NEW TRENDS IN QUANTUM SYSTEMS
IN CHEMISTRY AND PHYSICS

Progress in Theoretical Chemistry and Physics

VOLUME 6

The titles published in this series are listed at the end of this volume.

New Trends in Quantum Systems in Chemistry and Physics

Volume 1
Basic Problems and Model Systems
Paris, France, 1999

Edited by

Jean Maruani
CNRS, Paris, France

Christian Minot
UPMC, Paris, France

Roy McWeeny
Università di Pisa, Italy

Yves G. Smeyers
CSIC, Madrid, Spain

and

Stephen Wilson
Rutherford Appleton Laboratory, Oxfordshire, United Kingdom

SPRINGER-SCIENCE+BUSINESS MEDIA, B.V.

Library of Congress Cataloging-in-Publication Data

ISBN 978-1-4020-0411-7 ISBN 978-0-306-46951-0 (eBook)
DOI 10.1007/978-0-306-46951-0

Printed on acid-free paper

Progress in Theoretical Chemistry and Physics

A series reporting advances in theoretical molecular and material sciences, including theoretical, mathematical and computational chemistry, physical chemistry and chemical physics

Aim and Scope

Science progresses by a symbiotic interaction between theory and experiment: theory is used to interpret experimental results and may suggest new experiments; experiment helps to test theoretical predictions and may lead to improved theories. Theoretical Chemistry (including Physical Chemistry and Chemical Physics) provides the conceptual and technical background and apparatus for the rationalisation of phenomena in the chemical sciences. It is, therefore, a wide ranging subject, reflecting the diversity of molecular and related species and processes arising in chemical systems. The book series *Progress in Theoretical Chemistry and Physics* aims to report advances in methods and applications in this extended domain. It will comprise monographs as well as collections of papers on particular themes, which may arise from proceedings of symposia or invited papers on specific topics as well as initiatives from authors or translations.

The basic theories of physics – classical mechanics and electromagnetism, relativity theory, quantum mechanics, statistical mechanics, quantum electrodynamics – support the theoretical apparatus which is used in molecular sciences. Quantum mechanics plays a particular role in theoretical chemistry, providing the basis for the valence theories which allow to interpret the structure of molecules and for the spectroscopic models employed in the determination of structural information from spectral patterns. Indeed, Quantum Chemistry often appears synonymous with Theoretical Chemistry: it will, therefore, constitute a major part of this book series. However, the scope of the series will also include other areas of theoretical chemistry, such as mathematical chemistry (which involves the use of algebra and topology in the analysis of molecular structures and reactions); molecular mechanics, molecular dynamics and chemical thermodynamics, which play an important role in rationalizing the geometric and electronic structures of molecular assemblies and polymers, clusters and crystals; surface, interface, solvent and solid-state effects; excited-state dynamics, reactive collisions, and chemical reactions.

Recent decades have seen the emergence of a novel approach to scientific research, based on the exploitation of fast electronic digital computers. Computation provides a method of investigation which transcends the traditional division between theory and experiment. Computer-assisted simulation and design may afford a solution to complex problems which would otherwise be intractable to theoretical analysis, and may also provide a viable alternative to difficult or costly laboratory experiments. Though stemming from Theoretical Chemistry, Computational Chemistry is a field of research in its own right, which can help to test theoretical predictions and may also suggest improved theories.

The field of theoretical molecular sciences ranges from fundamental physical questions relevant to the molecular concept, through the statics and dynamics of isolated molecules, aggregates and materials, molecular properties and interactions, and the role of molecules in the biological sciences. Therefore, it involves the physical basis for geometric and electronic structure, states of aggregation, physical and chemical transformations, thermodynamic and kinetic properties, as well as unusual properties such as extreme flexibility or strong relativistic or quantum-field effects, extreme conditions such as intense radiation fields or interaction with the continuum, and the specificity of biochemical reactions.

Theoretical chemistry has an applied branch – a part of molecular engineering, which involves the investigation of structure–property relationships aiming at the design, synthesis and application of molecules and materials endowed with specific functions, now in demand in such areas as molecular electronics, drug design or genetic engineering. Relevant properties include conductivity (normal, semi- and supra-), magnetism (ferro- or ferri-), optoelectronic effects (involving nonlinear response), photochromism and photoreactivity, radiation and thermal resistance, molecular recognition and information processing, and biological and pharmaceutical activities, as well as properties favouring self-assembling mechanisms and combination properties needed in multifunctional systems.

Progress in Theoretical Chemistry and Physics is made at different rates in these various research fields. The aim of this book series is to provide timely and in-depth coverage of selected topics and broad-ranging yet detailed analysis of contemporary theories and their applications. The series will be of primary interest to those whose research is directly concerned with the development and application of theoretical approaches in the chemical sciences. It will provide up-to-date reports on theoretical methods for the chemist, thermodynamician or spectroscopist, the atomic, molecular or cluster physicist, and the biochemist or molecular biologist who wish to employ techniques developed in theoretical, mathematical or computational chemistry in their research programmes. It is also intended to provide the graduate student with a readily accessible documentation on various branches of theoretical chemistry, physical chemistry and chemical physics.

Contents

Part III. Relativistic Formulations and Effects

Part IV. Valence Theory

Part V. Nuclear Motion

Preface

These two volumes collect thirty-eight selected papers from the scientific contributions presented at the Fourth European Workshop on *Quantum Systems in Chemistry and Physics* (QSCP-IV), held in Marly-le-Roi (France) in April 22-27, 1999. A total of one hundred and fifteen scientists attended the workshop, 99 from Europe and 16 from the rest of the world. They discussed the state of the art, new trends, and future evolution of the methods and applications.

The workshop was held in the old town of Marly-le-Roi, which lies to the West of Paris between the historic centres of Saint-Germain-en-Laye and Versailles. Participants were housed at the National Youth Institute, where over sixty lectures were given by leading members of the scientific community; in addition, over sixty posters were presented in two very animated sessions. We are grateful to the oral speakers and to the poster presenters for making the workshop such an stimulating experience. The social programme was also memorable - and not just for the closing banquet, which was held at the French Senate House. We are sure that participants will long remember their visit to the 'Musée des Antiquités Nationales': created by Napoleon III at the birthplace of Louis XIV, this museum boasts one of the world finest collections of archeological artifacts.

The Marly-le-Roi workshop followed the format established at the three previous meetings, organized by Prof. Roy McWeeny at San Miniato Monastery, Pisa (Italy) in April, 1996 (the proceedings of which were published in the Kluwer TMOE series); Dr Steve Wilson at Jesus College, Oxford (United Kingdom) in April, 1997 (which resulted in two volumes in Adv. Quant. Chem.); and Prof. Alfonso Hernandez-Laguna at Los Alixares Hotel, Granada (Spain) in April, 1998 (for which proceedings appeared in the present series). These meetings, sponsored by the European Union in the frame of the Cooperation in Science and Technology (COST) chemistry actions, create a forum for discussion, exchange of ideas and collaboration on innovative theory and applications.

Quantum Systems in Chemistry and Physics encompasses a broad spectrum of research where scientists of different backgrounds and interests jointly place special emphasis on quantum theory applied to molecules, molecular interactions and materials. The meeting was divided into several sessions, each addressing a different aspect of the field: 1 - Density matrices and density functionals; 2 - Electron correlation treatments; 3 - Relativistic formulations and effects; 4 - Valence theory (chemical bond and bond breaking); 5 - Nuclear motion (vibronic effects and flexible molecules); 6 - Response theory (properties and spectra); 7 - Reactive collisions and chemical reactions, computational chemistry and physics; and 8 - Condensed matter (clusters and crystals, surfaces and interfaces).

Density matrices and density functionals have important roles in both the interpretation and the calculation of atomic and molecular structures and properties. The fundamental importance of electronic correlation in many-body systems makes this topic a central area of research in quantum chemistry and molecular physics. Relativistic effects are being increasingly recognized as an essential ingredient of studies on many-body systems, not only from a formal viewpoint but also for practical applications to molecules and materials involving heavy atoms. Valence theory deserves special attention since it improves the electronic description of molecular systems and reactions from the point of view used by most laboratory chemists. Nuclear motion constitutes a broad research field of great importance accounting for the internal molecular dynamics and spectroscopic properties.

Also very broad and of great importance in physics and chemistry is the topic of response theory, where electric and magnetic fields interact with matter. The study of chemical reactions and collisions is the cornerstone of chemistry, where traditional concepts like potential-energy surfaces or transition complexes appear to become insufficient, and the new field of computational chemistry finds its main applications. Condensed matter is a field in which progressive studies are performed, from few-atom clusters to crystals, surfaces and materials.

We are pleased to acknowledge the support given to the Marly-le-Roi workshop by the European Commission, the Centre National de la Recherche Scientifique (CNRS) and Université Pierre et Marie Curie (UPMC). We would like to thank Prof. Alfred Maquet, Director of Laboratoire de Chimie Physique in Paris, Prof. Alain Sevin, Director of Laboratoire de Chimie Théorique in Paris, and Dr Gérard Rivière, Secretary of COST-Chemistry in Brussels, for financial and logistic help and advice. Prof. Gaston Berthier, Honorary Director of Research, and Prof. Raymond Daudel, President of the European Academy, gave the opening and closing speeches. The supportive help of Ms Françoise Debock, Manager of INJEP in Marly-le-Roi, is also gratefully acknowledged. Finally, it is a pleasure to thank the work and dedication of all other members of the local organizing team, especially Alexandre Kuleff, Alexis Markovits, Cyril Martinsky and, last but not least, Ms Yvette Masseguin, technical manager of the workshop.

Jean Maruani and Christian Minot
Paris, 2000

Part I

Density Matrices and Density Functionals

ARE EXACT KOHN-SHAM POTENTIALS EQUIVALENT TO LOCAL FUNCTIONS?

R. K. NESBET
IBM Almaden Research Center
650 Harry Road, San Jose, California 95120-6099, USA

AND

R. COLLE
Dipartimento di Chimica Applicata, Bologna
and Scuola Normale Superiore, Pisa, Italy

Abstract. In Kohn-Sham density functional theory, equations for the occupied orbital functions of a model state can be derived by minimizing the exact ground-state energy functional of Hohenberg and Kohn. It has been assumed for some time that the effective potentials in exact Kohn-Sham equations are equivalent to local potential functions. Specializing this theory to the exchange-only problem in a Hartree-Fock model, for which exact solutions are known, this assumption is tested in a situation relevant to real atoms. It is shown that the assumption fails.

1. Introduction

Density functional theory (DFT) is based on a proof [1] that the external potential acting on an N-electron system is uniquely associated with the electronic ground-state density function. The ground-state energy is a functional of the spin-indexed electron density $\rho(\mathbf{r})$, and this energy functional is minimized by the ground-state density function. Introducing an orbital model or reference state [2], the spin-indexed reference-state density $\rho = \sum_i n_i \phi_i^* \phi_i$ is expressed as a sum of densities of orthonormal spin-indexed orbital functions weighted by reference-state occupation numbers n_i, which for nondegenerate ground states have values $0, 1$ only. Unless explicitly varied, these occupation numbers are considered to be constants such that $\sum_i n_i = N$. This and other equations are simplified here by omit-

J. Maruani et al. (eds.), New Trends in Quantum Systems in Chemistry and Physics, Volume 1, 3–11.

ting spin indices for orbital functions, densities, and potential functions. Sums over spin indices are implied in integrals and summations. Exact wave functions Ψ are spin eigenstates, but single-determinant model or reference states Φ have broken spin symmetry except for closed-shell singlet states. An exact Kohn-Sham (KS) theory is defined by varying these occupied orbital functions so as to minimize the Hohenberg-Kohn (HK) ground-state energy functional. Alternatively, if it is assumed that the variational Euler-Lagrange equations for the orbital functions can be expressed in terms of purely local effective potential functions (the locality hypothesis), the same result should be achieved by minimizing only the mean kinetic energy of the reference state over all sets of occupied orbital functions that produce the ground-state density [2]. This procedure defines the Kohn-Sham construction. It is commonly assumed that the locality hypothesis is valid, so that exact KS theory and the KS construction should be equivalent when the same Hohenberg-Kohn energy functional is used.

If the locality hypothesis were valid as a general consequence of variational theory, it should apply to both exchange-correlation and kinetic energies. In their original paper, Kohn and Sham (1965) propose a model theory in which the kinetic energy is expressed by the linear operator $-\frac{1}{2}\nabla^2$ of Schrödinger. They do not resolve the question of locality of the exchange-correlation potential. Subsequent literature has assumed this potential to be a local function, not a linear operator. This would be consistent if the kinetic energy operator were equivalent to an effective local potential function $v_T(\mathbf{r})$ in correctly-formulated variational equations, as postulated in deriving the Thomas-Fermi equation [3]. It has recently been shown that this cannot generally be true [4].

Since the locality hypothesis fails for the kinetic energy operator, it cannot be assumed without proof for the exact ground-state exchange-correlation potential implied by DFT. We examine this issue here, compiling evidence from prior literature that appears to contradict the locality hypothesis, and adding new evidence from new test calculations. These calculations exploit the Hartree-Fock model of DFT, based on a demonstration that ground-state Hartree-Fock theory satisfies Hohenberg-Kohn theorems.

2. Kinetic energy and Thomas-Fermi theory

Given the exact HK energy functional $E[\rho]$ for external potential $v(\mathbf{r})$, infinitesimal variations of ρ induce variations of E within its range of definition. To avoid irrelevant mathematical complexities, it will be assumed here that only physically realizable density functions need be considered. In Kohn-Sham theory, this is assured by restricting orbital functions to the usual Hilbert space, requiring continuity with continuous gradients except for Kato cusp conditions at Coulomb potential singularities. The locality hypothesis assumes that such infinitesimal variations of E take the form

$$\delta E = \int d^3\mathbf{r} \frac{\delta E}{\delta \rho} \delta \rho = \int d^3\mathbf{r}\ v_E \delta \rho, \tag{1}$$

where $v_E(\mathbf{r})$ is a local function of $\mathbf{r}$. If $E[\rho]$ can be defined for unrestricted infinitesimal variations of ρ in a neighborhood of any physically realizable density, and a Lagrange multiplier μ is used to enforce the normalization constraint $\int \rho d^3\mathbf{r} = N$, the stationary condition is

$$\delta\{E - \mu(\int \rho d^3\mathbf{r} - N)\} = \int (\frac{\delta E}{\delta \rho} - \mu)\delta \rho d^3\mathbf{r} = 0. \tag{2}$$

For free variations of ρ this implies the Thomas-Fermi equation

$$\frac{\delta E}{\delta \rho} - \mu = 0. \tag{3}$$

If a kinetic energy functional can be defined, this derivation assumes that $\frac{\delta T}{\delta \rho} = v_T(\mathbf{r})$, a local effective potential, and that the residual exchange-correlation energy defines an exact local exchange-correlation potential, $\frac{\delta E_{xc}}{\delta \rho} = v_{xc}(\mathbf{r})$.

Kohn and Sham postulate an orbital decomposition of the density function. In order to examine evidence that Eq.(1) is not adequate for density-functional theory with this orbital structure, we consider an extended definition, in which infinitesimal variations of E are described in terms of a linear operator that acts on orbital wave functions. This generalized functional derivative is denoted by $\hat{v}_E$, and variations of E are given by

$$\delta E = \int d^3\mathbf{r} \sum_i \{\delta \phi_i^*(\mathbf{r}) n_i \hat{v}_E \phi_i(\mathbf{r}) + cc\}. \tag{4}$$

This generalization reduces to the standard form if $\hat{v}_E$ is equivalent to a local function $v_E(\mathbf{r})$ when acting on occupied orbitals of a Kohn-Sham model state.

Exact KS equations impose normalization by requiring the occupied orbitals of the model or reference state to be orthonormal. This introduces a matrix of Lagrange multipliers which can be diagonalized to give the canonical exact KS equations, for occupied orbitals $i \leq N$,

$$(\frac{\delta E}{\delta \rho} - \epsilon_i)\phi_i = 0, \tag{5}$$

where $\frac{\delta T}{\delta \rho}$ is replaced by the linear operator $\hat{v}_T = -\frac{1}{2}\nabla^2$.

If the Thomas-Fermi and exact KS equations were equivalent, they would give the same result when Eq.(3) is multiplied by ρ and integrated and Eq.(5) is multiplied by $n_i\phi_i^*$, integrated, and summed. Hence the equations are inconsistent unless $\sum_i n_i\epsilon_i = N\mu$. Because $\sum_i n_i = N$ and $\epsilon_i \leq \mu$, this equation cannot be true unless all orbital energies are equal. Except for unusual symmetry degeneracies, the exclusion principle rules this out for more than two electrons, and proves that Thomas-Fermi and exact KS equations are in general inconsistent [4].

This does not resolve the question of which of these equations is physically correct, but empirical evidence strongly supports Kohn and Sham in their choice of the linear operator $\hat{v}_T = -\frac{1}{2}\nabla^2$. Otherwise, Thomas-Fermi equations cannot incorporate Fermi-Dirac statistics [4] and do not describe atomic shell structure and chemical binding [3].

3. Natural definition of component functionals

The KS construction [2] defines a kinetic energy functional of the occupied orbitals of the model state Φ, constrained so that the model density ρ_Φ equals that of the true ground state Ψ. Variation of this orbital functional leads to the usual Schrödinger operator for kinetic energy in exact KS equations. In fact, this construction or any alternative rule that associates a single-determinant reference state with each N-electron wave function Ψ provides a natural definition of correlation energy and separately of each of the orbital functional components of the reference-state mean energy $(\Phi|H|\Phi)$ [5].

Using an unsymmetric normalization, $(\Phi|\Psi) = (\Phi|\Phi) = 1$, any energy eigenvalue of the N-electron Hamiltonian H is given exactly by an unsymmetrical formula $E = (\Phi|H|\Psi) = (\Phi|H|\Phi) + (\Phi|H|\Psi - \Phi) = (\Phi|H|\Phi) + E_c$. $\Psi - \Phi$ is orthogonal to Φ by construction. This defines $E_c = (\Phi|H|\Psi - \Phi)$ as an off-diagonal matrix element of the Hamiltonian between the reference state and its orthogonal complement. The leading term $(\Phi|H|\Phi)$ is an explicit orbital functional.

4. Orbital functional components of the energy functional

The N-electron Hamiltonian takes the form $H = T + U + V$, where T is the kinetic energy, U is the interelectronic Coulomb energy, and V is the external potential energy. These separate terms in $(\Phi|H|\Phi)$ define separate density functionals when the reference state Φ is determined by a wave function Ψ that is itself a density functional. Expressed as an explicit functional of the occupied orbitals of the reference state, the kinetic energy functional is

$$T[\rho] = (\Phi|T|\Phi) = \sum_i n_i(i| -\frac{1}{2}\nabla^2|i). \tag{6}$$

Similarly, the external potential energy is

$$V[\rho] = (\Phi|V|\Phi) = \sum_i n_i(i|v|i), \tag{7}$$

equal to the integral $\int v\rho d^3\mathbf{r}$, where ρ is the density function of the reference state. This definition is valid in the KS construction because $\rho_\Phi = \rho_\Psi$. Denoting the two-electron Coulomb interaction by u, Coulomb minus exchange by $\bar{u}$, the electronic interaction energy functional is

$$U[\rho] = (\Phi|U|\Phi) = \frac{1}{2}\sum_{i,j} n_i n_j (ij|\bar{u}|ij) = E_h[\rho] + E_x[\rho], \tag{8}$$

where $E_h[\rho] = \frac{1}{2}\sum_{i,j} n_i n_j (ij|u|ij)$ and $E_x[\rho] = -\frac{1}{2}\sum_{i,j} n_i n_j (ij|u|ji)$.

5. Functional derivatives and local potentials

When E_c is defined as indicated above, the component density functionals of $(\Phi|H|\Phi)$ are explicit orbital functionals. A consistent definition of orbital and density functional derivatives is required. If density functional derivatives were local functions, the variation of such a functional would be

$$\delta F = \int d^3\mathbf{r} \frac{\delta F}{\delta \rho}\delta\rho(\mathbf{r}) = \int d^3\mathbf{r}\; v_F(\mathbf{r})\delta\rho(\mathbf{r}). \tag{9}$$

This formula is not meaningful if the functional derivative is a linear operator that acts on orbital wave functions, because the notation $\hat{v}\delta\rho$ is not well-defined. A correct notation for such operators is

$$\delta F = \int d^3\mathbf{r} \sum_i \{\delta\phi_i^*(\mathbf{r}) n_i \hat{v}_F \phi_i(\mathbf{r}) + cc\}, \tag{10}$$

which reduces to Eq.(9) if $\hat{v}_F$ is equivalent to a local function $v_F(\mathbf{r})$. Variation of an explicit orbital functional takes the form

$$\delta F = \int d^3\mathbf{r} \sum_i \{\delta\phi_i^*(\mathbf{r}) \frac{\delta F}{\delta\phi_i^*(\mathbf{r})} + cc\}, \tag{11}$$

from which Schrödinger derived the kinetic energy operator $\hat{v}_T = -\frac{1}{2}\nabla^2$ used in the KS equations [2]. Consistency between Eqs.(10) and (11) implies the chain rule

$$\frac{\delta F}{\delta\phi_i^*(\mathbf{r})} = n_i \hat{v}_F \phi_i(\mathbf{r}). \tag{12}$$

From this chain rule, it follows for any functional derivative $\hat{v}_F$ that reduces to a local function $v_F(\mathbf{r})$ that

$$\sum_i \phi_i^* \frac{\delta F}{\delta \phi_i^*} = \sum_i n_i \phi_i^*(\mathbf{r}) \hat{v}_F \phi_i(\mathbf{r}) = v_F(\mathbf{r}) \rho(\mathbf{r}). \tag{13}$$

This sum rule determines the effective local potential if the locality hypothesis is valid.

6. Exchange energy in the Hartree-Fock model

It can be shown that the Hohenberg-Kohn theory is valid for an unrestricted Hartree-Fock model. The full N-electron Hamiltonian $H = T + U + V$ is used, but trial wave functions are limited to single normalized Slater determinants constructed from orthonormal spin-indexed orbital wave functions. Only closed-shell states will be considered in specific calculations discussed here. The variational energy functional is $(\Phi|H|\Phi)$ [6]. In this model, the universal HK functional is $(\Phi|T + U|\Phi)$, evaluated in any Hartree-Fock ground state, corresponding to the density ρ of the Hartree-Fock wave function Φ. The simplest definition of a reference state in this model is $\Phi = \Psi$, since the variational trial functions Ψ are limited to the form of single Slater determinants. With this definition, E_c vanishes exactly, and all other components of the HK energy functional are known from their values in Hartree-Fock ground states. This provides a model of DFT in which everything is known or can be computed accurately: density, energy, and wave function.

This model can be used to test the validity of the locality hypothesis for the effective exchange potential derived from the exchange energy functional.

7. Energy relationships in the Hartree-Fock model

In this model, exact KS equations are equivalent to ground-state HF (or UHF) equations and determine the same (unique) wave function Φ. The model defines the variational energy functional as $(\Phi|H|\Phi)$ for any trial function in the form of a single determinant. In the KS construction (KSC), the model function minimizes $(\Phi|T|\Phi)$ subject to a density constraint, and $E_{KSC} \geq E_{HF}$. In optimized effective potential theory (OEP) [7, 8], occupied orbital functions of a model state Φ are determined by equations of the same form as the KS equations, with a local exchange potential chosen to minimize the variational energy $(\Phi|H|\Phi)$. Since the OEP density is unconstrained except by normalization, while the KSC density is constrained to equal that of the HF ground-state, $E_{KSC} \geq E_{OEP}$. Hence the variational

TABLE 1. Variational energies for typical atoms (Hartree units)

Atom	E_{HF}	E_{OEP}	E_{KSC}
He	-2.8617	-2.8617	-2.8617
Be	-14.5730	-14.5724	-14.5724
Ne	-128.5471	-128.5455	-128.5454

energies of the three methods must be such that $E_{KSC} \geq E_{OEP} \geq E_{HF}$ [9].

If the locality hypothesis is valid, an exact local exchange potential exists that must be equivalent to the linear exchange operator of Fock when acting on the occupied ground-state orbital functions. Using this exchange potential, OEP must produce the HF ground state, and the KSC exchange potential must agree with it. Hence the wave functions, electronic densities, and variational energies should all be equal.

Evidence that $\rho_{OEP} \neq \rho_{HF}$ has been available for some time [10]. If $\rho_{OEP} \neq \rho_{HF}$, HK theory implies $E_{OEP} > E_{HF}$, and no local exchange potential can minimize the variational energy. This contradicts the locality hypothesis and implies that imposing locality is a variational constraint. Except for He, computations indicate a variational error of constraint ΔE in the OEP and KSC variational energies. Thus $\Delta E_{OEP} = E_{OEP} - E_{HF} > 0$ and $\Delta E_{KSC} \geq \Delta E_{OEP}$. Computed variational energies are shown for typical atoms in Table 1: E_{HF} [11], E_{OEP} [7, 8], and E_{KSC} [9, 12]. Computed energies are such that $E_{KSC} \geq E_{OEP} > E_{HF}$, in agreement with variational theory if there is a locality constraint in the OEP and KSC methods. These KSC results show that although solution of the KSC equations produces a local exchange potential, implying "noninteracting v-representability", this does not imply minimization of the variational energy functional, the "locality hypothesis".

8. Direct test of the locality hypothesis

A necessary condition for locality of a density functional derivative is provided by varying the nuclear charge in an atom [5]. If a local potential function $v_F(\mathbf{r})$ is equal to the functional derivative $\frac{\delta F}{\delta \rho(\mathbf{r})}$, then the definition of a functional derivative is tested by the criterion

$$P_F = \frac{\partial}{\partial Z} F[\rho] - \int d^3\mathbf{r}\, v_F(\mathbf{r}) \frac{\partial \rho}{\partial Z} = 0. \tag{14}$$

Only ground-state quantities are used in this formula.

TABLE 2. Results of the KS construction (Hartree units, signed integers indicate powers of 10)

At	P_h^{HF}	P_x^{HF}	P_x^{KS}	P_T^{HF}	P_T^{KS}
He	-0.347-9	0.174-9	0.174-9	-0.606-5	-0.606-5
Be	-0.316-9	0.126	-0.1-3	0.812	0.815
Ne	0.152-8	0.442	-0.3-4	6.859	6.862

We have carried out HF and KSC calculations, cross-checked between two different programs using different methods and procedures, and obtain total energies consistent with published HF [11] and KSC [9] results. Computed values of the criterion quantities P_F are shown in Table 2 for He, Be and Ne ground states. In this table, HF refers to local potentials computed for HF ground states using Eq.(13). The Hartree potential $v_h(\mathbf{r})$ satisfies the locality criterion to computational accuracy in all cases, while the effective kinetic potential $v_T(\mathbf{r})$ fails by a large margin for Be and Ne. P_x^{HF} tests the Slater local exchange potential computed using Eq.(13), and indicates that this test of locality fails. However, for the KSC local exchange potential, P_x^{KS} is nonzero, but at the margin of accuracy of these calculations. It has recently been shown that the integral equation of the OEP model implies $P_x^{OEP} = 0$ [5], although the OEP energy and density differ from the HF ground state [12]. Thus this necessary but not sufficient condition for locality of the exchange potential is satisfied exactly in the OEP model and to a close approximation in KSC.

9. Conclusions

The mathematical issue involved here is the validity of the locality hypothesis, that a correct variational derivation of exact Kohn-Sham equations implies the existence of local potentials equivalent to the density-functional derivatives that occur in this theory. This hypothesis has been shown to fail (for $N > 2$) for the kinetic energy [4], so it cannot be assumed to be valid for exchange and correlation energies without a separate proof. In the unrestricted Hartree-Fock model of density-functional theory, there is no correlation energy, and this hypothesis implies the existence of an exact local exchange potential. Here we have shown that it fails for typical atoms with more than two electrons. The UHF model has the same mathematical structure for all atoms and molecules. Thus it can be expected that an exact local exchange potential does not exist in this model of density-functional theory for quite general N-electron systems.

Kohn-Sham kinetic and exchange energies in a correlated system can be defined as orbital functionals with exactly the same form as in the UHF model [5], expressed as mean values in the model or reference state. In general, correlation energy cannot be expressed so directly as an orbital functional, much less as an explicit density functional. Except for correlation screening of continuum-electron exchange, it is not obvious how the mathematical character of these explicit orbital functionals could be qualitatively changed by cancellation against a necessarily more complex and numerically smaller correlation energy functional. We conjecture that the local exchange-correlation potential anticipated in exact Kohn-Sham theory does not exist for any real system with more than two electrons.

Acknowledgments

RKN is grateful to the Scuola Normale Superiore (Pisa) for funding a lectureship during which this work was initiated. RC gratefully acknowledges support of this work by the Italian MURST: Programmi di ricerca scientifica di rilevante interesse nazionale - Anno 1999.

References

1. Hohenberg, P. and Kohn, W. (1964) Inhomogeneous electron gas, *Phys.Rev.* **136**, B864-B871.
2. Kohn, W. and Sham, L.J. (1965) Self consistent equations including exchange and correlation effects, *Phys.Rev.* **140**, A1133-A1138.
3. March, N.H. (1957) The Thomas-Fermi approximation in quantum mechanics, *Adv.Phys.* **6**, 1-101.
4. Nesbet, R.K. (1998) Kinetic energy in density-functional theory, *Phys.Rev.A* **58**. R12-R15.
5. Nesbet, R.K. (1999) Exchange and correlation energy in density functional theory, *Int.J.Quantum Chem.*, in press.
6. Payne, P.W. (1979) Density functionals in unrestricted Hartree-Fock theory, *J.Chem.Phys.* **71**, 490-496.
7. Aashamar, K., Luke, T.M. and Talman, J.D. (1978) Optimized central potentials for atomic ground-state wavefunctions, *At.Data Nucl.Data Tables* **22**, 443-472.
8. Engel, E. and Vosko, S.H. (1993) Accurate optimized-potential-model solutions for spherical spin-polarized atoms: Evidence for limitations of the exchange-only local spin-density and generalized-gradient expansions, *Phys.Rev.A* **47**, 2800-2811.
9. Görling, A. and Ernzerhof, M. (1995) Energy differences between Kohn-Sham and Hartree-Fock wave functions yielding the same electron density, *Phys.Rev.A* **51**, 4501-4513.
10. Trickey, S.B. (1984) Comment on 'Electron removal energies in Kohn-Sham density-functional theory', *Phys.Rev.B* **30**, 3523.
11. Froese Fischer, C. (1977) *The Hartree-Fock Method for Atoms*, Wiley, New York.
12. Nesbet, R.K. and Colle, R. (1999) Tests of the locality of exact Kohn-Sham exchange potentials, *Phys.Rev.A*, submitted.

THEORY OF EXACT EXCHANGE RELATIONS FOR A SINGLE EXCITED STATE

Á. NAGY
Department of Theoretical Physics
University of Debrecen
H-4010 Debrecen, Hungary

Abstract. A recently proposed theory for a single excited state based on Kato's theorem is reviewed. The concept of adiabatic connection is extended and the validity of Kato's theorem along the adiabatic path is discussed. Exchange identities are derived utilizing the principle of adiabatic connection and coordinate scaling. A generalized 'Koopmans' theorem' is derived.

1. Introduction

Though the density functional theory was originally a ground-state theory [1], it has been extended to excited states, too. Nowadays, there are several methods to treat excited states of atoms, molecules and solids. Rigorous generalization for excited states was given by Theophilou [2] and by Gross, Oliveira and Kohn [3]. Several calculations were done with this method [4-11] This approach has the disadvantage that one has to calculate all the ensemble energies lying under the given ensemble energy to obtain the desired excitation energy. It is especially inconvenient to use it if one is interested in highly excited states. Recently, Görling [12] presented a new density fuctional formalism for excited states generalizing a recent perturbation theory [13]. An alternative theory, worth mentioning, is time-dependent density functional theory [14,15] in which transition energies are obtained from the poles of dynamic linear response properties. The work formalism proposed by Sahni and coworkers [16] has also been applied in excited-state density functional calculations [17]. Görling [18] has presented a generalized density fuctional formalism based on generalized adiabatic connection.

J. Maruani et al. (eds.), New Trends in Quantum Systems in Chemistry and Physics, Volume 1, 13–24.

Recently, a new approach of treating a single excited state has been presented [19,20]. It is based on Kato's theorem [21,22] and is valid for Coulomb external potential (i.e. free atoms, molecules and solids). It has the advantage that one can treat a single excited state.

2. Theory for a single excited state

It is well-known that the ground state electron density is sufficient in principle to determine all molecular properties. For Coulomb system this statement can be simply understood (Bright Wilson's [23] argument): Kato's theorem [21,22] states that

$$Z_\beta = -\frac{1}{2n(r)} \frac{\partial n(r)}{\partial r}\bigg|_{r=R_\beta} , \tag{1}$$

where the partial derivatives are taken at the nuclei β. So the cusps of the density tell us where the nuclei are (R_β) and what the atomic numbers Z_β are. On the other hand, the integral of the density gives us the number of electrons:

$$N = \int n(\mathbf{r}) d\mathbf{r} . \tag{2}$$

Thus from the density the Hamiltonian can be readily obtained from which every property can be determined. This argument is valid only for Coulomb systems, while the density functional theory is valid for any external potential.

Kato's theorem is valid not only for the ground state but also for the excited states. Consequently, if the density n_i of the i-th excited state is known, the Hamiltonian $\hat{H}$ is also in principle known and its eigenvalue problem

$$\hat{H}\Psi_k = E_k \Psi_k \qquad (k = 1, ..., i, ...) \tag{3}$$

can be solved, where

$$\hat{H} = \hat{T} + \hat{V} + \hat{V}_{ee} . \tag{4}$$

$$\hat{T} = \sum_{j=1}^{N} (-\frac{1}{2}\nabla_j^2) , \tag{5}$$

$$\hat{V}_{ee} = \sum_{k=1}^{N-1} \sum_{j=i+1}^{N} \frac{1}{|\mathbf{r}_k - \mathbf{r}_j|} , \tag{6}$$

and

$$\hat{V} = \sum_{k=1}^{N} \sum_{J=1}^{M} -Z_J/|\mathbf{r}_k - \mathbf{R}_J| , \tag{7}$$

are the kinetic energy, the electron-electron energy and the electron-nucleon operators, respectively.

Using the concept [24,25] of adiabatic connection Kohn-Sham-like equations can be derived. It is supposed that the density is the same for both the interacting and non-interacting systems, and there exists a continuous path between them. A coupling constant path is defined by the Schrödinger equation

$$\hat{H}_i^\alpha \Psi_k^\alpha = E_k^\alpha \Psi_k^\alpha , \tag{8}$$

where

$$\hat{H}_i^\alpha = \hat{T} + \alpha \hat{V}_{ee} + \hat{V}_i^\alpha . \tag{9}$$

The subscript i denotes that the density of the given excited state is supposed to be the same for any value of the coupling constant α. $\alpha = 1$ corresponds to the fully interacting case, while $\alpha = 0$ gives the non-interacting system:

$$\hat{H}_i^0 \Psi_k^0 = E_k^0 \Psi_k^0 . \tag{10}$$

For $\alpha = 1$ the Hamiltonian $\hat{H}_i^\alpha$ is independent of i. For any other values of α the 'adiabatic' Hamiltonian depends on i, we have different Hamiltonian for different excited states. Thus the non-interacting Hamiltonian ($\alpha = 0$) is different for different excited states.

Now, let us study Kato's theorem along the adiabatic path. First, the original form of Kato's theorem [21] is reviewed: 'The Hamiltonian is written as

$$\hat{H} = \hat{T} + \hat{W} , \tag{11}$$

where

$$\hat{W}(\mathbf{r}_1, ...\mathbf{r}_N) = W_0(\mathbf{r}_1, ...\mathbf{r}_N) + \sum_{j}^{N} \bar{V}_{0j}(\mathbf{r}_j) + \sum_{j<k}^{N} \bar{V}_{jk}(|\mathbf{r}_k - \mathbf{r}_j|) . \tag{12}$$

W_0 is a real-valued, measurable function bounded in the whole configuration space. For each j, k with $0 \leq j < k \leq N$ the $\bar{V}_{jk}$ are real-valued, measurable functions defined in the 3-dimensional space, which vanish identically outside some sphere and satisfy

$$\int |\bar{V}_{jk}(\mathbf{r})|^{\sigma} d\mathbf{r} < \infty \ , \tag{13}$$

where σ is a fixed constant such that $\sigma \geq 2$. These assumptions are satisfied by the Coulomb potential:

$$\bar{V}_{0j}(\mathbf{r}_j) = \sum_{J=1}^{M} -Z_J/|\mathbf{r}_j - \mathbf{R}_J| \ , \tag{14}$$

$$\bar{V}_{jk} = \frac{1}{|\mathbf{r}_k - \mathbf{r}_j|} \ . \tag{15}$$

Then Kato's theorem states:

$$\left.\frac{\partial \bar{\Psi}}{\partial r}\right|_{r=R_\beta} = -Z_\beta \left.\Psi\right|_{r=R_\beta} \ , \tag{16}$$

where the partial derivatives are taken at the nuclei β and $\bar{\Psi}$ is the average value of Ψ taken over the sphere $r = constant$ around the nucleus β, for fixed values of the remaining electron coordinates $\mathbf{r}_2, ..., \mathbf{r}_N$:

$$\bar{\Phi} = \frac{1}{4\pi} \int_{\omega_1} \Phi(\mathbf{r}_1, \mathbf{r}_2, ..., \mathbf{r}_N) d\omega_1 \ . \tag{17}$$

From this Eq. (1) can be derived [21].

Now let us turn to Kato's theorem along the adiabatic path. The Hamiltonian is given by Eq. (9). As the density n_i is fixed along the adiabatic path, its derivatives are also fixed thus we have Eq. (1) for any value of α. This gives a constraint for $\hat{V}_i^{\alpha}$. The Hamiltonian (9) has the form of (11) - (12):

$$\begin{aligned} \hat{W}^{\alpha,i}(\mathbf{r}_1, ...\mathbf{r}_N) &= W_0^{\alpha,i}(\mathbf{r}_1, ...\mathbf{r}_N) + \sum_j^N \bar{V}_{0j}^{\alpha,i}(\mathbf{r}_j) \\ &\quad + \alpha \sum_{j<k}^N \bar{V}_{jk}(|\mathbf{r}_k - \mathbf{r}_j|) \ . \end{aligned} \tag{18}$$

Though the second term is not the same as Eq. (14), in the vicinity of the nucleus R_β, the term $-Z_\beta/|\mathbf{r} - \mathbf{R}_\beta|$ should be the dominant in order to obtain (1).

In spite of the fact that Kato's theorem is valid along the adiabatic path, we do not know the potential $W^{\alpha,i}$ in (14) or V_i^{α} in (9). Moreover, it might as well happen that there are several potentials that give the same density. In the next section we show that for a given electron-configuration, the non-interacting effective potential can be uniquely constructed from the density of the excited state n_i [20].

3. Construction of excited-state effective potential

In the density functional theory there are several methods [26-34] to obtain the Kohn-Sham potential from the ground-state electron density. These methods can be generalized to excited states. We now show that for a given electron-configuration, the non-interacting effective potential can be given uniquely from the density of the excited state n_i. Here only nondegenerate case is considered. Then the non-interacting wave-function can be given as a Slater determinant of orbitals satisfying the one-electron equations

$$\left[-\frac{1}{2}\nabla^2 + v_i^0(\mathbf{r})\right]\phi_j^i = \varepsilon_j^i \phi_j^i. \tag{19}$$

The density n_i of the excited state is

$$n_i = \sum_{j=1}^{I} \lambda_j^i |\phi_j^i|^2 \,, \tag{20}$$

where the occupation numbers λ_j^i can be 0 or 1. I corresponds to the highest occupied orbital.

Now, we introduce the functions K_j^i with the following definition:

$$\phi_j^i = n_i^{\frac{1}{2}} K_j^i \,. \tag{21}$$

Substituting Eq. (21) into Eq.(19) we obtain

$$\left[\frac{1}{8}\left(\frac{\nabla n_i}{n_i}\right)^2 - \frac{1}{4}\frac{\nabla^2 n_i}{n_i}\right] K_j^i$$
$$-\frac{1}{2}\frac{\nabla n_i}{n_i}\nabla K_j^i - \frac{1}{2}\nabla^2 K_j^i + v_i^0(\mathbf{r}) K_j^i = \varepsilon_j^i K_j^i \,. \tag{22}$$

As it can be seen from the definition (21) the functions K_i are not all independent:

$$1 = \sum_j \lambda_j^i |K_j^i|^2 \,. \tag{23}$$

Let us suppose that the order of the orbitals corresponds to increasing eigenvalues: $\varepsilon_1^i \leq \varepsilon_2^i \leq \varepsilon_3^i$.... From Eq. (23) follows that

$$K_1^i = \frac{1}{\lambda_1^i}\left(1 - \sum_{j=2}^{I} |K_j^i|^2\right)^{1/2} . \tag{24}$$

(If $\lambda_1^i = 0$ then one should select another orbital which is occupied.) The effective potential $v_i^0(\mathbf{r})$ can then be eliminated from equations (22) as follows. From the first equation of (22)

$$v_i^0(\mathbf{r}) = -\left[\frac{1}{8}\left(\frac{\nabla n_i}{n_i}\right)^2 - \frac{1}{4}\frac{\nabla^2 n_i}{n_i}\right] + \frac{1}{2}\frac{\nabla n_i}{n_i}\frac{\nabla K_1^i}{K_1^i} + \frac{1}{2}\frac{\nabla^2 K_1^i}{K_1^i} + \varepsilon_1^i . \tag{25}$$

Then substitute this potential to the remaining $I-1$ equations of (22):

$$\frac{\nabla n_i}{n_i}\left(\frac{\nabla K_j^i}{K_j^i} - \frac{\nabla K_1^i}{K_1^i}\right) + \frac{\nabla^2 K_j^i}{K_j^i} - \frac{\nabla^2 K_1^i}{K_1^i} = \eta_j^i \ , \ j = 2, .., I \ , \tag{26}$$

where

$$\eta_j^i = 2(\varepsilon_1^i - \varepsilon_j^i) \ , \ j = 2, .., I \ . \tag{27}$$

So we arrive at a system of differential equations, $I-1$ equations for $I-1$ functions $K_j^i, j = 2, .., I$. The functions K_j^i should satisfy the proper boundary conditions and normalization

$$\int d\mathbf{r} n_i(\mathbf{r}) |K_j^i(\mathbf{r})|^2 = 1 \ , \ j = 2, .., I \tag{28}$$

and orthogonality

$$\int d\mathbf{r} n_i(\mathbf{r}) K_j^{i*}(\mathbf{r}) K_k^i(\mathbf{r}) = 0, \ j \neq k, j, k = 1, ..., I \tag{29}$$

conditions.

Now, we can integrate the system (26) of $I-1$ differential equations for arbitrarily chosen parameters η_j^i. Having specified the boundary conditions and satisfied the normalization and orthogonality conditions, the proper values of η_j^i can be determined. Though, this is a typical eigenvalue-eigenfunction problem, it may be questioned whether their solution exists because of their non-linear coupled character. This is analogous to the so-called non-interacting v-representability problem of the ground-state theory. In the ground-state theory it is generally supposed that the potential exists for true physical densities. Detailed investigations, though showed, that there are exceptions [35,36]. For example, Aryasetiawan and Stott [31] studied the problem of v-representability for several model systems and constructed densities that are not ground-state densities. Instead, these are excited state densities. (E. g. They determined the effective potentials corresponding to the ground- and the first excited-state densities in a model of three spinless fermions moving in the square-well potential.) So even if a given density is not a ground-state density, it is an excited-state density. It is straighforward to assume that physically acceptable excited-state densities are non-interacting v-representable.

4. Exact exchange relations

In this new theory, the exact form of the exchange energy functional for the excited state is unknown. For constructing approximations it is necessary to uncover constraints which are satisfied by the exact exchange energy and potential. With this in mind the following theorems are derived. (We mention in passing that these are straighforward generalizations of theorems valid for the ground state [37].)

The Kohn-Sham equations corresponding to the excited state i have the form

$$\left[-\frac{1}{2}\nabla^2 + v_i^0(\mathbf{r};[n_i])\right]\phi_j^i(\mathbf{r}) = \varepsilon_j^i \phi_j^i(\mathbf{r}) \,. \tag{30}$$

Here, we emphasize that the effective potential $v_i^0(\mathbf{r};[n_i])$ depends on the excited state density n_i, that is we have different effective potential for different excited states.

The excited state density n_i is given by

$$n_i = \sum_{k=1}^{M-1} \lambda_k^i |\phi_k^i|^2 + (\lambda_M^i - 1)|\phi_M^i|^2 + \lambda_J^i |\phi_J^i|^2 \tag{31}$$

$$J = M + j \tag{32}$$

and

$$\lambda_J^i = 1 \tag{33}$$

provided that an electron is excited to the level J from the level M. λ_k^i are the occupation numbers. We can also define a 'ground-state' density

$$n_0 = \sum_{k=1}^{M} \lambda_k^i |\phi_k^i|^2 \,, \tag{34}$$

$$\sum_{k=1}^{M} \lambda_k^i = N \,. \tag{35}$$

It is, of course, not the true ground-state density as it is constructed from the excited-state Kohn-Sham orbitals.

Now, the 'Koopmans' theorem' for the ith excited state reads

$$\langle \Psi_i^{\alpha=0} | \hat{H} | \Psi_i^{\alpha=0} \rangle - \langle \Psi_{i,0}^{\alpha=0,N-1} | \hat{H}^{N-1} | \Psi_{i,0}^{\alpha=0,N-1} \rangle = \varepsilon_J^i \,, \tag{36}$$

where $\Psi_i^{\alpha=0}$ is the single determinant of the non-interacting excited state i and $\Psi_{i,0}^{\alpha=0,N-1}$ is the non-interacting 'ground-state' determinant of $N-1$ electrons, i.e. $\Psi_{i,0}^{\alpha=0,N-1}$ is built of the lowest $N-1$ orbitals of the Hamiltonian (9) for $\alpha = 0$.

Another important theorem relates the orbital energy difference with the corresponding difference of the expectation values of the Hamiltonian $\hat{H}$.

$$\varepsilon_J^i - \varepsilon_{HOMO}^i = \langle \Psi_i^{\alpha=0} | \hat{H} | \Psi_i^{\alpha=0} \rangle - \langle \Psi_{i,0}^{\alpha=0} | \hat{H} | \Psi_{i,0}^{\alpha=0} \rangle \,, \tag{37}$$

where $\Psi_{i,0}^{\alpha=0}$ is the non-interacting 'ground-state' determinant of N electrons, i.e. $\Psi_{i,0}^{\alpha=0}$ is built of the lowest N orbitals of the Hamiltonian (9) for $\alpha = 0$. ε_{HOMO}^i is the energy of the 'highest occupied orbital' in determinant $\Psi_{i,0}^{\alpha=0}$.

A useful relation for exchange energy difference reads

$$\begin{aligned} E_x^i[n_i] - E_x^i[n_i - \Delta n_i^F] &= \int d\mathbf{r} v_x^i(\mathbf{r}; [n_i]) \Delta n_i^F(\mathbf{r}) \\ &+ \frac{1}{2} \int d\mathbf{r}_1 d\mathbf{r}_2 \frac{\Delta n_i^F(\mathbf{r}_1) \Delta n_i^F(\mathbf{r}_2)}{|\mathbf{r}_1 - \mathbf{r}_2|} \,, \end{aligned} \tag{38}$$

where $\Delta n_i^F(\mathbf{r})$ is the density of the highest occupied Kohn-Sham orbital:

$$\Delta n_i^F(\mathbf{r}) = |\phi_J^i(\mathbf{r})|^2 \tag{39}$$

This expression can also be considered as a recursion relation leading to

$$\begin{aligned} E_x^i[n_i] &= \frac{1}{2} \sum_{k=1}^{L} \lambda_k^i \int d\mathbf{r}_1 d\mathbf{r}_2 \frac{|\phi_k^i(\mathbf{r}_1)|^2 |\phi_k^i(\mathbf{r}_2)|^2}{|\mathbf{r}_1 - \mathbf{r}_2|} \\ &+ \sum_{k=1}^{L} \lambda_k^i \int d\mathbf{r} v_x^i(\mathbf{r}; [n_k^i]) |\phi_k^i(\mathbf{r})|^2 \end{aligned} \tag{40}$$

where in accordance with the notations of Eqs. (31), (32) and (34)

$$L = M + j \,, \tag{41}$$

$$\lambda_k^i = 0 \quad if \quad M+1 \le k \le L-1 \ , \tag{42}$$

$$n_j^i(\mathbf{r}) = \sum_{l=1}^{j} \lambda_l^i |\phi_l^i(\mathbf{r})|^2 \tag{43}$$

and

$$E_x^i[n_i] = \frac{1}{2} \sum_{k=1}^{K} \sum_{l=1}^{K} \lambda_k^i \lambda_l^i \int d\mathbf{r}_1 d\mathbf{r}_2 \frac{\phi_k^{i*}(\mathbf{r}_1)\phi_l^{i*}(\mathbf{r}_2)\phi_k^i(\mathbf{r}_2)\phi_l^i(\mathbf{r}_1)}{|\mathbf{r}_1 - \mathbf{r}_2|} \ . \tag{44}$$

Eq. (38) may be cast in the form of the following identity:

$$\int d\mathbf{r} \phi_J^{i*}(\mathbf{r}) \hat{v}_x^{HF} \phi_J^i(\mathbf{r}) = \int d\mathbf{r} \phi_J^{i*}(\mathbf{r}) v_x^i(\mathbf{r}; [n_i]) \phi_J^i(\mathbf{r}) \ , \tag{45}$$

where $\hat{v}_x^{HF}$ is the Hartree-Fock exchange potential with Kohn-Sham orbitals. One can immediatelly see that Eq. (45) has the consequence that the highest occupied orbital energy in the Hartree-Fock and Kohn-Sham theory equals. This statement is true both for the ground and the excited states. This theorem was derived in the ground-state Kohn-Sham theory [38-40] and utilized in the KLI approximation for the optimized potential method.

5. Proof of the theorem

Let us construct the Hamiltonian

$$\hat{H}_i^{\alpha} = \hat{T} + \alpha \hat{V}_{ee} + \sum_{k=1}^{N} v_i^{\alpha}(\mathbf{r}_k) \ , \tag{46}$$

where $v_i^{\alpha}(\mathbf{r}; [n])$ is defined such that the ith excitation density n_i remain independent of α.

From previous studies we know that the asymptotic decay of $n_i^{\alpha} = n_i$ is governed by $|E_i^{\alpha} - E_{i,0}^{\alpha,N-1}|$, where E_i^{α} is the ith excited-state energy of $\hat{H}_i^{\alpha}$. $E_{i,0}^{\alpha,N-1}$ is, on the other hand, the 'ground-state' energy of $\hat{H}_i^{\alpha}$ with one-electron removed. As n_i is independent of α, $|E_i^{\alpha} - E_{i,0}^{\alpha,N-1}|$ is also independent of α. Consequently,

$$\frac{\partial}{\partial \alpha} \left| E_i^{\alpha} - E_{i,0}^{\alpha,N-1} \right|_{\alpha=0} = 0 \ . \tag{47}$$

From Eqs. (46) and (47) we obtain

$$\langle \Psi_i^{\alpha=0} | \hat{V}_{ee} | \Psi_i^{\alpha=0} \rangle + \int d\mathbf{r} n_i \left. \frac{\partial v_i^\alpha}{\partial \alpha} \right|_{\alpha=0} = \langle \Psi_{i,0}^{\alpha=0,N-1} | \hat{V}_{ee} | \Psi_{i,0}^{\alpha=0,N-1} \rangle + \int d\mathbf{r} n_{i,0}^{N-1} \left. \frac{\partial v_i^\alpha}{\partial \alpha} \right|_{\alpha=0} , \qquad (48)$$

Now, following Görling and Levy [13] and Nagy [19] v_i^α can be expanded as

$$v_\alpha^i(\mathbf{r}; [n_i]) = v_i^0 - \alpha \left(u(\mathbf{r}; [n_i]) + v_x^i(\mathbf{r}; [n_i]) \right) + \ldots , \qquad (49)$$

where u and v_x^i are the Coulomb and exchange potentials, respectively. From Eqs. (48) and (49) it follows that

$$\langle \Psi_i^{\alpha=0} | \hat{H} | \Psi_i^{\alpha=0} \rangle - \langle \Psi_{i,0}^{\alpha=0,N-1} | \hat{H} | \Psi_{i,0}^{\alpha=0,N-1} \rangle = \langle \phi_J^i | -\frac{1}{2}\nabla^2 + v + u + v_x^i | \phi_J^i \rangle . \qquad (50)$$

Thus we obtained the 'Koopmans' theorem' for the excited state (Eq. (36)).

Now, we can apply the 'ground-state Koopmans' theorem' [37,38,40]:

$$\varepsilon_{HOMO}^i = \langle \Psi_{i,0}^{\alpha=0} | \hat{H} | \Psi_{i,0}^{\alpha=0} \rangle - \langle \Psi_{i,0}^{\alpha=0,N-1} | \hat{H} | \Psi_{i,0}^{\alpha=0,N-1} \rangle . \qquad (51)$$

Combining Eqs. (36) and Eq. (51) we arrive at Eq. (37). Now, returning to Eq. (50) and comparing the expectation values on both sides of the equation, we can immediatelly obtain identities (38) and (45). Recursion relation (40) can be readily derived from Eq. (45) by a simple substitution of Eqs. (41) - (44). In the theorems presented, we have the orbitals corresponding to the non-interacting excited-state potential, so these are useful constraints for this (unknown) potential.

Acknowledgements

The grant 'Széchenyi' from the Hungarian Ministry of Education is gratefully acknowledged. This work was supported by the grants OTKA No. T 029469 and T 025369.

References

1. Hohenberg, P. and Kohn, W., Phys. Rev. B **136**, 864 (1964).
2. Theophilou, A.K., J. Phys. C **12**, 5419 (1978).
3. Gross, E.K.U., Oliveira, L.N. and Kohn, W., Phys. Rev. A **37**, 2805, 2809, 2821 (1988).
4. Nagy, Á., Phys. Rev. A **42**, 4388 (1990).

5. Nagy, Á., J. Phys. **B 24**, 4691 (1991).
6. Nagy Á. and Andrejkovics, I., J. Phys. B **27**, 233 (1994).
7. Nagy, Á., Int. J. Quantum. Chem. **56**, 225 (1995).
8. Nagy Á., J. Phys. B **29**, 389 (1996).
9. Nagy, Á., Int. J. Quantum. Chem. S. **29**, 297 (1995).
10. Nagy , Á., Adv. Quantum. Chem. **29**, 159 (1997).
11. Nagy, Á., Phys. Rev. A **49**, 3074 (1994).
12. Görling, A., (1996) Phys. Rev. A **54**, 3912 (1996).
13. Görling, A. and Levy, M., Phys. Rev. A **47**, 13105 (1993); Phys. Rev. A **47**, 196 (1994); Int. J. Quantum. Chem. S. **29**, 93 (1995).
14. Gross, E. U. K., Dobson, J. F. and Petersilka, M., in *Density Functional Theory, (Topics in Current Chemistry)* (ed.) Nalewajski, R. Springer-Verlag, Heidelberg, vol. 181. p. 81 (1996).
15. Casida, M. F., in *Recent advances in the density functional methods, (Recent Advances in Computational Chemistry)* (ed) Chong, D. P., World Scientific, Singapore vol.1 p.155 (1996); Casida, M. E., Jamorski, Casida, K. C. and Salahub, D. R., J. Chem. Phys. **108**, 5134 (1998).
16. Harbola, M. K. and Sahni, V., Phys. Rev. Lett. **62**, 489 (1989); see also Sahni, V., in *Density Functional Theory (Topics in Current Chemistry)* (ed.) Nalewajski, R, Springer-Verlag, Berlin vol.182. p.1 (1996).
17. Sen, K. D., Chem. Phys. Lett. **168**, 510 (1992); Singh, R.and Deb, B. M., Proc. Indian Acad. Sci. **106**, 1321 (1994); J. Mol. Struc. Theochem **361**, 33 (1996); J. Chem. Phys. **104**, 5892 (1996); Roy, A. K., Singh, R. and Deb, B. M., J. Phys. B **30**, 4763 (1997); Int. J. Quantum. Chem. **65**, 317 (1997) ; Roy, A. K. and Deb, B. M., (1997) Phys. Lett. A **234**, 465 (1997).
18. Görling, A., Phys. Rev. A **59**, 3359 (1999).
19. Nagy, Á., Int. J. Quantum. Chem. **70**, 681, (1998).
20. Nagy, Á., in *Electron Correlations and Materials Properties*, (eds.) Gonis, A., Kioussis, N. and Ciftan, M.,Plenum, New York, 451 (1999).
21. Kato, T., Commun. Pure Appl. Math. **10**, 151 (1957).
22. Steiner, E. J., Chem. Phys. **39**, 2365 (1963); March, N. H., *Self-consistent fields in atoms, Pergamon*, Oxford (1975).
23. Handy, N. C., in *Quantum Mechanical Simulation Methods for Studying Biological Systems*, (eds) Bicout D. and Field, M., Springer-Verlag, Heidelberg p.1. (1996).
24. Gunnarsson, O. and Lundqvist, B.I., Phys. Rev. B **13**, 4274 (1976).
25. Harris J. and Jones, JR. O., Phys. F **4**, 1170 (1984); Harris, J., Phys. Rev. A **29**, 1648 (1984).
26. Nagy, Á., J. Phys. B **26**, 43 (1993) ; Phil. Mag. B **69**, 779 (1994).
27. Nagy, Á. and March, N. H., Phys. Rev. A **39**, 5512 (1989); Phys. Rev. A **40**, 5544 (1989).
28. Zhao, Q. and Parr, R. G., J. Chem. Phys. **98**, 543 (1993); Parr, R. G., Phil. Mag. B **69**, 737 (1994); Zhao, Q., Morrison, R. C. and Parr, R. G., Phys. Rev. A **50**, 2138 (1994). Morrison, R. C. and Zhao, Q., Phys. Rev. A **51**, 1980 (1995). Parr. R. G. and Liu, S., Phys. Rev. A **51**, 3564 (1995).
29. Almbladh, C. O. and Pedroza, A. P., Phys. Rev. A **29**, 2322 (1984).
30. Arysetiawan, F. and Stott, M. J., Phys. Rev. B **34**, 4401 (1986).
31. Arysetiawan, F. and Stott, M. J., Phys. Rev. B **38**, 2974 (1988).
32. Chen, J. and Stott, M. J., Phys. Rev. A **44**, 2816 (1991) ; Chen, J. Esquivel, R. O. and Stott, M. J., Phil. Mag. B **69**, 1001 (1994).
33. Görling, A., Phys. Rev. A **46**, 3753 (1992); Görling, A. and Ernzerhof, M., Phys. Rev. A **51**, 4501 (1995).
34. van Leeuwen, R. and Baerends, E. J., Phys. Rev. A **49**, 2421 (1994).
35. Levy, M., Phys. Rev. A **28**, 1200 (1982).

36. Lieb, E. H., in *Physics as a Natural Pliosophy,* (eds) Shimony, A. and Feshback, H., MIT Press, Cambridge, MA. (1982).
37. Levy, M. and Görling A., Phys. Rev. A **53**, 3140 (1996).
38. Perdew, J. P., in *Density Functional Methods in Physics*, NATO Advanced Study Institute, Series B: Physics, (ed.) Dreizler, R. M. and da Providencia, J., Plenum, New York, vol. 123 (1985) and references therein.
39. Sharp R. T. and Horton, G. K., Phys. Rev. **30**, 317 (1953); Talman, J. D. and Shadwick, W. F., Phys. Rev. A **14**, 36 (1976); Aashamar, K., Luke, T. M. and Talman, J. D., At. Data Nucl. Data Tables **22**, 443 (1978).
40. Krieger, J. B., Li, Y. and Iafrate, G. J., Phys. Lett. A **146**, 256 (1990); Int. J. Quantum. Chem. **41**, 489 (1992); Phys. Rev. A **45**, 101 (1992); Phys. Rev. A **46**, 5453 (1992).

CORRELATION ENERGY CONTRIBUTIONS FROM LOW-LYING STATES TO DENSITY FUNCTIONALS IN THE KLI APPROXIMATION

C. GUTLE AND A. SAVIN
Laboratoire de Chimie Théorique
CNRS and Université Pierre et Marie Curie
F-75252 Paris, France
gutle@lct.jussieu.fr, savin@lct.jussieu.fr

AND

J. B. KRIEGER
Physics Department Brooklyn College
CUNY Brooklyn, NY 11210, USA

Abstract. In this paper we consider a criterion to allow a new coupling between density functional theory and configuration interaction methods. We study as a possible criterion the ordering of the orbital energies produced by the exchange-only KLI potential. This idea arises from the observation that the KLI potential behaves as - 1/r for large r in agreement with the known properties of the exact Kohn-Sham potential. The KLI bound states can thus be classified into valence and Rydberg orbitals, the latter not expected to make an important contribution to the correlation energy. We verify this assumption for the first terms of the He and Be series, as in the former only dynamical correlation is supposed to be present, whereas in the latter peculiar near-degeneracy effects intervene. In addition, exact results are given for the Be series in the limit of infinite nuclear charge. Although the contribution to the correlation energy from the low lying virtual states are significantly different for the two series (saturating as Z increases for the He series and being proportional to Z for large Z for the Be series) the remaining contributions to the correlation energy for both series saturate suggesting the application of DFT to the calculation of this latter contribution.

J. Maruani et al. (eds.), New Trends in Quantum Systems in Chemistry and Physics, Volume 1, 25–44.

1. Motivation

While fitting to experimental data produces density functionals for the exchange-correlation energy with average errors approaching the 1 kcal/mol limit, some fundamental problems still remain. For example, the Kohn-Sham (KS) eigenstate is constructed from a single determinant, which may not even yield the correct density in the case of degeneracy ("non v-representability", see, e.g. [1]). Moreover, the nature of the KS determinant may rapidly change by the inflence of a vanishingly small perturbation. While the standard resolution of this problem is the ensemble treatment (see, e.g., [2, 3]) an alternative approach is to use the multi- (and not single-) determinant wave functions to describe the reference state. This approach has been initiated by Lie and Clementi 25 years ago [4, 5] and is based on the observation that a few Slater determinants might describe an important, system specific part of the correlation energy, the remaining part showing an easy transferable behavior from system to system. A general definition of this separation of the correlation energy (often called "static" and "dynamic", respectively) is not unique, as can be seen by the treatment given in the Lie and Clementi paper quoted above [5] (for more recent discussions, see, e.g. [6, 7]). In their first paper [4], Lie and Clementi treated diatomic molecules by considering the minimum number of Slater determinants to guarantee the proper dissociation of the molecule. The dynamic correlation energy was obtained by using an approximate density functional (DF). In the second one [5], it is shown that a few configurations remain important, and that they have to be added to the wave function calculation in order to achieve good results. Other methods to couple a multi-determinant treatment to density functionals have been proposed (see, e.g., Ref. [8, 9, 10, 11, 12]). Essentially, there are two ways of resolving the problem of the choice of the "important" determinants:

a) finding an appropriate separation into determinants yielding the "static" correlation energy,

b) using a criterion allowing an arbitrary number of Slater determinants.

The second approach is of course more flexible, as it allows the user to decide upon the most convenient separation. It also allows one to check the quality of the result by changing the separation, and to approach the exact value along with the reduction of the DF contribution and the increase of the wave function part. The former approach is, however, simpler for a systematic study and "black-box" applications; this is the approach we followed in the present paper. We will limit ourselves (for the present) to wave function calculations, but briefly mention some implications of adding approximate density functionals to correct for the remaining part of the correlation energy. The systems that we will study are deliberately simple:

the He and Be series, as in the latter the origin of the near-degeneracy is well understood. Linderberg and Shull [13] have shown that in the Be series (in contrast to the He series) a two-configuration ($1s^2 2s^2$ and $1s^2 2p^2$) treatment is required, due to the degeneracy of the 2s and 2p levels when the nuclear charge, Z, goes to infinity. A confirmation of the regular behavior of the correlation energy in the series can be seen in, e.g., the numerical study of Clementi [14], and is confirmed by more recent estimations of the exact correlation energy (cf. [15]). In fact, as Z becomes very large, the dominant term in the correlation energy becomes constant (Z-independent) in the He series, while in the Be series it is linear in Z, the proportionality constant being -0.011727 [13]. The difficulty of approximate DFs to describe this behavior was recognized long ago [16], and has not been solved to this day in approximate Kohn-Sham methods. We would like to recall that the careful treatment of quasi-degenerate states is also important in wave-function methods [17].

2. Method

The philosophy of our approach is based upon a simple observation: molecules normally possess an energy gap, while the uniform electron gas, used as a starting point in LDA and GGA's, does not. Furthermore we expect the states contributing to the dynamic correlation to lie in the continuum, based on the observation of Davidson [18] that the most important natural orbitals of He are significantly different from the Rydberg orbitals: the correlating orbitals have to be localized in the same spatial region as the strongly occupied ones, in order to describe the formation of the correlation hole. We thus expect that among the states below the ionization limit only a few (the valence states) contribute significantly to the correlation energy. To be more specific, we will consider the one-particle states to be eigenstates of some one-particle hamiltonian:

$$h_0 \phi_i = \varepsilon_i \phi_i \tag{1}$$

and treat the correlation by standard quantum mechanics methods in the space of the orbitals with negative eigenvalues ($\varepsilon_i < 0$). [1] The choice of h_0 can play a major role. It is well known that obtaining the Hartree-Fock wave function does not have any implication on the non-occupied (virtual) states. The freedom of choosing the virtual states has been used, for example, in

[1] The potentials, and the orbital energies are, of course, determined only up to an arbitrary constant, which does not modify the eigenstates. As we are interested only in a selection of the eigenstates, the choice of the constant is immaterial. In our case, we choose all states which are below ionization for the non-interacting systems, which corresponds to the negative eigenvalues for a potential which vanishes at infinity.

constructing improved virtual orbitals (see, e.g., [19]) or other types of orbitals (see, e.g., [20]). Conventional Hartree-Fock calculations very often show that the eigenvalues of the virtual orbitals are positive, and thus not suited for our purpose. We will consider in this paper a well-defined one-body hamiltonian, with a local potential: the Krieger-Li-Iafrate (KLI) [21] hamiltonian. This has been already done, e.g., by Engel et al.[22]. Using orbitals which do not originate from a Fock potential is, of course, not new (cf. the use of orbitals which originate from density functional calculations, e.g., in Ref. [23]). Fritsche [24] has argued that taking the Kohn-Sham (KS) determinant as the starting point of a wave function expansion guarantees that density is already correct at zeroth order. Görling and Levy [25] have constructed a perturbation theory which guarantees that the density is correct to each given order. Their second-order energy expression does not contain the correlation potential, and thus one may consider the KLI potential (as giving an excellent approximation to the exact exchange potential, in the DF sense) as a "best" starting point, to second order. Furthermore, the exact KS potential requires the knowledge of the exact density, while the KLI potential can be obtained with an effort comparable to that of an Hartree-Fock calculation. We thus write :

$$h_0 = -\frac{1}{2}\nabla^2 + V_{KLI} \tag{2}$$

Let us now analyse the behavior of V_{KLI} for $Z \to \infty$. Following the treatment of Linderberg and Shull [13], we will change to modified Hartree units ($E \to \tilde{E} = E/Z^2$, $r \to \tilde{r} = Zr$; E is the energy, r the distance of the electron from the nucleus in atomic units.) The exact hamiltonian in these units becomes for atomic systems:

$$\tilde{H} = \tilde{T} - \sum_{i=1}^{N} \frac{1}{\tilde{r}_i} + \frac{1}{Z}\sum_{i<j} \frac{1}{\tilde{r}_{ij}} \tag{3}$$

where T is the operator for the kinetic energy, N the number of electrons, r_i the distance of the ith electron from the nucleus and r_{ij} the distance between electrons i and j. For $Z \to \infty$, the system is that of N non-interacting electrons in the external potential of the hydrogen atom. As the electron-electron interaction has been turned off as $1/Z$ in $\tilde{H}$, it will also disappear in $\tilde{V}_{KLI}$ as $1/Z$ so that

$$V_{KLI}(r) = Z^2 \tilde{V}_{KLI}(\tilde{r}) \to -\frac{Z}{r} \quad as \quad Z \to \infty \tag{4}$$

In the modified Hartree units (cf. text preceding Eq. 3) we thus obtain

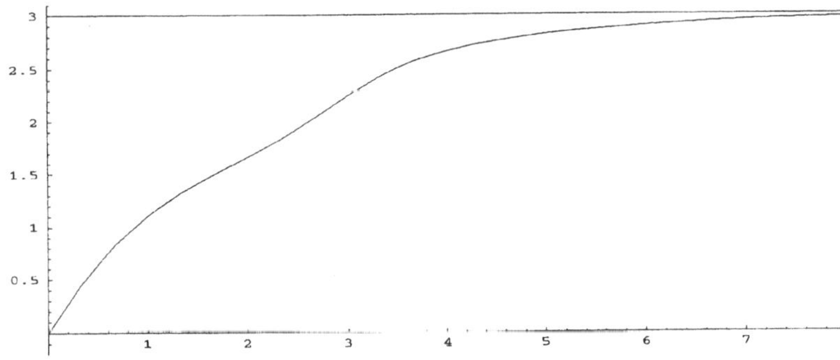

Figure 1. $\tilde{r}$ times $\tilde{V}^{(1)}_{KLI}(\tilde{r})$ as a function of $\tilde{r}$, asymptotically approaching $N-1=3$.

$$\tilde{V}_{KLI}(\tilde{r}) = -\frac{1}{\tilde{r}} + \frac{1}{Z}\tilde{V}^{(1)}_{KLI}(\tilde{r}) + \mathcal{O}\left(\frac{1}{Z^2}\right) \quad (5)$$

where $\tilde{V}^{(1)}_{KLI}$ is Z-independent. $\tilde{V}^{(1)}_{KLI}$ for the Be series is shown in Figure 1. Details about how to compute to first order the Coulomb and exchange parts of V_{KLI}, namely V_h and V_x in the limit of very large Z, are given in Appendix 1. The shape of $\tilde{r}$ times $\tilde{V}^{(1)}_{KLI}$ (cf. Fig. 1) is consistent with the asymptotic expression:

$$V_{KLI}(r) \rightarrow -\frac{Z-N+1}{r} \quad as \quad r \rightarrow \infty \quad (6)$$

which implies that Rydberg states exist both in the He and the Be series.

Let us now analyse the behavior of the KLI orbital energies as $Z \rightarrow \infty$. As for the potential expansion in Z, we write for the ith orbital energy, in modified Hartree units

$$\tilde{\varepsilon}_i = \tilde{\varepsilon}^{(0)}_i + \frac{1}{Z}\tilde{\varepsilon}^{(1)}_i + \mathcal{O}\left(\frac{1}{Z^2}\right) \quad (7)$$

Introducing Eq. 4 into Eq. 2 and applying perturbation theory, it turns out that the zeroth order is just the hydrogenic orbital energy

$$\varepsilon_{i,H} = \int \phi^*_{i,H}(\tilde{r})\left(-\frac{1}{2}\tilde{\nabla}^2 - \frac{1}{\tilde{r}}\right)\phi_{i,H}(\tilde{r})\, d^3\tilde{r} \quad (8)$$

($\phi_{i,H}$ being the ith orbital of the H atom), whereas we get the first-order change in the $\tilde{\varepsilon}_i$ to be

$$\tilde{\varepsilon}^{(1)}_i = \frac{1}{Z}\int |\phi_{i,H}(\tilde{r})|^2 \tilde{V}^{(1)}_{KLI}(\tilde{r})\, d^3\tilde{r} \quad (9)$$

in modified Hartree units (or $\sim Z$ in Hartree units). The difference between the $2s$ and $2p$ orbital energies is thus proportional to Z at most, for $Z \to \infty$. For $i = 2s$ and $2p$, we obtain for the integral, 0.530147 and 0.591788, respectively. We expect the KLI orbitals to have first-order changes with respect to $\phi_{i,H}$

$$\tilde{\phi}_i(\tilde{r}) = \phi_{i,H}(\tilde{r}) + \frac{1}{Z}\tilde{\phi}_i^{(1)}(\tilde{r}) + \mathcal{O}\left(\frac{1}{Z^2}\right) \qquad (i = 1s, 2s, 2p \cdots) \tag{10}$$

where $\tilde{\phi}_i^{(1)}$ is Z-independent. A similar expansion will arise for Slater determinants constructed from the $\tilde{\phi}_i$, $\tilde{\Phi}_I$ yielding

$$\tilde{\Phi}_I = \Phi_{I,H} + \frac{1}{Z}\tilde{\Phi}_I^{(1)} + \mathcal{O}\left(\frac{1}{Z^2}\right) \tag{11}$$

where $\Phi_{I,H}$ is constructed from $\phi_{i,H}$ (e.g., $i = 1s$, $2s$ for the Be $\Phi_{0,H}$), and $\tilde{\Phi}_I^{(1)}$ satisfies $\langle\Phi_{I,H}|\tilde{\Phi}_I^{(1)}\rangle = 0$, since it differs from $\Phi_{I,H}$ by the orbital $\tilde{\phi}_i^{(1)}$ which is orthogonal to $\phi_{i,H}$. These results allow us to calculate the behavior of the energy for $Z \to \infty$.

Starting with KLI determinant in modified Hartree units $\tilde{\Phi}_0$ and according to perturbation theory, we express the energy up to second order in $1/Z$ as[2]

$$\tilde{E}^{(0)} + \tilde{E}^{(1)} + \tilde{E}^{(2)} = \langle\tilde{\Phi}_0|\tilde{H}|\tilde{\Phi}_0\rangle + \sum_{I\neq 0} \frac{|\langle\tilde{\Phi}_0|\tilde{H}|\tilde{\Phi}_I\rangle|^2}{\sum\limits_{i\ in\ \tilde{\Phi}_0} 2\tilde{\varepsilon}_i - \sum\limits_{i\ in\ \tilde{\Phi}_I} 2\tilde{\varepsilon}_i} \tag{12}$$

i.e. in Hartree units for Be

$$E^{(0)} + E^{(1)} + E^{(2)} = -1.25Z^2 + 1.550111Z + \mathcal{O}\left(Z^0\right) \tag{13}$$

We have used the following formulas

$$\langle\tilde{\Phi}_0|\tilde{H}|\tilde{\Phi}_0\rangle = \langle\Phi_{0,H}|\sum_i -\frac{1}{2}\tilde{\nabla}_i^2 - \frac{1}{\tilde{r}_i}|\Phi_{0,H}\rangle + \frac{1}{Z}\langle\Phi_{0,H}|\sum_{i<j}\frac{1}{\tilde{r}_{ij}}|\Phi_{0,H}\rangle + \mathcal{O}\left(\frac{1}{Z^2}\right) = -1.25 + \frac{1}{Z}1.571000 + \mathcal{O}\left(\frac{1}{Z^2}\right) \tag{14}$$

[2]For the sake of simplicity we did not include single particle excitations into Eq. 12, as they contribute only to order $1/Z^2$ and higher. The reason is that the degenerate hydrogen orbitals, which might generate terms in $1/Z$, have different symmetry , and the matrix elements between orbitals of different symmetry are zero.

$$\langle\tilde{\Phi}_0|\tilde{H}|\tilde{\Phi}_I\rangle = \frac{1}{Z}\langle\Phi_{0,H}|\sum_{i<j}\frac{1}{\tilde{r}_{ij}}|\Phi_{I,H}\rangle + \mathcal{O}\left(\frac{1}{Z^2}\right)$$
$$= \begin{cases} \frac{1}{Z}0.050744 + \mathcal{O}\left(\frac{1}{Z^2}\right) & if\ \mathrm{I{=}1s^22}p^2 \\ \mathcal{O}\left(\frac{1}{Z}\right) & otherwise \end{cases} \qquad (15)$$

$$\sum_{i\ in\ \Phi_0} 2\tilde{\varepsilon}_i \quad \sum_{i\ in\ \Phi_I} 2\tilde{\varepsilon}_i = \frac{2}{Z}\int \tilde{V}^{(1)}_{KLI}(\tilde{r})\left(|\phi_{2s,H}(\tilde{r})|^2 - |\phi_{2p,H}(\tilde{r})|^2\right) d^3\tilde{r}$$
$$+\mathcal{O}\left(\frac{1}{Z^2}\right) = \begin{cases} -\frac{1}{Z}0.123282 + \mathcal{O}\left(\frac{1}{Z^2}\right) & if\ \mathrm{I{=}1s^22}p^2 \\ \mathcal{O}\left(Z^0\right) & otherwise \end{cases} \qquad (16)$$

where the term proportional to Z^2 in Eq. 13 is provided by $\tilde{\Phi}_0$ alone, while part of the term linear in Z comes from the second-order contribution. The correlation energy to order linear in Z is thus given by

$$E^{(2)} = -0.020887Z + \mathcal{O}\left(Z^0\right) \qquad (Be \quad series) \qquad (17)$$

The second order energy does not have a divergent term, in contrast to the one obtained using the hydrogen hamiltonian at zeroth order.
As the source of this term linear in Z is the $2s$-$2p$ degeneracy in the hydrogen atom, we can also diagonalize the hamiltonian matrix in this space [13]

$$\begin{pmatrix} \langle\tilde{\Phi}_s|\tilde{H}|\tilde{\Phi}_s\rangle & \langle\tilde{\Phi}_s|\tilde{H}|\tilde{\Phi}_p\rangle \\ \langle\tilde{\Phi}_p|\tilde{H}|\tilde{\Phi}_s\rangle & \langle\tilde{\Phi}_p|\tilde{H}|\tilde{\Phi}_p\rangle \end{pmatrix} = \qquad (18)$$
$$\begin{pmatrix} -1.25 + \frac{1}{Z}1.571000 + \mathcal{O}\left(\frac{1}{Z^2}\right) & \frac{1}{Z}0.050744 + \mathcal{O}\left(\frac{1}{Z^2}\right) \\ \frac{1}{Z}0.050744 + \mathcal{O}\left(\frac{1}{Z^2}\right) & -1.25 + \frac{1}{Z}1.778847 + \mathcal{O}\left(\frac{1}{Z^2}\right) \end{pmatrix}$$

where $\tilde{\Phi}_s$ is the $1s^22s^2$ configuration, and $\tilde{\Phi}_p$ is the $1s^22p^2$ configuration. To order $Z^{(0)}$, the expression for E is identical to that obtained when constructing the perturbation series with hydrogenic orbitals (instead of using KLI orbitals)[13]:

$$E = -1.25Z^2 + 1.559273Z + \mathcal{O}\left(Z^0\right) \qquad (19)$$

The resulting correlation energy is $-0.011727Z$ (using Eq. 14 for the KLI energy), which yields thus the exact leading term in the correlation energy. Notice that the KLI second-order energy overestimates the quasi-degeneracy effect: the coefficient of Z in Eq. 17 was nearly twice the exact

TABLE 1. **Energies in the He series**
$E^{(0)} + E^{(1)}$ is the expectation value of the hamiltonian with the KLI wave function; $E^{(2)}$ the correlation energy in second-order perturbation theory (Eq. 39); E_{exact} is taken from Ref. [15]. Total energies are given in order to facilitate the comparison with calculations using zeroth order wave functions different from KLI

Z	$E^{(0)} + E^{(1)}$	$E^{(0)} + E^{(1)} + E^{(2)}$	E_{exact}
2	-2.862	-2.907	-2.904
3	-7.236	-7.282	-7.280
4	-13.611	-13.656	-13.656
5	-21.986	-22.031	-22.031
6	-32.361	-32.406	-32.406
7	-44.736	-44.781	-44.781
8	-59.111	-59.156	-59.157
9	-75.486	-75.531	-75.532
10	-93.861	-93.906	-93.907

TABLE 2. **Energies in the Be series**
$E^{(0)} + E^{(1)}$ is the expectation value of the hamiltonian with the KLI wave function; $E^{(2)}$ the correlation energy in second-order perturbation theory (Eq. 39); E_{exact} is taken from Ref. [15]

Z	$E^{(0)} + E^{(1)}$	$E^{(0)} + E^{(1)} + E^{(2)}$	E_{exact}
4	-14.572	-14.693	-14.667
5	-24.237	-24.377	-24.349
6	-36.407	-36.566	-36.535
7	-51.081	-51.257	-51.223
8	-68.257	-68.448	-68.412
9	-87.933	-88.141	-88.101
10	-110.110	-110.334	-110.291

one. Of course, higher order terms compensate for this discrepancy. (We get for the third order energy a correction of $0.0035\frac{1}{Z}$ mH.)

3. Numerical results

Usually, the correlation energy is defined with respect to the non-relativistic Hartree-Fock energy. In density functional theory, the expectation value of

TABLE 3. **Energies in the He series in restricted subspaces**
$E_q^{(2)}$ is the correlation energy in second-order perturbation theory taking into account only the configurations $1s^2$, $2s^2$ and $2p^2$ built from KLI orbitals; E_q is the total energy obtained from a configuration interaction calculation in the same space; $E^{(2)}$ ($\varepsilon < 0$) is the correlation energy in second-order perturbation theory obtained in the space of the KLI orbitals with negative orbital energies

Z	$E^{(0)} + E^{(1)} + E_q^{(2)}$	E_q	$E^{(0)} + E^{(1)} + E^{(2)}$ ($\varepsilon < 0$)
2	-2.862	-2.862	-2.863
3	-7.237	-7.237	-7.238
4	-13.612	-13.612	-13.614
5	-21.987	-21.987	-21.989
6	-32.362	-32.362	-32.364
7	-44.737	-44.737	-44.739
8	-59.112	-59.112	-59.114
9	-75.487	-75.488	-75.489
10	-93.863	-93.862	-93.864

TABLE 4. **Energies in the Be series in restricted subspaces**
$E_q^{(2)}$ is the correlation energy in second-order perturbation theory taking into account only the configurations $1s^2 2s^2$ and $1s^2 2p^2$ built from KLI orbitals; E_q is the total energy obtained from a configuration interaction calculation in the same space; $E^{(2)}$ ($\varepsilon < 0$) is the correlation energy in second-order perturbation theory obtained in the space of the KLI orbitals with negative orbital energies; E_{MCHF} is the multiconfiguration Hartree-Fock energy taken from Ref. [15]

Z	$E^{(0)} + E^{(1)} + E_q^{(2)}$	E_q	$E^{(0)} + E^{(1)} + E^{(2)}$ ($\varepsilon < 0$)	E_{MCHF}
4	-14.619	-14.606	-14.627	-14.617
5	-24.308	-24.290	-24.315	-24.297
6	-36.498	-36.476	-36.504	-36.481
7	-51.189	-51.162	-51.195	-51.168
8	-68.380	-68.350	-68.387	-68.356
9	-88.072	-88.039	-88.080	-88.045
10	-110.265	-110.228	-110.273	-110.234

the KS determinant is used as a reference. As we start with a KLI determinant, we use still another definition of the correlation energy. In this paper, we will mean by 'correlation energy' the difference between the energy obtained at a given level and the expectation value of the hamiltonian obtained with the KLI determinant. In particular, we will compare the exact correlation energies, E_c, with those obtained within second-order perturbation theory, $E^{(2)}$ (cf. Tab. 1 and Tab. 2 for He and Be, respectively), $E_q^{(2)}$, the second-order contribution from the space of $1s$,$2s$ and $2p$ orbitals, $E_{c,q}$, the configuration interaction correlation energy in the same space[3], and finally the contribution to the second-order energy coming from the orbitals with negative orbital energies, $E^{(2)}$ $(\varepsilon_i < 0)$. The values for $E_q^{(2)}$, $E_{c,q}$, $E^{(2)}$ $(\varepsilon_i < 0)$ can be inferred from the data given in Tables 1 and 3 for He, and Tables 2 and 4 for Be. $E_q^{(2)}$, $E_{c,q}$, $E^{(2)}$ $(\varepsilon_i < 0)$ are compared to E_c in Figs. 2 and 3 (for He and Be respectively). Technicals details concerning our calculations are deferred to Appendix 2. We want to mention, however, that we estimate our second-order energies to be 1-2 mH too high, due to the basis sets limitations.

While for the He series, it turns out that the total energy is only slightly too low within second-order perturbation theory based upon KLI (cf. Tab. 1 and Fig. 2), a severe over-estimate of the correlation can be observed for the Be series (cf. Tab. 2 and Fig. 3). If we use only the orbitals with negative energies to obtain the correlation energy,$E^{(2)}$ $(\varepsilon_i < 0)$, we observe that in the He series, there is only a negligible contribution coming from these orbitals, while in the Be series it decreases linearly with Z, in accordance with the asymptotic behavior given by Eq. 17. Notice that the slope is more pronounced for $E^{(2)}(Z)$ and $E_q^{(2)}(Z)$ than for $E_c(Z)$ and $E_{c,q}(Z)$ (cf. Eqs. 17 and 19 and Fig. 3). Of course, higher order terms in the perturbation series will correct this effect. At infinite order in perturbation theory, using only the space of $1s^2 2s^2$ and $1s^2 2p^2$ configurations, we get the same result as that obtained by diagonalizing the hamiltonian in this space.

As the overestimation of the second-order correlation energy could be traced back to the quasi-degeneracy effect, a simple correction is the one suggested by Eggarter and Eggarter in their analysis of Hartree-Fock based second-order perturbation theory [26], namely that the correlation energy can be approximated by

$$E_c \approx E^{(2)} + \left(E_{c,q} - E_q^{(2)}\right) \tag{20}$$

This expression gives the correct linear term for $Z \to \infty$ as the erroneous linear term appearing in $E^{(2)}$ is compensated by that in $E_q^{(2)}$ (cf. Eq. 17). Using Eq. 20 the errors are reduced from maximally 43 to 13 mH in the Be series (cf. Tab. 4 and Fig. 4). For the He series a calculation in the space of determinants having the doubly occupied orbitals $1s$, $2s$ or $2p$ shows virtually no effect in comparison with a second-order energy calculation in the same space (cf. Tab. 3 and Fig. 4).

[3]We considered the $1s^2$,$2s^2$,$2p^2$ configurations in the He series, and the $1s^2 2s^2$ and $1s^2 2p^2$ configurations in the Be series.

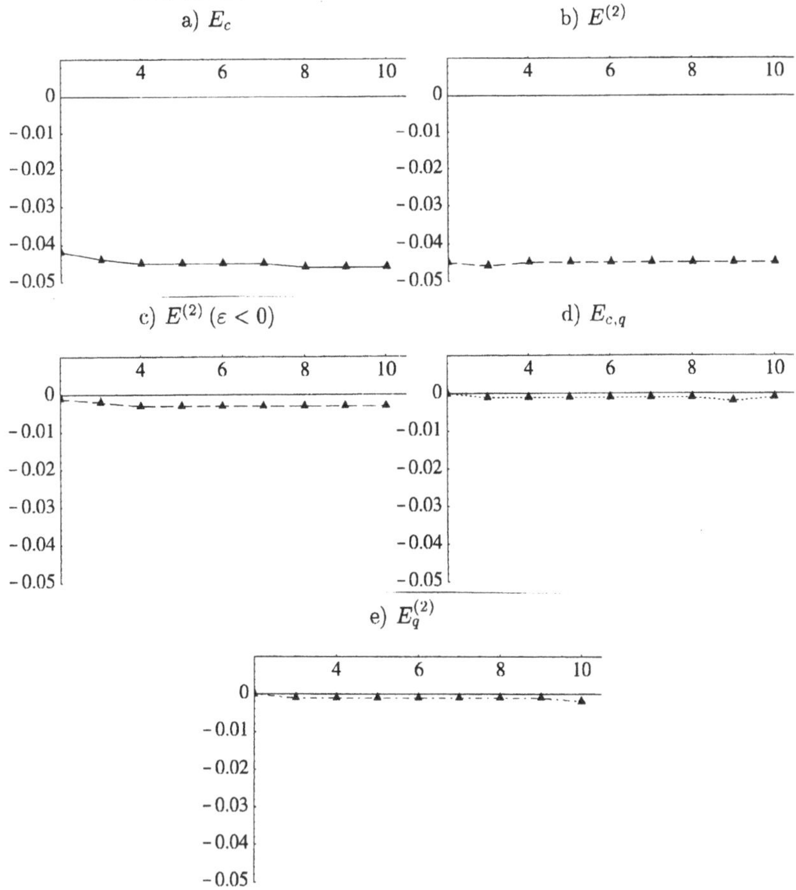

Figure 2. Correlation energies (i.e. differences to $\langle \Phi_0 | H | \Phi_0 \rangle$) in various approximations for the He series vs. the nuclear charge, Z: a) exact (taken from Ref.[15]); b) calculated up to second order in full space; c) calculated up to second order in the space of orbitals with negative energies; d) calculated with a configuration interaction in 1s,2s,2p subspace; e) calculated up to second order in 1s,2s,2p subspace.

In the Be series, we notice some difference between the second-order contributions in the $1s$, $2s$, $2p$ ($E_q^{(2)}$) space vs. $E^{(2)}(\varepsilon_i < 0)$ (cf. Fig. 3). (In the He series, both $E_q^{(2)}$ and $E^{(2)}(\varepsilon_i < 0)$ are very small, cf. Fig. 2.) Lower bounds for the energy obtainable in the space of the $1s$, $2s$, $2p$ orbitals are given by the multi-configuration Hartree-Fock (MCHF) calculations [15], which yield the lowest energies in the $1s$, $2s$, $2p$ space. We see from Tab. 4 that optimizing the orbitals in the MCHF calculation lowers the energy by roughly 10 mH in the Be series; this difference is similar to that obtained when comparing $E^{(2)}(\varepsilon_i < 0)$ with $E_q^{(2)}$. Thus, taking care of $E^{(2)}(\varepsilon_i < 0)$ might be used to compensate for orbital optimization.

We believe that a density functional might be used to obtain the total energies once the effect of quasi-degenerate states on the correlation energy is taken into account. In fact, introducing a gap in the uniform electron gas is possible. Several variants might be investigated, e.g.

a) shifting all virtual states by a gap

b) excluding all states above the Fermi level within a range for the remaining part of the correlation energy.

For both cases electron gas calculations were done, and the latter variant was already used in molecular calculations in a scheme combining multi-configuration wave-functions with density functionals (see, e.g., Ref. [10]). The uniform electron gas with a gap shift was also used in molecular calcu-

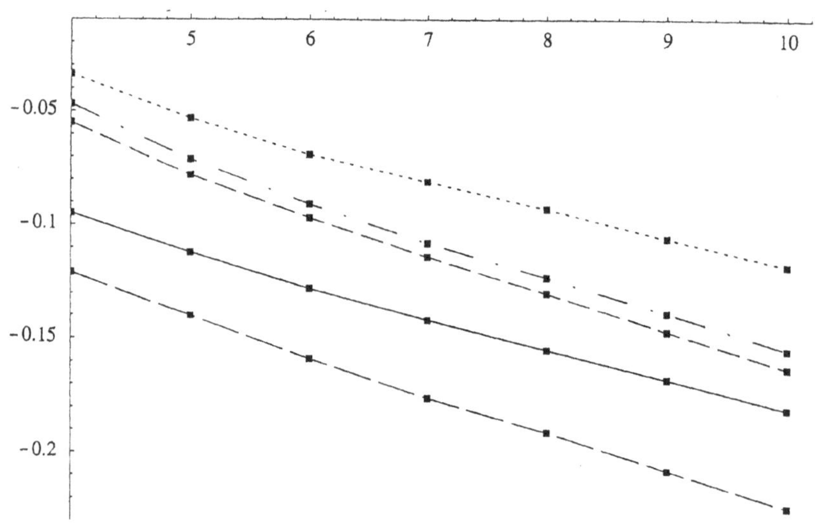

Figure 3. Correlation energies (i.e. differences to $\langle\Phi_0|H|\Phi_0\rangle$) in various approximations for the Be series vs. the nuclear charge, Z:
a) full line: exact taken from Ref.[15] (E_c);
b) long dashed line: calculated up to second order in full space ($E^{(2)}$);
c) short dashed line: calculated up to second order in the space of orbitals with negative energies ($E^{(2)}$ $(\varepsilon < 0)$);
d) dotted line: calculated with a configuration interaction in 1s,2s,2p subspace ($E_{c,q}$);
e) dashed dotted line: calculated up to second order in 1s,2s,2p subspace ($E_q^{(2)}$).

lations assuming that one can define a local gap $\frac{1}{8}\left(|\nabla n|/n\right)^2$ (where n is the electron density of the system considered) which asymptotically is equal to the ionization potential [28]. It turns out that one has to take into account also the self-interaction correction and a gradient correction. In connection with recent exchange density functionals, quite good results could be obtained [30]. Such a treatment neglects, in principle, all contributions coming from states within the gap. It thus seems natural to add the correlation energy of such a functional to that missing in a full calculation within the space of orbitals with negative energies. As shown above, this quantity is quite safely estimated from a configuration interaction within the space of quasi-degenerate states. Thus, the density functional should describe only $E_c - E_{c,q}$ which changes very little with Z (cf. Fig. 5; the fluctuations in the plot might be due to our numerical accuracy). To estimate the effect of the higher states with $\varepsilon_i < 0$ we use an equation similar to Eq. 20:

$$E_c\left(\varepsilon_i < 0\right) \approx E_{c,q} + \left(E^{(2)}\left(\varepsilon_i < 0\right) - E_q^{(2)}\right) \tag{21}$$

We plot in Fig. 5 the part of the correlation energy which has to be described by the density functional, viz., $E_c - E_c(\varepsilon_i < 0)$ which we approximate by

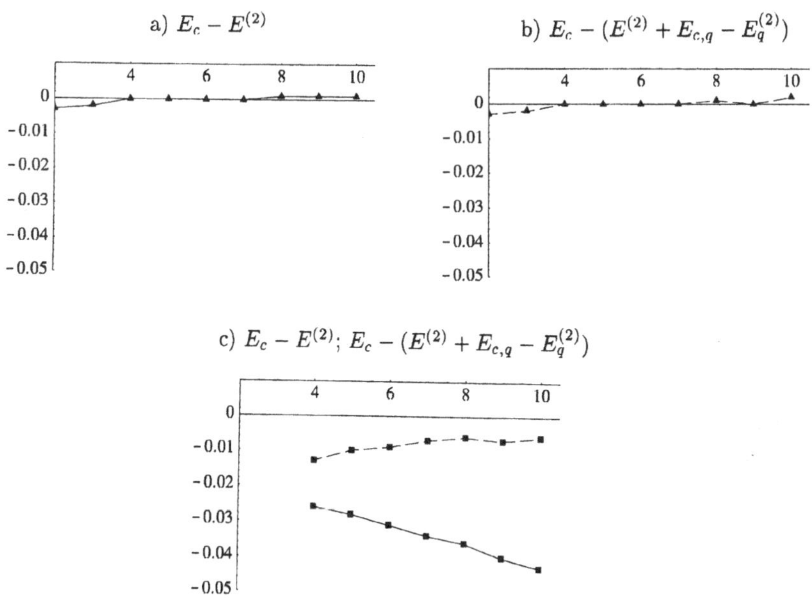

Figure 4. Errors of second-order energies and of Eq. 20 with respect to nuclear charge, Z:
a) He series, $E_c - E^{(2)}$;
b) He series, $(E_c - (E^{(2)} + E_{c,q} - E_q^{(2)}))$, cf. Eq. 20;
c) Be series, full line: $E_c - E^{(2)}$; broken line: $(E_c - (E^{(2)} + E_{c,q} - E_q^{(2)}))$, cf. Eq. 20.

E_c minus the r.h.s. of Eq. 20. We also show in Fig. 5 $E_c - E_{c,q}$ which differs only little from the preceding quantity. In both cases we notice that there is a very small change of this quantity with respect to Z, and this is in part due to our numerical inaccuracies. It would not be difficult to generate a functional having a behavior like that shown for $E_c - E_{c,q}$, or $E_c - E_c\,(\varepsilon_i < 0)$. It turns out, however, that using the functional of Ref. [30] for Be already yields a quite reasonable correlation energy, and using $E_{c,q}$ (cf. Fig. 6) plus this density functional for $E_c - E_{c,q}$ would yield a too large correlation energy. We also show the difference between E_c and $E^{(2)}(\varepsilon < 0)$ to show that the latter is not so easy to describe with the same type of Z-dependence, due to the error in the Z-dependent term, mentionned before.

4. Conclusion

Our results demonstrate that even when the nearly exact KLI exchange potential (cf. [21, 22]) is employed as the zeroth order KS potential, it is not possible to even qualitatively account for the correlation energy by employ-

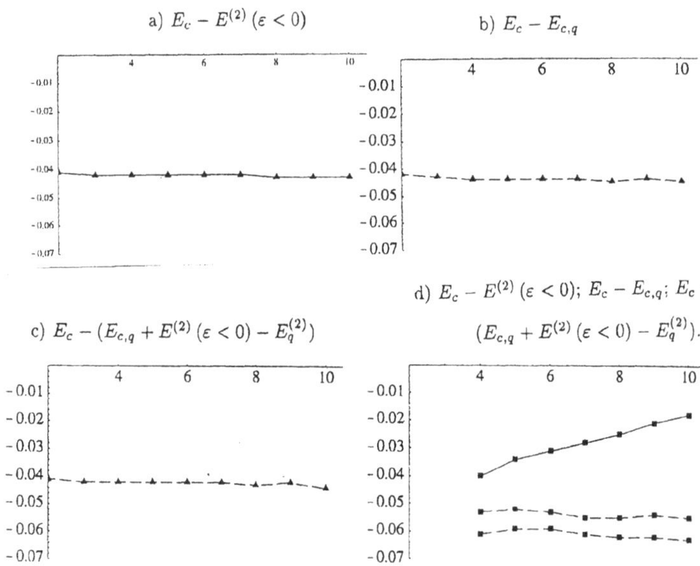

Figure 5. Correlation energy not described by states in the gap, with respect to nuclear charge, Z:
a) He series, $E_c - E^{(2)}\,(\varepsilon < 0)$;
b) He series, $E_c - E_{c,q}$;
c) He series, $(E_c - (E_{c,q} + E^{(2)}\,(\varepsilon < 0) - E_q^{(2)}))$, cf. Eq. 21;
d) Be series, full line: $E_c - E^{(2)}\,(\varepsilon < 0))$, long dashed line: $E_c - E_{c,q})$, short dashed line: $(E_c - (E_{c,q} + E^{(2)}\,(\varepsilon < 0) - E_q^{(2)}))$, cf. Eq. 21.

ing second order perturbation theory in a system such as the Be isoelectronic series in which virtual states exist which are quasi-degenerate with occupied orbitals. This follows from the fact that the coefficient of the term linear in Z, which dominates the correlation energy, is in error by nearly a factor of two when second order perturbation is employed. However, we find that diagonalizing the Hamiltonian in the space of quasi-degenerate orbitals by employing a linear combination of determinants leads to the correct linear dependence on Z of the correlation energy. Moreover, the remaining correlation energy due to contributions from all the other virtual orbitals in the system including those in the continuum (which actually make the largest contribution) rapidly saturate as Z increases as in the case of the He isoelectronic series in which there is no quasi-degeneracy of virtual with occupied orbitals. This encourages us to believe that it will be possible to construct density functionals for this component of the correlation energy since the latter tend to saturate in the high density limit when properly constructed to satisfy certain scaling conditions (25,28). The inclusion of quasi-degeneracy effects, although generally ignored in DF calculations, can have important effects on calculated ionization potentials, electron affinities and molecular atomization energies because the addition or subtraction of an electron or the creation of a chemical bond can easily change a system from one having a quasi-degeneracy to one that does not and vice-versa, leading to significant changes in the correlation energy.

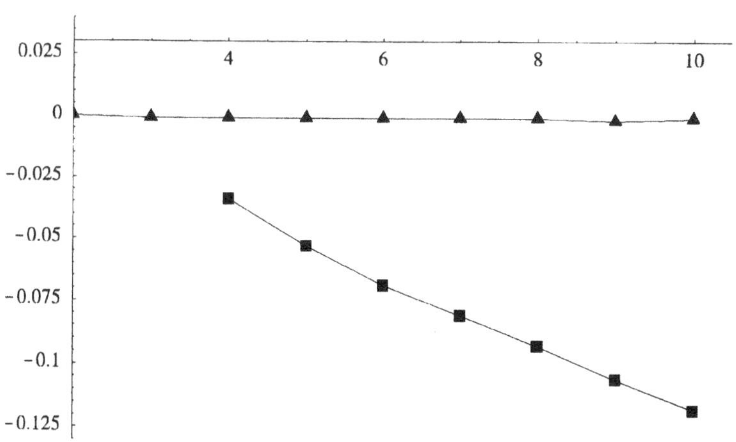

Figure 6. Difference between the energies in the space of Slater determinants constructed by doubly occupying the 1s, 2s, 2p orbitals, and the expectation values with the KLI determinants, $\langle\Phi_0|H|\Phi_0\rangle$, vs. the nuclear charge, Z. a) triangles connected by a full line: in the He series ($E_{c,q}$); b) squares connected by a full line: in the Be series ($E_{c,q}$).

We therefore anticipate that the inclusion of the considerations discussed above will make significant contributions to the accurate calculation of the ground state properties of atoms and molecules.

5. Appendix

5.1. $1/Z$ TERM OF $\tilde{V}_{KLI}$

In order to see what happens in order $1/Z$ in KLI, let us recall a few expressions.

We take for the Coulomb term:

$$\tilde{V}_h(\tilde{r}) = \frac{1}{Z}\int \frac{n_H(\tilde{r}')}{|\tilde{r}-\tilde{r}'|} d^3\tilde{r}' + \mathcal{O}\left(\frac{1}{Z^2}\right) \tag{22}$$

where

$$n_H(\tilde{r}) = \sum_{i=1s,2s} |\phi_{i,H}(\tilde{r})|^2 \tag{23}$$

and get the following expressions from mathematica[29]:

a) in the He series:

$$\tilde{V}_h(\tilde{r}) = \frac{1}{Z}\left(-\frac{2(1+\tilde{r})}{\tilde{r}}e^{-2\tilde{r}} + \frac{2}{\tilde{r}}\right) + \mathcal{O}\left(\frac{1}{Z^2}\right) \tag{24}$$

b) in the Be series:

$$\tilde{V}_h(\tilde{r}) = \frac{1}{Z}\left(-\frac{2(1+\tilde{r})}{\tilde{r}}e^{-2\tilde{r}} - \frac{8+6\tilde{r}+2\tilde{r}^2+\tilde{r}^3}{4\tilde{r}}e^{-\tilde{r}} + \frac{4}{\tilde{r}}\right) + \mathcal{O}\left(\frac{1}{Z^2}\right) \tag{25}$$

Using notations similar to those of [21], but dropping the spin index, as we restrict ourselves to closed-shells, we have for the exchange energy:

$$\tilde{E}_x = -\frac{1}{2}\sum_{ij}\left(\tilde{\phi}_i\tilde{\phi}_j|\tilde{\phi}_j\tilde{\phi}_i\right)\frac{1}{Z} \tag{26}$$

and its functional derivative, employing real wavefunctions, is

$$\frac{\delta\tilde{E}_x}{\delta\tilde{\phi}_i(\tilde{r})} = \frac{1}{Z}\sum_j \tilde{\phi}_j(\tilde{r})\int \frac{\tilde{\phi}_i(\tilde{r}')\tilde{\phi}_j(\tilde{r}')}{|\tilde{r}-\tilde{r}'|}d^3\tilde{r}' \tag{27}$$

From Eq. 12 of Ref. [21]

$$\tilde{v}_i(\tilde{r}) = -\frac{1}{Z}\frac{1}{\tilde{\phi}_i(\tilde{r})}\sum_j \tilde{\phi}_j(\tilde{r})\int \frac{\tilde{\phi}_i(\tilde{r}')\,\tilde{\phi}_j(\tilde{r}')}{|\tilde{r}-\tilde{r}'|}d^3\tilde{r}' \tag{28}$$

From Eq. 48 of Ref. [21] and developping the $\tilde{\phi}_i$ according to Eq. 10, we get for the local exchange part of KLI potential:

a) in the He series:

$$\tilde{V}_x(\tilde{r}) = -\frac{1}{Z}\int \frac{\phi_{1s,H}(\tilde{r}')\,\phi_{1s,H}(\tilde{r}')}{|\tilde{r}-\tilde{r}'|}d^3\tilde{r}' + \mathcal{O}\left(\frac{1}{Z^2}\right) \tag{29}$$

b) in the Be series:

$$\tilde{V}_x(\tilde{r}) = \{\sum_{i=1s,2s} |\phi_{i,H}(\tilde{r})|^2\tilde{v}_i(\tilde{r}) + |\phi_{1s,H}(\tilde{r})|^2\tilde{C}\}\frac{1}{n_H(\tilde{r})} + \mathcal{O}\left(\frac{1}{Z^2}\right) \tag{30}$$

and the constant $\tilde{C}$ is determined by the requirement (Eq. 18 of Ref. [21]):

$$\int [\tilde{V}_x(\tilde{r}) - \tilde{v}_{1s}(\tilde{r})]|\tilde{\phi}_{1s}(\tilde{r})|^2 d^3\tilde{r} = \tilde{C} \tag{31}$$

Substituting into the above equation the expression of $\tilde{V}_x(\tilde{r})$ and integrating over $\tilde{r}$, we get for the Be series:

$$\begin{aligned}
\tilde{C}\ &\left(1 - \int \frac{|\phi_{1s,H}(\tilde{r})|^2}{n_H(\tilde{r})}d^3\tilde{r} + \mathcal{O}\left(\frac{1}{Z}\right)\right) = \\
-\ &\frac{1}{Z}\sum_{i,j=1s,2s}\int \frac{|\phi_{1s,H}(\tilde{r})|^2}{n_H(\tilde{r})}\phi_{i,H}(\tilde{r})\,\phi_{j,H}(\tilde{r})\frac{1}{|\tilde{r}-\tilde{r}'|} \\
\times\ &\phi_{i,H}(\tilde{r}')\,\phi_{j,H}(\tilde{r}')\,d^3\tilde{r}d^3\tilde{r}' + \frac{1}{Z}\sum_{i=1s,2s}(\phi_{1s,H}\phi_{i,H}|\phi_{i,H}\phi_{1s,H})
\end{aligned} \tag{32}$$

From the following values for the integrals involving hydrogenic orbitals:

$$\begin{aligned}
(\phi_{1s,H}\phi_{1s,H}|\phi_{1s,H}\phi_{1s,H}) &= 0.624999 \\
(\phi_{2s,H}\phi_{2s,H}|\phi_{2s,H}\phi_{2s,H}) &= 0.150390 \\
(\phi_{1s,H}\phi_{1s,H}|\phi_{2s,H}\phi_{2s,H}) &= 0.209876 \\
(\phi_{1s,H}\phi_{2s,H}|\phi_{1s,H}\phi_{2s,H}) &= 0.021948
\end{aligned} \tag{33}$$

we obtain with *Mathematica*

$$\tilde{C} = -0.0360971\frac{1}{Z} + \mathcal{O}\left(\frac{1}{Z^2}\right) \tag{34}$$

and

$$\tilde{v}_{1s}(\tilde{r}) = \frac{1}{Z}\left(-\frac{1}{\tilde{r}} + \frac{3\tilde{r}^2 - 4\tilde{r} - 4}{54}e^{-\tilde{r}} + \frac{1+\tilde{r}}{\tilde{r}}e^{-2\tilde{r}}\right) + \mathcal{O}\left(\frac{1}{Z^2}\right) \tag{35}$$

$$\tilde{v}_{2s}(\tilde{r}) = \frac{1}{Z}\left(-\frac{1}{\tilde{r}} + \frac{8 + 6\tilde{r} + 2\tilde{r}^2 + \tilde{r}^3}{8\tilde{r}}e^{-\tilde{r}} + \frac{16(2+3\tilde{r})}{27(\tilde{r}-2)}e^{-2\tilde{r}}\right) + \mathcal{O}\left(\frac{1}{Z^2}\right) \tag{36}$$

5.2. TECHNICAL DETAILS

The KLI potentials were obtained with the numerical program [31].

The zeroth, first and second-order energies were calculated according to the formulas:

$$E_{KLI}^{(0)} = \sum_a^{occ} \langle \phi_a | \hat{T} + \hat{V}_{ne} + \hat{V}_{KLI} | \phi_a \rangle \tag{37}$$

$$\begin{aligned} E_{KLI}^{(1)} &= \sum_a^{occ} \left(-2\langle \phi_a | \hat{V}_{KLI} | \phi_a \rangle + \langle \phi_a \phi_a | \phi_a \phi_a \rangle\right) \\ &+ \sum_{a<b}^{occ} \left(4\langle \phi_a \phi_b | \phi_a \phi_b \rangle - 2\langle \phi_a \phi_b | \phi_b \phi_a \rangle\right) \end{aligned} \tag{38}$$

$$\begin{aligned} E_{KLI}^{(2)} = 2\sum_a^{occ}\sum_r^{virt} & \frac{\left| -\langle \phi_a | \hat{V}_{KLI} | \phi_r \rangle + \sum_b^{occ} \left(2\langle \phi_a \phi_b | \phi_r \phi_b \rangle - \langle \phi_a \phi_b | \phi_b \phi_r \rangle\right) \right|^2}{\varepsilon_a - \varepsilon_r} \\ +4\sum_{a<b}^{occ}\sum_{r<s}^{virt} & \frac{|\langle \phi_a \phi_b | \phi_r \phi_s \rangle|^2 + |\langle \phi_a \phi_b | \phi_s \phi_r \rangle|^2 - \langle \phi_a \phi_b | \phi_r \phi_s \rangle \langle \phi_a \phi_b | \phi_s \phi_r \rangle}{\varepsilon_a + \varepsilon_b - \varepsilon_r - \varepsilon_s} \\ & + \sum_a^{occ}\sum_r^{virt} \frac{|\langle \phi_a \phi_a | \phi_r \phi_r \rangle|^2}{2\varepsilon_a - 2\varepsilon_r} + 2\sum_{a<b}^{occ}\sum_r^{virt} \frac{|\langle \phi_a \phi_b | \phi_r \phi_r \rangle|^2}{\varepsilon_a + \varepsilon_b - 2\varepsilon_r} \\ & + 2\sum_a^{occ}\sum_{r<s}^{virt} \frac{|\langle \phi_a \phi_a | \phi_r \phi_s \rangle|^2}{2\varepsilon_a - \varepsilon_r - \varepsilon_s} \end{aligned} \tag{39}$$

where ϕ_a, ϕ_b (respectively ϕ_r,ϕ_s) are occupied (respectively virtual) KLI orbitals with energies ε_a, ε_b (respectively ε_r, ε_s).

The correlation energies in the space of the configurations generated by doubly occupying the 1s, 2s and 2p KLI orbitals were obtained with *Mathematica* [29].

The KLI orbitals were obtained in a Slater type even-tempered basis set (available on request).

The second-order energies, using all orbitals, can be compared for the He series to the more accurate ones obtained by Engel et al [22], which quote 48.2 and 46.8 for He and Ne^{8+}, respectively for their correlated OPM results (cf.Tab.5 of [22]) from which we estimate our errors to be $\lesssim 2mH$, which may well be attributed to the missing f, g, ... functions in our basis set. As the 4f, 5g, ... states are not expected to yield a significant contribution to the correlation energy, we believe that our results for $\varepsilon_i < 0$ should be accurate to the same order of magnitude.

References

1. F. Aryasetiawan, M.J. Stott, Phys. Rev. B **34**, 4401 (1986).
2. M. Levy, Phys. Rev. A **26**, 1200 (1982).
3. E.H. Lieb, Int. J.Quantum Chem. **24**, 243 (1983).
4. G.C.Lie, E.Clementi, J. Chem. Phys. **60**, 1274 (1974).
5. G.C.Lie, E.Clementi, J. Chem. Phys. **60**, 1278 (1974).
6. E. Valderrama, E.V. Ludeña, J. Hinze, J. Chem. Phys. **106**, 9227 (1997).
7. D.K.W. Mok, R. Neumann, N.C. Handy, J. Phys. Chem. **100**, 6225 (1996).
8. F. Moscardo, E. San-Fabian, Phys. Rev. A **44**, 1549 (1991).
9. R.Colle, O. Salvetti, Theor. Chim. Acta **53**, 55 (1979).
10. A.Savin, Int. J. Quantum Chem. S **22**, 59 (1988).
11. A.Savin, H.J. Flad, Int. J. Quantum Chem. **56**, 327 (1995).
12. N.O.J. Malcolm, J.J.W. McDouall, Chem. Phys. Lett. **282**, 121 (1998).
13. J. Linderberg and H. Shull, J. Mol. Spectry **5**, 1 (1960).
14. E.Clementi, J. Chem. Phys. **42**, 2783 (1965).
15. S.J. Chakravorty, J. Phys. Chem. **100**, 6172 (1996), http://php.indiana.edu/ davidson/atom/paper.html.
16. J.P. Perdew, E.R. McMullen, A. Zunger, Phys. Rev. A **23**, 2785 (1982).
17. S. Salomonson, I.Lindgren, A.M. Martenson, Physica Scripta **21**, 351 (1980).
18. E.R. Davidson, Rev. Mod. Phys. **44**, 451 (1972).
19. S.Huzinaga, D. McWilliams, A.A. Cantu, Adv. Quantum Chem. **7**, 187 (1973).
20. E.R. Davidson, J. Chem. Phys. **57**, 1999 (1972).
21. J. Krieger, Y. Li, G. Iafrate, in *Density Functional Theory*, E.K.U. Gross, R.M. Dreizler, eds., Plenum Press, New York, 1995, p.191.
22. E. Engel, R.M. Dreizler, J. Comp. Chem **20**, 31 (1999).
23. S. Shankar, P.T. Narasimhan, Phys. Rev. A **29**, 58 (1984).
24. L. Fritsche, Phys. Rev. B **33**, 3976 (1986).
25. A. Görling and M. Levy, Phys. Rev. B **47**, 13105 (1993).
26. T. P. Eggarter and E. Eggarter, J. Phys. B: Atom. Molec. Phys.
27. C. Gutle, A.Savin, J. Chen, J.Krieger, Int. J. Quantum Chem. **75**, 885 (1999).
28. J. Krieger, J. Chen, G. Iafrate, A. Savin, in: *Electron Correlation and Material Properties*, A. Gonis and N. Kioussis, eds., Plenum, New-York, to be published.

29. S. Wolfram, *The Mathematica Book* (Wolfram Media/Cambridge University Press, Cambridge, 1996).
30. S. Kurth, J. P. Perdew, P. Blaha, Int. J. Quantum Chem. **75**, 889 (1999).
31. Y. Li, unpublished.

ORBITAL LOCAL-SCALING TRANSFORMATION APPROACH: FERMIONIC SYSTEMS IN THE GROUND STATE

YA. I. DELCHEV *, A. I. KULEFF *, #, P. TZ. YOTOV *
J. MARUANI #, AND R. L. PAVLOV *

* *Institute for Nuclear Research and Nuclear Energy, Bulgarian Academy of Sciences, 72 Tzarigradsko Chaussée, 1784 Sofia, Bulgaria (ydelchev@inrne.bas.bg)*

Laboratoire de Chimie Physique, UPMC and CNRS, 11 rue Pierre-et-Marie-Curie, 75005 Paris, France (maruani@ccr.jussieu.fr)

Abstract. The local-scaling transformation (LST) variational method is used to derive an energy functional which depends on R dynamical variables, the LST functions realizing the transformation of the R orbitals involved in the expansion of the model N-fermion wave function. The solution of the Euler-Lagrange system with respect to these variables leads to the determination of approximate ground-state properties for the considered system. A set of atomic and molecular problems is discussed.

1. Introduction

A local-scaling transformation (LST) is a special type of mapping of the three-dimensional vector space onto itself: $\vec{r} \to \vec{f}(\vec{r})$, which maintains the direction $\vec{r}_0 = \vec{r}/r$ of the original vector $\vec{r}$. This transformation is a generalization of the standard scaling transformation: $\vec{r} \to \lambda\vec{r}$. LSTs have been applied in quantum chemistry (e.g., Refs [1-8]). Petkov and Stoitsov [9] and Petkov and coworkers [10] have systematically developed the LST variational method, using a single transformation for all particle coordinates $\vec{r}_i$. On the basis of this method a rigorous version of DFT was formulated [11-20].

In the present paper we propose an extension of the LST method, introducing independent LSTs for different single-particle orbitals. This orbital local-scaling transformation (OLST) method is applicable to various multifermionic systems.

J. Maruani et al. (eds.), New Trends in Quantum Systems in Chemistry and Physics, Volume 1, 45–61.

In the next section the main features of the LST method are recalled.

The third section is devoted to the analysis of the effect of independent LSTs for the different one-particle orbitals. It is shown that the nodal structure of the initial orbitals, and that of the new (locally-scaled) ones, lead to a weakening of the constraints on the Jacobian with respect to the case of a uniform transformation of the orbitals. The OLST operators which realize the transformation of the reference set of orbitals are introduced. The set manifold thus generated forms an orbit whose elements are, by construction, in one-to-one correspondence with the wave functions built through these elements. The set of all these functions forms a class in the N-particle Hilbert space L_N.

In the fourth section is formulated a variational method based on the OLSTs defined in this class, $\mathcal{L}_N^R \subset L_N$. The energy of the system becomes a functional of the LST functions, $f_i(\vec{r})$, that realize the transformation of the orbitals $\varphi_i(\vec{r})$. The variation of this functional with respect to the new (locally-scaled) dynamical variables $f_i(\vec{r})$ leads to an Euler-Lagrange system. Using the solution of this system one can derive the approximate wave function and then all ground-state properties.

In the last section we discuss some possible implementations of the proposed OLST method for atoms and molecules, and also some technical points for its realization.

2. Local-scaling transformation of the real three-dimensional vector space

We consider the three-dimensional vector space

$$\mathbf{R}^3 \equiv \{\vec{r} = (r, \vec{r}_0);\, r \in \mathbf{R}^1 \equiv [0, \infty), \vec{r}_0 \in S^2\}, \tag{1}$$

where S^2 is the sphere of unit vectors $\vec{r}_0 = \vec{r}/r$ centered on the origin of coordinates. The local-scaling transformation is the following continuous transformation of $\mathbf{R}^3$:

$$\begin{aligned} &\vec{f} : \mathbf{R}^3 \to \mathbf{R}^3 \\ &\vec{r} \to \vec{f}(\vec{r}) = f(r, \vec{r}_0)\vec{r}/r. \end{aligned} \tag{2}$$

Obviously, the mapping $\vec{r} \to \vec{r}\,' = \vec{f}(\vec{r}) \in \mathbf{R}^3$ ascertains a one-to-one correspondence between every vector $\vec{r} \in \mathbf{R}^3$ and the vector $\vec{r}\,'$, with the same direction but changed length: $r' = f(r, \vec{r}_0) \in \mathbf{R}^1$.

We assume that the set of all scalar functions $f(r, \vec{r}_0)$ satisfies the conditions:

$$\begin{aligned}&(i)\quad f(\vec{r})\in C^1(\mathbf{R}^3),\\&(ii)\quad f(r_1,\theta,\varphi)<f(r_2,\theta,\varphi),\quad \forall r_1<r_2 \textit{ and } \forall \vec{r}_0\equiv(\theta,\varphi),\end{aligned}\tag{3}$$

and that the functional determinant (Jacobian of the LSTs) obeys the conditions:

$$(iii)\quad J(\vec{f}(\vec{r});\vec{r})\neq 0 \quad \textit{and} \quad J(\vec{f}(\vec{r});\vec{r})<\infty,\quad \forall\vec{r}\in\mathbf{R}^3. \tag{4}$$

The Jacobian can be written as:

$$J(\vec{f}(\vec{r});\vec{r})=\frac{1}{3r^3}\vec{r}.\nabla f^3(\vec{r}), \tag{5}$$

In spherical polar coordinates, for every fixed pair (θ, φ), Eqn (5) takes the form

$$J(\vec{f}(\vec{r});\vec{r})=\frac{f^2(r,\theta,\varphi)}{r^2}\frac{\partial f(r,\theta,\varphi)}{\partial r}. \tag{6}$$

The properties of the local-scaling transformation set fulfilling the conditions (3) and (4) can be formulated in the following propositions [9, 10], the proofs of which are given in Ref. [9].

Lemma 1: The set of LSTs defined in Eqn (2) forms the so-called LST group $\mathcal{F}$ with respect to the operation "sequential application":

$$\vec{f}_1\circ\vec{f}_2\rightarrow\vec{f}_{12}(\vec{r})=\vec{f}_2(\vec{f}_1(\vec{r})). \tag{7}$$

This means that:

1) The binary operation is associative for any three elements $\vec{f}_1(\vec{r}),\vec{f}_2(\vec{r}),\vec{f}_3(\vec{r})$ belonging to $\mathcal{F}$.

2) There exists a unit element, $\vec{f}_e(\vec{r})=\vec{r}\in\mathcal{F}$.

3) For every $\vec{f}(\vec{r})\in\mathcal{F}$ there exists an inverse element, $\vec{f}^{-1}(\vec{r})\in\mathcal{F}$, such that

$$\vec{f}(\vec{f}^{-1}(\vec{r}))=\vec{f}^{-1}(\vec{f}(\vec{r}))=\vec{f}_e(\vec{r}),\quad\forall\vec{r}\in\mathbf{R}^3.$$

Let L_N be the antisymmetric Hilbert space of an N-fermion system described by a Hamiltonian $\hat{H}$ independent of the spin variables, and let $\Phi(\vec{r}_1,...,\vec{r}_N)$ be an arbitrary normalized wave function belonging to L_N. To every element of the set $\mathcal{F}$ one may associate a unitary linear operator $\mathcal{U}_{\vec{f}}^N$ defined through the rule:

$$\mathcal{U}_{\vec{f}}^{N}\Phi(\vec{r}_1,\ldots,\vec{r}_N):=\prod_{i=1}^{N}[J(\vec{f}(\vec{r}_i);\vec{r}_i)]^{1/2}\Phi(\vec{f}(\vec{r}_1),\ldots,\vec{f}(\vec{r}_N))\equiv\Psi([f];\vec{r}_1,\ldots,\vec{r}_N)\,. \quad (8)$$

We denote the set of operators $\mathcal{U}_{\vec{f}}^{N}$ corresponding to the elements $\vec{f}(\vec{r})\in\mathcal{F}$ by

$$\mathcal{U}_{\mathcal{F}}^{N}=\{\mathcal{U}_{\vec{f}}^{N}\mid\vec{f}(\vec{r})\in\mathcal{F}\}\,.$$

Lemma 2: The set $\mathcal{U}_{\mathcal{F}}^{N}$ is a group with respect to the operator product and is a unitary representation of $\mathcal{F}$ in the space L_N.

Lemma 3: The groups $\mathcal{F}$ and $\mathcal{U}_{\mathcal{F}}^{N}$ are isomorphic.

One may also define the action of the operator $\mathcal{U}_{\vec{f}}^{N}$ as N repeated actions of the operator $\mathcal{U}_{\vec{f}}$:

$$\mathcal{U}_{\vec{f}}^{N}\Phi(\vec{r}_1,\ldots,\vec{r}_N):=\underbrace{\mathcal{U}_{\vec{f}}\ldots\mathcal{U}_{\vec{f}}}_{N\ times}\Phi(\vec{r}_1,\ldots,\vec{r}_N)\,. \quad (9)$$

The i-th action of $\mathcal{U}_{\vec{f}}$ realizes the LST of the wave function Φ with respect to the i-th coordinate, which leads to Eqn (9) [9, 10] when all transformations are performed. The set of unitary linear operators $\mathcal{U}_{\vec{f}}$ forms a group isomorphic to $\mathcal{F}$: $\mathcal{U}_{\mathcal{F}}\equiv\{\mathcal{U}_{\vec{f}}\mid\vec{f}(\vec{r})\in\mathcal{F}\}$. The group $\mathcal{U}_{\mathcal{F}}^{N}$ can be represented as the N-th direct diagonal product of N replicas of the group $\mathcal{U}_{\mathcal{F}}$ [11]:

$$\mathcal{U}_{\mathcal{F}}^{N}=[\times]^{N}\mathcal{U}_{\mathcal{F}}\,. \quad (10)$$

3. Independent local-scaling transformations of the single-particle orbitals

We consider the space of square-integrable single-particle functions L_1. Let us assume that $\{\varphi_i(\vec{r},\sigma)\}_i\equiv\{\varphi_i(\vec{r})\chi_i(\sigma)\}_i\equiv\{\varphi_i(\vec{x})\}_i$ is a given orthonormal (or simply normalized) linearly independent sequence containing, in general, an infinite number of spin-orbitals, and that L_1^R is a subspace of L_1 spanned onto a finite ordered set $\{\varphi_i(\vec{x})\}_{i=1}^{R}$ of spin-orbitals. We transform them through independent LSTs acting on every spin-orbital by a different operator $\mathcal{U}_{\vec{f}_i}$:

$$\psi_i(\vec{r}) \equiv \psi_i([f_i];\vec{r}) = \mathcal{U}_{\vec{f}_i}\varphi_i(\vec{r}) = [J(\vec{f}_i(\vec{r});\vec{r})]^{1/2}\varphi_i(\vec{f}_i(\vec{r})) = \\ = \frac{f_i(r,\theta,\varphi)}{r}\left(\frac{\partial f_i(r,\theta,\varphi)}{\partial r}\right)^{1/2}\varphi_i(\vec{f}_i(\vec{r})), \quad i=1,\ldots,R, \tag{11}$$

where the second line of Eqn (11) is written in spherical coordinates for every fixed pair (θ, φ), writing $\varphi_i(\vec{f}_i(\vec{r})) = \varphi_i(f_i(r,\theta,\varphi),\theta,\varphi)$, $\psi_i(\vec{r}) = \psi_i(r,\theta,\varphi)$, and $f_i(\vec{r}) \equiv f_i(r,\theta,\varphi)$. The corresponding "spin-orbital density distributions" are:

$$\rho_i^{\psi}(\vec{r}) \equiv \rho_i^{\psi}([f_i];\vec{r}) = \psi_i^*(\vec{r})\psi_i(\vec{r}) = J(\vec{f}_i(\vec{r});\vec{r})\varphi_i^*(\vec{f}_i(\vec{r}))\varphi_i(\vec{f}_i(\vec{r})) = \\ = \frac{f_i^2(r,\theta,\varphi)}{r^2}\frac{\partial f_i(r,\theta,\varphi)}{\partial r}\rho_i^{\varphi}(\vec{f}_i(\vec{r})), \quad i=1,\ldots,R, \tag{12}$$

and it is supposed that the scalar functions $f_i(r,\vec{r}_0)$, which are continuous and differentiable, satisfy the conditions (3). These conditions define $\{f_i(\vec{r})\}_{i=1}^{R}$ as monotonically increasing for every fixed pair $\Omega_0=(\theta_0,\varphi_0)$. Therefore, there exists a one-valued function $f_i^{-1}(r,\theta_0,\varphi_0) \equiv r(f_i,\theta_0,\varphi_0)$, inverse to $f_i(r,\theta_0,\varphi_0)$. It follows from Eqs (12) and (3) that every function $\rho_i^{\psi}(\vec{r})$ must be non-negative.

Let $\varphi_i(\vec{r})$ be an arbitrary orbital of the set $\{\varphi_i(\vec{r})\}_{i=1}^{R}$, $f_i(\vec{r})$ an LST function, and the transformation (12) generate a continuous function $\rho_i^{\psi}(\vec{r})$. Integrating Eqn (12) one obtains:

$$\langle\psi_i|\psi_i\rangle = \int \rho_i^{\psi}([f_i(\vec{u})];\vec{u})d\vec{u} = \int \rho_i^{\varphi}(\vec{v})d\vec{v} = \langle\varphi_i|\varphi_i\rangle = 1. \tag{13}$$

The equalities (13) show that the considered functions $\psi_i(\vec{r})$ must also be normalized and, as we have the relation $\langle\psi_i|\psi_i\rangle = \langle\varphi_i|\mathcal{U}_{\vec{f}_i}^{+}\mathcal{U}_{\vec{f}_i}|\varphi_i\rangle$, the operators $\mathcal{U}_{\vec{f}_i}$ are unitary. Thus, fixing the LST function $f_i(\vec{r})$ - which obeys Eqs (3) - and $\rho_i^{\varphi}(\vec{r})$ - which is a normalized, non-negative and continuous distribution, one establishes that the function $\rho_i^{\psi}(\vec{r})$ must also be normalized.

Let now $\rho_i^{\varphi}(\vec{r})$ and $\rho_i^{\psi}(\vec{r})$ be fixed functions. Then Eqn (12) can be brought into the form of an equation with separate variables, whose particular solution

$$\int_0^r \rho_i^{\psi}(u)u^2du - \int_{f_i(0)}^{f_i(r)} \rho_i^{\varphi}(v)v^2dv = 0, \quad f_i(0) \geq 0 \tag{14}$$

always exists [20]. Solving Eqn (14) for all pairs (θ,φ) one obtains the total LST function $f_i(\vec{r}) = f_i(r,\theta,\varphi)$.

As the action of these operators does not affect the spin functions, $\chi_i(\sigma)$, we have omitted the spin coordinates in Eqs (11) and (12), and shall use the term orbitals to cover both orbitals and spin-orbitals. We shall study the existence and definiteness of transformations (11) for both orthonormal and just normalized initial sets $\{\varphi_i(\vec{r})\}_{i=1}^{R}$, assuming the orbitals $\varphi_i(\vec{r})$ are continuous.

Let us assume further that orbitals $\varphi_i(\vec{r})$ and $\psi_i(\vec{r})$ have some nodal points. Then, the corresponding one-orbital densities $\rho_i^{\varphi}(\vec{r})$ and $\rho_i^{\psi}(\vec{r})$ are equal to zero and have minima for these nodal points. For a chosen direction, $\Omega_0 = (\theta_0, \varphi_0)$, Eqn (12) can be put in the form of a first-order differential equation:

$$\frac{df_i(r)}{dr} = \frac{r^2}{f_i^2(r)} \frac{\rho_i^{\psi}(r)}{\rho_i^{\varphi}(f_i(r))}, \quad f_i(0) \geq 0. \tag{15}$$

As the functions $\rho_i^{\psi}(r)$, $\rho_i^{\varphi}(r)$ and $f_i(r)$ are continuous, they are locally limited. Thus, besides the points where $\rho_i^{\varphi}(f_i(r))$ becomes zero, the right-hand side of Eqn (15):

$$F(r,f) = \frac{r^2}{f_i^2(r)} \frac{\rho_i^{\psi}(r)}{\rho_i^{\varphi}(f_i(r))}, \tag{16}$$

is a locally limited function.

Let the nodal points of the orbital $\varphi_i(r)$ be located at $r = \tilde{r}_\alpha^{\varphi_i}$, $\alpha = 1, \ldots, k$. Then the function $\rho_i^{\varphi}(f_i(r))$ in the denominator of the fraction $F(r,f)$ vanishes for $r = r_\alpha^{\varphi_i} = f_i^{-1}(\tilde{r}_\alpha^{\varphi_i})$, $\alpha = 1, \ldots, k$. At these points, the conditions for the existence theorem in Eqn (15) are not fulfilled [21]. The derivative $(df_i(r)/dr) \to \infty$ when $r \to r_\alpha^{\varphi_i}$, and one cannot obtain values of the LST function $f_i(r)$ by solving Eqn (15).

Let now the density distribution $\rho_i^{\psi}(f_i(r))$ become zero at $r = r_\beta^{\psi_i}$ ($\beta = 1, \ldots, l$), and

$$r_\alpha^{\varphi_i} \neq r_\beta^{\psi_i}, \quad \forall \alpha, \beta. \tag{17}$$

Then the derivative $df_i(r)/dr$, for $r = r_\beta^{\psi_i}$, is parallel to the r-axis. For the points $r_\alpha^{\varphi_i}$ where the conditions for the existence theorem in Eqn (15) are not fulfilled, the equation

$$\frac{dr(f_i)}{df_i} = \frac{1}{F(r,f)} = \frac{f_i^2}{r^2} \frac{\rho_i^{\varphi}(f_i(r))}{\rho_i^{\psi}(r)} \tag{18}$$

is well defined for all $r \neq r_\beta^{\psi_i}$. Determining the solution $r(f_i)$ of Eqn (18), one can now find the $f_i(r = r_\alpha^{\varphi_i})$, $\alpha = 1, \ldots, k$, as solutions of the equation

$$r(x) = r_\alpha^{\varphi_i} . \tag{19}$$

Here Eqn (15) has no meaning and, hence, the function $f_i(r, \theta, \varphi)$, for the fixed direction Ω_0.

The above statements can be summarized as follows.

Proposal 1: A set of continuous, linearly independent, normalized orbitals $\{\varphi_i(\vec{r})\}_{i=1}^R$ can be transformed uniquely into a normalized set $\{\psi_i([f_i(r)];\vec{r})\}_{i=1}^R$, locally scaling the initial set by the operators (11) and imposing the following conditions on the distribution $\rho_i^\psi(\vec{r})$:

$$0 \leq \rho_i^\psi(\vec{r}) < \infty, \quad i = 1, \ldots, R . \tag{20a}$$

For those points $\vec{r}_\alpha^{\varphi_i}$ (if they exist) that are related to the nodal points $\vec{r} = \tilde{\vec{r}}_\alpha^{\varphi_i}$ of the initial orbitals $\varphi_i(\vec{r})$ by the equation

$$\vec{r}_\alpha^{\varphi_i} = \vec{f}_i^{-1}(\tilde{\vec{r}}_\alpha^{\varphi_i}), \tag{20b}$$

the Jacobian goes to infinity:

$$J(\vec{f}_i(\vec{r});\vec{r}) \to \infty, \quad \vec{r} \to \vec{r}_\alpha^{\varphi_i}, \tag{21a}$$

for every chosen direction. The positions $\vec{r} = \vec{r}_\beta^{\psi_i}$ (if they exist) where the derivative $(\partial f_i(\vec{r})/\partial r) = 0$, i.e., when the Jacobian of the LSTs annihilate:

$$J(\vec{f}_i(\vec{r});\vec{r}) = 0, \tag{21b}$$

for every chosen direction, are nodal points of the functions $\psi_i([f_i(\vec{r})];\vec{r})$. Anywhere else, the following conditions:

$$J(\vec{f}_i(\vec{r});\vec{r}) \neq 0 \quad \textit{and} \quad J(\vec{f}_i(\vec{r});\vec{r}) < \infty \tag{21c}$$

must be fulfilled.

As an illustration, we display below the functions $f_i(r)$ and $\partial f_i(\vec{r})/\partial r$ that realize the LSTs of the orthogonal, hydrogen-like orbitals $\varphi_{1,00}^H = R_{1s}^H(r)Y_{00}(\theta,\varphi)$ and

$\varphi_{2,00}^{H} = R_{2s}^{H}(r)Y_{00}(\theta,\varphi)$ into the analytical HF orbitals of Clementi and Roetti [22] $\varphi_{1,00}^{CR} = R_{1s}^{CR}(r)Y_{00}(\theta,\varphi)$ and $\varphi_{2,00}^{CR} = R_{2s}^{CR}(r)Y_{00}(\theta,\varphi)$ for lithium.

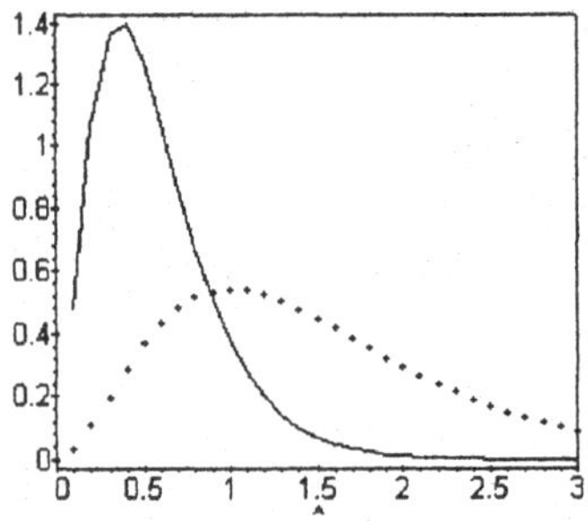

Figure 1a. Graphics of the orbital densities $\rho_{1s}^{H}(r)$ (solid line) and $\rho_{1s}^{CR}(r)$ (dotted line) in atomic units.

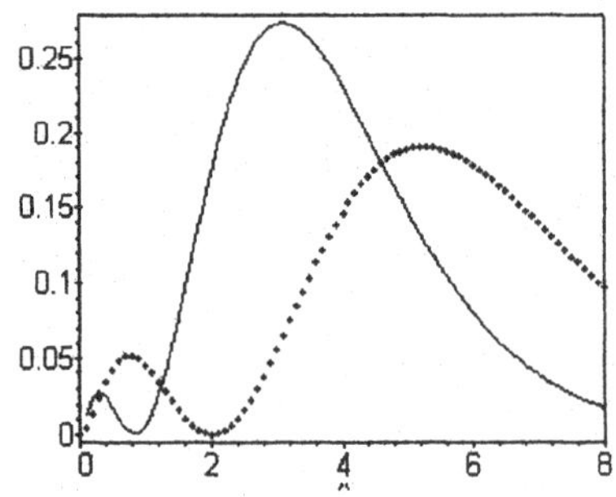

Figure 1b. Graphics of the orbital densities $\rho_{2s}^{H}(r)$ (solid line) and $\rho_{2s}^{CR}(r)$ (dotted line) in atomic units.

Using Eqn (14) we have calculated in a closed analytical form the LST functions $f_{1s}(r)$ and $f_{2s}(r)$ and their derivatives $\partial f_{1s}(r)/\partial r$ and $\partial f_{2s}(r)/\partial r$. In Figures (1a) and (1b) graphics of the one-orbital densities $\rho_{1s}^{H}(r)$, $\rho_{1s}^{CR}(r)$ and $\rho_{2s}^{H}(r)$, $\rho_{2s}^{CR}(r)$, respectively, are displayed. Figures (2a) and (2b) show the LST functions $f_{1s}(r)$, $f_{2s}(r)$ and the first derivative of the latter. As is seen in Fig. (2b) $f'_{2s}(r) \rightarrow$ infinity for one of the two inflexion points of $f_{2s}(r)$ (which corresponds to the zero of $\rho_{2s}^{H}(f_{2s}(r))$. The other inflexion point (for which $f'_{2s}(r) = 0$) corresponds to the zero of $\rho_{2s}^{CR}(r)$.

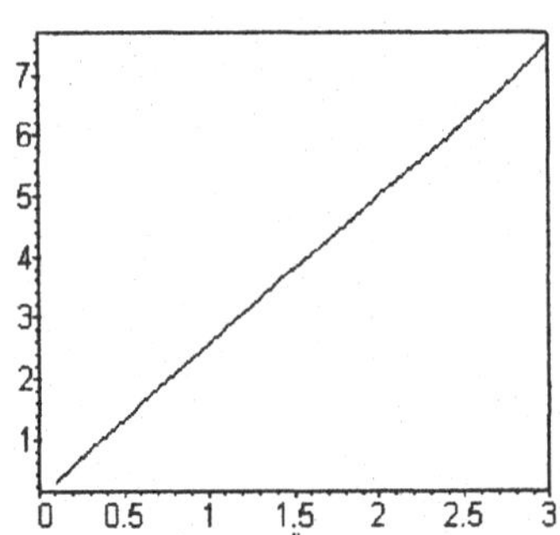

Figure 2a. Graphic of the LST function $f_{1s}(r)$ in atomic units.

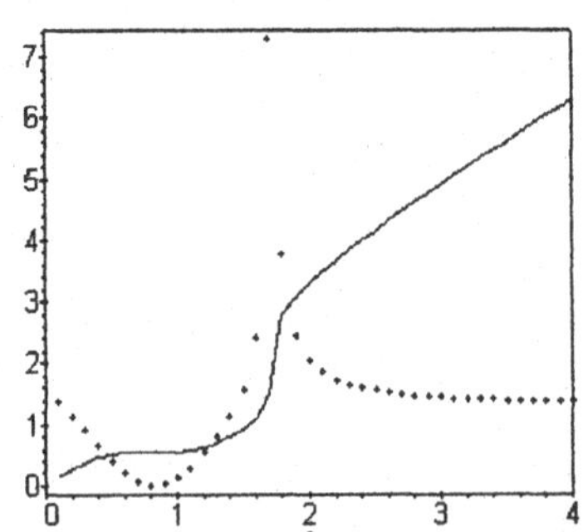

Figure 2b. Graphics of the LST function $f_{2s}(r)$ (solid line) and its first derivative (dotted line) in atomic units.

It can be proven that the set of all LSTs, $\{\vec{f}_i\}$, fulfilling the conditions (20) and (21) forms a group $\mathcal{F}_i \equiv \{\vec{f}_i\}$ (see Section 2). Every element $\vec{f}_i$ of this group transforms the i-th reference orbital $\varphi_i(\vec{r})$ into the function $\psi_i([f_i(\vec{r})];\vec{r})$, and distorts the corresponding distribution $\rho_i^{\varphi}(\vec{r})$ into the distribution $\rho_i^{\psi}([f_i(\vec{r})];\vec{r})$, keeping the angles with the original direction $\vec{r}_0 = \vec{r}/r$ unchanged. The set of unitary linear operators $\mathcal{U}_{\vec{f}_i}$ forms a group $\mathcal{U}_{\mathcal{F}_i} \equiv \{\mathcal{U}_{\vec{f}_i} \mid \vec{f}_i \in \mathcal{F}_i\}$, isomorphic to the group $\mathcal{F}_i$.

We form the set $\mathcal{F}^R$ of all elements

$$F_{\vec{f}_1,\vec{f}_2,\ldots,\vec{f}_R} := (\vec{f}_1(\vec{r}),\vec{f}_2(\vec{r}),\ldots,\vec{f}_R(\vec{r})) \equiv \{\vec{f}_i(\vec{r})\}_{i=1}^R, \tag{22}$$

where every $\vec{f}_i(\vec{r})$ is a fixed LST function belonging to $\mathcal{F}_i$. We can write the set

$$\mathcal{F}^R \equiv \{F_{\vec{f}_1,\vec{f}_2,\ldots,\vec{f}_R} \mid \vec{f}_i \in \mathcal{F}_i, i = 1,\ldots,R\} \tag{23}$$

as a direct product of the groups $\mathcal{F}_i$:

$$\mathcal{F}^R = \mathcal{F}_1 \otimes \mathcal{F}_2 \otimes \ldots \otimes \mathcal{F}_R. \tag{24}$$

Since the group $\mathcal{F}^R$ is a set of tensor products of elements of the LST groups $\mathcal{F}_i$, through which the orbital transformations are realized, we call $\mathcal{F}^R$ the orbital LST (OLST) group.

To every element $F_{\vec{f}_1,\vec{f}_2,\ldots,\vec{f}_R} \in \mathcal{F}^R$ we associate an operator $\mathcal{U}_{\vec{f}_1,\vec{f}_2,\ldots,\vec{f}_R}$ acting on the set $\{\varphi_i(\vec{r})\}_{i=1}^R$ according to the rule:

$$\begin{aligned}
\mathcal{U}_{\vec{f}_1,\vec{f}_2,\ldots,\vec{f}_R}\{\varphi_i(\vec{r})\}_{i=1}^R &:= (\mathcal{U}_{\vec{f}_1},\mathcal{U}_{\vec{f}_2},\ldots,\mathcal{U}_{\vec{f}_R})(\varphi_1(\vec{r}),\varphi_2(\vec{r}),\ldots,\varphi_R(\vec{r})) \equiv \\
&\equiv (\mathcal{U}_{\vec{f}_1}\varphi_1(\vec{r}),\mathcal{U}_{\vec{f}_2}\varphi_2(\vec{r}),\ldots,\mathcal{U}_{\vec{f}_R}\varphi_R(\vec{r})) = \\
&= (\psi_1([f_1(\vec{r})];\vec{r}),\psi_2([f_2(\vec{r})];\vec{r}),\ldots,\psi_R([f_R(\vec{r})];\vec{r}) \equiv \\
&= \{\psi_i([f_i(\vec{r})];\vec{r})\}_{i=1}^R,
\end{aligned} \tag{25}$$

where the action of the one-particle operators $\mathcal{U}_{\vec{f}_i}$ is defined in Eqn (11).

The operators $\mathcal{U}_{\vec{f}_1,\vec{f}_2,\ldots,\vec{f}_R}$ that transform the initial set $\{\varphi_i\}_{i=1}^R \in L_1^R$ into a normalized set $\{\psi_i\}_{i=1}^R$ form a group $\mathcal{U}_{\mathcal{F}}^R \equiv \{\mathcal{U}_{\vec{f}_1,\vec{f}_2,\ldots,\vec{f}_R}^R \mid \{\vec{f}_i\}_{i=1}^R \in \mathcal{F}^R\}$, by virtue of their definition (25). The normalized sets $\{\psi_i\}_{i=1}^R$ generated by the action of the operators $\mathcal{U}_{\vec{f}_1,\vec{f}_2,\ldots,\vec{f}_R} \in \mathcal{U}_{\mathcal{F}}^R$ are not linearly independent. Here we formally redefine the groups $\mathcal{F}^R$ and $\mathcal{U}_{\mathcal{F}}^R$, excluding the elements $F_{\vec{f}_1,\vec{f}_2,\ldots,\vec{f}_R} \in \mathcal{F}^R$ (and the corresponding ones $\mathcal{U}_{\vec{f}_1,\vec{f}_2,\ldots,\vec{f}_R}$) that produce linearly dependent sets, for which the Grammian is equal to zero:

$$\Gamma(\mathcal{U}_{\vec{f}_1}\varphi_1, \mathcal{U}_{\vec{f}_2}\varphi_2, \ldots, \mathcal{U}_{\vec{f}_R}\varphi_R) = \det\left[\left\langle \mathcal{U}_{\vec{f}_i}\varphi_i \middle| \mathcal{U}_{\vec{f}_j}\varphi_j \right\rangle\right] = 0. \tag{26}$$

The groups $\bar{\mathcal{F}}^R$ and $\bar{\mathcal{U}}_{\mathcal{F}}^R$ reduced in this manner are mutual isomorphic.

Let us fix a normalized set of R linearly independent orbitals $\{\varphi_i(\vec{x})\}_{i=1}^R \in L_1^R$ (we restore the spin variables in the notation). The manifold of all orbital sets $\{\psi_i\}_{i=1}^R$ induced by the operators $\bar{\mathcal{U}}_{\vec{f}_1,\vec{f}_2,\ldots,\vec{f}_R} \in \bar{\mathcal{U}}_{\mathcal{F}}^R$ - Eqn (25) - forms an orbit:

$$\vartheta[\{\varphi_i\}_{i=1}^R] \equiv \left\{\bar{\mathcal{U}}_{\vec{f}_1,\ldots,\vec{f}_R}\{\varphi_i(\vec{r})\}_{i=1}^R \mid \bar{\mathcal{U}}_{\vec{f}_1,\ldots,\vec{f}_R} \in \bar{\mathcal{U}}_{\mathcal{F}}^R\right\}, \tag{27}$$

where $\{\varphi_i(\vec{x})\}_{i=1}^R$ is the orbit-generating set. For this orbit $\vartheta[\{\varphi_i\}_{i=1}^R] \equiv \vartheta_\varphi^R$, holds the inclusion relation $\vartheta_\varphi^R \subset L_1$. Hence ϑ_φ^R is that manifold in L_1 which consists of all sets of linearly independent orbitals $\{\psi_i([\vec{f}_i(\vec{r})];\vec{x})\}_{i=1}^R$ induced by the elements of the group $\bar{\mathcal{U}}_{\mathcal{F}}^R$.

Let $R = N$ and

$$\phi(\vec{x}_1, \vec{x}_2, \ldots, \vec{x}_N) \equiv \varphi_1(\vec{x}_1)\varphi_2(\vec{x}_2)\ldots\varphi_N(\vec{x}_N) \tag{28}$$

be the N-particle product for a chosen spin-orbital set $\{\varphi_i(\vec{x})\}_{i=1}^N \in L_1^N$. Antisymmetrizing the expression (28) yields the corresponding Slater determinant:

$$\Phi^S(\vec{x}_1, \vec{x}_2, \ldots, \vec{x}_N) \equiv A_N[\varphi_1(\vec{x}_1)\varphi_2(\vec{x}_2)\ldots\varphi_N(\vec{x}_N)] \equiv [\varphi_i(\vec{x}_j)]_{ij} \in S_N \subset L_N, \tag{29}$$

where A_N is the antisymmetrizer, L_N the antisymmetric N-particle Hilbert space, and S_N the subclass of single Slater determinants in L_N. In the same way we build the Slater determinants for the spin-orbital set belonging to the orbit ϑ_φ^N:

$$\begin{aligned}\Psi^S_{f_1,f_2,\ldots,f_N}(\vec{x}_1, \vec{x}_2, \ldots, \vec{x}_N; \phi) &\equiv \\ &\equiv A_N[\psi_1([f_1];\vec{x}_1)\psi_2([f_2];\vec{x}_2)\ldots\psi_N([f_N];\vec{x}_N)] \equiv [\psi_i([f_i];\vec{x}_j)]_{ij}.\end{aligned} \tag{30}$$

The class $S_N[\bar{\mathcal{F}}^N]$ of determinantal wave functions (30) originates from the model determinant (29). The very mode of generation of the determinants (30) sets up a correspondence of every spin-orbital set of the orbit ϑ_φ^N with one and only one function of the class (30): $\{\psi_i[\vec{f}_i]\}_{i=1}^N \in \vartheta_\varphi^N \Leftrightarrow \Psi^S_{\vec{f}_1,\ldots,\vec{f}_N} \in S_N[\bar{\mathcal{F}}^N]$, and vice versa. This means that the determinant class $S_N[\mathcal{F}^N]$ is well defined both for the OLSTs, which belong to the group $\bar{\mathcal{F}}^N$, and on the orbit ϑ_φ^N : $S_N[\bar{\mathcal{F}}^N] \equiv S_N[\vartheta_\varphi^N]$. Since, for a given prototype determinant Φ, one could not generate all N-particle Slater determinants through OLSTs - Eqs. (11) and (12), $S_N[\mathcal{F}^N] \subset S_N$.

Every N-particle wave function, $\Psi_0(\{\vec{x}_i\}_{i=1}^N)$, can be approximated through a unique, multidimensional expansion belonging to the subspace $L_N^R \subset L_N$, spanned on $M = \binom{R}{N}$ Slater determinants $\{\Phi_I \equiv A[\varphi_{i_1}(\vec{x})\ldots\varphi_{i_N}(\vec{x})]\}_{I=1}^M$:

$$\begin{aligned}\Phi(\vec{x}_1,\ldots,\vec{x}_N) &= \sum_{i_1<\ldots<i_N} C(i_1,\ldots,i_N)A[\varphi_{i_1}(\vec{x}_1)\ldots\varphi_{i_N}(\vec{x}_N)] = \\ &= \sum_{I=\{i_1,\ldots,i_N\}} C_I\Phi_I^S(\vec{x}_1,\ldots,\vec{x}_N).\end{aligned} \tag{31}$$

In the above expansion, the determinants are built from the spin-orbital sets $\{\varphi_{i_k}(\vec{x})\}_{i_k=1}^N$ of N indexes: $I = \{i_k\}$, $i_1<i_2<\ldots<i_N$. Then, using Eqn (30), one gets:

$$\Psi_{f_1,\ldots,f_R}(\vec{x}_1,\ldots,\vec{x}_N;\Phi) = \sum_{\{i_k\}}^{M} C\{i_k\}[\psi_{i_k}([f_{i_k}];\vec{x}_j)]_{i_k j} \in \mathcal{L}_N^R(\mathcal{F}^R). \tag{32}$$

Every function of the class $\mathcal{L}_N^R[\mathcal{F}^R]$ of configuration-interaction (CI)-like wave functions is built from a single spin-orbital set $\{\psi_i([f_i(\vec{r})];\vec{x})\}_{i=1}^R$. This leads, by construction, to a one-to-one correspondence between the sets $\{\psi_i[f_i]\}_{i=1}^R \in \vartheta_\varphi^R$ and the approximate CI-like wave functions (32):

$$\{\psi_i[f_i]\}_{i=1}^R \in \vartheta_\varphi^R \Leftrightarrow \Psi_{\vec{f}_1,\ldots,\vec{f}_R} \in \mathcal{L}_N^R[\mathcal{F}^R], \tag{33}$$

and, hence, $\mathcal{L}_N^R[\mathcal{F}^R] \equiv \mathcal{L}_N^R[\vartheta_\varphi^R]$. It is obvious that, as in the one-determinant case, one also has: $\mathcal{L}_N^R \subset L_N^R$.

4. Variational method based on the OLSTs

Using a class of trial N-particle wave functions, $\Psi \in L_N$, to find an approximation to the ground-state energy, E_0, by minimization of the functional

$$E[\Psi] = \frac{\langle \Psi | \hat{H} | \Psi \rangle}{\langle \Psi | \Psi \rangle},$$

one performs an application of the variational method based on the LSTs of the one-particle spin-orbitals, which we call the OLST method.

We consider an N-particle fermionic system in its ground state, described by the Hamiltonian

$$\hat{H}(\vec{r}_1, \ldots, \vec{r}_N) = \sum_{i=1}^{N} (-\tfrac{1}{2}\Delta_i + V(\vec{r}_i)) + \sum_{i>j} g(\vec{r}_i, \vec{r}_j). \tag{34}$$

We choose a CI-like prototype wave function, Eqn (32), as a generating function of the class $\mathcal{L}_N^R \subset L_N^R$. For all states $\Psi_{\bar{f}_1,\ldots,\bar{f}_R} \in \mathcal{L}_N^R[\bar{\mathcal{F}}^R]$, the expectation value of the energy:

$$\begin{aligned} E[\Psi_{f_1,\ldots,f_R}] &= \frac{\langle \Psi_{f_1,\ldots,f_R} | \hat{H} | \Psi_{f_1,\ldots,f_R} \rangle}{\langle \Psi_{f_1,\ldots,f_R} | \Psi_{f_1,\ldots,f_R} \rangle} = E[f_1,\ldots,f_R;\Phi] \equiv \\ &\equiv E[f_1,\ldots,f_R;\{C_I\}_{I=1}^M;\{\varphi_i\}_{i=1}^R] \end{aligned} \tag{35}$$

takes the form of a functional of both the OLST functions $\{f_i\}_{i=1}^R$ and the CI-like reference wave function Φ. As this latter is expressed through the expansion coefficients $\{C_I\}_{I=1}^M$ and the initial set $\{\varphi_i\}_{i=1}^R$, the above expression can be viewed as a functional of $\{f_i\}_{i=1}^R$, depending on $\{C_I\}_{I=1}^M$ and $\{\varphi_i\}_{i=1}^R$.

The determination of the ground-state energy, which is performed through minimization of the functional $E[f_1,\ldots,f_R;\{C_I\}_{I=1}^M;\Phi]$, represents a variational approach defined in the class $\mathcal{L}_N^R \subset L_N$. In terms of the variational method, this means that the OLSTs define a class of trial functions (32) of L_N, and the global minimum of the energy functional $E[f_1,\ldots,f_R;\{C_I\}_{I=1}^M;\Phi]$ in this class leads to an upper limit to E_0, if it exists.

This orbital local-scaling transformation procedure, which allows to find approximate values for the ground-state properties of a many-electron system, consists of the following steps:

1) Selection of a proper initial set of spin-orbitals, $\{\varphi_i(\vec{x})\}_{i=1}^{R}$, and then of the model wave function $\Phi \in L_N$ which generates the class of trial wave functions $\Psi_{\vec{f}_1,\ldots,\vec{f}_R}$, according to Eqn (32).

2) Building of the functional $E[f_1,\ldots,f_R;\{C_I\}_{I=1}^{M};\Phi]$, using the Hamiltonian (34) and Eqn (35).

3) Finding the solution of the variational system of coupled equations:

$$\frac{\delta E}{\delta f_i} = 0, \quad \frac{\delta E}{\delta C_I} = 0, \quad \textit{for all } i = 1,\ldots,R \textit{ and } I = 1,\ldots,M, \textit{as } \{\vec{f}_i\}_{i=1}^{R} \in \mathcal{F}^R, \tag{36}$$

assuming the existence of a solution in the class $\mathcal{L}_N^R \subset L_N$.

4) Finding an approximate wave function, $\Psi_{f_1^{opt},\ldots,f_R^{opt}}(x_1,\ldots,x_N)$, using the resulting optimal sets of coefficients $\{C_I\}_{I=1}^{M}$ and functions $\{f_i^{opt}\}_{i=1}^{R}$, and then all properties of the system.

Let us construct the functional

$$E[\Psi_{f_1,\ldots,f_R}] = \left\langle \Psi_{f_1,\ldots,f_R} \middle| \hat{H} \middle| \Psi_{f_1,\ldots,f_R} \right\rangle \equiv E[f_1,\ldots,f_R;\{C_I\}_{I=1}^{M};\{\varphi_i\}_{i=1}^{R}], \tag{37}$$

where the $\Psi_{f_1,\ldots,f_R}$ are built from determinants Ψ_I^S, with $I=\{i_k\}$ and $i_1<i_2<\ldots<i_N$, corresponding to the orthogonal transformed set $\{\psi_i[f_i]\}_{i=1}^{R}$, Eqn (30). We form the auxiliary functional ε, adding to Eqn (37) the following constraint:

$$\mathcal{E}[f_1,\ldots,f_R;\{C_I\}_{I=1}^{M};\{\varphi_i\}_{i=1}^{R}] = E[f_1,\ldots,f_R;\{C_I\}_{I=1}^{M};\{\varphi_i\}_{i=1}^{R}] - \varepsilon\left(\sum_{I=1}^{M}\left|C_I\right|^2 - 1\right). \tag{38}$$

The second term on the right hand side of this equation ensures that the solution of the Euler-Lagrange system, Eqs (36), leads to the normalized wave functions, under the constraint $\left\langle \Psi_{f_1^{opt},\ldots,f_R^{opt}} \middle| \Psi_{f_1^{opt},\ldots,f_R^{opt}} \right\rangle = 1$. The spin-orbital set, $\{\psi_i[f_i^{opt}]\}_{i=1}^{R}$, corresponding to the stationary value of the energy will be orthonormal. The optimal energy, $E_{opt} > E_0$, is equal to the resulting value of the Lagrange multiplier ε: $E[\Psi^{opt}] = \varepsilon_0$. Obviously, the explicit form of the functional (38) will be rather simpler than that of the functional (35), which is constructed assuming the variational set $\{\psi_i[f_i]\}_{i=1}^{R}$ to be non-orthogonal.

5. Implementation of the OLST method in quantum chemistry

5.1. ATOMIC CASE

We choose the initial single-electron orbitals as products of radial, angular and spin functions:

$$\varphi_i(\zeta) = \overline{R}_i(r) Y_{l_i m_i}(\theta, \varphi) \chi_{s_i}(\sigma), \tag{39}$$

and impose the following constraint on the LST functions:

$$f_i(r, \theta, \varphi) = f_i(r). \tag{40}$$

Thus the LSTs of the orbitals do not affect the angular variables:

$$\varphi_i(\zeta) \xrightarrow{LST} \frac{f_i(r)}{r} \sqrt{\frac{df_i(r)}{dr}} \overline{R}_i(f_i(r)) Y_{l_i m_i}(\theta, \varphi) \chi_{s_i}(\sigma). \tag{41}$$

After integration over the angle and spin variables the energy (37) takes the form of a functional of LST functions:

$$E = E[f_1(r), \ldots, f_R(r); \{C_I\}_{I=1}^{M}; \{\varphi_i(r)\}_{i=1}^{R}]. \tag{42}$$

The Euler-Lagrange system will then consist of the habitual integro-differential equations with respect to the functions $f_i(r)$ completed by an additional set of equations for the coefficients C_I.

The analysis of the variational equations for the nonrelativistic Hamiltonian in the Born-Oppenheimer approximation shows that $f_i(r) \sim r$ when $r \to 0$ or $r \to \infty$. Using these boundary conditions one can solve numerically the Euler-Lagrange system, getting the optimal energy and then all ground-state properties of the N-fermion system.

Let the atomic orbitals be chosen in the form (39) and $\mathcal{R}_i(r)$ be the radial factors obtained by solving the multiconfigurational SCF equations corresponding to the actual expansion: $\Phi = \sum_I C_I \Phi_I$. Let us choose as initial radial functions in Eqn (39) the corresponding nodeless - and hence non-orthogonal - Slater radial functions. Then (as it was shown in Section 3) there exists a unique set of LST functions, $\{f_i(r)\}_{i=1}^{R}$, that bring the initial set of radial factors $\{\overline{R}_i\}_{i=1}^{R} \equiv \{R_i^{(S)}(r)\}_{i=1}^{R}$ into the set $\{\mathcal{R}_i(r)\}$ of the optimized atomic-orbital radial factors.

This shows that, in the approximation of a separated angular dependence in the one-particle atomic orbitals (39), the proposed OLST method is equivalent to the MCSCF approach. One may expect that, for light atoms at the HF level, the functions $f_i(r)$ may be parameterized in a suitable way involving a limited number of parameters. This follows from the simplicity of the behavior of the

LST functions $f_i(r)$: they must be monotonic and possess inflexion points, which generate nodes in the transformed radial functions:

$$R_i([f_i],r) = \frac{f_i(r)}{r}\sqrt{\frac{df_i(r)}{dr}}R_i^{(S)}(f_i(r)), \tag{43}$$

and also from the flexibility of the Slater functions $R_i^{(S)}(r)$. Then, the energy functional will be reduced to a function depending on these parameters, and the variational problem, to a Ritz minimization procedure.

The resolution of the Euler-Lagrange system in the OLST method could be performed by using algorithms based on the spline methods [23, 24]. This may permit to increase the accuracy of numerical MCSCF calculations [25]. One may anticipate that the use of analytical approximations for the LST functions $f_i(r)$, instead of the radial factors $R_i(r)$, with suitable core-splines [23] or B-splines [24], will improve the convergence characteristics of the numerical procedure in actual updating of the radial functions (43).

5.2. MOLECULAR CASE

The positions $\{R_J\}_{J=1}^{K}$ of the K nuclei forming the molecular skeleton in the Born-Oppenheimer approximation are assumed to be fixed. Since every transformation $\varphi_i(\vec{r}) \xrightarrow{\hat{u}_{\vec{f}_i}} \psi_i([f_i];\vec{r})$ is generated by a mapping $\vec{r} \to \vec{r}\,'$ of $\mathbf{R}^3$ onto itself, the positions of the nuclei must remain unaltered after the orbitals distorsion:

$$\vec{f}_i(\vec{R}_J) = \vec{R}_J, \quad \forall i = 1,\ldots,R \text{ and } \forall J = 1,\ldots,K. \tag{44}$$

Therefore, only those $\vec{f}_i \in \mathcal{F}_i$ that fulfill Eqn (44) have to be considered. The set of these $\vec{f}_i$ forms a group, which is subgroup of $\mathcal{F}_i$.

Let us take the initial molecular orbitals as linear combinations of predefined basis functions:

$$\varphi_i = \sum_{\mu=1}^{n} C_{\mu i}\chi_\mu , \tag{45}$$

where the $C_{\mu i}$ are expansion coefficients. The basis functions $\chi_1,\ldots,\chi_n$ are taken as normalized and centered on the atomic nuclei. We choose the basis functions χ_μ, $\mu = 1, \ldots, n$, as atomic orbitals of the form (39).

Then, for LST functions of the type (40), the locally-scaled molecular orbital (45) turns into an orbital depending on the function $f_i(r)$ and its derivative:

$$\varphi_i(r,\theta,\varphi) \xrightarrow{LST} \sum_{\mu} C_{\mu i} \frac{f_i(r)}{r} \sqrt{\frac{df_i(r)}{dr}} \chi_\mu(f_i(r),\theta,\varphi),$$

and the energy, as in the atomic case, takes the form of a functional of the OLST functions (42). Choosing the expansion coefficients $C_{\mu i}$ in a proper way, one can then determine completely this functional.

Let us fix the coefficients $\{C_I\}$. Then, solving the Euler-Lagrange system:

$$\frac{\delta}{\delta f_i} E[f_1,\ldots,f_R;\{C_I^{fix}\}_{I=1}^{M};\{\varphi_i\}_{i=1}^{R}] = 0, \quad \forall i = 1,\ldots,R, \tag{46}$$

one can find the radially optimized molecular orbitals $\psi_i([f_i^{opt}];\vec{r})$: this variational solution does not depend on the initial values of the coefficients $C_{\mu i}$ or on the number n of basis functions χ_μ in Eqn (45). However, since we do not optimize the angular dependence of the initial molecular orbitals $\varphi_i(\vec{r})$, the accuracy of the resulting one-particle functions $\psi_i([f_i^{opt}];\vec{r})$ will depend, in this respect, on the choice of the basis $\{\chi_\mu\}_{\mu=1}^{n}$. Persuing the variational process with respect to the coefficients $\{C_I\}$ and the orbitals $\{\psi_I\}$, one would obtain, in principle, the ground-state characteristics of the system in the OLST approach.

Because the optimization of the energy functional (42) must be realized by solving numerically a system of differential equations, the applicability of the OLST method is, in practice, limited to simple molecules, considering a strongly reduced CI space. But one could perform calculations of main properties with a high level of accuracy, using one-electron functions satisfying the nuclear cusp condition and possessing the correct long-range behavior.

An important feature of this model is that it makes it possible to compare the accuracy of analytical, Roothaan-Hall molecular calculations with that of numerical ones. Using, for instance, optimized analytical functions $\chi_i(\vec{r})$, one could further improve them in the framework of the OLST method. One can thus test the capacity of different finite basis sets to reproduce various properties.

In conclusion, the mathematical treatment of the variational problem based on the OLST method is more general in the molecular case than in the atomic case, since now any initial set of appropriately defined functions $\varphi_i(\zeta)$ may be used. A next step in our investigations will be an extension of the OLST method to approaching excited states.

Acknowledgements

This work was supported in part by a twinning convention between Universities Pierre et Marie Curie (Paris) and Saint Kliment Ohridski (Sofia) and European

COST-D9/0010/98 contract. One of us (A.I.K.) would like to thank the French Government for a predoctoral fellowship. Professor Mateyi Mateev is gratefully acknowledged for useful discussions.

References

1. G.G. Hall, Proc. Phys. Soc. (London) **75**, 575 (1960).
2. M.J.T. Hoor, Int. J. Quant. Chem. **33**, 563 (1988).
3. S.T. Epstein and J.O. Hirschfelder, Phys. Rev. **A 123**, 1495 (1961).
4. S.T. Epstein, *The Variational Method in Quantum Chemistry*, Academic Press, New York, 1974.
5. N.M. Witriol, Int. J. Quant. Chem. S **6**, 145 (1972).
6. T. Koga, Y. Yamamoto and E.S. Kryachko, J. Chem. Phys. **91**, 4758 (1989).
7. I.Zh. Petkov and M.V. Stoitsov, Sov. J. Nucl. Phys. **37**, 692 (1984).
8. L.S. Georgiev, Ya.I. Delchev, R.L. Pavlov and J. Maruani, Theochem **433**, 35 (1998).
9. I.Zh. Petkov and M.V. Stoitsov, Teor. Mat. Fiz. **55**, 407 (1983) (in Russian).
10. I.Zh. Petkov, M.V. Stoitsov and E.S. Kryachko, Int. J. Quant. Chem. **29**, 149 (1986).
11. E.S. Kryachko and E.V. Ludeña, *Energy Density Functional Theory of Many-Electron Systems*, Kluwer, Dordrecht, 1990.
12. E.S. Kryachko and E.V. Ludeña, Phys. Rev. **A 43**, 2179 (1991).
13. E.S. Kryachko, I.Zh. Petkov and M.V. Stoitsov, Int. J. Quant. Chem. **32**, 467 (1987).
14. E.S. Kryachko, I.Zh. Petkov and M.V. Stoitsov, Int. J. Quant. Chem. **32**, 473 (1987).
15. E.S. Kryachko and E. . Ludeña, Phys. Rev. **A 35**, 957 (1987).
16. E.S. Kryachko and E.V. Ludeña, Phys. Rev. **A 43**, 2194 (1991).
17. I.Zh. Petkov and M.V. Stoitsov, *Nuclear Density Functional Theory,* Oxford University Press, New York, 1991.
18. E.S. Kryachko and E.V. Ludeña, Int. J. Quant. Chem. **43**, 769 (1992);
T. Koga, Y. Yamamoto and E. Ludeña, Phys. Rev. **A 43**, 5814 (1991).
19. E.V. Ludeña, R. Lopez-Boada, J.E. Maldonado, E. Valderrama, E.S. Kryachko, T. Koga, and J. Hinze, Int. J. Quant. Chem. **56**, 285 (1995).
20. R.L. Pavlov, J. Maruani, Ya.I. Delchev and R. McWeeny, Int. J. Quant. Chem. **65**, 241 (1997);
R.L. Pavlov, F.E. Zakhariev, Ya.I. Delchev and J. Maruani, ibid. **65**, 257 (1997).
21. A.N. Tihonov, A.B. Vasil'eva and A.G. Sveshnicov, *Differential Equations*, Nauka, Moscow, 1980 (in Russian).
22. E. Clementi and C. Roetti, Atomic and Nuclear Data Tables **14**, 177 (1974).
23. L. Alexandrov and D. Karadjov, J. Comp. Math. & Math. Phys. **20**, 923 (1980) (in Russian).
24. C. Froese-Fischer and W. Guo, J. Comp. Phys. **90**, 486 (1990).
25. C. Froese-Fischer, W. Guo and Z. Shen, Int. J. Quant. Chem. **42**, 849 (1992).

REDUCED DENSITY-MATRIX TREATMENT OF SPIN-ORBIT INTERACTION TERMS IN MANY-ELECTRON SYSTEMS

R. L. PAVLOV *, #, A. I. KULEFF *, #, P. TZ. YOTOV *
J. MARUANI #, AND YA. I. DELCHEV *

* *Institute for Nuclear Research and Nuclear Energy, Bulgarian Academy of Sciences, 72 Tzarigradsko Chaussée, 1784 Sofia, Bulgaria (ropavlov@inrne.bas.bg)*

Laboratoire de Chimie Physique, UPMC and CNRS, 11 rue Pierre-et-Marie-Curie, 75005 Paris, France (maruani@ccr.jussieu.fr)

Abstract. In the formalism of reduced density matrices and functions, using the irreducible tensor-operator technique and the space-spin separation scheme, the matrix elements of one of the main spin-relativistic corrections of the Breit-Pauli Hamiltonian, the spin-orbit interactions, are expressed in a form suitable for numeric implementation. A comparison with other methods is made and the advantages of such an approach are discussed.

1. Introduction

In the formalism of reduced density matrices and functions (RDM & RDF) [1-3] the matrix elements and expectation values of the different types of spin-involving operators take the form of a product of space and spin factors [1, 4]. The spin part is determined by the spin symmetry and reduces to 3j-symbols and the spatial part is determined by the action of space operators on the spin distribution or correlation matrices or functions [1, 4, 5]. The spin distribution and correlation functions are built from the spatial part of the RDMs of first or second order, respectively. In this approach, the space-spin separation results from the possibility of separating the space and spin variables in the RDMs [6-9] and of the resulting expressions for the matrix elements of the perturbation terms in the Breit-Pauli Hamiltonian [10].

In terms of the RDMs and RDFs and in the scheme of the space-spin separation there are presented in Refs [1, 4] the matrix elements and expectation values

J. Maruani et al. (eds.), New Trends in Quantum Systems in Chemistry and Physics, Volume 1, 63–74.

of the various spin-involving operators, which are relativistic corrections of the Breit-Pauli Hamiltonian. The matrix elements and expectation values of the operators corresponding to the different types of spin interactions are reduced to products of a multiplier determined by the spin symmetry, characterizing a spin state or a transition between two states in a given spin multiplet, and a space part which depends neither on the spin state nor on a transition between two states. This space part is expressed by the action of the considered space operator on the space part of the relevant spin distribution or correlation matrix or function.

In the present work we present a further stage in the treatment of the matrix elements of one of the main relativistic corrections, the spin-orbit interactions, which should be amenable to direct numerical implementation. Using the irreducible tensor-operator technique and applying the Wigner-Eckart theorem, the matrix elements of the spin-orbit interactions are presented as products, or sums of products, of multipliers determined by the spin symmetry, multipliers characterizing the orbital symmetry, and a spatial part determined by the action of the symmetrized space tensor-operators on the normalized spin distribution or correlation matrices or functions. The action of these space operators, which is the same for a given spin multiplet and is independent on the investigated splitting or transition, is reduced by a standard procedure. The expectation values of the operators of spin-same orbit and spin-other orbit couplings, giving the amount of splitting of the energy levels, are expressed in an analytical form suitable for numerical implementation. We also consider the transition matrix elements of these operators, giving the contribution of spin-orbit interactions to the corresponding spin transitions.

2. Matrix elements in the RDM formalism

The RDM of order s (s-RDM) of an N-electron system ($1 \leq s \leq N$) in the state K, described by a wave function $\Psi_K(\tau_1,\ldots,\tau_N)$, $\tau_i = (\vec{r}_i, \sigma_i)$, eigenfunction of the operators S^2 and S_Z, has the form [1-3, 11-14]:

$$\begin{aligned}\rho(KK|\tau_1,\ldots,\tau_s;\tau_1',\ldots,\tau_s') &= N(N-1)\ldots(N-s+1)\times\\ &\times\int\Psi_K(\tau_1,\ldots,\tau_s;\tau_{s+1},\ldots,\tau_N)\Psi_K^*(\tau_1',\ldots,\tau_s';\tau_{s+1},\ldots,\tau_N)d\tau_{s+1}\ldots d\tau_N.\end{aligned} \quad (1)$$

The corresponding RDF of order s (s-RDF) is defined by the expression:

$$\rho(KK|\tau_1,\ldots,\tau_s) = \rho(KK|\tau_1,\ldots,\tau_s;\tau_1',\ldots,\tau_s')\big|_{\tau_i'=\tau_i} = \rho(KK|\tau_1,\ldots,\tau_s;\tau_1,\ldots,\tau_s). \quad (2)$$

The generalized, transition s-RDM between states K and K' described by Ψ_K and $\Psi_{K'}$ has the form [1-3]:

$$\rho(KK'|\tau_1,\ldots,\tau_s;\tau_1',\ldots,\tau_s') = N(N-1)\ldots(N-s+1)\times \\ \times\int \Psi_K(\tau_1,\ldots,\tau_s,\tau_{s+1},\ldots,\tau_N)\Psi^*_{K'}(\tau_1',\ldots,\tau_s';\tau_{s+1},\ldots,\tau_N)d\tau_{s+1}\ldots d\tau_N. \quad (3)$$

For the transition s-RDF we have a similar expression, following from Eq. (2).

The expectation value of a given s-particle operator, $\hat{F}(i_1,i_2,\ldots,i_s)$, can be written as:

$$\left\langle \sum_{\{i_s\}} \hat{F}(i_1,\ldots,i_s)\right\rangle = Sp_{i_1,i_2,\ldots,i_s}\sum_{\{i_s\}}\hat{F}(i_1,\ldots,i_s)\rho(\tau_1,\ldots,\tau_N;\tau_1',\ldots,\tau_N') = \\ = Sp_{i_1,i_2,\ldots,i_s}\hat{F}(i_1,\ldots,i_s)\rho(\tau_1,\ldots,\tau_s;\tau_1',\ldots,\tau_s'). \quad (4)$$

In the above expression, $\sum_{\{i_s\}}$ means a summation over all possible sets (i_1, i_2, ..., i_N) and $Sp_{i_1,i_2,\ldots,i_s}$ is an operator of integration over particle coordinates labelled with the corresponding numbers, after identification of the primed and unprimed coordinates.

In many-electron theory, the expectation values and matrix elements of symmetric sums of identical operators are of main importance. Then, for the expectation values of a symmetric sum of identical s-particle operators, $\hat{F}(i_1,i_2,\ldots,i_s)$, assuming the following order of numbers i_k, $1 \le i_1 < i_2 < \ldots < i_s \le N$, we have:

$$\left\langle \sum_{\{i_s\}} \hat{F}(i_1,i_2,\ldots,i_s)\right\rangle = \left\langle \Psi_K(\tau_1,\ldots,\tau_N)\right|\sum_{\{i_s\}}\hat{F}(i_1,i_2,\ldots,i_s)\left|\Psi_K(\tau_1,\ldots,\tau_N)\right\rangle = \\ = \frac{1}{s!}Sp\hat{F}(i_1,i_2,\ldots,i_s)\rho(K|\tau_{i_1},\ldots,\tau_{i_s};\tau_{i_1}',\ldots,\tau_{i_s}'). \quad (5)$$

This is a general expression. Usually, only RDMs or RDFs of 1st or 2nd order are relevant. Higher-order matrices or functions are used only in specific cases.

After separation of the space and spin variables, the 1-RDM takes the form:

$$\rho(\tau_1;\tau_1') = \sum_{\gamma,\gamma'=\alpha,\beta}\rho^{\gamma,\gamma'}(\vec{r}_1;\vec{r}_1')\gamma(s_1)\gamma'^*(s_1'), \quad (6)$$

where the $\rho^{\gamma,\gamma'}(\vec{r}_1;\vec{r}_1')$ are the space components and the $\gamma(\sigma)$ ($\sigma = \alpha, \beta$) the spin one-electron wave functions. The space components form the charge and spin distribution matrices:

$$\rho(\vec{r}_1;\vec{r}_1') = \rho^{\alpha,\alpha}(\vec{r}_1;\vec{r}_1') + \rho^{\beta,\beta}(\vec{r}_1;\vec{r}_1'), \quad (7)$$

$$q(KK|\vec{r}_1;\vec{r}_1') = (M/S)\, q(\underline{KK}\,|\vec{r}_1;\vec{r}_1') = M D(\vec{r}_1;\vec{r}_1'), \tag{8}$$

where K is the index of the spin state corresponding to $<S_Z> = M$ ($M = S, S - 1, \ldots, - S$), $\underline{K}$ corresponds to the maximal value: $M = S$, $D(\vec{r}_1;\vec{r}_1')$ is the normalized spin distribution matrix (which is the same within a given spin multiplet), and

$$q(\underline{KK}|\vec{r}_1;\vec{r}_1') = (1/2)\,[\rho^{\alpha,\alpha}(\underline{KK}\,|\vec{r}_1;\vec{r}_1') - \rho^{\beta,\beta}(\underline{KK}\,|\vec{r}_1;\vec{r}_1')]. \tag{9}$$

Similarly, the 2-RDM can be written in the form:

$$\rho(\tau_1,\tau_2;\tau_1',\tau_2') = \sum_{\gamma,\gamma',\gamma'',\gamma'''=\alpha,\beta} \rho^{\gamma\gamma',\gamma''\gamma'''}(\vec{r}_1,\vec{r}_2;\vec{r}_1',\vec{r}_2')\gamma(s_1)\gamma'(s_2)\gamma''^{*}(s_1')\gamma'''^{*}(s_2'), \tag{10}$$

where the $\rho^{\gamma\gamma',\gamma''\gamma'''}(\vec{r}_1,\vec{r}_2;\vec{r}_1',\vec{r}_2')$ are the space components. There are altogether 16 components in Eq. (10), but only 6 are independent space components [1, 2, 6]. These latter form the spin correlation matrix.

The spin-orbit interaction matrix can then be written in the form:

$$q_{SO}(KK|\vec{r}_1,\vec{r}_2;\vec{r}_1',\vec{r}_2') = (M/S)\,q_{SO}(\underline{KK}|\vec{r}_1,\vec{r}_2;\vec{r}_1',\vec{r}_2') = MD_{SO}(\vec{r}_1,\vec{r}_2;\vec{r}_1',\vec{r}_2'), \tag{11}$$

where $D_{SO}(\vec{r}_1,\vec{r}_2;\vec{r}_1',\vec{r}_2')$ is the normalized spin-orbit interaction matrix (the same for all states in a given spin multiplet), and $q_{SO}(\underline{KK}|\vec{r}_1,\vec{r}_2;\vec{r}_1',\vec{r}_2')$ is given by

$$\begin{aligned} q_{SO}(\vec{r}_1,\vec{r}_2;\vec{r}_1',\vec{r}_2') = \frac{1}{2}[&\rho^{\alpha\alpha,\alpha\alpha}(\vec{r}_1,\vec{r}_2;\vec{r}_1',\vec{r}_2') + \rho^{\alpha\beta,\alpha\beta}(\vec{r}_1,\vec{r}_2;\vec{r}_1',\vec{r}_2') - \\ &- \rho^{\beta\alpha,\beta\alpha}(\vec{r}_1,\vec{r}_2;\vec{r}_1',\vec{r}_2') - \rho^{\beta\beta,\beta\beta}(\vec{r}_1,\vec{r}_2;\vec{r}_1',\vec{r}_2')], \end{aligned} \tag{12}$$

constructed on four of the components of Eq. (10).

3. Spin-orbit interactions

3.1. THE FUNCTIONAL OF THE *SPIN-SAME ORBIT* INTERACTION

For an N-electron system in the field of M nuclei, the operator of the *spin-same orbit* interaction in the effective Breit-Pauli Hamiltonian has the form:

$$\hat{h}_{SO}(i) = -\sum_{\lambda} Z_\lambda \frac{g_0\alpha^2}{r_{\lambda i}^3}\mathbf{S}_i.\mathbf{L}_{i/\lambda}, \tag{13}$$

where $\mathbf{L}_{i/\lambda}$ is the orbital momentum of the i-th electron with respect to nucleus λ, with radius-vector $\vec{R}_\lambda$ and charge Z_λ, $\vec{r}_{\lambda i} = \vec{R}_\lambda - \vec{r}_i$, $\alpha = 1/2c$ is the fine-structure constant, and g_0 is the g-factor of the free electron [15]. We consider the transition matrix element between two states K and K' in the same spin multiplet. The transition between different spin multiplets is impossible from a symmetry point of view because it is not possible to express elements of an irreducible representtion of the group SO(3) with elements of a different irreducible representation of the same group. We obtain the transition matrix element of the operator (13) by expressing the scalar product in terms of irreducible tensor operators, using the expressions of the transition matrix elements of the spin-involving operators [1, 4], in the form:

$$\langle K\,SM_S|\sum_i \hat{h}_{SO}(i)|K'\,SM_S'\rangle = -Zg_0\alpha^2\sum_m(-1)^m \int_{\vec{r}_1'=\vec{r}_1} r_1^{-3}L_{-m}^1(1)q_S^f(KK'|\vec{r}_1;\vec{r}_1')_m^1 d\vec{r}_1 =$$

$$= -Zg_0\alpha^2\sum_m(-1)^m\langle SM_S|S_m^1|SM_S'\rangle \int_{\vec{r}_1'=\vec{r}_1} r_1^{-3}L_{-m}^1(1)D_S^f(K|\vec{r}_1;\vec{r}_1')d\vec{r}_1,$$

where S_m^1 and $L_{-m}^1(1)$ are the components of the symmetrized operators of the total spin and total orbital momentum and

$$q_S^f(KK'|\vec{r}_1;\vec{r}_1')_m^1 = \int_{s_1'=s_1} s_m^1(1)\rho_f(KK'|\tau_1;\tau_1')ds_1.$$

Applying the Wigner-Eckart theorem to the matrix elements of S_m^1 and expressing the angular momentum operator (a vector product of $\mathbf{r}$ and $\mathbf{p}$) in terms of a product of irreducible tensor operators, one obtains

$$\langle K\,SM_S|\sum_i \hat{h}_{SO}(i)|K'\,SM_S'\rangle = \\ = -iZg_0\alpha^2\sqrt{3}\sum_m(-1)^m C_{MmM'}^{S1S}\sum_{q,p}C_{qp-m}^{111}\int_{\vec{r}_1'=\vec{r}_1} r_1^{-3}C_{-q}^1(1)(\nabla^*)_p^1 D_S^f(K|\vec{r}_1;\vec{r}_1')d\vec{r}_1, \quad (14)$$

where

$$C_q^p(1) = C_q^p(\theta_1,\varphi_1) = [4\pi/(2p+1)]^{1/2}Y_q^p(\theta_1,\varphi_1)$$

are the normalized spherical functions, $(\nabla^*)_p^1$ are the symmetrized components of the angular part of the ∇ operator [8, 16], and $C_{MmM'}^{S1S}$ are 3j-symbols. Expression (14) is valid also when a magnetic field is applied to the system.

For a same state, Eqn (14) takes the form:

$$\langle K\,LS\,JM_J\,|\sum_i \hat{h}_{SO}(i)|\,K\,LS\,JM_J\rangle = \\ = -iZg_0\alpha^2\sqrt{3}M_S\sum_q C^{111}_{q-q0}\int_{\vec{r}_1'=\vec{r}_1} r_1^{-3}C^1_{-q}(1)(\nabla^*)^1_q D^f_S(K|\vec{r}_1;\vec{r}_1')d\vec{r}_1. \tag{15}$$

This term gives the width of the spectral line splitting.

For a system with fixed nuclei, Z_λ, the angular momentum takes the form:

$$\mathbf{L}_{i/\lambda} = \mathbf{L}^\lambda(i) = \mathbf{r}_{\lambda i}\times\mathbf{p}(i) = \mathbf{r}_i\times\mathbf{p}(i) - \mathbf{r}_\lambda\times\mathbf{p}(i).$$

From the general formula of the matrix element of a product of two tensor operators:

$$\langle K|f\cdot g|K\rangle = \sum_{K'}\langle K|f|K'\rangle\langle K'|g|K\rangle, \tag{16}$$

it follows that $\mathbf{r}_\lambda\times\mathbf{p}(i) = 0$, because the matrix element of $\mathbf{r}_\lambda$ differs from zero only when $M_L(i) = M'_L(i)$ while that of $\mathbf{p}(i)$ differs from zero only when $M'_L(i) = M_L(i) + 1$ [17]. Then Eqn (15) takes the form:

$$\langle K\,LS\,JM_J\,|\sum_i \hat{h}_{SO}(i)|\,K\,LS\,JM_J\rangle = \\ = -iZg_0\alpha^2\sqrt{3}M_S\sum_\lambda Z_\lambda\sum_q C^{111}_{q-q0}\int_{\vec{r}_1'=\vec{r}_1} r_1^{-3}C^1_{-q}(1)(\nabla^*)^1_q D^f_S(K|\vec{r}_1;\vec{r}_1')d\vec{r}_1. \tag{17}$$

Expressions (15) and (17) are very suitable for direct numerical implementation, due to the separation of space and spin variables and their simplicity.

3.2. THE FUNCTIONAL OF THE *SPIN-OTHER ORBIT* INTERACTION

The *spin-other orbit* interaction operator in the effective Breit-Pauli Hamiltonian can be expressed as the two-particle symmetric sum:

$$\frac{1}{2}\sum_{i,j} g_{SO}(i,j) = -\frac{g_0\alpha^2}{r_{ij}^3}\left[\mathbf{S}_i\cdot\left(\mathbf{L}_{i/j} + 2\mathbf{L}_{j/i}\right) + \mathbf{S}_j\cdot\left(\mathbf{L}_{j/i} + 2\mathbf{L}_{i/j}\right)\right]. \tag{18}$$

There the angular momentum of the j-th electron with respect to the i-th electron has the form:

$$\mathbf{L}_{i/j} = \mathbf{L}^j(i) = \mathbf{r}_{ji}\times\mathbf{p}(i) = \mathbf{r}_i\times\mathbf{p}(i) - \mathbf{r}_j\times\mathbf{p}(i). \tag{19}$$

According to Eqn (16), for a given state, the expectation value of $\mathbf{r}_j\times\mathbf{p}(i)$ is equal to zero. Then, the expectation value of the operator $\mathbf{L}_{i/j}$ takes the form:

$$<\mathbf{L}_{i/j}> = <\mathbf{r}_i\times\mathbf{p}(i)> \equiv <\mathbf{L}(i)>.$$

Consequently, for the expectation value of Eqn (18), one has:

$$\left\langle \frac{1}{2}\sum_{i,j} g_{SO}(i,j)\right\rangle = -g_0\alpha^2\left[\left\langle \sum_{i,j}\frac{\mathbf{S}(i).(\mathbf{L}(i)+\mathbf{L}(j))}{r_{ij}^3}\right\rangle + \left\langle \sum_{i,j}\frac{\mathbf{S}(i).\mathbf{L}(j)}{r_{ij}^3}\right\rangle\right] + [i \leftrightarrow j]. \quad (20)$$

Applying the Wigner-Eckart theorem for the expectation value of the operator **S** yields, for the first term in Eqn (20):

$$\left\langle K\,LS\,JM_J \left|\sum_{i,j}\frac{\mathbf{S}(i).(\mathbf{L}(i)+\mathbf{L}(j))}{r_{ij}^3}\right| K\,LS\,JM_J\right\rangle = (3/2)^{1/2} M$$

$$\left\{ \int_{\vec{r}_1'=\vec{r}_1} r_{12}^{-3} L_0^1(1) D_{SO}^f(K|\vec{r}_1,\vec{r}_2;\vec{r}_1',\vec{r}_2)d\vec{r}_1 d\vec{r}_2 + \int_{\vec{r}_2'=\vec{r}_2} r_{12}^{-3} L_0^1(1) D_{SO}^f(K|\vec{r}_1,\vec{r}_2;\vec{r}_1,\vec{r}_2')d\vec{r}_1 d\vec{r}_2 \right\}.$$

Expressing the angular momentum as a vector product of the operators **r** and **p** in terms of irreducible tensor operators, one obtains:

$$\left\langle K\,LS\,JM_J \left|\sum_{i,j}\frac{\mathbf{S}(i).(\mathbf{L}(i)+\mathbf{L}(j))}{r_{ij}^3}\right| K\,LS\,JM_J\right\rangle =$$

$$= i(3/2)^{1/2} M \sum_q C_{q-q0}^{111}\left\{ \int_{\vec{r}_1'=\vec{r}_1} r_{12}^{-3} C_q^1(1)(\nabla_1^{*})_{-q}^1 D_{SO}^f(K|\vec{r}_1,\vec{r}_2;\vec{r}_1',\vec{r}_2)d\vec{r}_1 d\vec{r}_2 + \right. \quad (21)$$

$$\left. + \int_{\vec{r}_2'=\vec{r}_2} r_{12}^{-3} C_q^1(1)(\nabla_2^{*})_{-q}^1 D_{SO}^f(K|\vec{r}_1,\vec{r}_2;\vec{r}_1,\vec{r}_2')d\vec{r}_1 d\vec{r}_2 \right\}.$$

This expression determines the contribution of the first terms in Eqn (20) to the width of the spectral line splitting.

The last terms in Eqn (20) can be written in the form, well known in atomic spectroscopy [17]:

$$\left\langle K\,LS\,JM_J \left|\sum_{i,j}\frac{\mathbf{S}(i).\mathbf{L}(j)}{r_{ij}^3}\right| K\,LS\,JM_J\right\rangle = \left\langle K\,LS\,JM_J \left| a(\vec{r})[\mathbf{S}.\mathbf{L}]\right| K\,LS\,JM_J\right\rangle \approx$$

$$\approx \langle r^{-3}\rangle C\left[J(J+1) - L(L+1) - S(S+1)\right], \quad (22)$$

where $a(r)$ is a one-particle tensor operator, and the constant C in the last expression must be determined for every different case.

For different states within a spin multiplet, the component $\mathbf{r}_j\times\mathbf{p}(i)$ of the angular momentum (19) is non-zero. This follows from the general form of (16):

$$\langle K|f\cdot g|K'\rangle=\sum_{K''}\langle K|f|K''\rangle\langle K''|g|K'\rangle .$$

In this case, using the Wigner-Eckart theorem yields, for the matrix element of the first terms in Eqn (18):

$$\left\langle K\,SM_S\left|\sum_{i,j}r_{ij}^{-3}\left[\mathbf{S}(i).\left(\mathbf{L}^j(i)+2\mathbf{L}^i(j)\right)\right]\right|K'\,SM_S'\right\rangle=(3/2)^{1/2}\sum_m(-1)^m C^{S1S}_{MmM'}$$
$$\int_{\substack{\vec r_1'=\vec r_1\\ \vec r_2'=\vec r_2}} r_{12}^{-3}\left[\left[L^{(2)}(1)\right]^1_{-m}+2\left[L^{(1)}(2)\right]^1_{-m}\right]D^f_{SO}(K|\vec r_1,\vec r_2;\vec r_1',\vec r_2')d\vec r_1 d\vec r_2 .$$

The expression in square brackets on the left-hand side can be written as:

$$[\mathbf{S}(i).\mathbf{L}_{i/j}+2\,\mathbf{S}(i).\mathbf{L}_{i/j}]=[\mathbf{L}(i)+\mathbf{L}(j)].\mathbf{S}(i)-[\mathbf{r}_j\times\mathbf{p}(i)+\mathbf{r}_i\times\mathbf{p}(j)].\mathbf{S}(i)+\mathbf{L}^{(i)}(j).\mathbf{S}(i). \quad (23)$$

Applying the Wigner-Eckart theorem for the operator **S** yields, for the matrix element of the first terms in Eqn (23):

$$\left\langle K\,SM_S\left|\sum_{i,j}\frac{\mathbf{S}(i).(\mathbf{L}(i)+\mathbf{L}(j))}{r_{ij}^3}\right|K'\,SM_S'\right\rangle=(3/2)^{1/2}\sum_m(-1)^m C^{S1S}_{MmM'}$$
$$\int_{\substack{\vec r_1'=\vec r_1\\ \vec r_2'=\vec r_2}} r_{12}^{-3}\left[L(1)^1_{-m}+L(2)^1_{-m}\right]D^f_{SO}(K|\vec r_1,\vec r_2;\vec r_1',\vec r_2')d\vec r_1 d\vec r_2 .$$

Expressing the angular momentum as a vector product of the operators **r** and **p** in terms of irreducible tensor operators yields, for the matrix element of the transition $M_S\rightarrow M'_S$:

$$\left\langle K\,LS\,JM_J\left|\sum_{i,j}\frac{\mathbf{S}(i).\left(\mathbf{L}(i)+\mathbf{L}(j)\right)}{r_{ij}^3}\right|K'\,LS\,JM_J'\right\rangle=i(3/2)^{1/2}\sum_m(-1)^m C^{S1S}_{MmM'}\sum_{q,p}C^{111}_{qp-m}$$
$$\left\{\int_{\vec r_1'=\vec r_1} r_{12}^{-3}C^1_{-q}(1)(\nabla_1^\bullet)^1_p D^f_{SO}(K|\vec r_1,\vec r_2;\vec r_1',\vec r_2)d\vec r_1 d\vec r_2+\right. \quad (24)$$
$$\left.+\int_{\vec r_2'=\vec r_2} r_{12}^{-3}C^1_{-q}(1)(\nabla_2^\bullet)^1_p D^f_{SO}(K|\vec r_1,\vec r_2;\vec r_1,\vec r_2')d\vec r_1 d\vec r_2\right\}.$$

Using the irreducible tensor-operator technique one can express the second terms in Eqn (23) in the form:

$$\left[\mathbf{r}_j \times \mathbf{p}(i) + \mathbf{r}_i \times \mathbf{p}(j)\right].\mathbf{S}(i) = i\sqrt{2}\sum_m (-1)^m \sum_{q,\lambda} C^{111}_{q\lambda-m} \{ r_j \frac{\partial}{\partial r_i} C^1_q(j) C^1_\lambda(i) + $$
$$+ r_i \frac{\partial}{\partial r_j} C^1_q(i) C^1_\lambda(j) + \frac{r_j}{r_i} C^1_q(j)(\nabla^*_i)^1_{-\lambda} + \frac{r_i}{r_j} C^1_q(i)(\nabla^*_j)^1_{-\lambda} \} S^1_m(i).$$

Then, for the matrix element we obtain:

$$\left\langle K SM_S \left| \sum_{i,j} r_{ij}^{-3} \left[\mathbf{r}_j \times \mathbf{p}(i) + \mathbf{r}_i \times \mathbf{p}(j)\right].\mathbf{S}(i) \right| K' SM'_S \right\rangle = i\sqrt{3}\sum_m (-1)^m C^{S1S'}_{MmM'} \sum_{q,\lambda} C^{111}_{q\lambda-m}$$
$$\left\{ \int_{\substack{\vec{r}_1'=\vec{r}_1 \\ \vec{r}_2'=\vec{r}_2}} r_{ij}^{-3} \{ r_2 \frac{\partial}{\partial r_1} C^1_q(2) C^1_\lambda(1) + r_1 \frac{\partial}{\partial r_2} C^1_q(1) C^1_\lambda(2) \} D^f_{SO}(K|\vec{r}_1,\vec{r}_2;\vec{r}_1',\vec{r}_2') d\vec{r}_1 d\vec{r}_2 + \right. \tag{25}$$
$$\left. + \int_{\substack{\vec{r}_1'=\vec{r}_1 \\ \vec{r}_2'=\vec{r}_2}} r_{ij}^{-3} \{ \frac{r_2}{r_1} C^1_q(2)(\nabla^*_1)^1_{-\lambda} + \frac{r_1}{r_2} C^1_q(1)(\nabla^*_2)^1_{-\lambda} \} D^f_{SO}(K|\vec{r}_1,\vec{r}_2;\vec{r}_1',\vec{r}_2') d\vec{r}_1 d\vec{r}_2 \right\}.$$

The third term in Eqn (23) can be written in the form:

$$\mathbf{L}^{(i)}(j) \,.\, \mathbf{S}(i) = \mathbf{L}(j) \,.\, \mathbf{S}(i) - [\mathbf{r}_i \times \mathbf{p}(j)] \,.\, \mathbf{S}(i). \tag{26}$$

The matrix element of the first term in Eqn (26), analogous to (22), has a form that is well known in atomic spectroscopy. Using the irreducible tensor-operator technique we can express the second term as:

$$\left[\mathbf{r}_i \times \mathbf{p}(j)\right].\mathbf{S}(i) = i\sqrt{2} r_i \sum_m (-1)^m \sum_{q,\lambda} C^{111}_{q\lambda-m} \{ C^1_\lambda(j) \frac{\partial}{\partial r_i} C^1_q(i) S^1_m(i) + \tag{27}$$
$$+ r_j^{-1} (\nabla^*_j)^1_\lambda C^1_q(i) S^1_m(i) \}.$$

Then, the corresponding matrix element takes the form:

$$\left\langle K SM_S \left| \sum_{i,j} \left[\mathbf{r}_i \times \mathbf{p}(j)\right].\mathbf{S}(i) \right| K' SM'_S \right\rangle = i\sqrt{2}\sum_m (-1)^m C^{S1S}_{MmM'} \sum_{q,\lambda} C^{111}_{q\lambda-m}$$
$$\int_{\substack{\vec{r}_1'=\vec{r}_1 \\ \vec{r}_2'=\vec{r}_2}} r_1 \{ C^1_\lambda(2) \frac{\partial}{\partial r_1} C^1_q(1) S^1_m(1) + r_2^{-1} (\nabla^*_2)^1_\lambda C^1_q(1) S^1_m(1) \} D^f_{SO}(K|\vec{r}_1,\vec{r}_2;\vec{r}_1',\vec{r}_2') d\vec{r}_1 d\vec{r}_2 . \tag{28}$$

We can rewrite Eqn (27) as:

$$[\mathbf{r}_i \times \mathbf{p}(j)].\mathbf{S}(i) = i\sqrt{2}\sum_m (-1)^m \sum_{q,\lambda} C^{111}_{q\lambda-m} \{C^1_\lambda(j) r_i \frac{\partial}{\partial r_i} V^{11}_{qm}(i) + $$
$$+ r_i r_j^{-1} (\nabla^*_j)^1_\lambda V^{11}_{qm}(i)\},$$

where the $V^{kl}(i)$ are direct products of the operators $\mathbf{C}^k(i)$ and $\mathbf{S}^1(i)$. The numerical values of the matrix elements of the operators $V^{11}_{qm}(i)$ are measured in atomic spectroscopy [17]. In this case Eqn (29), and also Eqn (22), do not take part in the minimization.

4. Discussion and Conclusion

The results presented here allow, not only to include the spin-orbit interactions in the energy functional for a spin-polarized, many-electron systems, in terms of density matrices [1, 2, 18-20], but also to set up a numerical minimization procedure, in the frame of the variational approach for density matrices.

It is possible to include the *spin-same orbit* interactions in the Hartree-Fock scheme but there are basic difficulties with the *spin-other orbit* interactions, connected with the one-particle representation of the two-body matrix for spin-orbit interactions. These difficulties can be overcome if one constructs the many-electron wave function in the vector-model scheme [16, 17, 21] or in the framework of the valence-bond method [1]. But even for the consideration of the *spin-same orbit* interactions in the Hartree-Fock scheme one must impose additional conditions of preservation of spin symmetry.

In atomic spectroscopy [17] the irreducible tensor-operator method allows the calculation of the width of the spectral line splitting induced by the spin-orbit interactions, determining the constant in expressions such as Eqn (22). The calculations are realized *after* minimization of the energy functional constructed in the vector-model scheme. In contrast to this, the results of the present work give the possibility to determine the width of the line splitting *through* a minimization of the energy functional of the density matrix, constructed with a wave function corresponding to a given quantum mechanical approach. This is realized with the inclusion of the terms (15), (21), (22) in the energy functional. In contrast to the case in atomic spectroscopy, we add Eqs (15) and (21) to expressions such as Eqn (22). These terms precise the width of the spectral line splitting. In addition, while the term (22) is determined by the global characteristics of the system (total spin and orbital angular momentum), the terms (15) and (21) are determined by individual characteristics, such as the particular spin and space geometry of the system (3j-symbols).

The matrix elements (14), (24) and (25), resp. (28), describing the contributions of the spin-orbit interactions to the transitions in a given spin multiplet, are also determined by specific characteristics of the system. The inclusion of these terms in the energy functional gives the possibility to determine their contributions in the transition through the minimization procedure. The constraint for the orbital quantum numbers, which follows from the symmetry properties of the 3j-symbols, results from the fact that the inclusion of these interactions, expressed by a spin-involving operator, removes the orbital degeneracy.

In the Barth-Hedin construction [19, 22, 23], the most widely used in Kohn-Sham-type calculations for spin-polarized systems, the energy functional is defined in terms of the first-order density matrix. This does not allow the description of relativistic corrections (including the spin-orbit interactions), which are formed with two-body density matrices. Then it is only possible to determine the influence of an external magnetic field on the ground state [19].

The formalism presented here can be used for the determination of all relativistic corrections in the Breit-Pauli Hamiltonian as well as for calculation of the influence of an external magnetic field, not only for the ground state but also for an arbitrary state of the spin multiplet. In practice, the problem is reduced to building the corresponding spin distribution and correlation matrices and functions within of a suitable quantum mechanical approximation, and performing the elementary mathematical operations presented in this work. The use of a suitable minimization procedure, preserving automatically the space and spin symmetry (e.g., applying the local-scaling transformation method [24-26] or, more precisely, its formulation for spin-polarized systems [20, 27, 28]) allows a direct minimization of the density-matrix energy functional, including arbitrary relativistic correction terms. This is the aim of our future investigations.

Acknowledgements

This work was supported in part by a twinning convention between Universities Pierre et Marie Curie (Paris) and Saint Kliment Ohridski (Sofia) and European COST-D9/0010/98 contract. One of us (A.I.K.) would like to thank the French Government for a predoctoral fellowship. Professor Mateyi Mateev is gratefully acknowledged for useful discussions.

References

1. R. McWeeny, *Methods of Molecular Quantum Mechanics*, Academic Press, New York, 1989.
2. M. M. Mestetchkin, *Methods of Reduced Density Matrices in Quantum Molecular Theory*, Naukova Dumka, Kiev, 1977 (in Russian).
3. K. Husimi, Proc. Phys. Math. Soc. Japan **22** (1940) 264.
4. R. McWeeny, J. Chem. Phys. **42** (1965) 1717.

5. J. Maruani, in P. Becker (ed.), *Electron and Magnetization Densities in Molecules and Crystals*, Plenum Press, New York, 1980, p. 633.
6. V. A. Fock, J.E.T.P. **10** (1940) 961.
7. R. McWeeny and Y. Mizuno, Proc. Roy. Soc. **A259** (1961) 554.
8. R. McWeeny, Mol. Eng. **7** (1997) 7.
9. R. McWeeny, Adv. Quant. Chem. **31** (1999) 15.
10. A. P. Yutsis, I. B. Levinson and V. V. Vanagas, *Mathematical Formalism of Angular Momentum Theory*, Gospolizdat, Vilnius, 1968 (in Russian).
11. R. McWeeny, Proc. Roy. Soc. **A223** (1954) 63.
12. P.-O. Löwdin, Phys. Rev. **97** (1955) 1474.
13. C. N. Yang, Rev. Mod. Phys. **34** (1962) 694.
14. R. McWeeny, Rev. Mod. Phys. **32** (1960) 355; P.-O. Löwdin, Rev. Mod. Phys. **32** (1960) 328; A. J. Coleman, Rev. Mod. Phys. **35** (1963) 668.
15. A. I. Akhiezer and V. B. Berestetzkii, *Quantum Electrodynamics*, Nauka, Moskow, 1969 (in Russian).
16. A. P. Yutsis and A. A. Bandzaytis, *Angular Momentum Theory in Quantum Mechanics*, Mokslas, Vilnius, 1977 (in Russian).
17. I. I. Sobelman, *Introduction to Atomic Spectra Theory*, Nauka, Moscow, 1977 (in Russian).
18. E. S. Kryachko and E. V. Ludeña, *Energy Density Functional Theory of Many-Electron Systems*, Kluwer, Dordrecht, 1990.
19. U. Barth and L. von Hedin, J. Phys. **C5** (1972) 1629.
20. R. L. Pavlov, J. Maruani, Ya. I. Delchev and R. McWeeny, Int. J. Quant. Chem. **65** (1997) 241.
21. L. C. Biedenharn and J. D. Louck, *Angular Momentum in Quantum Physics*, Addison Wesley, Massachusetts, 1981.
22. M. M. Pant and A. K. Rajagopal, Solid State Comm. **10** (1972) 1157.
23. A. K. Rajagopal and J. Gallaway, Phys. Rev. **B7** (1973) 1912.
24. I. Zh. Petkov and M. V. Stoitsov, T.M.F. **55** (1983) 407; Yad. Fisica **37** (1983) 1167 (in Russian).
25. I. Zh. Petkov, M. V. Stoitsov and E. S. Kryachko, Int. J. Quantum Chem. **29** (1986) 140.
26. I. Zh. Petkov and M. V. Stoitsov, *Nuclear Density Functional Theory*, Oxford University Press, Oxford, 1991.
27. Ya. I. Delchev, R. L. Pavlov and C. I. Velchev, Proc. Bulg. Acad. Sci. **44** (1991) 35.
28. R. L. Pavlov, F. E. Zakhariev, Ya. I. Delchev and J. Maruani, Int. J. Quant. Chem. **65** (1997) 257.

Part II

Electron Correlation Treatments

MANY-ELECTRON STURMIANS APPLIED TO ATOMS AND IONS IN STRONG EXTERNAL FIELDS

JOHN AVERY AND CECILIA COLETTI
H.C. Ørsted Institute, University of Copenhagen
DK-2100 Copenhagen, Denmark

Abstract. Methods are introduced for constructing sets of antisymmetrized many-electron Sturmian basis functions using the nuclear attraction potential of an atom or ion as the basis potential. When such basis sets are used, the kinetic energy term disappears from the secular equation, the Slater exponents are automatically optimized, convergence is rapid, and a solution to the many-electron Schrödinger equation, including correlation, is found directly, without the use of the SCF approximation. This technique is applied to atomic ions in external fields so strong that treatments based on perturbation theory are not possible.

1. Many-electron Sturmians for atoms

Suppose that we are able to find a set of functions $\phi_\nu(\mathbf{x})$ which are solutions to the many-electron equation:

$$\left[-\frac{1}{2}\Delta + \beta_\nu V_0(\mathbf{x}) - E\right]\phi_\nu(\mathbf{x}) = 0 \tag{1}$$

In equation (1), Δ is the generalized Laplacian operator

$$\Delta \equiv \sum_{j=1}^{N}\left(\frac{\partial^2}{\partial x_j^2} + \frac{\partial^2}{\partial y_j^2} + \frac{\partial^2}{\partial z_j^2}\right) \equiv \sum_{j=1}^{N}\Delta_j \tag{2}$$

while $\mathbf{x}$ represents the set of coordinate vectors for all the electrons in the system:

$$\mathbf{x} = \{\mathbf{x}_1, \mathbf{x}_2, ..., \mathbf{x}_N\} \tag{3}$$

and

$$\mathbf{x}_j \equiv \{x_j, y_j, z_j\} \qquad j = 1, 2, ..., N \tag{4}$$

J. Maruani et al. (eds.), New Trends in Quantum Systems in Chemistry and Physics, Volume 1, 77–93.

In the present paper we shall discuss the case where

$$V_0(\mathbf{x}) = -\sum_{j=1}^{N} \frac{Z}{r_j} \tag{5}$$

represents the attractive potential of the nucleus of an N-electron atom or ion, but the method discussed here can also be applied to molecules, as we have shown in previous papers [13,14]. In the molecular case, $V_0(\mathbf{x})$ would represent the attractive potential of all the nuclei in the system. The constants β_ν in equation (1) are especially chosen so that all the solutions to the zeroth-order many-electron wave equation (1) correspond to the same value of the energy E, regardless of the quantum numbers ν. Thus the set of functions $\phi_\nu(\mathbf{x})$ may be regarded as a set of generalized Sturmian basis functions [1-11]. But can we obtain a set of solutions to equation (1); and can we find a suitable set of weighting factors for the potential so that all these solutions correspond to the same energy? We shall now try to demonstrate that such a set of solutions can indeed be constructed.

Let

$$\phi_\nu(\mathbf{x}) = \chi_\mu(\mathbf{x}_1)\chi_{\mu'}(\mathbf{x}_2)...\chi_{\mu''}(\mathbf{x}_N) \tag{6}$$

where

$$\begin{aligned} \chi_{nlm,+1/2}(\mathbf{x}_j) &= R_{nl}(r_j)Y_{lm}(\theta_j,\phi_j)\alpha(j) \\ \chi_{nlm,-1/2}(\mathbf{x}_j) &= R_{nl}(r_j)Y_{lm}(\theta_j,\phi_j)\beta(j) \end{aligned} \tag{7}$$

and

$$\begin{aligned} R_{nl}(r_j) &= \mathcal{N}_{nl}(2k_\mu r_j)^l e^{-k_\mu r_j} F(l+1-n|2l+2|2k_\mu r_j) \\ \mathcal{N}_{nl} &= \frac{2k_\mu^{3/2}}{(2l+1)!}\sqrt{\frac{(l+n)!}{n(n-l-1)!}} \end{aligned} \tag{8}$$

The functions $\chi_\mu(\mathbf{x}_j)$ are just the familiar hydrogenlike atomic spin-orbitals, except that the orbital exponents k_μ are left as free parameters which we shall determine later by means of subsidiary conditions. In equations (6)-(8), μ represents the set of quantum numbers $\{n,l,m,s\}$. The functions $\chi_\mu(\mathbf{x}_j)$ satisfy the relationships:

$$\left[-\frac{1}{2}\Delta_j + \frac{1}{2}k_\mu^2 - \frac{nk_\mu}{r_j}\right]\chi_\mu(\mathbf{x}_j) = 0 \tag{9}$$

$$\int d\tau_j |\chi_\mu(\mathbf{x}_j)|^2 \frac{1}{r_j} = \frac{k_\mu}{n} \tag{10}$$

and

$$\int d\tau_j |\chi_\mu(\mathbf{x}_j)|^2 = 1 \tag{11}$$

We now introduce the subsidiary conditions

$$k_\mu^2 + k_{\mu'}^2 + k_{\mu''}^2 + ... = -2E \tag{12}$$

and

$$nk_\mu = n'k_{\mu'} = ... = n''k_{\mu''} = Z\beta_\nu \tag{13}$$

Provided that the subsidiary conditions are satisfied, the product shown in equation (6) will be a solution to (1), since

$$\begin{aligned}\left[-\frac{1}{2}\Delta - E\right]\phi_\nu(\mathbf{x}) &= \left[-\frac{1}{2}\Delta_1 + \frac{1}{2}k_\mu^2 - \frac{1}{2}\Delta_2 + \frac{1}{2}k_{\mu'}^2 + ...\right]\chi_\mu(\mathbf{x}_1)\chi_{\mu'}(\mathbf{x}_2)... \\ &= \left[\frac{nk_\mu}{r_1} + \frac{n'k_{\mu'}}{r_2} + ..\right]\chi_\mu(\mathbf{x}_1)\chi_{\mu'}(\mathbf{x}_2)... \\ &= \beta_\nu\left[\frac{Z}{r_1} + \frac{Z}{r_2} + ...\right]\chi_\mu(\mathbf{x}_1)\chi_{\mu'}(\mathbf{x}_2)... \\ &= -\beta_\nu V_0(\mathbf{x})\phi_\nu(\mathbf{x})\end{aligned} \tag{14}$$

It can easily be seen that an antisymmetrized function of the form

$$\begin{aligned}\phi_\nu(\mathbf{x}) &\equiv |\chi_\mu\chi_{\mu'}...\chi_{\mu''}| \\ &\equiv \frac{1}{\sqrt{N!}}\begin{vmatrix}\chi_\mu(\mathbf{x}_1) & \chi_{\mu'}(\mathbf{x}_1) & ... & \chi_{\mu''}(\mathbf{x}_1) \\ \chi_\mu(\mathbf{x}_2) & \chi_{\mu'}(\mathbf{x}_2) & ... & \chi_{\mu''}(\mathbf{x}_2) \\ \vdots & \vdots & & \vdots \\ \chi_\mu(\mathbf{x}_N) & \chi_{\mu'}(\mathbf{x}_N) & ... & \chi_{\mu''}(\mathbf{x}_N)\end{vmatrix}\end{aligned} \tag{15}$$

will also satisfy (1), since the antisymmetrized function can be expressed as a sum of terms, each of which has the form shown in equation (6). When we use the set of functions $\phi_\nu(\mathbf{x})$ as a basis set to build up the wave function of an N-electron atom or ion, we shall of course use the antisymmetrized functions shown in equation (15), since we wish the wave function to satisfy the Pauli principle.

Sturmian basis sets satisfy potential-weighted orthonormality relations, and similarly, our generalized many-electron Sturmian basis functions satisfy an orthonormality relation where the weighting factor is the potential $V_0(\mathbf{x})$ [12-14]. To see this, we consider two different solution to equation (1). From (1), it follows that they satisfy

$$\int dx \phi_{\nu'}^*(\mathbf{x})\left[\frac{1}{2}\Delta + E\right]\phi_\nu(\mathbf{x}) = \beta_\nu \int dx \phi_{\nu'}^*(\mathbf{x})V_0(\mathbf{x})\phi_\nu(\mathbf{x}) \tag{16}$$

and

$$\int dx \phi_\nu^*(\mathbf{x}) \left[\frac{1}{2}\Delta + E\right] \phi_{\nu'}(\mathbf{x}) = \beta_{\nu'} \int dx \phi_\nu^*(\mathbf{x}) V_0(\mathbf{x}) \phi_{\nu'}(\mathbf{x}) \tag{17}$$

If we take the complex conjugate of (17), subtract it from (16), and make use of the Hermiticity of the operator $\frac{1}{2}\Delta + E$, we obtain:

$$(\beta_\nu - \beta_{\nu'}) \int dx \phi_{\nu'}^*(\mathbf{x}) V_0(\mathbf{x}) \phi_\nu(\mathbf{x}) = 0 \tag{18}$$

where the constants β_ν are assumed to be real. From equation (18) it follows that if $\beta_\nu - \beta_{\nu'} \neq 0$, then

$$\int dx \phi_{\nu'}^*(\mathbf{x}) V_0(\mathbf{x}) \phi_\nu(\mathbf{x}) = 0 \tag{19}$$

For the case where $\nu = \nu'$, we have

$$\begin{aligned} \int dx V_0(\mathbf{x}) |\phi_\nu(\mathbf{x})|^2 &= -\sum_{\mu \subset \nu} Z \int d\tau_j |\chi_\mu(\mathbf{x}_j)|^2 \frac{1}{r_j} \\ &= -\sum_{\mu \subset \nu} Z \frac{k_\mu}{n} \\ &= -\frac{1}{\beta_\nu} \sum_{\mu \subset \nu} k_\mu^2 = \frac{2E}{\beta_\nu} \end{aligned} \tag{20}$$

Combining (19) and (20), and making use of the orthonormality of the spin functions and the spherical harmonics, we obtain the potential-weighted orthonormality relations:

$$\int dx\ \phi_{\nu'}^*(\mathbf{x}) V_0(\mathbf{x}) \phi_\nu(\mathbf{x}) = \delta_{\nu',\nu} \frac{2E}{\beta_\nu} \tag{21}$$

where ν stands for the set of quantum numbers $\{\mu, \mu', ..., \mu''\}$.

2. The secular equation

Having constructed a set of many-electron Sturmian basis functions by the method just described, we would like to use this basis set to solve the Schrödinger equation for an N-electron atom in a strong external field. Using atomic units, we can write the Schrödinger equation in the form:

$$\left[-\frac{1}{2}\Delta + V(\mathbf{x}) - E\right] \psi(\mathbf{x}) = 0 \tag{22}$$

where

$$V(\mathbf{x}) = V_0(\mathbf{x}) + V'(\mathbf{x}) + V''(\mathbf{x}) \tag{23}$$

Here $V_0(\mathbf{x})$ is the nuclear attraction potential shown in equation (5), while

$$V'(\mathbf{x}) = \sum_{i>j}^{N} \sum_{j=1}^{N} \frac{1}{r_{ij}} \tag{24}$$

is the interelectron repulsion potential and

$$V''(\mathbf{x}) = \sum_{j=1}^{N} \mathcal{E} \cdot \mathbf{x}_j \tag{25}$$

is the potential due to the applied field. In equation (25) we show the potential due to a constant electric field whose field strength $\mathcal{E}$ is expressed in atomic units. Expanding the wave function as series of generalized Sturmian basis functions, and making use of equation (1), we obtain:

$$\sum_{\nu} \left[-\frac{1}{2}\Delta + V(\mathbf{x}) - E \right] \phi_\nu(\mathbf{x}) B_\nu =$$
$$\sum_{\nu} \left[-\beta_\nu V_0(\mathbf{x}) + V(\mathbf{x}) \right] \phi_\nu(\mathbf{x}) B_\nu = 0 \tag{26}$$

Multiplying (26) from the left by a conjugate basis function, integrating over the coordinates, and making use of the potential-weighted orthonormality relation (21), we obtain:

$$\sum_{\nu} \left[\int dx \phi^*_{\nu'}(\mathbf{x}) V(\mathbf{x}) \phi_\nu(\mathbf{x}) - 2E\delta_{\nu',\nu} \right] B_\nu = 0 \tag{27}$$

We now introduce a parameter, p_0, which is related to the energy E of the system by

$$p_0^2 \equiv -2E \tag{28}$$

and we also introduce the definition:

$$T_{\nu',\nu} \equiv -\frac{1}{p_0} \int dx \phi^*_{\nu'}(\mathbf{x}) V(\mathbf{x}) \phi_\nu(\mathbf{x}) \tag{29}$$

We shall see below that if $V(\mathbf{x})$ is a potential produced by Coulomb interactions, then the matrix $T_{\nu',\nu}$, defined in this way, is independent of p_0. In terms of these new parameters, the secular equation, (27), can be written in the form:

$$\sum_{\nu} \left[T_{\nu',\nu} - p_0 \delta_{\nu',\nu} \right] B_\nu = 0 \tag{30}$$

The total potential, V, consists of a nuclear attraction part, V_0, an interelectron repulsion part, V', and a part due to the applied field, V''. The matrix $T_{\nu',\nu}$ can also be divided into three parts

$$T_{\nu',\nu} = T^0_{\nu',\nu} + T'_{\nu',\nu} + T''_{\nu',\nu} \tag{31}$$

corresponding respectively to nuclear attraction, interelectron repulsion and applied field. From the potential-weighted orthonormality relation (21) and the subsidiary conditions, (12) and (13), it follows that:

$$\begin{aligned} T^0_{\nu',\nu} &= -\frac{1}{p_0}\int dx \phi^*_{\nu'}(\mathbf{x}) V_0(\mathbf{x}) \phi_\nu(\mathbf{x}) \\ &= -\frac{2E}{p_0\beta_\nu}\delta_{\nu',\nu} = \frac{p_0}{\beta_\nu}\delta_{\nu',\nu} = \frac{1}{\beta_\nu}\left(\sum_{\mu\subset\nu} k_\mu^2\right)^{1/2} \delta_{\nu',\nu} \\ &= Z\left(\sum_{\mu\subset\nu}\frac{1}{n^2}\right)^{1/2} \delta_{\nu',\nu} \end{aligned} \tag{32}$$

Thus $T^0_{\nu',\nu}$ is independent of p_0, as we mentioned above. If there is no external field, and if the interelectron repulsion is neglected, the matrix $T_{\nu',\nu} = T^0_{\nu',\nu}$ is diagonal, and the Sturmian secular equation, (30), simply requires that

$$p_0 = Z\left(\sum_{\mu\subset\nu}\frac{1}{n^2}\right)^{1/2} \tag{33}$$

The zeroth-order energy of the system is then given by

$$E = -\frac{p_0^2}{2} = -\frac{Z^2}{2}\sum_{\mu\subset\nu}\frac{1}{n^2} \tag{34}$$

which is the correct energy of a system of N noninteracting electrons in the attractive field of the atom's nucleus. (For simplicity we use the approximation where the motion of the nucleus is neglected.)

3. The interelectron repulsion matrix

Let us now turn to the evaluation of the interelectron repulsion matrix,

$$T'_{\nu',\nu} \equiv -\frac{1}{p_0}\int dx \phi^*_{\nu'}(\mathbf{x}) \sum_{i>j}^{N}\sum_{j=1}^{N}\frac{1}{r_{ij}}\phi_\nu(\mathbf{x}) \tag{35}$$

We shall see that this matrix is also independent of p_0. In order to evaluate $T'_{\nu',\nu}$, we need to calculate 2-electron integrals of the form:

$$\begin{aligned} J = \int_0^\infty dr_1 r_1^{2+j_1} e^{-\zeta_1 r_1} \int_0^\infty dr_2 r_2^{2+j_2} e^{-\zeta_2 r_2} \times \\ \int d\Omega_1 W_1(\hat{\mathbf{x}}_1) \int d\Omega_2 W_2(\hat{\mathbf{x}}_2) \frac{1}{r_{12}} \end{aligned} \tag{36}$$

where $W_1(\hat{\mathbf{x}}_1)$ and $W_2(\hat{\mathbf{x}}_2)$ are products of spherical harmonics. This can be done using Fourier-transforms and contour integration, as described in some of our previous papers. Alternatively we can expand $1/r_{12}$ in a series of Legendre polynomials:

$$\frac{1}{r_{12}} = \sum_{l=0}^{\infty} \frac{r_<^l}{r_>^{l+1}} P_l(\hat{\mathbf{x}_1} \cdot \hat{\mathbf{x}}_2) \tag{37}$$

Then

$$J = \sum_{l=0}^{\infty} a_l I_l \tag{38}$$

where

$$a_l \equiv \int d\Omega_1 W_1(\hat{\mathbf{x}}_1) \int d\Omega_2 W_2(\hat{\mathbf{x}}_2) P_l(\hat{\mathbf{x}_1} \cdot \hat{\mathbf{x}}_2) \tag{39}$$

and

$$I_l \equiv \int_0^{\infty} dr_1 r_1^{j_1+2} e^{-\zeta_1 r_1} \int_0^{\infty} dr_2 r_2^{j_2+2} e^{-\zeta_2 r_2} \frac{r_<^l}{r_>^{l+1}} \tag{40}$$

Usually very few of the angular coefficients a_l are non-zero, so the sum in equation (38) involves only a few terms. The radial integrals I_l can be evaluated by means of the relationship

$$\begin{aligned} &\int_0^{\infty} dr_1\, r_1^{j_1+2} e^{-\zeta_1 r_1} \int_0^{\infty} dr_2\, r_2^{j_2+2} e^{-\zeta_2 r_2} \frac{r_<^l}{r_>^{l+1}} \\ &= \frac{\Gamma(j_1+j_2+5)}{(\zeta_1+\zeta_2)^{j_1+j_2+5}} \left[\frac{{}_2F_1(1, j_1+j_2+5; j_2+l+4; \zeta_2/(\zeta_1+\zeta_2))}{j_2+l+3} \right. \\ &\quad \left. + \frac{{}_2F_1(1, j_1+j_2+5; j_1+l+4; \zeta_1/(\zeta_1+\zeta_2))}{j_1+l+3} \right] \end{aligned} \tag{41}$$

where

$${}_2F_1(a, b; c; x) \equiv 1 + \frac{ab}{c} x + \frac{a(a+1)b(b+1)}{c(c+1)} \frac{x^2}{2!} + \ldots \tag{42}$$

is a hypergeometric function. For example, suppose that we are considering a 2-electron atom or ion and that

$$\phi_\nu(\mathbf{x}) = |\chi_{1s}\chi_{\bar{1}s}| \equiv \frac{1}{\sqrt{2}}[\chi_{1s}(1)\chi_{\bar{1}s}(2) - \chi_{1s}(2)\chi_{\bar{1}s}(1)] \tag{43}$$

where

$$\begin{aligned} \chi_{1s}(1) &= \left(\frac{k_\mu^3}{\pi}\right)^{1/2} e^{-k_\mu r_1} \alpha(1) \\ \chi_{\bar{1}s}(1) &= \left(\frac{k_\mu^3}{\pi}\right)^{1/2} e^{-k_\mu r_1} \beta(1) \end{aligned} \tag{44}$$

so that

$$\phi_\nu(\mathbf{x}) = \frac{k_\mu^3}{\pi} e^{-k_\mu(r_1+r_2)} \frac{1}{\sqrt{2}} [\alpha(1)\beta(2) - \beta(1)\alpha(2)] \tag{45}$$

Then the diagonal element of the interelectron repulsion matrix involving the configuration $\nu = \{1, 0, 0, \frac{1}{2}; 1, 0, 0, -\frac{1}{2}\}$ is given by

$$\begin{aligned} T'_{\nu,\nu} &= -\frac{1}{p_0} \int dx \phi_\nu^*(\mathbf{x}) \frac{1}{r_{12}} \phi_\nu(\mathbf{x}) \\ &= -\frac{1}{p_0} \left(\frac{k_\mu^3}{\pi}\right)^2 \int d^3x_1 \int d^3x_2 e^{-2k_\mu(r_1+r_2)} \frac{1}{r_{12}} \\ &= -\frac{1}{p_0} \left(\frac{k_\mu^3}{\pi}\right)^2 (4\pi)^2 \int_0^\infty dr_1 r_1^2 e^{-2k_\mu r_1} \int_0^\infty dr_2 r_2^2 e^{-2k_\mu r_2} \frac{1}{r_>} \end{aligned} \tag{46}$$

The radial integral can be evaluated by means of equation (41), and the result is:

$$T'_{\nu,\nu} = -\frac{5}{8} \frac{k_\mu}{p_0} = -\frac{1}{\sqrt{2}} \frac{5}{8} \tag{47}$$

which is a pure number, independent of p_0. The matrix elements of the interelectron repulsion potential, $T'_{\nu',\nu}$, always prove to be pure numbers, and they are always independent of p_0. The reason for this is that the subsidiary relations (12) and (13) require that

$$\frac{k_\mu}{p_0} = \frac{1}{n\sqrt{\frac{1}{n^2} + \frac{1}{n'^2} + \ldots + \frac{1}{n''^2}}} \tag{48}$$

Thus the ratios k_μ/p_0 are always pure numbers; and from equations (8), (35) and (41) it follows that $T'_{\nu',\nu}$ can always be expressed in terms of these ratios. The simple example which we have considered here already allows us to find the energies of the 2-electron isoelectronic series of atoms and ions in the rough approximation where our basis set consists only of a single 2-electron Sturmian basis function - that shown in equation (45). In that case, the secular equation reduces to the requirement that

$$p_0 = T^0_{\nu,\nu} + T'_{\nu,\nu} = \sqrt{2}\left(Z - \frac{5}{16}\right) \tag{49}$$

where we have made use of equations (32) and (47). In this rough approximation, the energies of the atoms and ions in the 2-electron isoelectronic series are given by

$$E = -\frac{p_0^2}{2} = -\left(Z - \frac{5}{16}\right)^2 \tag{50}$$

More accurate energies could of course be obtained by using more basis functions; but the crude 1-basis-function result shown in equation (50) already is in good agreement with Clementi's Hartree-Fock energies [15], as is shown in Figure 1.

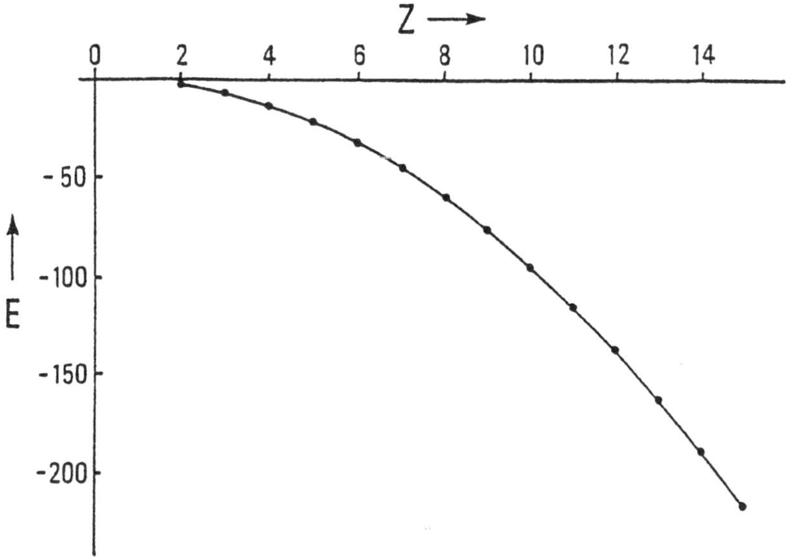

Figure 1. This figure shows the energies as a function of Z for the ions in the two electron isoelectronic series calculated in the crude approximation where only one configuration is used (equation (50)). The line shows the energies calculated from equation (50), while the dots represent Clementi's Hartree-Fock values [15].

4. The 3-electron isoelectronic series in a strong electric field

We are now in a position to calculate the properties of an atom or atomic ion in a strong external field. For example, let us consider the 3-electron isoelectronic series, Li, Be^+, B^{2+}, C^{3+}, N^{4+}, O^{5+},... etc subjected to a constant external electric field so strong that treatments based on perturbation theory must fail. We saw above that $T^0_{\nu,\nu}$ and $T'_{\nu,\nu}$ are pure numbers, independent of p_0. By contrast, the matrix

$$\begin{aligned} T''_{\nu',\nu} &= -\frac{1}{p_0}\int dx\phi^*_{\nu'}(\mathbf{x})\sum_{j=1}^{N}\mathcal{E}\cdot\mathbf{x}_j\phi_\nu(\mathbf{x}) \\ &= -\frac{\mathcal{E}}{p_0^2}\int dx\phi^*_{\nu'}(\mathbf{x})\sum_{j=1}^{N}p_0 r_j\cos\theta_j\phi_\nu(\mathbf{x}) \end{aligned} \qquad (51)$$

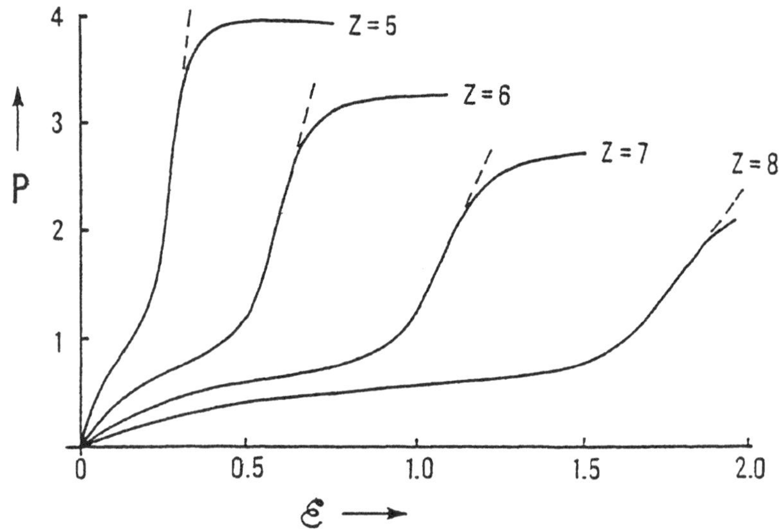

Figure 2. Figure 2 shows the polarization of ions in the three-electron isolectronic series. P is the induced dipole moment in electron-Bohrs, while $\mathcal{E}$ is the external field in Hartrees/electron-Bohr. It can be seen that a pre-ionization anomaly occurs when the induced dipole moment reaches approximately 1 electron-Bohr. This pre-ionization anomaly is associated with an avoided crossing between the first excited state and the ground state (figure 3), where the character of the ground state changes abruptly and begins to include a large proportion of configurations corresponding to high values of n (figure 4). The dashed lines indicate how we believe the calculated induced dipole moments would behave if our basis sets were more complete.

depends on p_0; but it can be separated into a pure number (the integral in the second part of equation (51)) multiplied by the factor $\eta = \mathcal{E}/p_0^2$, where $\mathcal{E}$ is the electric field strength in atomic units. Thus we can write:

$$T''_{\nu',\nu} \equiv \eta \mathcal{V}_{\nu',\nu} \tag{52}$$

where

$$\begin{aligned} \mathcal{V}_{\nu',\nu} &\equiv -\int dx \phi^*_{\nu'}(\mathbf{x}) \left(\sum_{j=1}^{N} p_0 r_j \cos\theta_j \right) \phi_\nu(\mathbf{x}) \\ \eta &\equiv \frac{\mathcal{E}}{p_0^2} \end{aligned} \tag{53}$$

For simplicity we shall confine our basis set to the following 9 configurations:

$$\begin{aligned} \phi_{nS} &= |\chi_{ns}\chi_{1s}\chi_{\bar{1s}}| \qquad n = 2, 3, 4 \\ \phi_{nP} &= |\chi_{np_0}\chi_{1s}\chi_{\bar{1s}}| \qquad n = 2, 3, 4 \end{aligned}$$

$$\begin{aligned}\phi_{nD} &= |\chi_{nd_0}\chi_{1s}\chi_{\bar{1}s}| \qquad n=3,4 \\ \phi_{nF} &= |\chi_{nf_0}\chi_{1s}\chi_{\bar{1}s}| \qquad n=4\end{aligned} \tag{54}$$

Each of these configurations consists of a Slater determinant involving a helium-like inner shell and an outer spin-up orbital. If we choose z-axis in the direction of the applied field, the z component of total angular momentum is a good quantum number; and we consider here only the $M=0$ states of the system. We would like the configurations of equation (52) to be 3-electron Sturmian basis functions - that is to say, they should all satisfy equation (1), and they should all correspond to the same value of the energy, $E=-p_0^2/2$. As we saw above, this can be achieved by building the configurations of orbitals of the form shown in equations (7) and (8), with Slater exponents k_μ satisfying the subsidiary conditions (12) and (13). It then follows that

$$T^0_{\nu',\nu} = Z\left(2+\frac{1}{n^2}\right)^{1/2} \tag{55}$$

and

$$\frac{k_\mu}{p_0} = \left\{\frac{1}{n\sqrt{2+\frac{1}{n^2}}}, \frac{1}{\sqrt{2+\frac{1}{n^2}}}, \frac{1}{\sqrt{2+\frac{1}{n^2}}}\right\} \tag{56}$$

Thus if we take matrix elements between configurations involving different values of n, we do not have orthonormality between the atomic orbitals of one configuration and those of the other. As a consequence we must either use the generalized Slater-Condon rules [16-18] or in some other way take into account the absence of orthonormality between configurations involving different values of n. Apart from this complication, the evaluation of the matrix elements $T'_{\nu',\nu}$ and $\mathcal{V}_{\nu',\nu}$ is straightforward; and the results are shown in Tables 1 and 2. The interelectron repulsion matrix is block-diagonal, each block corresponding to a value of the total orbital angular momentum quantum number L. For the matrix elements of the applied field, the only non-zero values are those corresponding to $\Delta L = \pm 1$, as shown in Table 2. In order to obtain the wave functions, energies and polarizations for the 3-electron series in the external field, we first pick a value of Z and a value of the parameter η. We then solve the secular equation

$$\sum_\nu \left[T'_{\nu',\nu} + \eta\mathcal{V}_{\nu',\nu} + \left(Z\sqrt{2+\frac{1}{n^2}} - p_0\right)\delta_{\nu',\nu}\right] B_\nu = 0 \tag{57}$$

and by repeating the diagonalization for many other values of η we obtain the ground state and excited state energies, wave functions, and polarizations of the system as functions of this parameter. Since we also obtain a spectrum of p_0 values with each diagonalization, we can use the relationship

$\mathcal{E} = \eta p_0^2$ to find these properties as functions of the external field. For example, Figure 2 shows the polarizations of the ground states of the B^{2+}, C^{3+}, N^{4+} and O^{5+} ions as functions of the strength of the applied field. In this figure, atomic units are used, both for the polarization and for the applied electric field strength. It can be seen from this figure that when the polarization reaches a value of approximately 1 atomic unit (1 electron-Bohr), the polarizability increases sharply. This seems to be a pre-ionization anomaly corresponding to an avoided crossing in the spectrum (Figure 3), where the character of the ground state changes from being dominated by the $n=2$ configurations and begins to include a large proportion of configurations corresponding to higher values of n, as shown in Figure 4. Our basis set is too poor to represent the true behavour of the polarization following the avoided crossing. In fact, avoided crossings of higher excited states occur at much lower values of the applied field, so that when the first excited state meets the ground state for an avoided crossing, it is already rich in configurations corresponding to high values of n. This is reflected in Figure 2 by dotted lines, which show how we believe the calculated polarizations would continue to increase with increasing field if our basis set were richer.

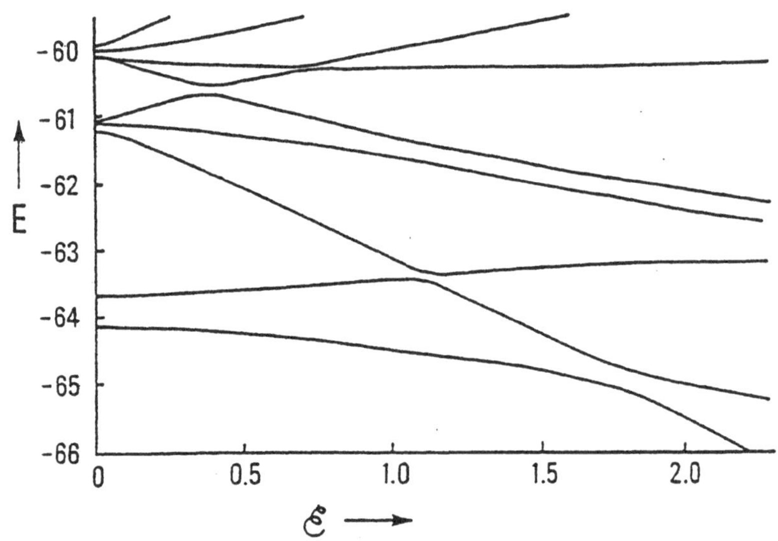

Figure 3. This figure shows the $M = 0$ energy levels in Hartrees for O^{5+} in a very strong external electric field. An avoided crossing between the first excited state and the ground state occurs when the applied field is approximately 1.7 Hartrees/electron-Bohr, and this can be seen to correspond to the field where the pre-ionization anomaly for $Z = 8$ occurs in figure 2. Since an avoided crossing between the first excited state and the second excited state has already occured at a lower value of the field the ground state becomes rich in configurations corresponding to high values of n.

TABLE 1. Blocks of $T'_{\nu',\nu}$ for the 3-electron isoelectronic series

SS block	$n=2$	$n=3$	$n=4$
$n'=2$	0.6818701	0.0408137	0.0214969
$n'=3$	0.0408137	0.5631285	0.0282403
$n'=4$	0.0214969	0.0282403	0.5138263

PP block	$n=2$	$n=3$	$n=4$
$n'=2$	0.7290174	0.0542473	0.0290072
$n'=3$	0.0542473	0.5768334	0.0342767
$n'=4$	0.0290072	0.0342767	0.5195789

DD block	$n=3$	$n=4$
$n'=3$	0.5828068	0.0306734
$n'=4$	0.0306734	0.5220717

FF block	$n=4$
$n'=4$	0.5222305

TABLE 2. Blocks of $\mathcal{V}_{\nu',\nu}$

PS block	$n = 2$	$n = 3$	$n = 4$
$n' = 2$	4.5000000	0.9954544	0.4018616
$n' = 3$	2.8707321	10.6770782	2.2577101
$n' = 4$	1.1309399	4.8345780	19.2678488

DP block	$n = 2$	$n = 3$	$n = 4$
$n' = 3$	3.7546458	7.5498344	1.0989138
$n' = 4$	1.2472158	5.8493764	15.4142790

FD block	$n = 3$	$n = 4$
$n' = 4$	7.5923117	11.5607093

5. Discussion

The use of many-electron Sturmian basis functions derived from the actual nuclear attraction potential of an atom or molecule offers a number of advantages:

1. The matrix representation of the nuclear attraction potential in this

basis is diagonal.
2. When such basis functions are used, the kinetic energy term vanishes from the secular equation.
3. The Slater exponents of the basis set are automatically optimized.
4. Convergence is rapid.
5. A solution to the many-electron Schrödinger equation is obtained directly, without the use of the SCF approximation.
6. Excited states are obtained with good accuracy (as is illustrated in Table 3).

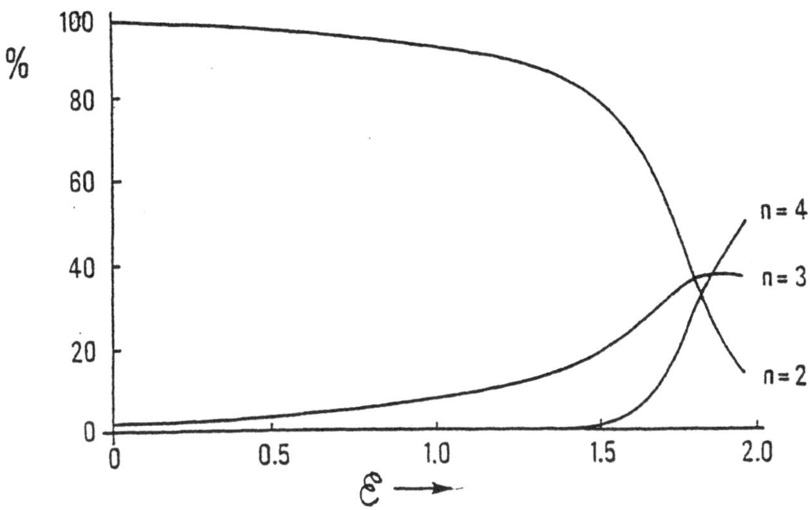

Figure 4. This figure shows the changes in the character of the ground state of O^{5+} at the avoided crossing between the ground state and the first excited state. The percentage of the wavefunction corresponding to the $n = 2$ configuration, equation (54), drops sharply, to be replaced by configurations corresponding to higher values of n. Since the ground state wavefunction then contains d and f functions the polarizability increases.

The method can be applied both to atoms and to molecules. Only atoms are considered in the present paper, but in previous papers [13,14] we have shown that the momentum-space methods of Fock, Shibuya and Wulfman are very suitable for constructing molecular many-electron Sturmian basis sets.

In the present paper, we have illustrated the method by considering atomic ions of the 3-electron isoelectronic series in very strong external electric fields. Our basis set consisted of only 9 generalized Sturmian basis functions (equation (54)); but it would be possible to represent the ground states with a still smaller set, with no loss of accuracy, if the basis functions

TABLE 3. Excited ^2S-states of O^{5+} and F^{6+} in Hartrees. The basis set used consisted of the 5 configurations, $\phi_{nS} = |\chi_{ns}\chi_{1s}\chi_{\bar{1s}}|$ $n = 2, ..., 6$. Experimental values are taken from Moore's tables [22]

O^{5+}	$3s\ ^2S$	$4s\ ^2S$	$5s\ ^2S$	F^{6+}	$3s\ ^2S$	$4s\ ^2S$	$5s\ ^2S$
calc.	61.232	60.267	59.832	*calc.*	78.356	77.057	76.467
expt.	61.219	60.250	59.813	*expt.*	78.342	77.040	76.450
err.	0.021%	0.028%	0.032%	*err.*	0.017%	0.022%	0.022%

were constructed from parabolic hydrogenlike orbitals of the type studied by Aquilanti and co-workers [19-21]. These alternative hydrogenlike orbitals, separated in parabolic coordinates, are in fact the most suitable building blocks for representing the wave function of an atom or ion in an external electric field.

Our basis set is so small that it is insufficient to represent accurately the neutral lithium atom; but convergence improves with increasing Z, and the ions in the 3-electron isoelectronic series are described increasingly well as Z becomes large. Once the matrix elements shown in Tables 1 and 2 have been calculated, solutions for all values of Z and $\mathcal{E}$ can be obtained with almost no calculational effort by diagonalizing the 9×9 matrix shown in equation (57).

In conclusion, the method of many-electron Sturmian basis functions seems to offer an interesting and promising alternative to the usual SCF-CI methods.

References

1. Epstein, P.S., Proc. Natl. Acad. Sci. (U.S.A.) **12**, 637 (1926).
2. Podolski, B., Proc. Natl. Acad. Sci. (U.S.A.) **14**, 253 (1928).
3. Pauling, L. and Wilson, E.B., *Introduction to Quantum Mechanics*, Chapter VIII, McGraw Hill, New York and London (1935).
4. Shull, H. and Löwdin, P.O., J. Chem. Phys. **30**, 617 (1959).

5. Rotenberg, M., Ann. Phys. (New York) **19**, 262 (1962).
6. Rotenberg, M., Adv. At. Mol. Phys. **6**, 233 (1970).
7. Goscinski, O., *Preliminary Research Report No. 217*, Quantum Chemistry Group, Uppsala University (1968).
8. Gazeau, J.P. and Maquet, A., J. Chem. Phys. **73**, 5147 (1980).
9. Duchon, Ch., Dumont-Lepage, M. Cl., and Gazeau, J.P., J. Chem. Phys. **76**, 445 (1982).
10. Avery, J., *Hyperspherical Harmonics; Applications in Quantum Theory*, Kluwer Academic Publishers, Dordrecht, Netherlands (1989).
11. Avery, J. and Herschbach, D.R., Int. J. Quantum Chem. **41**, 673 (1992).
12. Aquilanti, V. and Avery, J., Chem. Phys. Letters **267**, 1 (1997).
13. Avery, J, J. Math. Chem. **21**, 285 (1997) .
14. Avery, J, Advances in Quantum Chemistry **31**, 201 (1999).
15. Clementi E., J. Chem. Phys. **38**, 996 (1963).
16. Löwdin, P.O., J. Appl. Phys. Suppl. **33**, 251 (1962).
17. Amos, A.T. and Hall, G.G., Proc. Roy. Soc. (London) **A263**, 483 (1961).
18. King, H.F., Stanton, R.E., Kim, H., Wyatt, R.E., and Parr, R.G., J. Chem. Phys. **47**, 1936 (1967).
19. Aquilanti V., Cavalli S., Coletti C., Grossi G., Chem. Phys. **209**, 405 (1996).
20. Aquilanti V., Cavalli S., Coletti C., Chem. Phys. **214**, 1 (1997).
21. Aquilanti V., Cavalli S., Coletti C., Phys. Rev. Lett. **80**, 3209 (1998).
22. Moore, C.E., *Atomic Energy Levels, Circular of the National Bureau of Standards 467*, Superintendent of Documents, U.S. Government Printing Office, Washington 25 D.C. (1949).

AN IMPLEMENTATION OF THE CONFIGURATION-SELECTING MULTI-REFERENCE CONFIGURATION-INTERACTION METHOD ON MASSIVELY PARALLEL ARCHITECTURES

P. STAMPFUSS AND W. WENZEL
Universität Dortmund, Institut für Physik
D-44221 Dortmund, Germany

Abstract. We report on a scalable implementation of the configuration-selecting multi-reference configuration interaction method for massively parallel architectures with distributed memory. Based on a residue driven evaluation of the matrix elements this approach, which was adapted to allow the selective treatment of triple and quadruple excitations with respect to the reference space, allows the routine treatment of Hilbert spaces of well over 10^9 determinants. We demonstrate the scalability of the method for up to 128 nodes on the IBM-SP2 and for up to 256 nodes on the CRAY-T3E. We elaborate on the specific adaptation of the transition residue-based matrix element evaluation scheme that ensures the scalability and load-balancing of the method.

1. Introduction

For many years the multi-reference configuration interaction method (MRCI) [1, 2, 3] has been one of the benchmark tools for accurate investigation into the electronic structure of atoms and molecules. Ever since the development of the direct CI algorithm [1] highly efficient implementations [4] have been used for a wide variety of molecules. The generic lack of extensivity of the MRCI method has at least been partially addressed with a number of *a posteriori* [5, 6] corrections and through direct modification of the CI energy-functional [7, 8, 9, 10, 11, 12].

Due to its high computational cost, however, application of MRCI remain constrained to relatively small systems. For this reason the configuration-selective version of the MRCI-method (MRD-CI), introduced by Buenker and Peyerimhoff [13, 14, 15], has arguably become one of its most

J. Maruani et al. (eds.), New Trends in Quantum Systems in Chemistry and Physics, Volume 1, 95–113.

widely used versions. In this variant only the most important configurations of the interacting space of a given set of primary configurations are chosen for the variational wavefunction[16], while the energy contributions of the remaining configurations are estimated on the basis of second-order Rayleigh-Schrödinger perturbation theory [17, 18]. Since the variationally treated subspace of the problem is much smaller than the overall Hilbert space, the determination of its eigenstates requires far less computational effort. Indeed, for typical applications the overwhelming majority of the computational effort is concentrated in the expansion loop, where the energy contribution of candidate configurations is computed.

Even within this approximation, the cost of MRCI calculations remains rather high. The development of efficient configuration-selecting CI codes [19, 20, 18, 25, 21, 22, 23, 26] is inherently complicated by the sparseness and the lack of structure of the selected state-vector. In order to further extend the applicability of the method, it is thus desirable to employ the most powerful computational architectures available for such calculations. Here we report on the progress of the first massively parallel, residue-driven implementation of the MRD-CI method for distributed memory architectures. While efforts to parallelize standard MR-SDCI (all single and double excitations) on distributed memory architectures face significant difficulties rooted in the need to distribute the CI vectors over many nodes [24, 27, 28, 29] — a parallel implementation of MRD-CI can capitalize on the compactness of its state representation. In our implementation the construction of the subset of nonzero matrix elements is accomplished by the use of a residue-based representation of the matrix elements that was originally developed for the distributed memory implementation of MR-SDCI [29]. This approach allows us to efficiently evaluate the matrix elements both in the expansion loop as well as during the variational improvement of the coefficients of the selected vectors.

In order to attain an efficient implementation of the MRCI family of methods on massively parallel machines with distributed memory mechanisms must be devised that distribute the data among the nodes of the machine such that all computations can be accomplished using only *local* data. In our determinant-based implementation the residue-tree, as discussed in section 2, plays the role of the organizing principle. Going somewhat beyond the standard MRD-CI, our implementation was specifically optimized to estimate the importance of triple- and quadruple excitations of the reference configuration. The energy arising from such configurations yields the overwhelming contribution to the energy difference between FCI and MR-SDCI and is thus of paramount importance for the development approximately extensive versions of the MRCI method [7, 8, 9, 10]. Since the number of higher-than-doubly excited configuration rises so quickly with system size,

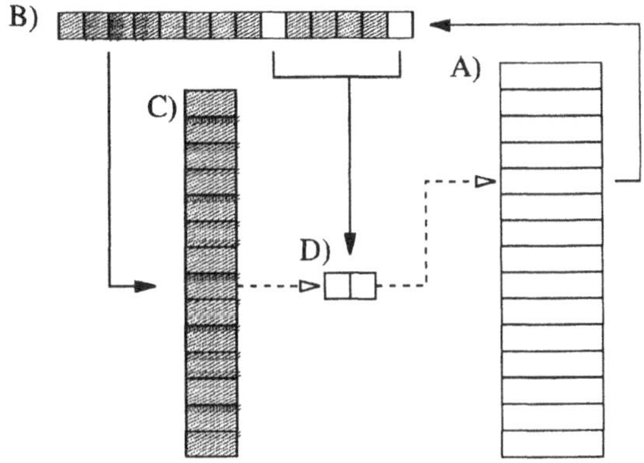

Figure 1. Schematic representation of the two-particle residue-tree. For each element of the configuration-list (A) all possible two-particle residues are constructed. In the configuration illustrated in (B) each box represents one occupied orbital, the shaded region corresponds to the residue and the two white boxes to the orbital pair. The $(n_e - 2)$-electron residue configuration is looked up in the residue-tree (C), where an element (D) is added that encodes the orbitals that where removed, information regarding the permutation required and the index of the original configuration in the configuration list. Solid arrows in the figure indicate logical relations ships, dotted arrows indicate pointers incorporated in the data structure. The residue-list, along with all elements must be rebuilt once after each expansion loop, the effort to do so is proportional to product of n_e^2 with the number of configurations. The number of matrix elements encoded in a single element of the residue-tree is proportional to the *square of the number of entries* of type (D).

FCI as well as CI-SDTQ calculations are prohibitively expensive for all but the smallest systems. In addition it is possible to modify the treatment of the TQ excitations, such as to provide explicit extensive dressings of the CI matrix elements for incomplete primary spaces. Configuration selecting CI provides a particularly effective, maybe the only viable, compromise between computational efficiency and accuracy for the treatment of the TQ space. Here we report on the key principles of the the implementation of the method and provide timings for benchmark applications that demonstrate the scalability of the method for up to 128 nodes of an IBM-SP2 and up to 256 nodes of a CRAY-T3E for Hilbert spaces of dimension up to 5×10^9 of which up to 5×10^6 elements were selected for the variational wavefunction.

2. Methodology

In the following we will describe the key ingredients for the residue based parallel implementation of the configuration selecting MR-CI method. We begin with a description of the orbital partitioning scheme that allows a flexible treatment of the triple and quadruple excitations with regard to the active space. We illustrate the principle of the residue-based matrix-element evaluation that is at the heart of our algorithm. Next we present the results of benchmarks of the method for O_2, NO_2 and benzene as a function of the number of nodes used.

2.1. MATRIX ELEMENT EVALUATION

Virtually all computational effort of the configuration selective CI method is concentrated in two steps. First, the many body field

$$q_i = \langle \phi_i | H | \Psi \rangle \tag{1}$$

must be computed for all non-selected configurations $|\phi_i\rangle$ to asses there importance. Here

$$|\Psi\rangle = \sum_j c_j \, |\phi_i\rangle \tag{2}$$

designates either the set of all previously selected configurations or a suitably chosen reference set. Secondly, matrix elements of the same form as equation (1) must be evaluated repeatedly for all selected configurations with respect to the CI vector in the variational subspace to determine its eigenstates. In configuration selecting CI these operations are complicated by the lack of structure in the selected Hilbert space even on single-processor machines.

Several parallelization strategies have been advocated to implement complex algorithms on distributed memory machines, *data locality* becomes a paramount issue. One widely used approach is the use of client-server models, where one central node distributes the data among the client nodes on demand. This model is very versatile and has been used for a number of applications. However, in complex algorithms involving large amounts data, communication bottlenecks can easily arise as the communication patterns vary widely with the size of the active space, the number of electrons and the size of the orbital basis. In addition, one of the key challenges in data-management for CI application arises from the fact that inter-node communication is always slower than a typical computation. As a result, as parallel implementation must insure that all data transmitted between nodes can be used many times before it is discarded.

In our implementation we have therefore chosen an alternate communication scheme, where all operations and data are distributed *a priori* among the nodes of the machine accoriding to an organizing principle that ensures an equal distribution of the work among the nodes. The advantage of this approach is that the scalability of the algortihm is a mathematical necessicity, however, the concepts employed are quite narrow and differ already appreaciable between closely related problems such as the configuration selecting and the non-selecting versions of MRCI. Hence, much effort must be devoted to optimize the organizing principle for each particular application of the scheme.

To compute the matrix elements of the Hamilton operator we exploit an enumeration scheme in which each matrix element between two determinants (or configuration state functions) $|\phi_1\rangle$ and $|\phi_2\rangle$ is associated with the subset of orbitals that occur in both the target and the source determinant. This unique subset of orbitals is called the *transition residue* mediating the matrix element and serves as a sorting criterion to facilitate the matrix element evaluation on distributed memory architectures. For a given many-body state, we consider a tree of all possible transition residues as illustrated in Figure (1). For each such residue we build a list of *residue-entries*, composed of the orbital-pairs (or orbital for a single-particle residue) which combine with the residue to yield a selected configuration and a pointer to that configuration. For configuration selecting CI the reduction in the number of selected configurations combined with the large total memory of modern distributed memory machines allows us to build the residue tree for the selected configuration, provided that only the required section of the residue tree are stored in the different stages of the computation.

Once the residue tree is available the evaluation of the matrix elements is very efficient. In the *expansion step*, one must evaluate $q_i = \langle\phi_i|HP|\Psi\rangle$, where P projects on the part of the Hilbert space in which only inactive and active and low orbitals are occupied. For SD (single & double excitations only) calculations one needs to consider only residues and orbital pairs that contain no d-type orbitals. For TQ (single & double, triple and quadruple excitations) at most two orbitals of type h and l are allowed in the residue and none in the orbital pair. This portion of the residue tree contains but a fraction of the overall residue tree and is easily accommodated on all nodes. For each $|\phi_i\rangle$ we determine the required single- and two-particle residues, which are then searched for in the residue tree. In a SD calculation one can eliminate the search step by constructing the allowed excitations directly from the internal residues. If a match is found the information in the tree enables us to immediately compute *all matrix elements* associated with the given residue. As a result the overall numerical effort scales strictly linear with the number of configurations Φ_i for which matrix elements must be evaluated and the number of non-trivial operations per configuration is proportional to n_e^2. Again, there is a significant difference between configuration-selecting and non-selecting CI calculations. In non-selecting CI the number of non-zero matrix elements scales as $n_e^2N^2$ for each configuration, the step to determine the transition residues is thus

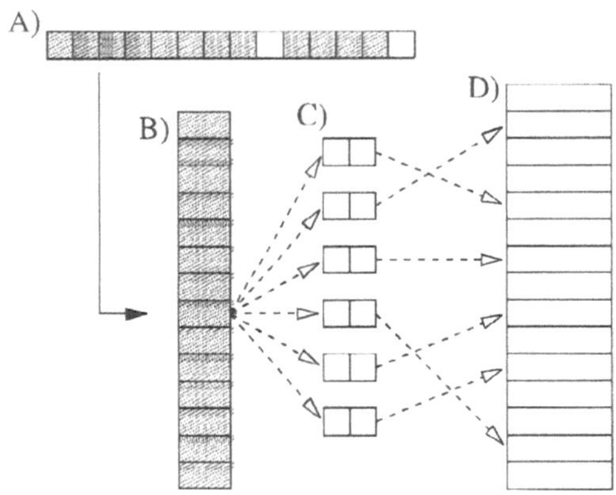

Figure 2. Schematic representation of the computation of two-particle matrix-elements in the expansion step using the residue-tree. For a given configuration (A) we form all two-particle residues, which are looked up in the residue tree. In the configuration illustrated in (A) each box represents one occupied orbital, the shaded region corresponds to the transition residue and the two white boxes to the orbital pair. The $(n_e - 2)$-electron residue configuration is looked up in the residue-tree (B). Each orbital pair (C) associated with the residue encodes a matrix element with an element of the configuration list (D). The orbital indices of the required integral are encoded in the orbital pairs in (C), the coefficient of the source configuration is looked up directly in (D). Only one lookup operation is required to compute all matrix elements associated with the given transition residue and only the subset of matrix elements that lead to selected source-configurations are constructed.

not dominant. In configuration-selecting CI, most of the possible matrix elements do not lead to selected configurations, thus the number of entries per element of the transition residue-tree is much smaller than $O(N^2)$. Here the residue-based implementation avoids the explicit enumeration of matrix elements that do not lead to selected configurations. For this reason, a residue-based implementation of MRD-CI on a massively parallel architecture offers a good balance between the computational demands and storage requirements.

In the *iteration phase* the full residue tree for all selected configurations must be built, but a single copy of the tree can be distributed across all nodes. All matrix elements associated with a given transition residue can be locally evaluated if the associated orbital pairs are present on a unique node. We note that the residue tree itself (part B in Figure (2)) is not required at all, only the set of connected orbital pairs is needed. As a result no lookup operations are required in this step and one can simply loop over the locally available section of the orbital pair segments to evaluate

TABLE 1. List of the distinct computational steps in the parallel implementation of the configuration-selecting MRCI procedure. There are three phases, associated with the initialization of the program and the expansion and iteration of the state-vector respectively. The details of the phases are discussed in the text. The fourth row of the table details which set of nodes is involved in each step, while the fifth row indicates the type of operation that dominates the step. Almost all the computational effort is concentrated in steps E and M, the next leading contribution arising from the logic steps D2,R2,G2. Only standard high-level communication routines were used to make the program as portable as possible.

	Phase		Type
	Expansion & Logic Steps		
D1	distribution of initial state	All nodes	Broadcast
R1	build section of restricted residue tree	All nodes	Computation
G1	gather and distribute residue tree	All nodes	Pairwise
E	expansion loop	All nodes	Computation
D2	distribute selected configurations	All nodes	All-to-All
R2	build one section of full residue tree	All nodes	Computation
G2	gather the residue tree	All nodes	All-to-All
	Iteration Steps		
D3	Distribute New Coefficients	Node 0	Broadcast
M	Evaluate matrix elements	All nodes	Computation
G3	Gather many body field	Node 0	Gather
X	Davidson iteration	Node 0	Iteration

all matrix elements that can be constructed for the present orbital sets. Since each matrix element is uniquely identified by its transition residue, the contributions to the many-body field can be simply collected at the end of this step on a single node to perform the Davidson iteration. This mechanism allows a rapid evaluation of all matrix elements while using the available core memory to its fullest extent.

In order to facilitate the explicit treatment of triple and quadruple excitation from a given reference set, we partition the orbital space into five segments according to a partition scheme that proved promising in an earlier investigation [30]. The population of these segments allows the gradual inclusion of triple and quadruple (TQ) excitations in cases where even the enumerative search of such configurations in the expansion loop is prohibitively expensive.

2.2. PARALLEL IMPLEMENTATION

In a truly scalable implementation great care must be take to divide all work equally across the participating nodes. A remaining non-scalable portion of 1% of the computational effort of a single processor application translates into a 100% overhead if the same task is distributed across 100 nodes. Our massively parallel algorithm for configurations-selecting MRCI is therefore based on a client server model that strictly separates the calculation from the communication steps. The latter were chosen to require only global communication directives of the underlying MPI communication library which can be expected to execute efficiently on most modern parallel architectures.

According to the above outline the overall work can be broken into two distinct phases that require the same order of magnitude of computational effort. Table Table (1) summarizes the most important steps of the configurations-selecting CI procedure. In an *expansion step* we begin with the distribution of the current state-vector to all nodes (D1). Each nodes then builds the restricted residue tree for all the set of configurations it received in step (D1). The effort per node involved in this step (R1) is strictly proportional to $n_e^2\ N_{\text{conf}}/N_k$, it will therefore scale well with the number of nodes N_k. Next (step G1) the residue tree must be distributed to all nodes. Since orbital pairs belonging to the same transition residue have been created on several different nodes this is a nontrivial operation. Using a hashing mechanism we first assign a unique node to each transition residue and then gather all orbital pairs belonging to that residue on the appropriate node. Then each node builds its unique section of the residue tree. The hashing mechanism ensures the balancing of the computational effort across all nodes. Finally the information of all the nodes is distributed via an all-to-all communication across the entire machine. Now (step E) each node can run through a predetermined section of the search space to evaluate the energy contributions and to select the configurations for the variational subspace. Step E dominates the overall computational effort of the configuration-selection by a large margin.

The next three steps prepare the variational subspace for the iterations. Since the distribution of the selected configurations on the different nodes can be rather uneven, we first redistribute the configurations among the nodes(D2). Then each node constructs its portion of the full residue tree (R2). These contributions are gathered in analogy to step (G1) across all nodes, such that each node has all orbital pairs for its assigned transition residues. In contrast to step (G1), however, the entries are not distributed across the machine, but remain on the nodes for the local matrix element evaluation. The computational effort in the matrix elements evaluation step

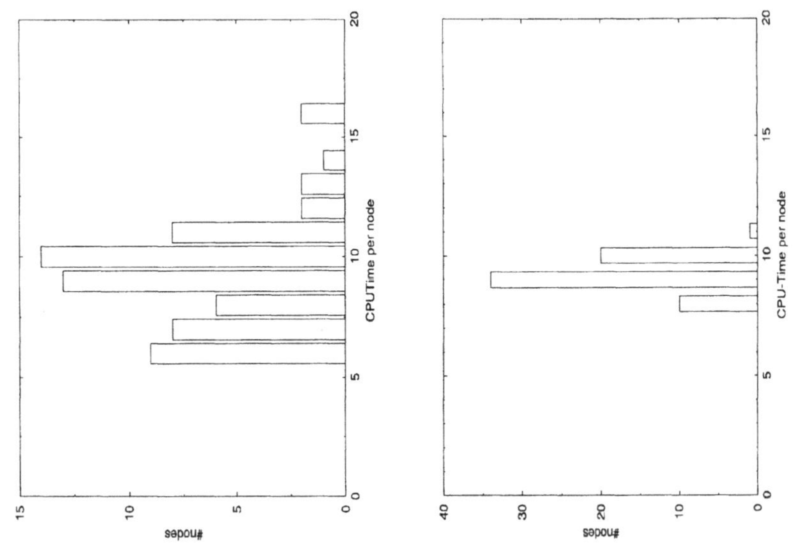

Figure 3. Histogram of the CPU time in the iteration step for the O_2 benchmark calculation including triple and quadruple excitations described in the text (a) in the absence of load-balance through the exchange of residue-entries between nodes and (b) with such load-balancing. The reduction in the width of the distribution improves the scalability of the algorithm.

is proportional to the expectation value of the square of the number of orbital pairs over the transition residues. The hashing mechanism we used to assign transition residues to nodes, however, ensures only that there are approximately the same number of residues on each node. Figure (3)(a) demonstrates that the computational effort can nevertheless vary quite significantly among the nodes, an effect that worsens with an increasing number of nodes. Such an imbalance in the work-distribution leads to the loss of scalability of the algorithm. It is therefore important to redistribute the workload among the nodes to achieve better performance. To this end we gather discretized histograms of the transition residue distribution on the server node, which uses this information to assign approximately even work-loads to all nodes. Based on this technique the theoretical deviations in the variation of the work-load can be reduced from over 50% to less than 12% of the overall average computational effort. Figure (3) shows histograms of the distribution of the work-load without and without active load-balancing to demonstrate this observation.

After the redistribution of the transition-residue table, the program can proceed with the iteration steps to converge the variational subspace. The four steps comprising an iteration are executed many times after each expansion loop, but require no further logic information. Almost all the work is concentrated in step M. We note in passing that in the expansion step only a fraction of all possible integrals are required on the nodes, a fact that will be exploited in future versions of the code.

3. Benchmark Calculations

In order to demonstrate the scalability of the implementation we have conducted benchmark calculations for two typical applications of the program. The first example is concerned with the evaluation of the importance of the triple and quadruple excitations for the potential energy surfaces of the oxygen molecule and its anion. Previous work has established that the accurate calculation of the electron affinity of O_2 remains a formidable challenge even to present day quantum chemical techniques. At the level of a CAS-SCF description the adiabatic electron affinity of the oxygen molecules is predicted with the wrong sign even in the basis set limit. A careful study[31] concluded that strong differential dynamical correlation effects are most likely entirely responsible for the source of this discrepancy. In MRCI-SD calculations the correct sign for the electron affinity can barely be reached using aug-QZP quality basis sets. A semiquantitative agreement between experiment and theory was reached, when the multi-reference generalization of the Davidson correction[6] was applied to estimate the effect of higher excitations to the MRCI wavefunction.

O_2 is therefore one of the simplest molecules which challenges one of the central paradigms of modern quantum chemical correlation methods that rest on the assumption the explicit treatment of single and double excitations of a chemically motivated reference set of configurations is sufficient to quantitatively account for dynamical correlation effects. This observation, as well as the desire to explicitely test approximations for extensivity corrections to MR-SDCI[7, 8, 9, 10] motivated the development of the present code. Since the CAS+SDTQ Hilbert space of O_2 in a aug-QZP basis has dimension 32×10^9, this problem cannot be treated with any of the presently available MR-SDCI or MRD-CI implementations, but provides a suitable challenge for our parallel implementation. The calculations were performed in a (sp)-augmented cc-pVTZ, cc-pVQZ and cc-pV5Z basis set in D_{2h} symmetry at the experimental geometries. In Hilbert spaces of dimension up to 5×10^9 containing triple and quadruple excitations we selected up to 5×10^6 determinants as a function of the threshold for the coefficients ranging from 10^{-3} to 10^{-6}.

In the second example, more within the traditional applications of confi-

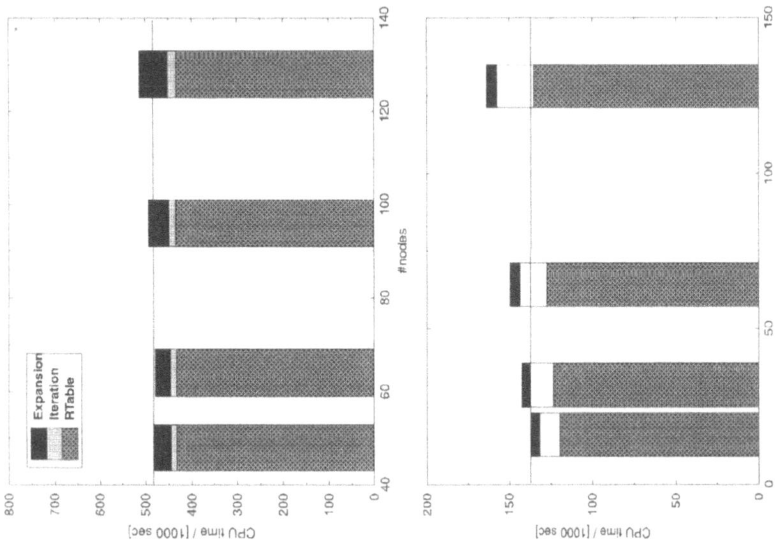

Figure 4. Total CPU time in (sec) for the fully converged calculation of the ground state of the two benchmark calculations described in the text as a function of the number of nodes of the IBM-SP2. A straight line indicates perfect scaling of the computational effort with the number of nodes. The shaded areas in the bars, from top to bottom, indicate the contributions of the matrix element evaluation (steps D3,M,G3,X in Table (1)), logic (D1,R1,G1,D2,R2,G2) and the expansion loop (step E).

guration-selecting CI, we have computed the ground state energy of benzene in a cc-pVDZ basis set using active spaces of 6 and 12 active orbitals. The latter calculation was motivated by the desire to test the program for very large Hilbert spaces, but the smaller active space is sufficient to adequately describe the chemistry of benzene. The calculation was performed in C_{6v} symmetry resulting in Hilbert spaces of up to 3×10^9 determinants of which up to 2×10^6 were selected for the variational subspace.

3.1. SCALABILITY

The most important consideration in the evaluation of the performance of a parallel program is its scalability with the number of processors used for a given calculation. For scaling purposes we selected a typical run with 10^9 determinants (1.8×10^6 selected) for O_2 and another with 1.3×10^9 determinants (1.6×10^6selected) for benzene respectively. For these cases we performed benchmark runs on the 256-node IBM-SP2 of the Karlsruhe supercomputer center. We also tested the program on the on the 256-node and 512-node CRAY-TE3's of the supercomputer center (HLRZ) of the Research Center Jülich and using the maximally available number of processors for standard runs, i.e. 256 on the CRAY-TE3 and 128 on the IBM-SP2 respectively. The runs on the CRAY T3E with its larger data types but smaller core memory per node forced us to use a somewhat smaller threshold than on the IBM-SP2 for the scaling runs in order to be able to finish the calculation even for a small number of processors. Since all data except integrals and state-vector is distributed across the machines the size of the maximally treatable Hilbert space grows significantly with the number of nodes. Unfortunately the T3E consists of two machines of different physical characteristics: the smaller cluster (128 nodes) permits runs ranging from 16 to 64 nodes, the larger one allows runs requiring 65-256 nodes. The interpretation of the scaling data will have to take this "break" into account.

Figure (4) shows the total computational effort (excluding the time to read the integral file) of the aforementioned scaling runs on the IBM-SP2 as a function of the number of nodes. In these plots, the computational effort for all logic-steps sections (D1,R1,G1,D2,R2,G2) are subsumed in one category, the expansion loop (E) and the iteration loop (D3,M,G3,X) constitute the other main components of the program. This division is motivated by the fact that the relative importance of these three main computational steps varies with the type of calculation performed and a different scaling behavior of these steps will result in an overall different performance for different calculations. The number of times the expansion loop is executed, for instance (in the test calculations twice for the SD and once for the TQ segment), will significantly affect the overall performance. Varying the threshold leads to a redistribution of the effort from expansion to matrix elements. Increasing the number of states (only one state per symmetry was computed in the test calculations) will reduce the importance of the logic section, as does an increase in the desired accuracy for the state-vector in the iteration steps. Since the expansion step is easier to parallelize than iteration and logic the test calculations we have selected less-than-ideal runs to test the program by using only the minimal number of expansion steps with a relatively low accuracy in the iteration step that somewhat overemphasizes the importance of the logic step — the most difficult part of the program.

For benzene we find almost perfect scaling from 48 to 128 nodes for the IBM-SP2. The total computational effort in the expansion loop, which dominates the overall computational effort, is constant to within 0.4% in going from the smallest to the largest number of nodes. In contrast, the effort associated with logic and communication grows somewhat with the

TABLE 2. Total CPU times for the benchmark calculations described in the text on the IBM-SP2 and the CRAY T3E as a function of the number of nodes. Given is the time in sec/node for the expansion and convergence of a single state in each calculation. The fractional computational loss between two test runs is defined as the ratio of the CPU-times per node divided by the perfect speedup factor given by the ratio of the nodes. The loss-data in the table always refer to successive entries. The calculation for benzene on the IBM-SP2 employed 12 active orbitals, that on the CRAY-T3E used a realistic active space of six orbitals. Note that the sensible limit for the latter calculations lies around 64 nodes, where less than 5 minutes are required to converge the calculation.

IBM-SP2				
	O_2		C_6H_6	
number of nodes	time (s/node)	loss	time (s/node)	loss
16	8308			
32	4374	5%		
48			10122	
64	2480	13%	7510	0%
96			5160	3%
128	1410	14%	4012	4%
CRAY-T3E				
	O_2		C_6H_6	
16	5334			
32	2887	8%	319	
64	1539	10%	172	8%
65	2062		244	
128	1107	6%	147	20%
256	620	12%		

number of nodes. This is to be expected, since the communication cost grows with the number of nodes and a total of 3.7 / 9.1 GB of data have to be transmitted across the machine for the small and large residue tables respectively. The overall speedup factor from 64 to 128 nodes is 1.86 (see Table (2)).

For the benchmark calculation of O_2 a more pronounced increase in the

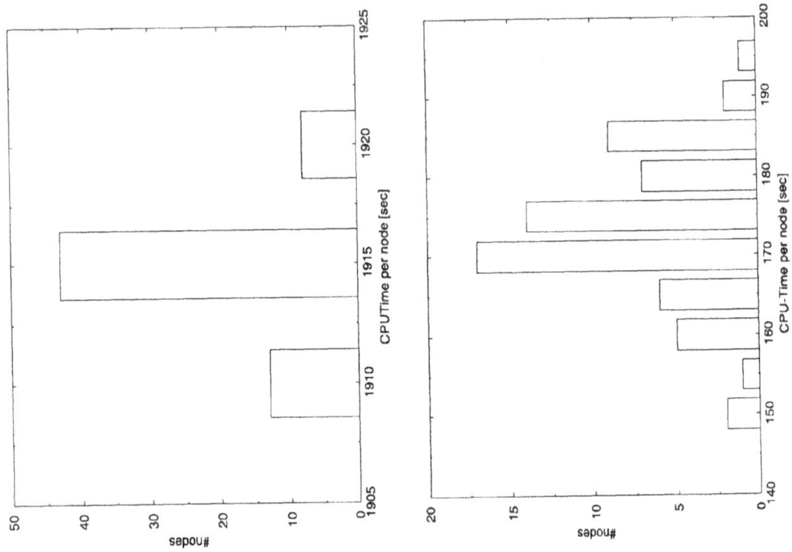

Figure 5. Histogram of the CPU-time distribution of (a) an expansion step including TQ excitations and (b) an iteration step for the O_2 benchmark calculation described in the text on 64 nodes of the IBM-SP2. The computational effort in the expansion step is almost perfectly distributed that of the iteration step varies with a standard deviation of approximately 4% resulting in a loss of computational efficiency as "fast" nodes have to wait for the "slow" nodes to finish. Without the use of load-balancing the width of distribution for the iteration step increases significantly. Since the number of transition residues/node decreases with the number of nodes, fluctuations in the computational effort become more difficult to balance for a large number of processors.

overall computational effort is observed in going from 32 to 128 nodes, in particular in with the last doubling from 64 to 128 processors. Again, the computational effort in the expansion loop is relatively stable, increasing only 6.7% (6.3%) in going from 32 to 64 (64 to 128) nodes respectively. This is the result of the near perfect load-balancing in evidence in Figure (5), which shows a histogram of the total CPU time/node for a run on 64 nodes. The presence of triple and quadruple excitations significantly complicates the overhead associated with the generation of the residue trees. Given the relatively wide distribution of the computational effort per transition residue that results from the presence of the TQ excitations, it becomes more and more difficult to balance the computational load in the matrix element step. This results in a larger variation in the load among the processors, which decreases the performance of the iteration step, since all processors must wait for the last node to finish. Because large amounts of data are required for the evaluation of the matrix elements it is difficult to

TABLE 3. Performance data for selected MRD-CI SD calculations. All calculations were performed on the CRAY-T3E of the HLRZ Jülich. The calculation on O_2 was performed in D_{2h} symmetry at the experimental geometry using a (2s2p) complete active space and a coefficient threshold of $\lambda = 10^{-5}$. The calculations on NO_2 were perfomed on the X $^1A_1$ ground state in C_{2v} symmetry at the experimental geometry using a (1211) complete active space and $\lambda = 3 \times 10^{-5}$. The calculation on C_6H_6 were performed on the X $^1A_1$ ground state in D_{2h} symmetry using a C_{6v} geometry with $R = 2.79$Åand a (00002121) active space with $\lambda = 3 \times 10^{-5}$. N_{det} designates the size of the Hilbert space, N_{sel} that of the selected subspace, the next column indicates the number of nodes used and the last the turnaround time till convergence in seconds. $^{(*)}$ The calculation in the cc-pVQZ basis requires twice the number of iterations of the calculation in the cc-pVTZ basis, which is responsible for the disproportionate increase in CPU time.

	Basis	N_{det}	N_{sel}	Nodes	CPU (s)
O_2^-	aug-cc-pVDZ	1×10^5	5.1×10^4	8	24
	aug-cc-pVTZ	1×10^6	3.5×10^5	65	61
	aug-cc-pVQZ	3×10^7	7.5×10^5	65	226
NO_2	cc-pVDZ	3^5	9^4	8	80
	cc-pVTZ	2^6	4^5	64	127
	cc-pVQZ	9^6	5^5	64	$1225^{(*)}$
C_6H_6	cc-pVDZ	4×10^7	2×10^5	64	172

go beyond the present implementation and to dynamically adjust the load while the iteration is in process.

The data for the test runs is summarized in detail in Table (2). For the IBM SP-2 we find near-perfect speedups for benzene. For O_2 the speedup factors associated with the doubling of the nodes are somewhat worse, but still warrant the use of a large number of nodes to perform the calculation in most circumstances. MRCI calculations require nontrivial communication steps on parallel machines, so that some loss of computational efficiency is unavoidable. On the CRAY-T3E we find a similar situation: For benzene we report calculations with the more realistic six-orbital space. On 128 nodes this calculation requires less than three minutes total turnaround time and the residue table in the iteration step is spread so thinly that it becomes impossible to balance. This explains the somewhat large loss of 20% efficiency in going from 65 to 128 nodes. Note that the data for 64 (65) nodes where obtained on the small (large) cluster of the T3E described above. The time differences are indicative of the relative performance of these two machines.

3.2. PERFORMANCE

Having verified the scalability of the method we now turn to the overall performance of the implementation to give some impression on the total CPU requirement for some well studied molecules. A great deal of caution is required in the interpretation of this data in a rapidly changing hardware landscape, as modifications in processor speed and architecture, caching capabilites and memory structure make comparisons between test calculations very difficult. Keeping these shortcomings in mind, however, overall performance data provides the prospective user with an order-of-magnitude estimate of the computational requirements. The benchmarks were performed on both IBM-SP2 with 120 Mhz RISC2 processors and 512 MB/node and the CRAY TE3 with 300 Mhz DEC Alpha Processors and 128 MB/node. Even though there is almost a factor 3 between the cycle rates of these machines the IBM processor has the larger floating point throughput. Neither of these processors is at the high end of present-day computational performance. Table (3) summarizes total CPU-times for calculations on O_2, NO_2 and benzene for a number of basis sets to give an impression of the total turnaround-times that can be expected for standard calculations.

4. Discussion

Accurate benchmark methods for the treatment of dynamical correlation effects, such as MRCI, have made a significant impact in the development of quantum chemistry. Since their computational effort rises rapidly (as n_e^6) with the number of electrons, only the use of the most powerful computational architectures ensures their continued relevance to the field. Because massively parallel architectures with distributed memory will yield the highest computational throughput in the foreseeable future, it is worthwhile to pursue the use of these machines for quantum chemical benchmark calculations. The development of the first scalable implementation of one of the most popular variants of the MRCI method family on such architectures is one important step in this direction. The present implementation allows the treatment of Hilbert spaces and systems that are larger than those that can be treated on traditional architectures, while significantly reducing the turnaround time for more moderate applications. With the ability to routinely treat Hilbert space exceeding 10 billion determinants many questions that require a delicate balance of dynamical and non-dynamical correlation effects, e.g in transition metal chemistry, become amenable to the MRCI method.

A spin-adapted implementation of the residue-based MRCI algorithm, both for MR-SDCI[29] and its configuration-selecting variant is presently under way. We report here on the progress of the determinantal program, a spin-adapted version is presently under development. However, it is unclear whether a significant computational advantage can be expected from this code, in particular for the configuration-selecting algorithm including TQ excitations. The reason for this expectation lies in the fact that the number of unpaired spins rises rapidly for complex molecules and highly excited configurations. The size of the representation matrices of the symmetric or unitary groups that are required in spin-adapted implementations grows exponentially with the number of unpaired electrons. In non-selecting MR-SDCI the fact that all possible configuration state functions (CSF) are present allows an efficient implementation, in configuration-selecting MRD-CI the sparseness of the state-vector leads to a significant computational overhead in the generation of the representation matrices of the line-up permutations that are necessary to evaluate the matrix elements[23]. We note that FCI codes have long ago abandoned CSF based implementations[32, 33] in favor of a determinantal approach, even though the number of electrons is strongly limited in FCI. With larger Hilbert spaces and an increasing electron number a similar trend may appear for configuration-selecting MRD-CI.

One of the intrinsic bottlenecks of our present implementation configuration-selecting MRD-CI is the difficulty to develop efficient integral-driven matrix element evaluation methods in the iteration step, which ultimately limits the size of the basis that can be employed for these calculations. While many interesting calculations will be possible with the existing code, we are presently exploring the possibility of non-local integral storage for basis sets exceeding 400 orbitals[1]. In addition, it is worthwhile to investigate approximations, such as multi-reference second-order Brillouin-Wigner perturbation theory [17, 34, 35, 36], that eliminate the selected variational subspace in MRD-CI altogether. We note that the selection step in MRD-CI scales with $n_e^2 N^2$ (where N is the number of orbitals), as opposed to the iteration step which principally scales as $n_e^2 N^4$. The MR-BWPT approximation rests on the assumption that the presence of the vast majority of the selected configurations in MRD-CI is required not because their individual energy contribution cannot be estimated perturbatively, but because they generate a many-body field on the primary configurations that alters the relative importance of the latter.

Acknowledgments

This work was supported by DFG Grant KEI-164/11-2 and computational resources at the HLRZ Jülich and the HRZ Karlsruhe. We greatfully ac-

[1] Note added in proof: The integral driven version of the program has been completed.

knowledge stimulating discussions with H. Lischka and I. Shavitt in the course of this work.

References

1. B. O. Roos. *Chem. Phys. Letters*, 15:153, 1972.
2. B. O. Roos and P. E. M. Siegbahn. The direct configuration interaction method. In H.F. Schaefer III, editor, *Methods of Electronic Structure Theory*, page 189. Plenum, New York, 1994.
3. I. Shavitt. In H. F. Schaefer III, editor, *Modern Theoretical Chemistry*. Plenum, New York, 1977.
4. R. Sheppard, I. Shavitt, R. M. Pitzer, D.C. Comeau, M. Pepper, H. Lischka, P. G. Szalay, R. Ahlrichs, F. B. Brown, and J. Zhao. *Int. J. Quantum Chem. Symp.*, 22:149, 1988.
5. S. R. Langhoff and E. R. Davidson. *Int. J. Quantum Chem.*, 8:61, 1974.
6. W. Butscher, S. Shih, R. J. Buenker, and S. D. Peyerimhoff. *Chem. Phys. Letters*, 52:457, 1977.
7. J. Cižek. *J. Chem. Phys.*, 45:4256, 1966.
8. R. J. Bartlett and I. Shavitt. *Chem. Phys. Letters*, 50:190, 1977.
9. R. Gdanitz and R. Ahlrichs. *Chem. Phys. Letters*, 143:413, 1988.
10. P. Szalay and R. J. Bartlett. *J. Chem. Phys.*, 103:3600, 1995.
11. J.-L. Heully and J. P. Malrieu. *Chem. Phys. Letters*, 199:545, 1993.
12. J. P. Daudey, J.-L. Heully, and J. P. Malrieu. *J. Chem. Phys.*, 99:1240, 1993.
13. R. J. Bunker and S. Peyerimhoff. *Theor. Chim. Acta.*, 12:183, 1968.
14. R. J. Buenker and S. D. Peyerimhoff. *Theor. Chim. Acta*, 35:33, 1974.
15. R. J. Buenker and S. D. Peyerimhoff. *Theor. Chim. Acta*, 39:217, 1975.
16. F. Illas, J. Rubio, J.M. Ricart and P.S. Bagus. *J. Chem. Phys.*, 95:1877, 1991.
17. Z. Gershgorn and I. Shavitt. *Int. J. Quantum Chem.*, 2:751, 1968.
18. B. Huron, J.P Malrieu, and P. Rancurel. *J. Chem. Phys.*, 58:5745, 1973.
19. R. J. Buenker and S. D. Peyerimhoff. *New Horizons in Quantum Chemistry*. Reidel, Dordrecht, 1983.
20. J. L. Whitten and M. Hackmeyer. *J. Chem. Phys.*, 51:5548, 1969.
21. R. J. Harrison. *J. Chem. Phys.*, 94:5021, 1991.
22. S. Krebs and R. J. Buenker. *J. Chem. Phys.*, 103:5613, 1995.
23. M. Hanrath and B. Engels. New algorithms for an individually selecting mr-ci program. *Chem. Phys.*, 225:197, 1997.
24. M.Schuler, T. Kovar, H. Lischka, R. Sheppard, and R. J. Harrison. *Theor. Chim. Acta*, 84:489, 1993.
25. S. Evangilisti, J. P. Daudey, and J. P. Malrieu. *Chem. Phys.* 75:91, 1983.
26. C. Agneli and M. Persico. *Theor. Chim. Acc.* 98:117, 1998.
27. H. Lischka, F. Dachsel, R. Shepard, and R.J. Harrison. Parallel computing in quantum chemistry-message passing and beyond for a general ab initio program system. In W. Gentzsch and U. Harms, editors, *High-Performance Computing and Networking. International Conference and Exhibition Proceedings. Vol.1: Applications*, page 203. Spinger, Berlin, 1994.
28. H. Dachsel, H. Lischka, R. Shepard, J. Nieplocha, and R.J. Harrison. *J. Comp. Chem.*, 18:430, 1997.
29. F. Stephan and W. Wenzel. *J. Chem. Phys.*, 108:1015, 1998.
30. M. M. Steiner, W.Wenzel, J. W. Wilkins, and K. G. Wilson. *Chem. Phys. Letters*, 231:263, 1994.
31. R. Gonzáles-Luque, M. Merchán, M. P. Fülscher, and B. O. Roos. *Chem. Phys. Letters*, 204:323, 1993.
32. P. Knowles and N. Handy. *Chem. Phys. Letters*, 111:417, 1984.
33. J.Olsen, P. Jorgensen, and J.Simons. *Chem. Phys. Letters*, 169:463, 1990.

34. W. Wenzel and K. G. Wilson. *Phys. Rev. Letters*, 68:800, 1992.
35. W. Wenzel and M. M. Steiner. *J. Chem. Phys.*, 108:4714, 1998.
36. W. Wenzel. *Int. J. Quantum Chem.*, 70:613, 1998.

COMMENTS ON THE BASIS SETS USED IN RECENT STUDIES OF ELECTRON CORRELATION IN SMALL MOLECULES

S. WILSON
Rutherford Appleton Laboratory,
Chilton, Oxfordshire OX11 0QX, England

D. MONCRIEFF
Supercomputer Computations Research Institute,
Florida State University, Tallahassee,
FL 32306-4130, U.S.A.

AND

J. KOBUS
Instytut Fizyki, Uniwersytet Mikołaja Kopernika,
ul. Grudziądzka 5, 87-100 Toruń, Poland

Abstract. Some comments are made on the basis sets employed in recent studies of electron correlation energies for small molecules with particular reference to calculations for the ground states of the nitrogen and water molecules. For diatomic systems, the use of finite difference and finite basis set approximations in generating the Hartree-Fock reference function is compared. The *d*istributed, *u*niversal, *e*ven-*t*empered basis sets (for which we introduce the acronym *duet*) and correlation consistent basis sets are compared for both the Hartree-Fock model and in treatments of correlation effects. The use of correlation treatments based on many-body perturbation theory and on coupled cluster expansions are discussed. The systematic approximation of the molecular integral supermatrix corresponding to *duet* basis sets is addressed, as are applications to molecular systems containing heavy atoms.

1. Introduction

The famous 1927 paper of Heitler and London [1] on the ground state of the hydrogen molecule not only established the burgeoning field of quantum chemistry but also introduced the basis set into molecular electronic

J. Maruani et al. (eds.), New Trends in Quantum Systems in Chemistry and Physics, Volume 1, 115–132.

structure calculations - an approximation which is ubiquitous in almost all practical applications - and with it the basis set truncation error. This error has plagued practical applications through to the present time. Indeed, interest in reduction of the basis set truncation errors arising in studies of the simplest of molecules, the ground state of the hydrogen molecule, has continued though to the end of the twentieth century ([2]-[6]).

Three years ago, in 1996, we reported ([7],[8]) highly correlated calculations for the ground state of the nitrogen molecule and for the ground state of the water molecule. These calculations employed systematically developed *d*istributed *u*niversal *e*ven-*t*empered primitive spherical harmonic Gaussian basis sets in conjunction with a correlation treatment based on second order many-body perturbation theory. Over the past three years, a number of other studies of these two systems have been reported ([9]-[14]) using a variety of basis sets and methods for handling the correlation problem. One of the primary purposes of this paper is to compared our methods and the results they support with this more recent work with particular emphasis on the basis sets employed.

In section 2, we comment on the basis sets used in the various studies of electron correlation effects in small molecules cited above. We briefly discuss the systematic sequences of *d*istributed *u*niversal *e*ven-*t*empered primitive spherical harmonic Gaussian basis sets employed in our calculations in subsection 2.1. The correlation consistent basis sets developed by Dunning and his collaborators ([15]-[22]) and used in more recent studies ([9]-[14]) of N_2 and H_2O are also described. In subsection 2.2, we specifically consider recent calculations for the ground state of the nitrogen molecule whilst in subsection 2.3 we turn our attention to the ground state of the water molecule. The systematic approximation of the molecular integral supermatrix corresponding to *d*istributed *u*niversal *e*ven-*t*empered basis sets of primitive spherical harmonic Gaussian functions is considered in subsection 2.4. Section 3 contains a summary.

2. Comments on the basis sets used in recent studies of electron correlation in small molecules

Most molecular electronic structure calculations are carried out in two distinct stages ([23],[24]); the first involving an independent particle model (usually the Hartree-Fock model) and the second taking account of electron correlation effects (usually by means of a "many-body" technique, such as many-body perturbation theory or a cluster expansion). For atoms the use of spherical polar coordinates allows the problem to be factorize into an angular part which can be handled analytically and a one-dimensional radial part which can be solved numerically by introducing a suitable grid.

For molecules, there is, in general, no suitable coördinate system and thus the use of finite basis set expansions has become ubiquitous. This leads to the matrix Hartree-Fock equations and the so-called algebraic approximation in which the integro-differential Hartree-Fock equations become a set of algebraic equations for the expansion coefficients defining single particle state functions or orbitals. The solution of the matrix Hartree-Fock equations yields a set of orbitals, both occupied and unoccupied, which can be used to develop a description of electron correlation effects.

Diatomic molecules are an exception, in that the use of elliptical coordinates facilitates the factorization of the problem into a part, depending on the azimuthal angle, φ, which can be solved analytically and a part, depending on the prolate spheroidal coordinates[1] λ and μ, which is handled numerically. This two-dimensional grid gives rise to computational demands which are not inconsiderable and it is perhaps not surprising that, although the atomic Hartree-Fock problem could be already handled numerically using finite difference techniques in the 1930s and handled with ease by the 1950s when digital computing machines became available, such molecular problems were not regarded as tractable until the early 1980s [25] (for a recent review see [26], which might be termed the dawn of the "supercomputer age". Even today, finite difference diatomic molecule calculations are restricted to the Hartree-Fock model.

For many years, the development of basis sets for molecular electronic structure calculations was seen as a compromise between, on the one hand, the need to use a basis set that was sufficiently flexible to support the level of accuracy required in a particular investigation, and, on the other hand, the size of the basis set is small enough to allow the calculation to be carried within a particular time scale (See, for example, ([27],[28])). In much contemporary work, this is often the balance that has to be struck in specific applications of quantum chemical techniques. But this is an approach which is not without its pitfalls. Too often the assumed accuracy of the results depends on a fortuitous cancellation of errors arising from truncation of the basis sets with those associated with, for example, the neglect of certain correlation effects. Such errors may not be at all apparent to the casual user of quantum chemical computer programs. However, the need to develop methods for *systematically* refining basis sets has been recognized as an essential ingredient of any techniques for handling molecular electronic structure which aims to achieve high precision.

[1] $\lambda = (r_A + r_B)/R$, $\mu = (r_A - r_B)/R$

2.1. SYSTEMATIC SEQUENCES OF DISTRIBUTED UNIVERSAL EVEN-TEMPERED PRIMITIVE SPHERICAL-HARMONIC GAUSSIAN BASIS SETS

In recent years, we have reported ([2],[29]-[37]) a number of studies in which detailed comparisons of finite difference Hartree-Fock calculations for diatomic molecules with the corresponding finite basis set calculations using Gaussian basis sets have been reported. Such comparisons have been made for closed-shell systems, first for systems containing light atoms ([30]-[33]) and then for molecules containing heavy atoms ([34],[35]). More recently, comparisons have been made for some prototypical open-shell diatomic molecules, again, first for systems containing only light atoms [36] and then, very recently, for molecules containing heavy atoms [37]. In all cases the finite basis set total energies supported by the large and flexible basis sets employed approached the finite difference results at an accuracy approaching the sub-μHartree level. These "high precision" basis sets have subsequently been used in studies of polyatomic molecules and in studies of electron correlation effects. In particular, in 1996 highly correlated calculations for the ground state of the nitrogen molecule [7] and for the ground state of the water molecule [8] were reported. Over the past three years, a number of other studies of these two systems have been reported ([9]-[14]) using a variety of basis sets and methods for handling the correlation problem. As we have already said, one of the primary purposes of this paper is to compared our results published in 1996 with this more recent work.

We introduce the acronym *duet* - *d*istributed *u*niversal *e*ven-*t*empered to describe the basis sets employed in our calculations. These basis sets combined a number of features:-

(i) they are *distributed*. Although finite basis set approximations to molecular wavefunctions can be formally developed in terms of a one-centre expansion [28], it is well known that this approach is often poorly convergent, especially when off-centre heavy atoms are present. Most often a molecular basis set is constructed from subsets centred on each of the component atoms. These distributed basis sets with atom-centred subsets are employed in the vast majority of contemporary quantum chemical calculations. However, in general, the distribution may involve other expansion centres. For example, in the case of diatomic molecules the introduction of a subset centred on the bond mid-point has been demonstrated to be beneficial in "high precision" Hartree-Fock studies in that such functions lead to more rapid convergence of the energy expectation value.

(ii) they are *universal*. The parameters defining the basis set are not selected with a view to describing one particular atom, property or molecular environment. The subsets of the *duet* basis set can, therefore, be centred on any atomic nucleus or other point where significant electron density may accumulate. Furthermore, *duet* basis sets are therefore available for molecules

containing any atom of the Periodic Table and specific applications do not depend on the availability of tabulated atomic sets.

(iii) they are *even-tempered.* The exponents are taken to form a geometric series. The *duet* basis sets can therefore be easily enlarged if higher accuracy is required. Computational linear dependence, which has to be carefully control in "high precision" studies, can be routinely controlled.

(iv) they form a *systematic sequence.* The *duet* basis sets form a systematically constructed sequence which can be explicitly demonstrated to approach a complete set for a given expansion centre.

(v) they are *primitive spherical-harmonic Gaussian functions.* The use of spherical-harmonic Gaussian basis functions avoids the linear dependence problems which can arise with cartesian Gaussian basis functions when higher harmonics are included. The use of primitive Gaussian is essential for the formal mathematical completeness of the sets employed in the limit of an infinite number of basis functions in the one-centre case. For the multicentre case the approach is heuristic.

(vi) they are *not "energy biased".* The parameters defining the *duet* basis sets are not specifically chosen to support the lowest expectation of the energy. The *duet* basis sets might, therefore, be expected to deliver "high precision" for a range of expectation values. Indeed, a current study ([37],[38]) of multipole moments using finite difference and finite basis set approaches supports this view.

The more recent publications with which we compare our 1996 calculations for N_2 and H_2O have exclusively used the correlation consistent basis sets introduced by Dunning [15] in 1989. It is useful to compare the characteristics of the correlation consistent basis sets with those of the *duet* basis sets. For the purposes of the present study, the essential features of the correlation consistent basis sets are

(i) they are *atom-centred.* Indeed, correlation consistent basis sets are developed for atomic systems and then used in the synthesis of molecular basis sets.

(ii) they are *atom specific.* Tabulations of correlation consistent basis sets are available. However, they are not available for all atoms of the Periodic table and for different molecular environments.

(iii) they form a *sequence.* Empirical extrapolation procedures have been proposed to investigate the "infinite basis set" limit for calculations carried out with correlation consistent basis set sequences. However, the present authors are not aware of any proof that this limit corresponds to a complete basis set in the one-centre case. For the multicentre case this approach is also heuristic.

(iv) they are *contracted.* A property that would appear to preclude any formal demonstration of completeness.

(v) they are *energy biased.* Correlation consistent basis set are specifically designed to recover electron correlation energy.

2.2. A COMPARISON OF RECENT STUDIES OF THE N_2 GROUND STATE

Almost one half of the electronic binding energy of the nitrogen molecule ground state is attributable to electron correlation effects and it is not surprising that interest in this system continues unabated.

In 1993, Kobus published [39] a finite difference Hartree-Fock energy for the nitrogen molecule ground state at a nuclear separation of 2.068 bohr. Use of a grid of 169×193 points supported a finite difference Hartree-Fock energy of $-108.993\ 825\ 7$ Hartree. A systematically developed distributed basis set of even-tempered Gaussian functions [29] supported an energy of $-108.993\ 824\ 5$ Hartree which deviates by just 1.2 μHartree from the finite difference value. The basis set employed in this calculation consisted of primitive Gaussian-type functions. The functions were distributed on three centres - the two nuclei and the bond mid-point. It has been established that within the Hartree-Fock model this distribution of expansion centres leads to more rapid convergence that the more widely used distribution in which functions are centred on the nuclei only. On each centre subsets of functions of successively higher symmetry are added in order to explore the convergence pattern. For each symmetry type on a given centre even-tempered sequences were employed to reduce the associated basis set truncation error. The final basis set employed in our study of the Hartree-Fock energy of the nitrogen molecule can be written [$30s15p15d15f$; $27s12p10d10f$ *bc*], which consists of a $30s15p15d15f$ set centred on each of the nuclei and a $27s12p10d10f$ on the bond centre. The latter basis set was obtained from the $30s15p15d15f$ set by deleting the most diffuse functions, which, when positioned on the bond centre, give rise to computational linear dependence.

The total molecular electronic energy for the ground state of the nitrogen molecule is $-109.587\ 82$ Hartree. Quiney, Moncrieff and Wilson [41] have defined the post Hartree-Fock energy for the system XY as

$$\Delta E_{XY} = E_{XY} - E_{XY}^{HF} \qquad (1)$$

where E_{XY} is the total molecular ground state energy, which is available from experiment, and E_{XY}^{HF} is the exact Hartree-Fock energy, which is available from finite difference and finite element calculations, and from finite basis set studies in which convergence with respect to basis set has been monitored. The values of the *post*-Hartree-Fock energy for the nitrogen molecule ground state at a nuclear separation of 2.068 bohr has been estimated to be -594.00 mHartree. The accuracy with which the *post*-Hartree-Fock energy can be determined depends both on the accuracy of the finite difference Hartree-Fock energy and that of the total molecular energy derived from experiment. Errors in the experimentally derived total molecular energy are the dominant source of error in the *post*-Hartree-Fock energies.

Almost 91% of the post-Hartree-Fock energy of the nitrogen molecule is associated with non-relativistic electron correlation effects. An empirical estimate of the correlation energy is −539.52 *m*Hartree [41]. In 1996, we employed [7] the basis set described above as a starting point for the treatment of correlation effects using second order many-body perturbation theory. Using a basis set containing functions of s, p, d, f, g and h symmetry on three centres, the nuclei and the bond centre, we recovered a correlation energy of −530.43 *m*Hartree. This represents some 98.3% of the empirical estimate of the total correlation energy. Extrapolation of the correlation energies supported by a systematic sequence of basis set including successively higher symmetries gives an estimated correlation of −535.4 *m*Hartree, which is some 99.2% of the empirical estimate. This also represent 90.1% of the *post*-Hartree-Fock energy. The resulting correlation energies were also assessed by comparison with Klopper's "MP2-R12" method for which an application to the nitrogen molecule was reported in 1995 [42]. Klopper reports an estimated second-order correlation energy component of −536.14 Hartree for N_2 with a nuclear separation of 2.07 bohr. Thus second order many-body perturbation theory accounts for over 99% of the empirical correlation energy. The remaining error in the correlation energy is therefore an order of magnitude smaller than the remaining error in the *post*-Hartree-Fock energy which is mainly associated with relativistic effects.

In 1997, A.K. Wilson and Dunning [9] presented "MP2" calculations for the nitrogen molecule ground state using the correlation consistent basis sets. In a study which included only 'valence' correlation effects they reported a correlation energy of −408.53 *m*Hartree from their basis set designated *cc*-pV5Z and −413.23 *m*Hartree from their largest basis set, the *cc*-pV6Z set. Klopper reported a valence correlation energy of −420.37 *m*Hartree for this system. For the *cc*-pV5Z basis set, Moncrieff and Wilson [7] reported a total correlation energy estimate of −477.70 *m*Hartree for a nuclear separation of 2.068 bohr, whilst in subsequent work [4] for the same geometry they employed the correlation consistent basis set designated *aug-cc*-pCV5Z to obtain a total correlation energy estimate of −523.9 *m*Hartree. This represents about 97% of the empirical correlation energy estimate. In the same year, Dunning's group [10] also employed the correlation consistent basis sets in coupled cluster calculations for a number of homonuclear diatomic molecules, including the nitrogen molecule.

In 1998, Klopper and Helgaker [13] examined the convergence of "CCSD-R12" calculations for a sequence of correlation consistent basis sets. This explicitly correlated coupled-cluster doubles model involves non-standard two-, three-, four- and even five-electron integrals in addition to the standard one- and two-electron integrals. These non-standard, many-electron integrals are reduced to, at most, two-electron integrals by inserting a resolution of the identity [43]. However, there is no unique way of inserting this resolution of the identity and, furthermore, its accuracy is dependent on the quality of the basis set employed. In the same year as Klopper and Helgaker's work was published, Halkier et al [12] made a comparison of second order perturbation theory energies and coupled cluster correlation energies for a sequence of correlation consistent basis sets. For the nitrogen molecule ground state with a nuclear separation of 109.77 pm (2.0744 *bohr*) they reported Hartree-Fock, "MP2", "CCSD" and "CCSD(T)" calculations using *cc* pCVnZ correlation consistent basis sets, $n = D, T, Q, 5, 6$. They also reported a numerical Hartree-Fock energy of $-108.993\,188$ Hartree for the same nuclear geometry. The errors in the matrix Hartree-Fock calculations carried out with correlation consistent basis sets were found to be (in μHartree)

cc-pCVDZ	38271 $\mu Hartree$
cc-pCVTZ	8788 $\mu Hartree$
cc-pCVQZ	1876 $\mu Hartree$
cc-pCV5Z	360 $\mu Hartree$
cc-pCV6Z	88 $\mu Hartree$

The *cc*-pCV5Z basis set can thus support a total Hartree-Fock energy at the sub-μHartree level. The error in the matrix Hartree-Fock energy corresponding to the *cc*-pCV6Z set should be compared with the error of 1.2 μHartree associate with the *duet* basis set. Turning to the correlation energy calculations reported by Halkier et al [12], the "MP2" studies recover the following percentages of the empirical correlation energy estimate

cc-pCVDZ	71%
cc-pCVTZ	89%
cc-pCVQZ	95%
cc-pCV5Z	97%
cc-pCV6Z	98%

whilst the "CCSD" estimates of the electron correlation energy were

cc-pCVDZ	72%
cc-pCVTZ	89%
cc-pCVQZ	94%
cc-pCV5Z	96%
cc-pCV6Z	96%

By comparison, the *duet* basis set supported 98.3% of the empirical correlation energy and 99.2% after extrapolation. For the correlation consistent basis sets, extrapolation leads to 99.6% and 97.5% of the empirical correlation energy for the "MP2" and "CCSD" methods, respectively.

Halkier et al [12] also reported the correlation energies given by the "CCSD(T)" method when supported by correlation consistent basis sets. Whereas perturbative analysis [23] indicates that "CCSD" is a third order theory in that all terms through third order in the energy perturbation expansion are included, "CCSD(T)" is a fourth order approach. Single, double and quadruple replacements are included by means of the coupled cluster ansatz. Triple replacements are included perturbatively. The "CCSD(T)" method recovers the following percentages of the empirical correlation energy when supported by the correlation consistent basis sets

cc-pCVDZ	74.2%
cc-pCVTZ	92.3%
cc-pCVQZ	98.0%
cc-pCV5Z	99.8%
cc-pCV6Z	100.6%

The slight overestimate of the correlation energy for the largest basis set in the sequence may be attributable to error in the empirical estimate of the correlation energy.

2.3. A COMPARISON OF RECENT STUDIES OF THE H_2O GROUND STATE

The ground state of the water molecule is the second molecular system for which the results of recent studies can be usefully compared. The geometry employed in all of the calculations discussed here was R_{O-H}=1.80885 bohr and H–O–H=104.52. For this polyatomic system finite difference results which allow the precise assessment of the Hartree-Fock energies are not available. However, the systematic development of a distributed basis set for this molecule by Moncrieff and Wilson [8] is believed to lead to an energy which is within a few μHartree of the Hartree-Fock limit. In Table 1, the matrix Hartree-Fock energies supported by the correlation consistent basis sets are compared with the distributed basis set result. Note that the error associated with the largest correlation consistent basis set is comparable with that measured for the N_2 ground state for a basis set of the same type.

TABLE 1. Comparison of matrix Hartree-Fock energies for the water ground state supported by correlation consistent basis sets and by universal basis set sequences. All energies are given in Hartree. δ is the difference between the total matrix Hartree-Fock energy for a correlation consistent basis set and the corresponding *duet* basis set energy. δ is given in μHartree.

Method/Basis Set	$E_{(m)HF}$	Reference	δ
cc-pCVDZ	−76.027 204	a	40284
cc-pCVTZ	−76.057 358	a	10130
cc-pCVQZ	−76.064 948	a	2540
cc-pCV5Z	−76.067 105	a	383
cc-pCV6Z	−76.067 404	a	84
duet	−76.067 488	b	

a A. Halkier, T. Helgaker, P. Jorgensen, W. Klopper, H. Koch, J. Olsen and A.K. Wilson, *Chem. Phys. Lett.* **286**, 243 (1998).

b D. Moncrieff and S. Wilson, *J. Phys. B: At. Mol. & Opt. Phys.* **29**, 6009 (1996).

For the previously defined H_2O molecular geometry, the estimated empirical correlation energy of the water molecule in its ground state is $-370\pm$ 3 mHartree. The "MP2-R12" energy indicates that some 97.8 % of the empirical correlation energy is supported by the second-order expansion for the energy. In Table 2, the second order energies supported by the correlation consistent basis sets are compared with the *duet* basis set result. The largest correlation consistent basis set (*cc*-pCV6Z) recovers some 98.5% of the second order correlation energy component which is increased to 99.7% by extrapolation. The *duet* basis set accounts for some 98.6% of the estimated exact second order energy or some 99.96% after extrapolation. "CCSD" recovers about 96.4% of the empirical correlation energy whilst "CCSD(T)" accounts for 99.2% of the empirical correlation energy.

2.4. SYSTEMATIC APPROXIMATION OF THE MOLECULAR INTEGRAL SUPERMATRIX CORRESPONDING TO DUET BASIS SETS

Because of their flexibility, *duet* basis sets are necessarily large. Large basis sets of Gaussian functions are not a problem in quantum chemical studies provided that care is taken to avoid computational linear dependence. However, the number of two-electron integrals which arise formally increases as the fourth power of the number of basis functions. The traditional approach to basis set construction of using a set of as smaller size as possible whilst supporting the desired accuracy is motivated by the requirement that the two-electron integral list be kept as short as possible. The large number of two-electron integrals which arise in most quantum chemical calculations has motivated many studies of reliable schemes for approximating them.

Indeed, the whole of semi-empirical quantum chemistry can be regarded as the development of schemes for approximating the two-electron integrals which arise in *ab initio* molecular electronic structure calculations. Schemes are required for the systematic approximation of the molecular integral supermatrix corresponding to duet basis sets. In this section, we briefly examine some possibilities.

TABLE 2. Comparison of second order energy components supported by correlation consistent basis sets and universal basis set sequences with the approximate explicitly correlated energies obtained by Klopper. δ is the difference between the energy supported by a given basis set and that given by Klopper. δ is given in *mHartree*.

Method/Basis Set	E_2	δ	Reference
MP2-R12	−0.362 01		a
cc-pCVDZ	−0.241 326	120.68	b
cc-pCVTZ	−0.317 497	44.51	b
cc-pCVQZ	−0.342 631	19.38	b
cc-pCV5Z	−0.352 283	9, 73	b
cc-pCV6Z	−0.356 407	5.60	b
Estimated limit	−0.361 (1)	0.9	b
duet	−0.356 828	5.18	c
Estimated limit	−0.361 850	0.16	c

a W. Klopper, *J. Chem. Phys.* **102**, 6168 (1995).

b A. Halkier, T. Helgaker, P. Jorgensen, W. Klopper, H. Koch, J. Olsen and A.K. Wilson, *Chem. Phys. Lett.* **286**, 243 (1998).

c D. Moncrieff and S. Wilson, *J. Phys. B: At. Mol. & Opt. Phys.* **29**, 6009 (1996).

2.4.1. *The two-electron integral supermatrix*

The two-electron integrals can be arranged as a symmetric, positive definite supermatrix with rows and columns labeled by the index

$$(ij), \ j = 1, 2, ..., i; \ i = 1, 2, ..., n \tag{2}$$

so that the supermatrix takes the form

	11	21	22	31	32	33 ...
11	[1111]					
21	[2111]	[2121]				
22	[2211]	[2221]	[2222]			
31	[3111]	[3121]	[3122]	[3131]		
32	[3211]	[3221]	[3222]	[3231]	[3232]	
33	[3311]	[3321]	[3322]	[3331]	[3332]	[3333]
...						

A unique index can be assigned to each row/column by means of the formula

$$(ij) = [i(i-1)]/2 + j \tag{3}$$

Systematic approximation of the molecular integrals arising in a particular calculation should address the problem of approximating the whole of this supermatrix rather than the individual elements.

2.4.2. *Application of the Schwartz inequality*

Application of the Schwartz inequality to the two-electron repulsion integrals gives [44]

$$[pq \mid rs] \leq \sqrt{[pq \mid pq]} . \sqrt{[rs \mid rs]} \tag{4}$$

The two-centre integrals $[pq \mid pq]$ and $[rs \mid rs]$ can be evaluated rapidly and the above inequality might facilitate the elimination of large numbers of electron repulsion integrals without sacrificing accuracy of calculated properties. Introducing a threshold τ_1, quantum chemical algorithms can be constructed in which all integrals for which

$$J_{pqrs} < \tau_1 \tag{5}$$

where

$$J_{pqrs} = \sqrt{[pq \mid pq]} . \sqrt{[rs \mid rs]} \tag{6}$$

are neglected for a suitably chosen τ_1. A further refinement might be to define a second threshold τ_2 so that integrals for which

$$\tau_1 < J_{pqrs} < \tau_2 \tag{7}$$

can be approximated and only integrals for which

$$\tau_2 < J_{pqrs} \tag{8}$$

are explicitly evaluated. In the direct self-consistent field procedure ([45]-[48]) and in other algorithms which are both direct and iterative the thresholds might be changed from iteration to iteration and reduced as convergence is approached.

For large, that is extended, systems, the total number of two-electron integrals which have a magnitude less than some chosen tolerance may be very large and the cumulative effect of neglecting these integrals may be quite significant. Indeed, the Schwartz inequality discussed above can result in an unacceptable loss of accuracy for very large molecular systems even though the threshold may be set quite tightly.

2.4.3. *Fast Gaussian methods*

Much recent work in *ab initio* quantum chemistry has been directed towards the development of "fast Gaussian" methods which scale linearly $[O(n)]$ or as $O(n \log n)$ with the number of electrons ([49]-[57]). Indeed, in studies of the many-body problem across a range of applications for both the classical and the quantum formulation ([58]-[67]), a variety of "fast hierarchical" methods have been introduced which significantly reduce the complexity of the N-body problem by subdividing space into a fixed hierarchy of cells and exploiting a tree-like data structure. In "many-body tree" approaches the root (parent) problem is recursively subdivided into smaller cells (children). The complexity of the algorithm is reduced by approximating the distribution of particles within a cell as a series (multipole) expansion which converges rapidly in the far field. Unlike the classical many-body problem the quantum mechanical formulation involves continuous distributions. Molecular integrals that involve charge distributions which do not penetrate may be handled via a multipole approach whilst integrals with overlapping charge distributions are near-field and must be evaluated. The efficiency of these hierarchical multipole methods in quantum chemical application is critically dependent on the effective partition of near- and far-field interactions. This partition is complicated by the fact that it depends not only on the extent and separation of the charge distributions but also on the tolerance imposed and the order of multipole expansion employed.

Carlson and Rushbrooke [68] introduced a (real arithmetic) two-centre multipole expansion in 1950. The expansion expresses the electron-electron interaction as a sum of products of three terms - the first depending on the coordinates of electron 1 with respect to a centre $\mathbf{P}$, the second depending on the distance $\mathbf{P} - \mathbf{Q}$, and the third depending on the coordinates of electron 2 with respect to centre $\mathbf{Q}$. Explicitly, the multipole expansion may be written

$$\begin{aligned}\frac{1}{|\mathbf{r}-\mathbf{r}'|} &= \sum_{\ell_p}^{\infty}(-1)^{\ell_p}\sum_{\ell_q}^{\infty}\sum_{m_p=-\ell_p}^{\ell_p}\sum_{m_q=-\ell_q}^{\ell_q} \\ &\quad \Big(C_{\ell_p}^{m_p}[\mathbf{r}-\mathbf{P}]\; C_{\ell_p+\ell_q}^{m_p+m_q}[\mathbf{P}-\mathbf{Q}]\; C_{\ell_q}^{m_q}[\mathbf{r}'-\mathbf{Q}]\end{aligned}$$

$$+S_{\ell_p}^{m_p}[\mathbf{r}-\mathbf{P}]\ \mathcal{S}_{\ell_p+\ell_q}^{m_p+m_q}[\mathbf{P}-\mathbf{Q}]\ S_{\ell_q}^{m_q}[\mathbf{r}'-\mathbf{Q}]$$
$$-S_{\ell_p}^{m_p}[\mathbf{r}-\mathbf{P}]\ \mathcal{C}_{\ell_p+\ell_q}^{m_p+m_q}[\mathbf{P}-\mathbf{Q}]\ S_{\ell_q}^{m_q}[\mathbf{r}'-\mathbf{Q}]$$
$$+C_{\ell_p}^{m_p}[\mathbf{r}-\mathbf{P}]\ \mathcal{S}_{\ell_p+\ell_q}^{m_p+m_q}[\mathbf{P}-\mathbf{Q}]\ C_{\ell_q}^{m_q}[\mathbf{r}'-\mathbf{Q}]\Big) \qquad (9)$$

where

$$C_\ell^m[\mathbf{r}] = |\mathbf{r}|^\ell(\ell+m)!\ P_\ell^m(\cos\theta_\mathbf{r})\ \cos(m\phi_\mathbf{r}) \qquad (10)$$

and

$$S_\ell^m[\mathbf{r}] = |\mathbf{r}|^\ell(\ell+m)!\ P_\ell^m(\cos\theta_\mathbf{r})\ \sin(m\phi_\mathbf{r}) \qquad (11)$$

are multipole tensors, and

$$\mathcal{C}_\ell^m[\mathbf{R}] = |\mathbf{R}|^{-(\ell+1)}\frac{1}{(\ell-m)!}\mathcal{P}_\ell^m(\cos\theta_\mathbf{R})\cos(m\phi_\mathbf{R}) \qquad (12)$$

and

$$\mathcal{S}_\ell^m[\mathbf{R}] = |\mathbf{R}|^{-(\ell+1)}\frac{1}{(\ell-m)!}\mathcal{P}_\ell^m(\cos\theta_\mathbf{R})\sin(m\phi_\mathbf{R}) \qquad (13)$$

Computations are simplified by the following relations between the multipole tensors for positive and negative values of m:-

$$C_\ell^{-m}[\mathbf{r}] = (-1)^m C_\ell^m[\mathbf{r}] \qquad (14)$$

$$S_\ell^{-m}[\mathbf{r}] = (-1)^{(m+1)} S_\ell^m[\mathbf{r}] \qquad (15)$$

$$\mathcal{C}_\ell^{-m}[\mathbf{r}] = (-1)^m \mathcal{C}_\ell^m[\mathbf{r}] \qquad (16)$$

$$\mathcal{S}_\ell^{-m}[\mathbf{r}] = (-1)^{(m+1)} \mathcal{S}_\ell^m[\mathbf{r}] \qquad (17)$$

The reader is referred elsewhere for further details[49]-[57].

2.4.4. *Cholesky decomposition*

It has been shown in the previous section how the evaluation of integrals can be simplified in studies of extended molecules where there is no significant overlap between the two charge distributions involved for a significant fraction of the integrals. Accurate studies of small molecules necessitate the use of large basis sets for which the corresponding charge distributions do overlap significantly and the methods described above are not applicable. However, computational linear dependence amongst the charge distributions involved in the two-electron integral supermatrix can be exploited by means of a Cholesky decomposition. A brief outline of this approach is given below. Further details can be found elsewhere ([69]-[73]).

The two-electron supermatrix, $\mathbf{V} = \mathbf{V_{ij;kl}}$, is a symmetric, positive definite matrix may be written in the form

$$\mathbf{V} = \mathbf{L}\mathbf{L}^{\dagger} \tag{18}$$

where $\mathbf{L}$ is a lower triangular matrix and $\mathbf{L}^{\dagger}$ is its transpose. Explicitly, this matrix product may be written

$$V_{(ij),(kl)} = \sum_{(pq)} L_{(ij),(pq)} L_{(ij),(pq)} \tag{19}$$

where the index (pq) runs over all possible charge distributions. However, in the presence of computational linear dependence, an approximation to $\mathbf{V}$ may be written

$$V_{(ij),(kl)} = \sum_{(pq)}^{\nu} L_{(ij),(pq)} L_{(ij),(pq)} \tag{20}$$

where the effective numerical rank of the two-electron integral matrix, ν, is considerably less than the total number of charge distributions.

A very stable algorithm for the construction of $\mathbf{L}$ is known which most importantly does not require the construction of the full $\mathbf{V}$ supermatrix. For $(ij) = 1, 2, ..., [n(n+1)]/2$

$$L_{(ij),(ij)} = \sqrt{V_{(ij),(ij)} - \sum_{(kl)=1}^{(ij)-1} L_{(ij),(kl)}} \tag{21}$$

and

$$\begin{aligned} L_{(pq),(kl)} &= \frac{\left[V_{(mn),(ij)} - \sum_{(kl)=1}^{(ij)-1} L_{(pq),(kl)} L_{(ij),(kl)}\right]}{L_{(ij),(ij)}}, \\ (pq) &= [j(j+1)]/2, \; ..., \; [n(n+1)]/2 \end{aligned} \tag{22}$$

The summations are omitted when the upper index is zero.

3. Summary

We have critically compared the basis sets used in our 1996 studies of correlation effects in the ground states of the nitrogen molecule and of the water molecule with those used in some more recent studies. We have introduced the acronym *duet* for our *d*istributed *u*niversal *e*ven-*t*empered basis set. These *duet* basis sets have been compared with the more widely employed correlation consistent basis sets. The *duet* approach to basis set construction has been applied to systems containing heavy atoms, to open-shell systems, to properties other than the energy and to the relativistic electronic structure problem.

The *duet* basis sets have two principal advantages over more commonly used approaches to the construction of molecular basis sets:-

(i) accuracy - They can be systematically enlarged and formally approach a complete set in the limit of an infinite basis set for a given expansion centre.

(ii) lack of bias - The parameters defining the *duet* basis sets are not optimized for the calculation of one particular property, molecular environment or electronic state.

We have briefly discussed schemes for the systematic approximation of the molecular integral supermatrix corresponding to *duet* basis sets.

Acknowledgments

SW acknowledges the support of EPSRC under Grant GR/M74627. SW and JK acknowledge the support of the Komitet Badań Naukowych and the British Council through the Joint Research Collaboration Programme no. WAR/992/179. DM acknowledges the support of the Office of Energy Research, Office of Basic Energy Sciences, Division of Chemical Sciences US Department of Energy under grant DE-FG02-97ER-14758.

References

1. W. Heitler and F. London, *Z. Phys.* **44**, 455 (1927).
2. B.H. Wells and S. Wilson, *J. Phys. B: At. Mol. Opt. Phys.* **22**, 1285 (1989).
3. D. Moncrieff and S. Wilson, in *Quantum Systems in Chemistry and Physics. Trends in Methods and Applications*, p. 323, edited by R. McWeeny, J. Maruani, Y.G. Smeyers and S. Wilson, Kluwer Academic Publishers, Dordrecht (1997).
4. D. Moncrieff and S. Wilson, *J. Phys. B: At. Mol. Opt. Phys.* **31**, 3819 (1998).
5. D. Moncrieff and S. Wilson, in *Quantum Systems in Chemistry and Physics*, Part 1, edited by S. Wilson, P.J. Grout, J. Maruani, Y.G. Smeyers and R. McWeeny, Adv. Quantum Chem. **31**, p. 157 (1998).
6. F. Jensen, *J. Chem. Phys.* **110**, 6601 (1999).
7. D. Moncrieff and S. Wilson, *J. Phys. B: At. Mol. & Opt. Phys.* **29**, 2425 (1996).
8. D. Moncrieff and S. Wilson, *J. Phys. B: At. Mol. & Opt. Phys.* **29**, 6009 (1996).
9. A.K. Wilson and T.H. Dunning Jr., *J. Chem. Phys.* **106**, 8718 (1997).
10. K.A. Peterson, A.K. Wilson, D.E. Woon and T.H. Dunning Jr., *Theoret. Chem. Acc.* **97**, 251 (1997).
11. J. Noga, W. Klopper and W. Kutzelnigg, in *Recent Advances in Computational Chemistry* **3**, 1, ed. R.J. Bartlett, World Scientific, Singapore (1997).
12. A. Halkier, T. Helgaker, P. Jorgensen, W. Klopper, H. Koch, J. Olsen and A.K. Wilson, *Chem. Phys. Lett.* **286**, 243 (1998).
13. W. Klopper and T. Helgaker, *Theoret. Chem. Acc.* **99**, 265 (1998).
14. R. Gdanitz, *Chem. Phys. Lett.* **283**, 253 (1998).
15. T.H. Dunning, Jr.,*J. Chem. Phys.* **90**, 1007 (1989).
16. R.A. Kendall, T.H. Dunning, Jr., and R.J. Harrison, *J. Chem. Phys.* **96**, 6769 (1992).
17. D.E. Woon and T.H. Dunning, Jr., *J. Chem. Phys.* **98**, 1358 (1993).
18. D.E. Woon and T.H. Dunning, Jr., *J. Chem. Phys.* **100**, 2975 (1994).
19. D.E. Woon and T.H. Dunning, Jr., *J. Chem. Phys.* **103**, 103 (1995).
20. A.K. Wilson, T. van Mourik and T.H. Dunning, Jr., *J. Mol. Struct. (THEOCHEM)* **388**, 339 (1996).

21. T. van Mourik and T.H. Dunning, Jr., *J. Chem. Phys.* **107**, 2451 (1997).
22. T. van Mourik, A.K. Wilson, K.A. Peterson, D.E. Woon and T.H. Dunning, Jr., Adv. *Quantum Chem.* **31**, 105 (1998).
23. S. Wilson, *Electron correlation in molecules*, Clarendon Press, Oxford (1984).
24. R. McWeeny, *Methods of Molecular Quantum Mechanics*, Academic Press, London (1989).
25. L. Laaksonen, D. Sundholm, P. Pykk(ö), *Intern. J. Quantum Chem.* **23**, 309 (1983).
26. J. Kobus, L. Laaksonen and D. Sundholm, *Comput. Phys. Commun.* **98**, 346 (1996).
27. S. Wilson, in *Methods in Computational Molecular Physics*, edited by G.H.F. Diercksen and S. Wilson, p. 71, Plenum, New York (1982).
28. S. Wilson, Adv. Chem. Phys. **69**, 439 (1987).
29. D. Moncrieff and S. Wilson, *Chem. Phys. Lett.* **209**, 423 (1993).
30. D. Moncrieff and S. Wilson, *J. Phys. B: At. Mol. & Opt. Phys.* **26**, 1605 (1993).
31. J. Kobus, D. Moncrieff and S. Wilson, *J. Phys. B: At. Mol. & Opt. Phys.* **27**, 5139 (1994).
32. J. Kobus, D. Moncrieff and S. Wilson, *J. Phys. B: At. Mol. & Opt. Phys.* **27**, 2867 (1994).
33. D. Moncrieff, J. Kobus and S. Wilson, *J. Phys. B: At. Mol. & Opt. Phys.* **28**, 4555 (1995).
34. J. Kobus, D. Moncrieff and S. Wilson, *Molec. Phys.* **86**, 1315 (1995).
35. D. Moncrieff, J. Kobus and S. Wilson, *Molec. Phys.* **93**, 713 (1998).
36. J. Kobus, D. Moncrieff and S. Wilson, *Molec. Phys.* **96**, 1559 (1999).
37. J. Kobus, D. Moncrieff and S. Wilson, *Molec. Phys.* **98**, 401 (2000).
38. J. Kobus, D. Moncrieff and S. Wilson, *Phys. Rev. A accepted for publication.*
39. J. Kobus, Chem. Phys. Lett. **202**, 7 (1993).
40. D. Moncrieff and S. Wilson, *Progr. Theoret. Chem. & Phys.* **3**, 323 (2000).
41. H.M. Quiney, D. Moncrieff and S. Wilson, *Progr. Theoret. Chem. & Phys.* **2**, 127 (2000).
42. W. Klopper, *J. Chem. Phys.* **102**, 6168 (1995).
43. W. Kutzelnigg, Theor. chim. Acta. **68**, 445 (1985).
44. M. Häser and R. Ahlrichs, *J. Comput. Chem.* **10**, 104 (1989).
45. J. Almlof, K. Faegri and K. Korsell, *J. Comput. Chem.* **3**, 385 (1982).
46. M. Head-Gordon, J.A. Pople and M.J. Frisch, *Chem. Phys. Lett.* **153**, 503 (1988).
47. S. Saebo and J. Almlof, *Chem. Phys. Lett.* **154**, 83 (1989).
48. S. Wilson, in *Electron correlation in atoms and molecules*, ed. S. Wilson, Meth. Comput. Chem. **1** *p.* 301*ff* Plenum, New York (1987).
49. C.A. White, B. Johnson, P.M.W. Gill and M. Head-Gordon, *Chem. Phys. Letts.* **230**, 8 (1994).
50. C.A. White, B. Johnson, P.M.W. Gill and M. Head-Gordon, *Chem. Phys. Letts.* **253**, 268 (1996).
51. R. Kutteh and J.B. Nicholas, *Chem. Phys. Letts.* **238**, 173 (1996).
52. M. Challacombe, E. Schwegler and J. Almlof, in *Computational Chemistry: Review of Current Trends*, edited by J. Leczszynski, p. 53, World Scientific, Singapore (1996).
53. E. Schwegler and M. Challacombe, *J. Chem. Phys.* **105**, 2726 (1996).
54. M. Challacombe, E. Schwegler and J. Almlof, *J. Chem. Phys.* **104**, 4685 (1996).
55. D.L. Strout and G.E. Scuseria, *J. Chem. Phys.* **102**, 8448 (1995).
56. J.C. Burrant, G.E. Scuseria and M.J. Frisch, J. *Chem. Phys.* **105**, 8969 (1996).
57. M.C. Strain, G.E. Scuseria and M.J. Frisch, *Science* **271**, 51 (1996).
58. J. Barnes and P. Hut, *Nature* **324**, 446 (1986).
59. L. Hernquist, *Ap. J. Suppl.* **64**, 715 (1987).
60. J. Barnes and P. Hut, *Ap. J. Suppl.* **70**, 389 (1989).
61. S. Pfalzner and P. Gibbon, *Comput. Phys. Commun.* **79**, 24 (1994).
62. J.K. Salmon, *Intern. J. Super Appl.* **8**, 129 (1994).

63. M.S. Warren and J.K. Salmon, *Comput. Phys. Commun.* **87**, 266 (1995).
64. L Greengard and V. Rokhlin, *J. Comput. Phys.* **73**, 325 (1987).
65. K.E. Schmidt and M.A. Lee, *J. Stat. Phys.* **63**, 1223 (1991).
66. H.G. Petersen, D. Soelvason, J.W. Perram and E.R. Smith, *J. Chem. Phys.* **101**, 8870 (1994).
67. H.G. Petersen, E.R. Smith and D. Soelvason, *Proc. Roy. Soc.* **A 448**, 401 (1994).
68. B.C. Carlson and G.S. Rushbrooke, *Proc. Cambridge Philos. Soc.* **46**, 626 (1950).
69. N. Beebe and J. Linderberg, *Intern. J. Quantum Chem.* **12**, 683 (1977).
70. I. Roeggen and E. Wisloff-Nielsen, *Chem. Phys. Letts.* **132**, 154 (1986).
71. S. Wilson, *Meth. Comput. Chem.* **1**, 251 (1987).
72. S. Wilson, *Comput. Phys. Commun.* **58**, 71 (1990).
73. S. Wilson, in *The Effects of Relativity in Atoms, Molecules, and the Solid State*, p. 217, ed. S. Wilson, I.P. Grant and B.L. Gyorffy, Plenum, London (1991).

Part III

Relativistic Formulations and Effects

RELATIVISTIC QUANTUM MECHANICS OF ATOMS AND MOLECULES

H. M. QUINEY
School of Chemistry, University of Melbourne
Parkville, Victoria 3052, Australia

Abstract. An overview of relativistic electronic structure theory is presented from the point of view of quantum electrodynamics. The participation of the negative-energy states in practical calculations is described from complementary points of view, in order to illustrate how they enter into the operation of relativistic mean-field theories. Examples of our implementation of relativistic electronic structure theory are drawn from studies of gauge invariance, many-body perturbation theory, inner-shell processes, electron momentum spectroscopy, and relativistic density functional theory.

1. Introduction

The many layers of existing physical theories are characterised by a comparison of the energy of the processes which they describe to fundamental parameters. For example, Planck's constant, h, defines the granularity of the energy scale on which a system may be considered to be quantum mechanical, semi-classical, or classical. Formally, we can make a model quantum mechanical system appear to behave as a classical system by considering the limit $h \rightarrow 0$, a perpective which is particularly clear in the Lagrangian formulation of quantum theory [1, 2]. Of course we are not free to alter a fundamental constant in experiments, which are designed either to expose the shortcomings of an existing theory, or to extract new information within its known limits of validity.

Within atoms and molecules, electronic structure is strongly influenced by widely spaced discrete energy levels and quantum mechanical laws, but collisional processes are often well-described by classical models, and an effective continuum of translational energy states accessible to the sys-

J. Maruani et al. (eds.), New Trends in Quantum Systems in Chemistry and Physics, Volume 1, 135–173.

tem. The validity of physical models describing spectroscopy, reactivity, and chemical bonding is similarly determined by the relation between the mean speed of the electrons and the speed of light, c. If the mean electron speed is small compared with c, such as is the case with the light elements or in the valence shells of all but the heaviest elements, we may safely assume that $c \to \infty$ compared with the mean speed of the electrons, and adopt a non-relativistic formulation of quantum mechanics, and a classical theory of electrodynamics. Such an assumption is not warranted in the high-energy regime of inner-shell electrons, for which the special theory of relativity introduces significant modifications to non-relativistic quantum mechanics and to the electrodynamic interactions between charged particles. At the threshold of energies at which the creation of electron-positron pairs becomes possible, the scale of energy is set by the mass-energy relation, $E = mc^2$. Of course, c is another finite fundamental constant which we are not free to alter. It is a fundamental postulate of the theory of special relativity that light propagates *in vacuo* with speed c in all inertial frames of reference. The precise numerical value of c determines the structure of space-time, and the nature of transformations between frames. The extent to which we may assume that non-relativistic quantum mechanics will suffice in a given situation requires the same prudence as is required in the use of classical or semi-classical approximations in the description of low-energy phenomena.

Much of the current activity in relativistic quantum chemistry is motivated by the recognition of the influence of relativistic mechanics on the structural properties of molecules, and their interactions with electric and magnetic fields. The conventional treatment retains the non-relativistic basis of quantum chemistry to which are applied a series of "relativistic corrections". It is certainly the case that if attention is restricted to problems of this type, in which so-called "relativistic effects" are isolated in effective core potentials and spin-dependent perturbations, one is likely to obtain a reliable picture of the chemical properties of the heavy elements, provided that these properties depend mainly on the valence electron distribution. This approach is not the method of choice, however, in considerations of high-energy phenomena in the neighbourhood of heavy nuclei. In atomic physics, where the technical challenges of adopting relativistic quantum mechanics are easier to deal with, no practical or formal incentive remains which would persuade one to adopt non-relativistic methods in the regime in which relativistic effects are important. Given the simultaneous development of both computer hardware and numerical algorithms, and the increasing awareness of the important impact of special relativity on the properties of the heavy elements, it would appear to be just a matter of time until relativistic quantum electrodynamics (QED) finds similarly widespread application in chemistry.

A simple and general treatment emerges if a formulation derived from relativistic QED is adopted. The clear advantage of such an approach is that it is applicable to all electronic phenomena within atoms and molecules if it can be implemented consistently and to sufficient precision. This point of view places one in a position to study high-energy phenomena, particularly the radiative corrections to energy levels which consititute the Lamb shift, and the electroweak interactions between electrons and nucleons which violate parity inversion symmetry. In this latter case the energy scale is set by the Fermi constant, $G_F \simeq 2 \times 10^{-14}$ a.u., and the strength of the effect by the masses of the virtual W- and Z-bosons, M_W and M_Z, which mediate the effective P-odd interaction. One need not lose sight of non-relativistic and classical formulations, however, since all non-relativistic results are recovered in the parametric limit $c \to \infty$, and all classical results in the parametric limit $h \to 0$.

In this presentation, we discuss the formulation which underpins our treatment of electronic structures and electric and magnetic properties of atoms and molecules, and some technical details of how the method is implemented in BERTHA, a relativistic molecular electronic structure program. A detailed account of the program was given in the proceedings of the 1997 QSCP meeting in Oxford [3]; here we concentrate on some aspects of the treatment of electric and magnetic properties within our computational framework of finite basis set expansions.

2. Dirac equation

The manifestly covariant form of the Dirac equation [4] is

$$[\gamma_\mu(p^\mu + eA^\mu) - mc]\Psi(x) = 0 \tag{1}$$

where the space-time four-vector, x, is written

$$\begin{aligned} x &= (x^0, x^1, x^2, x^3) & (2)\\ &= (ct, \boldsymbol{r}). & (3) \end{aligned}$$

The four-momentum is defined by $p_\mu = i\partial/\partial x^\mu$, and A^μ is the classical four-potential, where

$$A^\mu = (\phi(\boldsymbol{r})/c, \boldsymbol{A}(\boldsymbol{r})). \tag{4}$$

The scalar potential is $\phi(\boldsymbol{r})/c$, and $\boldsymbol{A}(\boldsymbol{r})$ is the vector potential of the external electromagnetic field. The 4×4 matrices, γ_μ, satisfy the anticommutation relations

$$\{\gamma_\mu, \gamma_\nu\} = 2g_{\mu\nu} \tag{5}$$

where $g_{\mu\nu}$ is the Minkowski space metric, whose non-zero elements are $g_{00} = 1$ and $g_{11} = g_{22} = g_{33} = -1$. Repeated indices imply summation in the inner product $a_\mu b^\mu$.

Premultiplying by $c\gamma^0$, the Dirac equation is obtained in the non-relativistic variables $(\boldsymbol{x}, t)$,

$$\left\{ i\frac{\partial}{\partial t} + e\phi(\boldsymbol{r}) - c\boldsymbol{\alpha}\cdot(\boldsymbol{p} + e\boldsymbol{A}(\boldsymbol{r})) - \beta mc^2 \right\} \Psi(x) = 0 \tag{6}$$

The 4×4 matrices $\boldsymbol{\alpha}$ and β are given by

$$\boldsymbol{\alpha}_q = \begin{bmatrix} 0 & \boldsymbol{\sigma}_q \\ \boldsymbol{\sigma}_q & 0 \end{bmatrix}, \qquad \beta = \begin{bmatrix} \boldsymbol{I} & 0 \\ 0 & -\boldsymbol{I} \end{bmatrix} \tag{7}$$

where $q = \{x, y, z\}$, $\boldsymbol{\sigma}_q$ are the Pauli spin matrices, $\beta = \gamma^0$, and $\boldsymbol{I}$ is the 2×2 unit matrix.

If we assume that the external electromagnetic field consists only of a time-independent scalar potential, $V(\boldsymbol{r}) = -e\phi(\boldsymbol{r})$, the solutions are of the form

$$\Psi_k(x) = \psi_k(\boldsymbol{r}) e^{-iE_k t} \tag{8}$$

where $\psi_k(\boldsymbol{r})$ is a four-component function of position satisfying the spatial eigenvalue equation

$$\left\{ c\boldsymbol{\alpha} \cdot \boldsymbol{p} + \beta mc^2 + V(\boldsymbol{r}) \right\} \psi_k(\boldsymbol{r}) = E_k \, \psi_k(\boldsymbol{r}) \tag{9}$$

with eigenvalue, E_k. The solution of equations of this type forms the computational basis of the relativistic electronic structure theory of atoms and molecules. The solutions are classified as being of 'positive-energy' type for $E_k > 0$, and 'negative-energy' type for $E_k < 0$ (Figure 1). For attractive potentials, $V(\boldsymbol{r}) < 0$, of the type which most commonly occur in electronic structure theory, the positive-energy solutions are further classified as square-integrable bound states if $-mc^2 < E_k < mc^2$. All other solutions belong to a continuum of states representing scattering in the external field. In the case where there is no external field, the spectrum consists only of positive- and negative-energy continua.

3. The role of negative-energy states

The classical relativistic mass-energy expression for a free-particle

$$E^2 = c^2 p^2 + m^2 c^4 \tag{10}$$

clearly leads to two values of E, differing only in sign, for given values of p^2 and m. In classical relativistic mechanics the negative-energy solution

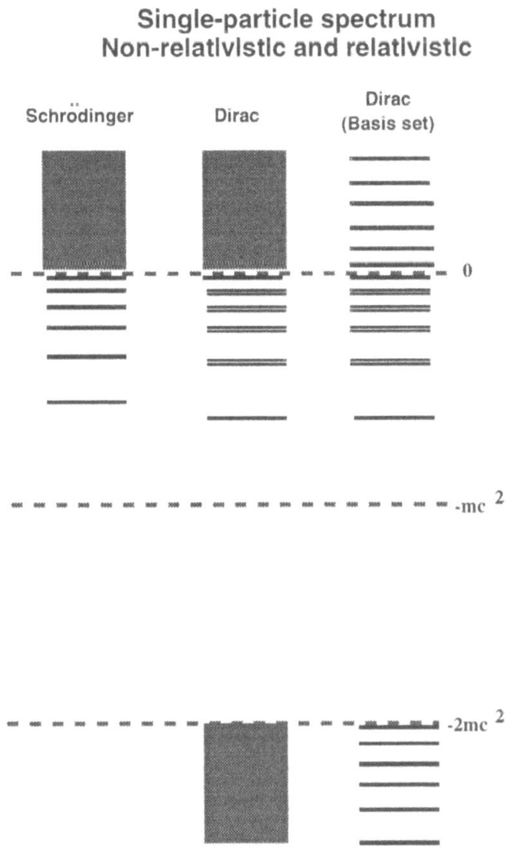

Figure 1. Schematic representation of the non-relativistic (Schrödinger) and relativistic (Dirac) spectra. The shaded areas represent continua, and the levels indicate discrete states. A finite basis set representation is included to indicate the manner in which the complete energy domain is spanned. Closely spaced levels are intended to suggest fine structure in the relativistic bound-state spectra.

is just discarded using physical arguments, and once a positive-energy solution has been selected, the particle will remain in that state unless an external perturbation is applied. The special principle of relativity requires that quantum mechanical free-particles also satisfy the mass-energy relation, Eq. (10), but now we may not discard so lightly the negative-energy solutions, because they form part of the complete set of eigenfunctions of Dirac's relativistic wave equation. In one-electron theory, which Dirac called c-number theory, a particle which is assumed initially to occupy a positive-energy state is unstable with respect to spontaneous radiative decay to a negative-energy state. In order to overcome this difficulty, it is necessary to reinterpret the solutions of the Dirac equation as operators which create or annihilate the quanta of a relativistic field, which in this case is the electron-positron field. Dirac denoted this approach as q-number theory, to distinguish the quantised creation and annihilation of particles from the classical prescriptions of c-number theory. Irrespective of how one performs this bookkeeping exercise, the use of the Dirac equation necessarily sacrifices any classical picture in which the number of particles is conserved, in favour of a more general conservation law in which the total charge is the conserved quantity. We shall examine how this works in practice, and the extent to which one need worry about such apparently exotic concepts in atomic physics and quantum chemistry, using some simple examples.

3.1. THE COMPOSITION OF A BOUND STATE

The only situation in which the labels "electron" and "positron" have simple meanings is for free-particles, and it is for this reason that we prefer to use the terminology "positive-energy" and "negative-energy" when discussing solutions of the Dirac equation. The application of an attractive external potential may generate positive-energy bound-state solutions of the Dirac equation, but expansion of any of these states in a complete basis of free-electron solutions includes contributions from both the positive- and negative-energy branches of the spectrum.

Taking the simplest case, we may find the expansion of the $1s_{1/2}$ bound-state of a hydrogenic ion, $\psi_{1s}(\boldsymbol{r})$, in a basis of positive- and negative-energy free-particle states, $\phi_p^{\pm}(\boldsymbol{r})$, in which $\boldsymbol{p}$ is the momentum associated with the state, and $p = |\boldsymbol{p}|$. If write this expansion in the form

$$\psi_{1s}(\boldsymbol{r}) = \int_0^{\infty} \left\{ c_p^+ \phi_p^+(\boldsymbol{r}) + c_p^- \phi_p^-(\boldsymbol{r}) \right\} dp$$

we find that

$$\begin{aligned} c_p^{\pm} &= (2Z)^{\gamma} \left\{ \sqrt{\frac{Z(1+\gamma)}{\Gamma(2\gamma+1)}} \sqrt{\frac{c^2 \pm E}{\pm\pi E}} \frac{\Gamma(\gamma+1)}{(p^2+Z^2)^{(\gamma+1)/2}} \sin(\gamma+1)\theta \right. \\ &\quad + \sqrt{\frac{Z(1-\gamma)}{\Gamma(2\gamma+1)}} \sqrt{\frac{-c^2 \pm E}{\pm\pi E}} \frac{(\gamma+2)\Gamma(\gamma)}{2p(p^2+Z^2)^{\gamma/2}} \\ &\quad \left. \times \left[\sin(\gamma\theta) - \frac{\gamma}{\gamma+2} \sin((\gamma+2)\theta) \right] \right\} \end{aligned}$$

where $\gamma = \sqrt{1-(Z/c)^2}$, $E = +c\sqrt{p^2+c^2}$, and

$$\theta = \sin^{-1}\left(\frac{p}{\sqrt{p^2+Z^2}} \right).$$

Although this expression is rather complicated, we can see that for finite values of c, the coefficients $\{c_p^-\}$ are non-vanishing, so that the external-field bound-state contains admixtures of the negative-energy solutions of the free-particle problem. If we consider the non-relativistic limit, $c \to \infty$, $\gamma \to 1$, and note that

$$\begin{aligned} \lim_{\gamma \to 1} \sin\left((\gamma+1)\theta\right) &= \frac{2pZ}{p^2+Z^2} \\ \sqrt{c^2p^2+c^4} &= c^2 + \frac{p^2}{2} + \frac{p^4}{8c^2} + \cdots \text{ for } c \gg p \end{aligned}$$

we recover the non-relativistic result [6]

$$c_p^{nr} = 4p\sqrt{\frac{2}{\pi}}\frac{Z^{5/2}}{(p^2+Z^2)^2}.$$

The fraction of the total bound-state density which is attributable to the negative-energy parts of its free-particle expansion is rather small for most elements of the Periodic Table, and it is worth wondering whether the negative-energy states are ever worth our serious consideration. We have identified two related cases in which the negative-energy states may play a crucial role in atomic and molecular problems: magnetic interactions [3], and the Lamb shift [5]. In second-order magnetic interactions, such as those encountered in nuclear magnetic resonance studies, the negative-energy states generated by an external electrostatic mean-field model may dominate the calculation of the interaction energy, though one may disguise this by modifying the effective potential which is used to construct the spinors. In the Lamb-shift, however, failure to treat the negative-energy parts correctly destroys the relativistic invariance of the problem to such an extent that that the renormalisation algorithm fails, leading to divergent, unphysical results. In the next section we shall see that, even in the simplest cases, negative-energy states enter into bound-state problems, though their influence may be indirect.

3.2. THE POSITRON AS A "HOLE"

In Dirac's "hole" theory of the positron, single-particle states whose energies are less than $-mc^2$ a.u. are assumed to be filled with electrons, according to the Pauli Exclusion Principle. The energy of the filled vacuum is subtracted from any physical model, on the grounds that it contributes a constant (albeit infinite) shift which is unobservable. In this model, a positron is interpreted as a hole in the density of negative-energy electrons, a description which Dirac constucted by analogy with the treatment of inner-shell processes, particularly those involving X-ray emission.

In Rayleigh-Schrödinger perturbation theory, the second-order correction to the energy, ε_2, of an occupied state, $|a\rangle$, whose zero-order energy is ε_a, is given by

$$\varepsilon_2 = \sum_v \frac{\langle a|\hat{O}|v\rangle\langle v|\hat{O}|a\rangle}{\varepsilon_a - \varepsilon_v}$$

where the perturbation is denoted by $\hat{O}$, and the unoccupied states of the complete spectrum are labelled $|v\rangle$, with eigenvalues ε_v.

In order to implement a valid formulation of bound-state quantum electrodynamics, we must incorporate the physical ideas of Dirac hole theory into any perturbative prescription for energy corrections to relativistic energy levels. We label negative-energy states by $|n\rangle$ with energies ε_n, occupied positive-energy states by $|p\rangle$ with energies ε_p, and unoccupied (virtual) states by $|v\rangle$ with energies ε_v. It is assumed that these states are solutions of the single-particle Dirac equation constructed from some time-independent external field.

In the most elementary relativistic treatment of this problem, the second-order correction to the total energy of this system consists of two parts:

1. The correction to the energy of the negative-energy vacuum electrons *and* the positive-energy electrons due to single-particle excitations involving virtual intermediate states.
2. The correction to the energy of the vacuum in the absence of occupied positive-energy states, which must be removed because it is unobservable.

Following this prescription, which involves only the occupancy of *electrons* in states of positive- or negative-energy, the relativistic second-order shift in the energy of the system, E_2^R, is

$$\begin{aligned} E_2^R &= \left\{ \sum_{nv} \frac{\langle n|\hat{O}|v\rangle\langle v|\hat{O}|n\rangle}{\varepsilon_n - \varepsilon_v} + \sum_{pv} \frac{\langle p|\hat{O}|v\rangle\langle v|\hat{O}|p\rangle}{\varepsilon_p - \varepsilon_v} \right\} \\ &- \left\{ \sum_{np} \frac{\langle n|\hat{O}|p\rangle\langle p|\hat{O}|n\rangle}{\varepsilon_n - \varepsilon_p} + \sum_{nv} \frac{\langle n|\hat{O}|v\rangle\langle v|\hat{O}|n\rangle}{\varepsilon_n - \varepsilon_v} \right\} \end{aligned}$$

In the second term, representing the correction to the "dressed" vacuum, the states $\{|p\rangle\}$ are unoccupied, and are consequently accessible as intermediate states in the energy correction to the vacuum.

Two of the summations cancel term-by-term, yielding

$$E_2^R = E_2^{R+} + E_2^{R-} \tag{11}$$

$$= \sum_{pv} \frac{\langle p|\hat{O}|v\rangle\langle v|\hat{O}|p\rangle}{\varepsilon_p - \varepsilon_v} - \sum_{np} \frac{\langle n|\hat{O}|p\rangle\langle p|\hat{O}|n\rangle}{\varepsilon_n - \varepsilon_p} \tag{12}$$

$$= \sum_{p\bar{p}} \frac{\langle p|\hat{O}|\bar{p}\rangle\langle\bar{p}|\hat{O}|p\rangle}{\varepsilon_p - \varepsilon_{\bar{p}}} \tag{13}$$

where $\{|\bar{p}\rangle\}$ is comprised of *all* those elements which are not in the set $\{|p\rangle\}$, irrespective of their positive- or negative-energy classifications.

While Eq. (11) conveys the impression that we should sum over all states outside of the set $\{|p\rangle\}$ as if they are virtual states, it is Eq. (12) which indicates the physical origin of the components of the second-order energy correction, within a framework drawn from quantum electrodynamics. The part of the summation involving positive-energy intermediate states is conventional, and describes the inclusion of singly-excited corrections to the wavefunction. The part of the summation involving negative-energy states, however, corresponds to a polarization of the vacuum charge-current density due to the blocking of $\{|p\rangle\}$ as accessible intermediate states, as a consequence of the Exclusion Principle. For a set $\{|p\rangle\}$ describing the bound single-particle states of a model atomic or molecular hamiltonian, the summation involving negative-energy states is a finite, observable component of the infinite, unobservable energy correction to the dressed vacuum.

3.3. THE POSITRON AS A "PARTICLE"

While the close analogy between electron-positron pair creation and core excitation phenomena in heavy elements is appealing, the manipulation involving subtraction of the divergent vacuum term is cumbersome. More elegant techniques have long been used in which the positron is treated as a particle, rather than as a hole.

We assume that the spatial part of the external-field Dirac field operator, $\psi_D(\mathbf{x})$, takes the form

$$\psi_D(\mathbf{x}) = \sum_m^{+} a_m \psi_m^{+}(\mathbf{x}) + \sum_n^{-} b_n^{\dagger} \psi_n^{-}(\mathbf{x})$$

where $a_m^{\dagger}$ and a_m are creation and annihilation operators for positive-energy states, and $b_m^{\dagger}$ and b_m are the corresponding operators for negative-energy states. The charge operator, Q, is

$$Q = -e \int \psi_D^{\dagger}(\mathbf{x})\psi_D(\mathbf{x})d\mathbf{x} + Q_0$$

where Q_0 is a constant associated with the vacuum charge, to be chosen later.

The hamiltonian in this representation is

$$h_D = \sum_m^{+} a_m^{\dagger} a_m E_m - \sum_n^{-} b_n^{\dagger} b_n E_n$$

and a one-electron positive-energy state, Ψ_a, is generated by operating on the external-field vacuum, $|0\rangle$,

$$\Psi_k = a_k^\dagger|0\rangle$$

If we wish to consider the lowest-order effect of virtual electron-positron pair creation, we must include in our one-electron bound-state trial wavefunction, Ψ, corrections arising from *three-particle* states which conserve the original total charge, $Q = -1$. These three-particle wavefunction corrections, Ψ' take the form

$$\Psi' = \sqrt{\frac{1}{3!}} a_m^\dagger a_{m'}^\dagger b_n^\dagger|0\rangle$$

revealing a key feature of relativistic quantum electrodynamics; one follows the total charge, and not the total number of particles.

This problem has been worked out in detail, and in greater generality by Sucher [7], who finds that the second-order energy correction due to virtual pair creation using this approach consists of two parts; a connected part corresponding precisely to E_2^{R-}, and a divergent, disconnected part corresponding to the vacuum correction term which was eliminated term-by-term in the treatment based on hole theory. The total observable second-order shift is identical to the one we deduced using hole theory, and the physical process giving rise to the term E_2^{R-} has been revealed to arise from the finite part generated by three-particle intermediate external-field states involving two electrons and one positron.

3.4. THE POSITRON AS AN ELECTRON PROPAGATING BACKWARDS IN TIME

According to the CPT theorem, the product of the symmetries of a physical system under charge- (C), parity- (P) and time-inversion (T) must be even. Electromagnetic interactions are parity conserving (P-even), and in a relativistic theory of mechanics we must consider the relative time ordering of events, rather than adopt the universal time frame assumed in Newtonian mechanics. Consequently, an electron moving backwards in time apparently exhibits the properties of a particle with the same mass as the electron, but whose charge has been reversed in sign. In a time-ordered representation, the particle-hole properties of electrons and positrons are replaced by a consideration of regions of space-time involving, respectively, times later and earlier than for a given space-time coordinate.

The prescriptions of Dirac hole theory were incorporated using the time-ordering of events by Feynman [8]. He wrote the propagation kernel,

$K(x_2, x_1)$, for relativistic electrons as

$$\begin{aligned} K(x_2,x_1) &= +\sum_p \psi_p(\mathbf{x}_2)\bar{\psi}_p(\mathbf{x}_1)\exp\left(-iE_p(t_2-t_1)\right) \text{ for } t_2 > t_1 \\ &\quad -\sum_n \psi_n(\mathbf{x}_2)\bar{\psi}_n(\mathbf{x}_1)\exp\left(-iE_n(t_2-t_1)\right) \text{ for } t_2 < t_1 \end{aligned}$$

so that the positive-energy states propagate an electron to later times, and the negative energy solutions propagate it to earlier times.

The relativistic second-order shift, E_2, due to a time-independent scalar perturbation, V, in this formulation is

$$\begin{aligned} E_2 &= i\int d^3\mathbf{x}_1\int d^3\mathbf{x}_2\int d(t_2-t_1)\bar{\psi}_0(\mathbf{x}_2)V(\mathbf{x}_2)K(x_2,x_1)V(\mathbf{x}_1)\psi_0(\mathbf{x}_1) \\ &= +i\int d^3\mathbf{x}_1\int d^3\mathbf{x}_2\int_0^\infty d(t_2-t_1)\bar{\psi}_0(\mathbf{x}_2)V(\mathbf{x}_2) \\ &\qquad \times\left(\sum_p \psi_p(\mathbf{x}_2)\bar{\psi}_p(\mathbf{x}_1)\right)V(\mathbf{x}_1)\psi_0(\mathbf{x}_1) \\ &\qquad \times\exp\left(-i(E_0-E_p)(t_2-t_1)\right) \\ &\quad -i\int d^3\mathbf{x}_1\int d^3\mathbf{x}_2\int_{-\infty}^0 d(t_2-t_1)\bar{\psi}_0(\mathbf{x}_2)V(\mathbf{x}_2) \\ &\qquad \times\left(\sum_n \psi_n(\mathbf{x}_2)\bar{\psi}_n(\mathbf{x}_1)\right)V(\mathbf{x}_1)\psi_0(\mathbf{x}_1) \\ &\qquad \times\exp\left(-i(E_0-E_n)(t_2-t_1)\right) \end{aligned}$$

We have assumed that the perturbation is a local function of the electronic coordinates which operates instantaneously, on the Dirac delta function $\delta(t_2-t_1)$. Integration over the time intervals $-\infty < (t_2-t_1) < 0$ and $0 < (t_2-t_1) < \infty$ is then trivial. Use of the identity

$$\frac{1}{\pi}\int_0^\infty \exp\left(-i\omega x\right)d\omega = \delta(x) + \frac{1}{i\pi x}$$

leads directly to the result

$$E_2 = \sum_p \frac{\langle 0|\hat{O}|p\rangle\langle p|\hat{O}|0\rangle}{\varepsilon_0-\varepsilon_p} - \sum_n \frac{\langle 0|V|n\rangle\langle n|V|0\rangle}{\varepsilon_n-\varepsilon_0} \tag{14}$$

$$+\sum_p \langle 0|\hat{O}|p\rangle\langle p|\hat{O}|0\rangle\delta(\varepsilon_0-\varepsilon_p) \tag{15}$$

$$= \sum_{\bar{p}} \frac{\langle 0|\hat{O}|\bar{p}\rangle\langle\bar{p}|\hat{O}|0\rangle}{\varepsilon_0-\varepsilon_{\bar{p}}} + i\pi\sum_p \langle 0|\hat{O}|p\rangle\langle p|\hat{O}|0\rangle\delta(\varepsilon_0-\varepsilon_p) \tag{16}$$

which is just a special case of Eq.(11) restricted to a one-electron problem. The second term depending on $\delta(\varepsilon_0 - \varepsilon_p)$ is necessarily imaginary, and represents a width, rather than the shift of the perturbation. The summation over $\bar{p}$ excludes the state ψ_0; the complete Gell-Mann-Low formalism [10] is required to demonstrate this feature.

3.5. THE VALIDITY OF RELATIVISTIC MEAN FIELD APPROXIMATIONS

These elementary exercises concerning the negative-energy states reveal a feature of relativistic many-electron theory which is of crucial importance in defining the place of the relativistic self-consistent field approximation within QED. Even if we take the trouble to ensure that the negative-energy states enter into a calculation only if real or virtual pair production processes are involved, the conclusions are indistinguishable from those of an unquantised theory, *provided that only one-body interactions are involved.* The formal equivalence between the results of the quantised and unquantised theories has been noted many times in the literature [8, 9], including in our own work on relativistic perturbation theory [12]. Consequently, we would make an error of interpretation if we were to imagine that the negative-energy states are empty in the vacuum state, and consequently available to serve as intermediate states in perturbation theory. On the basis of calculations performed using such a model, however, no serious problems would arise since a rigorous justification of them within QED can be presented using the principles outlined above. It is fair to suggest that much of the work performed in a relativistic atomic physics until the 1980's, beginning with the earliest formulation by Swirles in 1935 [11], was motivated by sound physical instinct and impressive agreement between theory and experiment, but not much thought was given to the potential problems posed by the negative-energy states; this simplified view was made possible by the concentration of this work on orbital models, mean-field potentials, and bound-states. The advent of successful relativistic finite basis parametrisations in the 1980's, explicit representations of the complete Dirac spectrum, and a rather active literature concerned with fundamental principles has served to bring these problems out into the open. An appreciation of the issues concerned with pair creation is necessary if one if develop numerical algorithms for relativistic calculations in atomic physics and quantum chemistry.

At each iteration of a relativistic self-consistent field calculation we define a one-body potential which supports a complete single-particle spectrum. As we have already seen, a hydrogenic bound-state includes admixtures of the negative-energy solutions of the free-particle Dirac equation field. More generally, the mean-field potentials at different stages of the iterative self-consistent field procedure may be quite different from one another, and any positive-energy spinor may contain admixtures of the negative-energy solutions generated in an earlier iteration. However, these mean-field potentials are of one-body type and, even if it is not apparent in the formalism, the orbital rotation process that takes us from one iteration to the next incorporates both excitation processes to positive-energy virtual states, *and* electron-positron pair creation processes in the spectral basis of earlier representations. Even in a multi-configurational formulation of mean field theory, the spinor rotation that updates the spinor $|i\rangle$ to a new approximation, $|i'\rangle$, takes the form

$$|i'\rangle = \sum_r |r\rangle U_{ri}$$

where $\boldsymbol{U}$ is unitary, and the sum over $\{|r\rangle\}$ includes all states of both positive- and negative-energy. In the relativistic generalisation of multi-configurational self-consistent field (MC-SCF) theory, the matrix $\boldsymbol{U}$ is constructed from the Hessian matrix, and is based on an expansion of the electronic energy to second-order, though formally we may infer the existence of a matrix $\boldsymbol{U}$ in any iterative SCF procedure. Failure to include the negative-energy states in the transformation destroys the quadratic convergence characteristics of relativistic Hessian-based MC-SCF theories [13]. On the face of it, such a procedure appears to involve the calculation of a highly-excited state, in which the negative-energy states play the role of virtual states on the same footing as unoccupied positive-energy levels, and in which half of the eigenvalues of the Hessian matrix are negative, and the other half are positive. But this appearance is quite illusory, since rotations involving the negative-energy states are equivalent to the virtual pair-creation terms which remain after the unobservable vacuum energy has been subtracted. Multi-configurational SCF theories involve the iterative improvement of one-body potentials, and so they fall within the same category as our model problem in perturbation theory, and under the protective umbrella of Furry bound-state QED theory.

It is only after we have defined a single-particle basis and a time-independent external field that we need be concerned about how to deal with truly many-body interactions. In this case the fortuitous equivalence between the formulae of quantised and unquantised perturbation theory suffers a catastrophic break down. If we restrict the formulae of many-body perturbation theory to include only excitation to positive-energy virtual states from positive-energy bound-states which are deemed initially to be occupied, we generate the "no-virtual pair" approximation. Implicit in this description is the understanding that no pairs are to be created or annihilated in the basis of configurations corresponding to the original external field, and the corresponding division of the spectrum into positive-

and negative-energy branches. The description does not imply, however, that no pair-creation processes at all are involved, because clearly the external field has associated with it a polarisation potential which distorts the spectrum corresponding to the case in which no fields are present. Similarly, the possibility of pair-creation processes is not precluded by the use of the no-virtual pair prescription, since the description is representation-dependent.

The rules for including pair creation processes in relativistic many-body theory and for avoiding the catastrophe often described as "Brown-Ravenhall disease" [14] have been presented several times, notably by Labzowsky [15] and Sapirstein [16]. They derive small many-body corrections to many-body perturbation theory, involving intermediate states in which the original N electrons are augmented by the creation of M pairs, violating the classical notion of the conservation of particle number, but respecting the conservation of the total charge of the system.

4. Relativistic finite basis set methods

4.1. SPINOR BASIS SETS

We write four-component atomic or molecular position-space amplitudes, $\psi_i(\boldsymbol{r})$, as an N-dimensional, multi-centre, finite basis set expansion

$$\begin{aligned}\psi_i(\boldsymbol{r}) &= \begin{bmatrix} \sum_{\mu=1}^{N} \frac{1}{r_\mu} c_\mu^L f_{\kappa_\mu}^L(r_\mu)\chi_{\kappa_\mu m_\mu}(\theta_\mu,\varphi_\mu) \\ \sum_{\mu=1}^{N} \frac{i}{r_\mu} c_\mu^S f_{\kappa_\mu}^S(r_\mu)\chi_{-\kappa_\mu m_\mu}(\theta_\mu,\varphi_\mu) \end{bmatrix} \\ &= \begin{bmatrix} \sum_{\mu=1}^{N} c_\mu^L M[L,\mu,\boldsymbol{r}] \\ \sum_{\mu=1}^{N} c_\mu^S M[S,\mu,\boldsymbol{r}] \end{bmatrix}\end{aligned}$$

where μ labels the basis function parameters, $T = \{L, S\}$, $f_{\kappa_\mu}^T(r)$ are radial functions and the spin-angular spinors are

$$\chi_{\kappa,m}(\theta,\varphi) = \begin{bmatrix} \left(\frac{j+m}{2j}\right)^{1/2} Y_{j-1/2}^{m-1/2}(\theta,\varphi) \\ \left(\frac{j-m}{2j}\right)^{1/2} Y_{j-1/2}^{m+1/2}(\theta,\varphi) \end{bmatrix} \qquad \kappa < 0,$$

$$\chi_{\kappa,m}(\theta,\varphi) = \begin{bmatrix} -\left(\frac{j+1-m}{2j+2}\right)^{1/2} Y_{j+1/2}^{m-1/2}(\theta,\varphi) \\ \left(\frac{j+1+m}{2j+2}\right)^{1/2} Y_{j+1/2}^{m+1/2}(\theta,\varphi) \end{bmatrix} \quad \kappa > 0.$$

A two-component basis function is denoted by $M[T,\mu,r]$. A number of classes of spinor basis set may be specified by a particular choice of radial function. Finite basis set calculations may exhibit catastrophic variational collapse problems, and we restrict our attention to choices of the radial functions $f_{\kappa_\mu}^T(r)$ for which such problems do not occur.

4.1.1. *Radial L- and S-spinors*

These are a suitable choice for one-centre, point-nuclear problems.

$$\begin{aligned} f_{\kappa,n}^T(r) &= C_{n\kappa} r^{\gamma_\kappa} \exp(-\lambda r) \\ &\times \left\{ -(1-\delta_{n0}) L_{n-1}^{2\gamma_\kappa}(2\lambda r) \pm \frac{N_n - \kappa}{(n+2\gamma_\kappa)} L_n^{2\gamma_\kappa}(2\lambda r) \right\} \end{aligned} \tag{17}$$

for fixed real λ, $n = 0, 1, 2, \cdots$ for $\kappa < 0$ and $n = 1, 2, 3, \cdots$ for $\kappa > 0$. The upper sign is chosen for $T = L$, and the lower sign for $T = S$. Elements of the S-spinor sets are generated from the L-spinor sets by fixing n (smallest value for fixed κ) and choosing exponent sets, $\{\lambda_{\kappa,i}\}$. The S-spinor set is the relativistic analogue of the non-relativistic Slater basis set. The advantage of L-spinors is that very large basis sets may be employed without introducing computational linear dependence. We are currently investigating the use of these functions in a relativistic implementation of the convergent close-coupled approximation to photoionisation.

4.1.2. *Radial G-spinors*

The G-spinor basis set is relativistic analogue of the Gaussian-type functions of quantum chemistry, and retains all of the advantages conferred by the Gaussian product theorem in the evaluation of multi-centre integrals. The form of the large- and small-component radial functions is fixed by the kinetic balance prescription, so that

$$f_\mu^L(r_{A_\mu}) = N_\mu^L \, r_{A_\mu}^{l_\mu+1} \exp(-\lambda_\mu r_{A_\mu}^2) \tag{18}$$

$$f_\mu^S(r_{A_\mu}) = N_\mu^S[(\kappa_\mu + l_\mu + 1) - 2\lambda_\mu r_{A_\mu}^2] \, r_{A_\mu}^{l_\mu} \exp(-\lambda_\mu r_{A_\mu}^2). \tag{19}$$

The exponent set $\{\lambda_\mu\}$ is chosen according to quantum chemical prescriptions.

4.2. KINETIC BALANCE

The necessity of a one-to-one mapping between the elements of the large- and small-component basis sets has been discussed in [17] and forms an integral part of the algorithms which have been implemented in BERTHA. Here, we generalise the treatment of [18] to examine the consequences of kinetic balance on the finite-dimensional representation of the complete Dirac spectrum, rather than just its positive-energy solutions. It is instructive to consider the free-particle Dirac equation

$$\begin{bmatrix} c^2 - \varepsilon & c\,\boldsymbol{\sigma}\cdot\boldsymbol{p} \\ c\,\boldsymbol{\sigma}\cdot\boldsymbol{p} & -c^2 - \varepsilon \end{bmatrix} \begin{bmatrix} \psi_\varepsilon^L \\ \psi_\varepsilon^S \end{bmatrix} = 0.$$

Straightforward algebra yields

$$(\boldsymbol{\sigma}\cdot\boldsymbol{p})\psi_\varepsilon^T = \eta_T \left(\frac{c^2 + \eta_T \varepsilon}{c} \right) \psi_\varepsilon^{\bar{T}},$$

where $\eta_T = 1$ if $T = L$, $\eta_T = -1$ if $T = S$, and $\bar{T} \neq T$. Further use of the operator identity

$$(\boldsymbol{\sigma}\cdot\boldsymbol{p})(\boldsymbol{\sigma}\cdot\boldsymbol{p}) = \hat{p}^2 \boldsymbol{I} \tag{20}$$

and the relativistic energy-momentum relation

$$\varepsilon^2 = c^2 p^2 + c^4$$

leads to separate equations for ψ_ε^T,

$$\hat{p}^2 \boldsymbol{I} \psi_\varepsilon^T = p^2 \psi_\varepsilon^T. \tag{21}$$

If we now consider the *matrix* representation of the free-particle Dirac equation

$$\begin{bmatrix} (c^2 - \varepsilon)\boldsymbol{S}_{LL} & c\boldsymbol{\Pi}_{LS} \\ c\boldsymbol{\Pi}_{SL} & -(c^2 + \varepsilon)\boldsymbol{S}_{SS} \end{bmatrix} \begin{bmatrix} \boldsymbol{c}_L \\ \boldsymbol{c}_S \end{bmatrix} = 0$$

the vectors of expansion coefficients, $\boldsymbol{c}_L$ and $\boldsymbol{c}_S$, form the solutions of a pair of simultaneous matrix equations

$$\boldsymbol{\Pi}_{\bar{T}T}\boldsymbol{c}_T = \eta_T \left(\frac{c^2 + \eta_T \varepsilon}{c} \right) \boldsymbol{S}_{\bar{T}\bar{T}} \boldsymbol{c}_{\bar{T}} \tag{22}$$

Equations involving either $\boldsymbol{c}_L$ or $\boldsymbol{c}_S$ may be obtained by elimination, yielding

$$\boldsymbol{\Pi}_{T\bar{T}} \boldsymbol{S}_{\bar{T}\bar{T}}^{-1} \boldsymbol{\Pi}_{\bar{T}T} \boldsymbol{c}_T^{\dagger} = p^2 \boldsymbol{S}_{TT} \boldsymbol{c}_T \tag{23}$$

Decoupling of the matrix representation of the Dirac equation, Eqn. (23), into valid representations of the Dirac free-particle operator equation, Eqn. (21) is possible only if the matrix identities

$$\boldsymbol{\Pi}_{T\bar{T}}\boldsymbol{S}_{\bar{T}\bar{T}}^{-1}\boldsymbol{\Pi}_{\bar{T}T} = \boldsymbol{p}_{TT}^{2} \tag{24}$$

are satisfied, where $\boldsymbol{p}_{TT}^2$ is the matrix representation of the operator $(\boldsymbol{\sigma}\cdot\boldsymbol{p})(\boldsymbol{\sigma}\cdot\boldsymbol{p}) = \hat{p}^2\boldsymbol{I}$ in the basis set $\{\varphi_k^T\}$. The matching of basis functions according to

$$\{\varphi_k^S\} = \{N_k^S\boldsymbol{\sigma}\cdot\boldsymbol{p}\varphi_k^L\} \tag{25}$$

generates representations which satisfy the Eqn. (24), because

$$\boldsymbol{\Pi}_{LS} = \boldsymbol{p}_{LL}^2 \tag{26}$$

$$\boldsymbol{\Pi}_{SL} = \boldsymbol{S}_{SS} \tag{27}$$

$$\boldsymbol{\Pi}_{LS} = \boldsymbol{\Pi}_{SL}^{\dagger}. \tag{28}$$

Without loss of generality we have chosen the arbitrary constants $\{N_k^S\}$ to be unity, but we need make no further assumptions about the functional form or normalization of the sets $\{\varphi_k^L\}$ and $\{\varphi_k^S\}$.

Substituting Eqns. (??) and (26) into Eqn. (23) yields two matrix representations of the free-particle Schrödinger equation for the large- and small-component functions

$$\boldsymbol{p}_{LL}^2\boldsymbol{c}_T = p^2\boldsymbol{S}_{LL}\boldsymbol{c}_T \qquad \text{for } T = L, S. \tag{29}$$

The large- and small-component expansion vectors, $\boldsymbol{c}_L$ and $\boldsymbol{c}_S$, are solutions of *identical* generalized matrix eigenvalue equations which, following these manipulations, can be written entirely in terms of $\boldsymbol{p}_{LL}^2$ and $\boldsymbol{S}_{LL}$, even though the large- and small-component basis functions are different. As a consequence of the identical values of p^2 which arise in the partitioned equations for $\boldsymbol{c}_T$, there is a one-to-one mapping of the positive and negative energy eigenvalues of the free-particle four-component equations of the form $E \to -E$ if strict kinetic balance is adopted. Substitution of Eqn. (26) into Eqn. (22) determines the simple relationship

$$\boldsymbol{c}_S = \frac{c}{c^2+\varepsilon}\boldsymbol{c}_L$$

between the large and small component vectors. The origin of the notation L and S for components is obvious if one considers the relative magnitudes of $\boldsymbol{c}_L$, for a *positive*-energy eigenvalue, $\varepsilon > c^2$. For negative-energy solutions, however, $\varepsilon < -c^2$ and the conventional notation is misleading. A $2M$-dimensional space of four-component solutions of the Dirac equation,

ψ_ε, is constructed for the eigenvalue spectrum $\varepsilon = \pm c^2\sqrt{(1+(p/c)^2)}$ from the M linearly independent solutions of the two-component equation Eqn. (29), each with eigenvalue p^2. It is straightforward to demonstrate that $\langle\psi_\varepsilon|\psi_{-\varepsilon}\rangle = 0$, establishing the linear independence of the spinor space.

The use of the unrestricted kinetic balance (UKB) prescription has been suggested in order to avoid the computational labour involved in implementation of the contraction of primitive radial functions implied by the definition of a small component G-spinor in Eq. (18). Since the RKB representation may be extended towards completeness and is linearly independent, however, the inflation of the small-component basis implied by the UKB basis must be a matter of considerable concern. It is found that UKB calculations suffer from serious linear dependence problems, and that negative-energy solutions are generated which have zero or near-zero kinetic energy, and vanishing large-components. We do not regard, therefore, the UKB prescription as a satisfactory relativistic finite basis set method for these reasons. A better solution is to eliminate these unphysical vectors with zero kinetic energy and small-component functions which are unmatched by large-component partners from the outset, by the use of a matched set of RKB G-spinor basis functions.

In practice, one may choose a non-orthogonal set of functions to solve the large-component free-particle equation, Eqn. (29), and use these two-component solutions to construct an orthonormal set of discrete four-component solutions of the Dirac equation. These four-component solutions span the same linear space as the union of the separate two-component kinetically balanced large- and small-component spaces, provided that *all* positive- and negative-energy solutions are included. From another point of view, we may use the prescriptions of this section to construct a $2N$-dimensional basis set of free-electron Dirac spinors starting with a two-component basis formed by solution of the free-electron Schrödinger equation, and the relationship between the expansion coefficients of a RKB set, Eq. (22). Introduction of an external field causes a rotation amongst the elements of the discrete free-electron spinor set, just as we found in the analytic decomposition of the $1s$ hydrogenic bound-state into its exact free-electron components.

4.3. REPRESENTATIONS OF THE CHARGE-CURRENT DENSITY

All quantities involved in relativistic calculations can be deduced from the components of the relativistic four-current, $j_\mu = (\varrho, \boldsymbol{j})$; this is true also for relativistic generalisations of density functional theories. We have chosen to work with these quantities directly, rather than construct them from scalar basis functions. For a four-component Dirac spinor,

$$\psi(\boldsymbol{r}) = \begin{bmatrix} \psi_1(\boldsymbol{r}) \\ \psi_2(\boldsymbol{r}) \\ \psi_3(\boldsymbol{r}) \\ \psi_4(\boldsymbol{r}) \end{bmatrix}$$

components of the four-current are defined to be

$$\begin{aligned} \varrho(\boldsymbol{r}) &= \psi^\dagger(\boldsymbol{r})\psi(\boldsymbol{r}) \\ j_q(\boldsymbol{r}) &= c\psi^\dagger(\boldsymbol{r})\boldsymbol{\alpha}_q\psi(\boldsymbol{r}) \end{aligned}$$

where $q = \{x, y, z\}$. Subsituting explicit expressions for the components of ψ and the matrices $\boldsymbol{\alpha}_q$, we find that the components or the four-current are real, and are given by

$$\begin{aligned} \varrho(\boldsymbol{r}) &= \psi_1^*(\boldsymbol{r})\psi_1(\boldsymbol{r}) + \psi_2^*(\boldsymbol{r})\psi_2(\boldsymbol{r}) + \psi_3^*(\boldsymbol{r})\psi_3(\boldsymbol{r}) + \psi_4^*(\boldsymbol{r})\psi_4(\boldsymbol{r}) \\ j_x(\boldsymbol{r}) &= c\{\psi_1^*(\boldsymbol{r})\psi_4(\boldsymbol{r}) + \psi_2^*(\boldsymbol{r})\psi_3(\boldsymbol{r}) + \psi_3^*(\boldsymbol{r})\psi_2(\boldsymbol{r}) + \psi_4^*(\boldsymbol{r})\psi_1(\boldsymbol{r})\} \\ j_y(\boldsymbol{r}) &= ic\{\psi_1^*(\boldsymbol{r})\psi_4(\boldsymbol{r}) - \psi_2^*(\boldsymbol{r})\psi_3(\boldsymbol{r}) + \psi_3^*(\boldsymbol{r})\psi_2(\boldsymbol{r}) - \psi_4^*(\boldsymbol{r})\psi_1(\boldsymbol{r})\} \\ j_z(\boldsymbol{r}) &= c\{\psi_1^*(\boldsymbol{r})\psi_3(\boldsymbol{r}) - \psi_2^*(\boldsymbol{r})\psi_4(\boldsymbol{r}) + \psi_3^*(\boldsymbol{r})\psi_1(\boldsymbol{r}) - \psi_4^*(\boldsymbol{r})\psi_2(\boldsymbol{r})\}. \end{aligned}$$

In our G-spinor representation, any component of the four-current may be reduced to a linear combinations of quantities derived from two-component objects, according to the rules:

Overlap charge density	$M^\dagger[T, \mu, \mathrm{r}]M[T, \nu, \mathrm{r}]$
Overlap current density	$M^\dagger[T, \mu, \mathrm{r}]\,\boldsymbol{\sigma}_q\, M[\bar{T}, \nu, \mathrm{r}]$

where T is either L or S, and $\bar{T} \neq T$. The operators, $\{\boldsymbol{\alpha}_q\}$ are constructed from the two-dimensional blocks $\{\boldsymbol{\sigma}_q\}$ so we are able to generate the components of the four-current without explicit consideration of zero-valued couplings between spinor components.

The Gaussian Product Theorem is invoked in order to write any of these four quantities in the general form

$$M^\dagger[T, \mu, \boldsymbol{r}_{A_\mu}]\,\boldsymbol{\sigma}_q\, M[T', \nu, \boldsymbol{r}_{A_\nu}] = \sum_{ijk} E_q[\mu, T; \nu, T'; i, j, k] H(\mu, \nu; i, j, k; \boldsymbol{r})$$

The label $q = 0$ refers to the overlap density, and $\sigma_0 = I$. The summation over $\{i, j, k\}$ terminates after $(\Lambda + 1)(\Lambda + 2)(\Lambda + 3)/6$ terms, where $\Lambda = \ell_\mu + \ell_\nu + k$, and $k = 0, 1,$ or 2.

This involves G-spinor expansion coefficients, E_q, and Hermite Gaussian functions, $H(\mu\nu; i, j, k; \boldsymbol{r})$. The Hermite Gaussian functions are the ones commonly used in quantum chemistry programs [19].

The important consequence of this formulation is that *ab initio* relativistic electronic structure atomic and molecular calculations may be performed using techniques borrowed from atomic physics and quantum chemistry, including the calculation of electric and magnetic properties. The "relativistic features" of the problem are largely buried in the E_q-coefficients, greatly simplifying the subsequent implementation a relativistic quantum mechanical description of electronic structure.

4.4. GAUGE INVARIANCE OF MATRIX METHODS

Gauge invariance of the Dirac equation implies that under the transformation

$$\boldsymbol{A} \rightarrow \boldsymbol{A}' = \boldsymbol{A} + \boldsymbol{\nabla}\Lambda(\boldsymbol{r}) \tag{30}$$

where $\Lambda(\boldsymbol{r})$ is an arbitrary differentiable function the Dirac hamiltonian transforms according to

$$\hat{h}_D \rightarrow \hat{h}'_D = \exp[-i\Lambda(\boldsymbol{r})]\ \hat{h}_D\ \exp[i\Lambda(\boldsymbol{r})]$$

which transforms the solution of the Dirac equation according to

$$\psi'(\boldsymbol{r}) = \exp[-i\Lambda(\boldsymbol{r})]\psi(\boldsymbol{r})$$

In order to illustrate the problems which may arise in the relativistic treatment of magnetic properties using finite basis set methods, we choose the simplest non-trivial example, $\Lambda(\boldsymbol{r}) = z$, which introduces an interaction hamiltonian $c\alpha_z$. The untransformed and transformed hamiltonians, $\hat{h}_D$ and $\hat{h}'_D$, respectively, are given by

$$\begin{aligned} \hat{h}_D &= c\boldsymbol{\alpha}\cdot\boldsymbol{p} + \beta mc^2 - \frac{Z}{r} \\ \hat{h}'_D &= c\boldsymbol{\alpha}\cdot\boldsymbol{p} + \beta mc^2 - \frac{Z}{r} + c\,\alpha_z \end{aligned}$$

and since they are related by an elementary gauge transformation, they generate identical eigenvalue spectra, with solution sets related by a complex phase factor.

The results in Table 1 represent a study of the convergence of the few lowest-energy bound-states of $\hat{h}'_D$ and $\hat{h}_D$ for H-like neon. Although the eigenvalue spectra of the two representations converge towards one another and towards the exact hydrogenic eigenvalues as the basis set is enlarged, it is necessary to saturate the space of G-spinors with $\ell = 4$ functions in order to achieve the level of agreement between the two representations considered here.

TABLE 1. Calculations with h_D and h'_D for H-like neon. The G-spinor basis set for all angular types is geometric with $\alpha = 0.01$ and $\beta = 1.9$. In order to highlight the selection rules operational in this problem, the original 30s 30p basis set has been augmented only by functions whose magnetic quantum number is $m_j = \pm 1/2$.

	$\hat{h}'_D$			$\hat{h}_D$
	30s 30p	30s 30p 30d	30s 30p 30d 30f 30g	30s 30p
1	50.066725	50.066720	50.066720	50.066718
2	-50.066725	-50.066720	-50.066720	-50.066718
3	-12.534053	-12.525472	-12.520899	-12.520860
4	-12.534053	-12.525472	-12.520899	-12.520860
5	-12.504163	-12.520163	-12.520867	-12.520856
6	-12.504163	-12.520163	-12.520867	-12.520856
7	-12.407097	-12.504293	-12.504163	-12.504163
8	-12.407097	-12.504293	-12.504163	-12.504163
9	-11.938560	-12.478914	-12.504151	-12.504163
10	-11.938560	-12.478914	-12.504151	-12.504163

There are other less obvious problems associated with the use of a non-zero gauge function $\Lambda(\boldsymbol{r})$, the most important of which is that the transformed Dirac hamiltonian can no longer be block-diagonalised in representations labelled by (κ, m_j). For a finite representation, this splits the $p_{3/2}$ states, for example, into two manifolds, characterised by $m_j = \pm 1/2$ and $m_j = \pm 3/2$. The eigenvalue $\varepsilon = -12.504163$ a.u. persists throughout all sets of calculations, because the only functions with $m_j = \pm 3/2$ included in the basis are of p-type, and consequently there are no non-zero off-diagonal matrix elements of $c\alpha_z$ involving these functions; this eigenvalue corresponds to the single-particle states $2p_{3/2,\pm 3/2}$. The degeneracy of Kramers pairs is preserved at all times because $c\alpha_z$ only couples basis states of the same m_j, but in the restricted subspace of $s-$ and $p-$type functions the two pairs of $p_{3/2}$ functions experience a physically meaningless splitting which is characteristic of an incomplete expansion. Moreover, the basis set representation of $\hat{h}'_D$ requires the use of complex arithmetic, while the representation of $\hat{h}_D$ may be constructed using real arithmetic, if the convention is adopted that the small-component amplitudes are chosen to be purely imaginary.

The reason for the complex nature of the representation of $\hat{h}'_D$ is easy to find; the inclusion of basis functions of higher angular momentum serves to construct the phase factor $\exp[-i\Lambda(\boldsymbol{r})]$ as a multipole expansion, which in this case takes the form

$$\exp[-i\Lambda(\boldsymbol{r})] = 1 - iz - \frac{z^2}{2!} + i\frac{z^2}{3!} + \cdots$$

We may accommodate this gauge transformation exactly by defining B-spinors, $B[T, \mu, \mathbf{r}_{A_\mu}]$, the two-component equivalents of London orbitals,

$$B[T, \mu, \mathbf{r}_{A_\mu}] = \exp[-i\Lambda(\boldsymbol{r})]M[T, \mu, \mathbf{r}_{A_\mu}]$$

where $M[T, \mu, \mathbf{r}_{A_\mu}]$ is a G-spinor. The phase factor $\exp[-i\Lambda(\boldsymbol{r})]$ has the effect of including basis functions with large angular momentum into the basis set.

5. Some practical considerations

5.1. THE NON-RELATIVISTIC LIMIT

It is a satisfying feature of the Dirac equation that its four-component spinor solutions may be re-interpreted as two-component spin-orbitals in the formal limit $c \to \infty$. From a computational point of view, however, fixing c to be more than two or three orders of magnitude larger than its natural value introduces numerical instability in the diagonalisation of the basis set representation of the Dirac equation; the practical upper limit is approximately $c \to 20000$ a.u. This method may not be used for high precision determinations of non-relativistic wavefunctions, since this value of c is unable to extinguish completely the fine structure in the core electronic levels of heavy elements.

In order to make direct comparisons between our relativistic formulation and the non-relativistic limit of quantum mechanics, we make use of the identity

$$(\boldsymbol{\sigma} \cdot \boldsymbol{p})(\boldsymbol{\sigma} \cdot \boldsymbol{p}) = p^2 \boldsymbol{I}$$

to write a two-component Schrödinger equation

$$\left(\frac{1}{2}(\boldsymbol{\sigma} \cdot \boldsymbol{p})(\boldsymbol{\sigma} \cdot \boldsymbol{p}) + V(\boldsymbol{r})\right)\psi^L(\boldsymbol{r}) = E\psi^L(\boldsymbol{r})$$

involving only the large-component two-spinor, $\psi^L(\boldsymbol{r})$, of a Dirac four-spinor, $\psi(\boldsymbol{r})$. For a given set of Gaussian orbital exponents, this formulation generates degenerate pairs of non-relativistic spin-orbitals in the relativistic large-component G-spinor basis set, with eigenvalues identical to those which would be generated in the usual scalar orbital formulation of quantum chemistry. These two-component solutions will, in general, involve unitary rotations amongst states of α and β spin, but can readily be transformed into states of pure spin if required.

All that is required to implement this non-relativistic procedure in addition to the elements of our relativistic electronic structure formalism, are the one-electron G-spinor integrals over the two-component form of the kinetic energy operator, $\frac{1}{2}(\boldsymbol{\sigma}\cdot\boldsymbol{p})(\boldsymbol{\sigma}\cdot\boldsymbol{p})$. Since the operator $\boldsymbol{\sigma}\cdot\boldsymbol{p}$ is translationally invariant, we may use the result from atomic structure theory

$$\boldsymbol{\sigma}\cdot\boldsymbol{p}\frac{f(r)}{r}\chi_{\kappa,m}(\vartheta,\varphi) = \frac{i}{r}\left(\frac{d}{dr}+\frac{\kappa}{r}\right)f(r)\chi_{-\kappa,m}(\vartheta,\varphi)$$

where $f(r)$ is an arbitrary function of r. Repeated application of this identity to a two-component G-spinor yields

$$(\boldsymbol{\sigma}\cdot\boldsymbol{p})\,(\boldsymbol{\sigma}\cdot\boldsymbol{p})\,\frac{f(r)}{r}\chi_{\kappa,m}(\vartheta,\varphi) = \frac{1}{r}\left(-\frac{d^2}{dr^2}+\frac{\kappa(\kappa+1)}{r^2}\right)f(r)\chi_{\kappa,m}(\vartheta,\varphi)$$

demonstrating that the operator p^2 returns the quantum numbers of the two-component spin-angular functions to their original values, so that the procedure introduces an elementary radial differential operator of the same general form as in non-relativistic atomic theory. In order to make the comparison complete, we need only note that for a given value of orbital angular momentum, $\ell > 0$, two fine-structure components may be formed, $\kappa = \ell$, and $\kappa = -\ell - 1$. In either case,

$$\kappa(\kappa+1) = \ell(\ell+1).$$

The special case where fine-structure is absent, corresponding to $\ell = 0$, also satisfies the equivalence between scalar and two-spinor non-relativistic formulations, since $\kappa = -1$ and $\kappa(\kappa+1) = 0$.

In order to construct Hartree-Fock wavefunctions using this prescription, all contributions to the relativistic Fock matrix which involve the small component amplitudes are simply ignored, since $\psi^S(\boldsymbol{r}) = 0$ for positive-energy solutions of the Dirac equation in the non-relativistic limit. For closed-shell systems, the solutions obtained using this prescription differ from restricted HF spin-orbital wavefunctions only by a unitary transformation. It can be also used as a basis for unrestricted Hartree-Fock wavefunctions for open-shell systems, with all the usual problems of spin contamination which accompany it.

5.2. INTEGRAL ECONOMISATION

We have described elsewhere our approach to the economisation of integral evaluation in direct implementations of Dirac-Hartree-Fock theory [3]. Many of the integral screening algorithms pioneered by Almlöf, Fægri and Korsell [20] may be carried over directly into the relativistic formulation;

this also forms the basis of the DIRAC program [21], whose developers were the first to implement this approach in a relativistic context. Visscher [22] has also made the important observation that the charge density associated with the small-component functions is highly localised, and that the interactions between such charge densities satisfy classical electrostatics.

In BERTHA, these insights have been combined into an operational algorithm, exploiting the localisation of charge characteristic of the small-component parts of Dirac spinors and our experience in constructing atomic solutions of the Dirac equation. Self-consistent field solutions of the Dirac equation are constructed in BERTHA according to the following scheme:

1. Initialise the molecular density matrix by first constructing a basis of atomic orbitals in the the chosen basis set. This involves an atomic structure calculation for each atom, and we have included the ability to select ionic states as the building block if this is appropriate. A minimal basis of atomic spinors is used to form a simple LCAS (linear combination of atomic spinors) molecular wavefunction. The core molecular spinors are only weakly perturbed atomic spinors, so an accurate and robust zero-order approximation to the total electron density is formed, particularly in the neighbourhood of a heavy nucleus. Deficiencies are concentrated in the valence region, and the procedure behaves rather like a non-relativistic calculation from this point onwards.
2. In the early SCF iterations, all interactions involving large-components are included without approximations, using conventional integral screening algorithms and dynamic choice of selection-rejection threshold. All one-centre contributions are treated exactly using Racah algebra methods borrowed from atomic physics. Direct two-centre charge-charge interactions involving small components are calculated using classical electrostatic methods, including the Almlöf J-matrix method [23], multipole expansions, and point-charge models [22]. All multi-centre small-component exchange contributions are neglected, on the assumption that small-component densities are localised to nearly spherical regions centred on the nuclei. This procedure is continued until the convergent region of the iterative algorithm has been entered.
3. Classes of neglected multi-centre small-component integrals are reintroduced exactly, using the integral screening algorithm to eliminated numerically insignificant contributions. The procedure is continued until convergence.

As one passes from one stage to the next, it is important to reinitialise the density matrix, so that the density *difference* matrix reflects changes due to a fixed subspace of the two-electron integral list.

In the early stages, the SCF procedure has all the characteristics of all-electron *ab initio* non-relativistic theory, and involves a similar cost. The one-centre two-electron integrals involve almost no overhead when calculated using methods derived from the decomposition of integrals into radial and angular contributions, so we calculate these at every iteration without bothering to include them in the screening algorithm. The approximate treatment of the multi-centre small-component contributions echoes semi-empirical techniques, in which the neglect of differential overlap is used to reject small contributions on physical grounds. It is important to note, however, that these approximations are used only as a computational tool which allows us to reach the final iterations of a complete Dirac-Hartree-Fock(-Breit) calculation by way of cheap iterations whose cost is essentially non-relativistic. In the end, nothing is neglected. The physical motivation which underlies neglect of integrals on the basis of differential overlap means that the approximations are well-tested and controllable, and that approximations which are intermediate between semi-empirical and *ab initio* limits may be constructed if required, by the simple expedient of switching off the evaluation of selected integral classes, and approximating the interactions using model charge distributions and classical electrostatics.

5.3. CONSTRUCTION OF THE RELATIVISTIC J-MATRIX

In relativistic single particle theories (DHF, DHFB, RDFT) it is convenient to separate the Fock matrix into Coulomb (J-matrix) and exchange-correlation (K-matrix) contributions. Electron density is a real, scalar quantity, and we may eliminate all spinor structure in J-matrix construction.

Following Almlöf [23], we define a scalar Hermite density matrix, $\mathcal{H}$,

$$\mathcal{H}[\alpha\beta;ijk] = \sum_{\mu\nu} \left\{ E^{LL}[\mu\nu;ijk]D^{LL}[\mu\nu] + E^{SS}[\mu\nu;ijk]D^{SS}[\mu\nu] \right\}$$

where the labels $\{\alpha\beta\}$ indicate origin locations and exponents, and sum over $\{\mu\nu\}$ includes all basis functions which share the labels $\{\alpha\beta\}$. The indices $\{ijk\}$ are Hermite polynomial indices. The large-component Hermite set is a subset of the small-component set as a consequence of kinetic balance and the matching of functions. The efficiency of this approach increases with increasing angular momentum (more density is accumulated in the sum over $\{\mu\nu\}$) and with the use of family or universal basis sets (in which many symmetries share the same sets $\{\alpha\beta\}$). Although the form of $\mathcal{H}[\alpha\beta;ijk]$ is the same as the non-relativistic case, the length is longer because $\{ijk\}$ is determined by the small component. For fixed angular momenta, ℓ and ℓ', the increase in length over the non-relativistic value is

$$\frac{(\Lambda+4)(\Lambda+5)}{(\Lambda+1)(\Lambda+2)}$$

where $\Lambda = \ell + \ell'$. For f-functions, this is results in factor of approximately two.

We employ the Hermite density to construct single-particle Coulomb matrix elements in a scalar Hermite basis set

$$[i'j'k';\alpha'\beta'|V_C] = \sum_{\alpha\beta}\sum_{ijk}[i'j'k';\alpha'\beta'|ijk;\alpha\beta]\mathcal{H}[\alpha\beta;ijk]$$

where $[i'j'k';\alpha'\beta'|ijk;\alpha\beta]$ is a two-electron electrostatic integral involving Hermite Gaussian functions. For $\{\alpha\beta\}$ corresponding to f-functions, this between 50 and 100 times faster than calculating two-electron G-spinor integrals to obtain the same quantity. For p-type functions the factor is about five, but the unit cost is much less. The improvements in practice are also geometry-dependent because of sparseness.

In BERTHA, we may readily evaluate the essential ingredients using the generalised Hermite charge-current density matrix

$$\mathcal{H}_q[\alpha\beta;ijk] = \sum_{\mu\nu}\left\{E_q^{T_1T_2}[\mu\nu;ijk]D^{T_1T_2}[\mu\nu] + E_q^{T_3T_4}[\mu\nu;ijk]D^{T_3T_4}[\mu\nu]\right\}$$

where the charge density requires $T_1 = T_2$, $T_3 = T_4$, $T_1 \neq T_3$, $q = 0$, and the components of the current density are obtained from $T_1 \neq T_2$, $T_3 \neq T_4$, $T_1 = T_4$, $q = \{x, y, z\}$. In terms of the Hermite Gaussian functions, $H[\alpha\beta;ijk;\boldsymbol{r}]$, the required quantities are

$$\begin{aligned}
\varrho(\boldsymbol{r}) &= \sum_{\alpha\beta}\sum_{ijk} H[\alpha\beta;ijk;\boldsymbol{r}]\mathcal{H}_0[\alpha\beta;ijk] \\
\frac{\partial\varrho(\boldsymbol{r})}{\partial x} &= -\sum_{\alpha\beta}\sum_{ijk} H[\alpha\beta;i+1,j,k;\boldsymbol{r}]\mathcal{H}_0[\alpha\beta;ijk] \\
j_q(\boldsymbol{r}) &= \sum_{\alpha\beta}\sum_{ijk} H[\alpha\beta;ijk;\boldsymbol{r}]\mathcal{H}_q[\alpha\beta;ijk]
\end{aligned}$$

All spinor structure has been absorbed in the modified densities, which are real, scalar quantities.

J-matrix elements in the G-spinor basis set are then constructed by repeated use of the scalar Hermite Gaussian integrals

$$J_{\mu'\nu'}^{TT} = \sum_{i'j'k'} E^{TT}[\mu'\nu';i'j'k'][i'j'k';\alpha'\beta'|V_C]$$

where the spinor elements $\{\mu'\nu'\}$ are spanned by the scalar labels (exponents and origins) $\{\alpha'\beta'\}$. This involves the cost of a nuclear attraction integral.

6. Applications

6.1. A HYDROGENIC MODEL PROBLEM

Some time ago we considered a simple model problem in relativistic perturbation theory, in which a hydrogenic ion with point nuclear charge Z is perturbed by an additional point nuclear charge, Z' [12]. When $Z' = 1$, this is a simple model for nuclear β-decay, and we shall assume that $Z' = 1$ throughout this section.

The energy levels of the perturbed and unperturbed systems may be determined exactly, as may the components of the perturbation expansion which describes the perturbed system,

$$\begin{aligned} E(Z+Z') &= \varepsilon_0 + \varepsilon_1 Z' + \varepsilon_2 Z'^2 + \cdots \\ &= E(z)|_{z=Z} + \left.\frac{\partial E(z)}{\partial z}\right|_{z=Z} Z' + \frac{1}{2}\left.\frac{\partial^2 E(z)}{\partial z^2}\right|_{z=Z} Z'^2 + \cdots \end{aligned}$$

For the $1s_{1/2}$ hydrogenic ground state, we find that the components of the perturbation expansion of the energy to second-order in Z' are given by

$$\begin{aligned} \varepsilon_0 &= (\gamma - 1)c^2 \\ \varepsilon_1 &= -\frac{Z}{\gamma} \\ \varepsilon_2 &= -\frac{1}{2}\left(\frac{1}{\gamma} + \frac{Z^2}{\gamma^3 c^2}\right) \end{aligned}$$

The calculations of the energy components of the model problem as a function of nuclear charge, Z, presented in Table 2 were performed using L-spinor basis sets, the dimensions of which were enlarged until the calculated value of ε_2 matched the exact value to the number of figures quoted. Both the single-particle eigenvalue, ε_0, and the first-order perturbation, ε_2 may be calculated exactly by the choice of the L-spinor exponent $\lambda = Z$, defined in Eq. (17). The second-order contributions, however, may be divided into contributions involving summations over positive- and negative-energy states, $\varepsilon_2^{(+)}$ and $\varepsilon_2^{(-)}$, respectively, as defined by Eq. (12).

The non-relativistic limit of ε_2 may be determined by setting $c \to \infty$ and $\gamma \to 1$, yielding $\varepsilon_2^{NR} = -1/2$, which is consistent with the result obtained directly from the Schrödinger equation; all non-relativistic energy components at third- and higher-order vanish identically. Expanding γ in a series in $(Z/c)^2$, we find that

$$\varepsilon_2 \simeq -\frac{1}{2} - \frac{3}{4}\frac{Z^2}{c^2} + O\left(\frac{Z^3}{c^3}\right).$$

TABLE 2. Calculations of the zero-order spinor energy, ε_0, positive- and negative- energy contributions, $\varepsilon_2^{(+)}$ and $\varepsilon_2^{(-)}$, respectively, to the second-order correction, ε_2, and the quasi-relativistic approximation, $\varepsilon_2^{qr} \simeq -1/2 - 3Z^2/4c^2$, as a function of the nuclear charge Z. The perturbation corresponds to $Z' = 1$ in all cases, and all energies are in atomic units. Numbers in parentheses indicate powers of ten.

Z	ε_0	$\varepsilon_2^{(+)}$	$\varepsilon_2^{(-)}$	$\varepsilon_2^{(+)} + \varepsilon_2^{(-)}$	$-\frac{1}{2} - \frac{3Z^2}{4c^2}$
10	-50.066742	-0.504126	1.05223(-4)	-0.504021	-0.503994
20	-201.076523	-0.517054	6.42479(-4)	-0.516412	-0.515975
30	-455.524907	-0.539992	1.76627(-3)	-0.538225	-0.535945
40	-817.807498	-0.575023	3.56735(-3)	-0.571455	-0.563902
50	-1294.626156	-0.625655	6.16022(-3)	-0.619495	-0.599846
60	-1895.682356	-0.697776	9.73442(-3)	-0.688042	-0.643779
70	-2634.846565	-0.801560	1.46176(-2)	-0.786943	-0.695699
80	-3532.192151	-0.955622	2.13944(-2)	-0.934228	-0.755607
90	-4617.757654	-1.197134	3.11666(-2)	-1.165967	-0.823502

We observe in Table 2 that the approximate formula reproduces the complete second order shift for small Z, but that the accuracy of the lowest-order expansion in powers of Z/c deteriorates as Z is increased. For large Z, the approximate formula is completely inadequate, and $\varepsilon_2^{(-)}$ varies approximately as $(Z/c)^3$, making a significant contribution to the total second-order energy shift.

6.2. INNER-SHELL PROCESSES

X-ray emission is a process in which single-particle relativistic effects dominate correlation effects in the calculation of transition energies, particularly for large Z. The relativistic contribution to the inner-shell ionisation energy, E_R, of an atom or molecule is given to a good approximation by

$$E_R = \{E_{DHF}[X] - E_{HF}[X]\} - \{E_{DHF} - E_{HF}\}$$

where E_{DHF} is the Dirac-Hartree-Fock energy of the neutral atom or molecule, E_{HF} is the Hartree-Fock energy, and [X] denotes a specified hole state.

Table 3 presents calculations of E_R and its constituent parts for selected atoms, X, and core-hole state ions, X[1s], while Table 4 extends the treatment to diatomic oxides, XO, and core-hole state molecular ions, X[1s]O. The dominance of the nuclear field is evident, since the effect of the chemical environment is rather small in all cases, and the value of E_R for the atomic system is a good first approximation to the corresponding value in

TABLE 3. Self-consistent field electronic energies of the elements, X, in Hartree atomic units. The notation [1s] denotes an atomic 1s hole localized at nucleus X; energies without this label refer to the ground electronic configuration. E_{DHF} denotes the Dirac-Hartree-Fock energy, and E_{HF} is the non-relativistic Hartree-Fock energy (in a.u.).

X	E_{DHF}	E_{HF}	$E_{DHF}[1s]$	$E_{HF}[1s]$	E_R
Be	-14.57589170	-14.57302261	-10.04264318	-10.04050268	0.00073
C	-37.67604073	-37.65969455	-26.75871771	-26.74712643	0.00475
O	-74.82498609	-74.76918802	-54.76742858	-54.72826630	0.01663
Mg	-199.9350663	-199.6145489	-151.7311565	-151.5021466	0.09150
Si	-289.4613374	-288.8344384	-221.4707322	-221.0195009	0.17567
Ca	-679.7101599	-676.7572540	-530.7855994	-528.6200637	0.78737
Ge	-2097.474048	-2075.331574	-1688.294489	-1671.937402	5.78604
Sr	-3178.079985	-3131.525998	-2584.449674	-2549.824143	11.92846
Sn	-6176.128513	-6022.842857	-5099.251964	-4983.968704	38.00240

TABLE 4. Self-consistent field energies of diatomic oxides, XO, in Hartree atomic units. The notation [X] denotes an atomic 1s hole localized at nucleus X; energies without this label refer to the ground electronic configuration. E_{DHF} denotes the Dirac-Hartree-Fock energy, and E_{HF} is the non-relativistic Hartree-Fock energy (in a.u.).

XO	E_{DHF}	E_{HF}	$E_{DHF}[X]$	$E_{HF}[X]$	E_R	λ
BeO	-89.5082	-89.4500	-84.9987	-84.9413	0.0008	0.67
CO	-112.8558	-112.7841	-101.9313	-101.8644	0.0168	0.830
MgO	-274.7558	-274.3809	-226.4938	-226.2103	0.0914	1.185
SiO	-364.5198	-363.8392	-296.4965	-295.9915	0.1758	1.270
CaO	-754.5440	-751.5430	-605.6562	-603.4438	0.7886	1.490
GeO	-2172.0787	-2150.0523	-1762.9201	-1746.6415	5.7417	1.708
SrO	-3252.7843	-3206.5430	-2659.3187	-2624.7110	11.9338	1.736
SnO	-6250.5354	-6097.6321	-5173.7115	-5058.7629	37.9547	1.860

the molecule. The atomic calculations were performed with both BERTHA [3] and GRASP [24], and found to be in agreement for the large basis sets employed here. The consistency of the results for E_R when atomic and molecular values are compared serves as a useful cross-check on our implementation of open-shell Dirac-Hartree-Fock and Hartree-Fock theory in a spinor basis set. It is likely that the basis set descriptions of the bonding region of the molecules and molecular ions would benefit from the addition of additional polarisation functions but that scarcely matters in the calculation of X-ray ionisation energies, since the basis set truncation errors evidently cancel to a good approximation when calculating the energy differences involving core-hole states.

For emission involving a 1s hole, it is convenient to adopt a relativistic hydrogenic model of mean-field screening in which

$$\begin{aligned} E_R &= c^2\sqrt{1-\frac{(Z-\lambda)^2}{c^2}} - c^2 + \frac{(Z-\lambda)^2}{2} \\ &\simeq \frac{(Z-\lambda)^4}{8c^2} + \frac{(Z-\lambda)^6}{16c^4} + \cdots \end{aligned}$$

where λ is a screening parameter.

The large value of the effective value of the screening constant, λ for high Z is indicative of the significant spatial contraction of the inner core spinor following ionisation.

6.3. RELATIVISTIC MOMENTUM-SPACE DISTRIBUTIONS

In the relativistic plane-wave impulse approximation, the cross-section for $(e-2e)$ ejection from ψ_i, σ, is defined by the differential relation [25]

$$\frac{d^5\sigma}{d\Omega_A d\Omega_B dE_A} = KS_i \int \psi_i^\dagger(\boldsymbol{p})\psi_i(\boldsymbol{p})d\Omega_p$$

where A labels the scattered electron, B the ejected electron. K and S_i are kinematic factors characteristic of the experiment.

The transformation of a G-spinor to a p-space representation is facilitated by the orthonormality of the spin-angular functions in both the r-space and p-space angular variables, $\chi(\vartheta_r, \varphi_r)$ and $\chi(\vartheta_p, \varphi_p)$, respectively. The transformation of a radial function, $f^T_{\kappa m}(r)$ to p-space form is achieved by evaluating spatial integrals over spherical wave amplitudes

$$f^T_{\kappa m}(p) = \sqrt{\frac{E \pm c^2}{\pi E}} \int_0^\infty f^T_{\kappa m}(r) j_{\ell_T}(pr)\, dr$$

where ℓ_T is the angular momentum associated with the two-spinor labelled by κ, $j_{\ell_T}(pr)$ is a spherical Bessel function, $E = +\sqrt{c^2p^2 + c^4}$, and the component labels take the values $T = L$ or $T = S$. The upper sign is chosen for $T = L$, and the lower sign for $T = S$. The radial integration may be performed analytically, for any $p \geq 0$. In the low-momentum and non-relativistic limits, $\sqrt{(E+c^2)/\pi E} \rightarrow \sqrt{2/\pi}$, and the small component p-space amplitudes vanish.

Using results from Racah algebra, we reduce two-centre G-spinor r-space overlap densities to angle-averaged p-space densities of the form

$$\begin{aligned}(\kappa, m; \kappa', m'; p)_{TT} &= f^T_{\kappa m}(p) f^T_{\kappa' m'}(p) \\ &\quad \times \sum_L (-1)^Q j_L(pR) C_L^{-Q}(\theta_R \varphi_R) \\ &\quad \times \langle \kappa, m | C_L^Q(\theta_p, \varphi_p) | \kappa', m' \rangle\end{aligned}$$

where $T = \{L, S\}$, $f^T_{\kappa m}(p)$ is the p-space transform of $f^T_{\kappa m}(r)$, $R = |\boldsymbol{R}_A - \boldsymbol{R}_B|$, $\boldsymbol{R} = (R, \theta_R, \varphi_R)$, $Q = m - m'$, and $\langle \kappa, m | C_L^Q(\theta_p, \varphi_p) | \kappa', m' \rangle$ is a two-spinor spherical tensor p-space matrix element. Figure 2 presents $p^2 \varrho(p)$ for the 3s orbital of Xe. The non-relativistic density shows a marked shift to smaller p, corresponding to a spatial expansion to larger r. The difference between the relativistic and non-relativistic profiles is readily accessible to experimental discrimination.

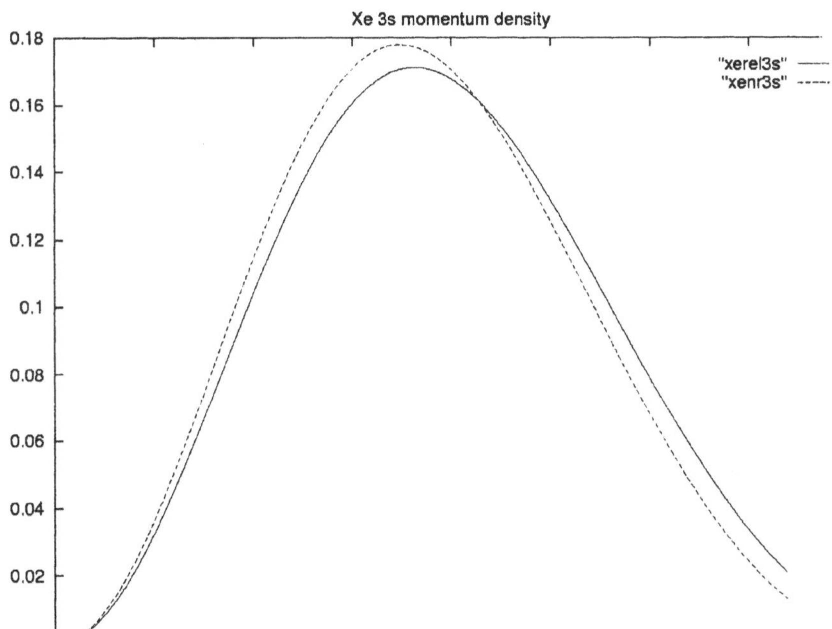

Figure 2. Free-particle momentum transform for the 3s level of neutral Xe. The plots are of angle-averaged values of $p^2 \varrho(p)$ in the Schrödinger and Dirac representations. The horizontal axis is in atomic units of momentum, and the vertical axis denotes the relative density of the non-relativistic and relativistic approximations.

6.4. CHEMICAL EFFECTS OF THE BREIT INTERACTION

The Breit interaction, b_{ij} is the lowest-order quantum electrodynamic correction to the electron-electron interaction

$$b_{ij} = b_{ij}^1 + b_{ij}^2 = -\frac{\boldsymbol{\alpha}_i \cdot \boldsymbol{\alpha}_j}{r_{ij}} + \frac{1}{2r_{ij}} \left[(\boldsymbol{\alpha}_i \cdot \boldsymbol{\alpha}_j) - \frac{(\boldsymbol{\alpha}_i \cdot \boldsymbol{r}_{ij})(\boldsymbol{\alpha}_j \cdot \boldsymbol{r}_{ij})}{r_{ij}^2} \right].$$

It comprises the Gaunt interaction, b_{ij}^1, and a correction characteristic of the choice of Coulomb gauge, b_{ij}^2.

The calculation of G-spinor integrals involving the Breit interaction necessarily involves more cost than a Coulomb integral. In Table 5 we see that the effect of the Breit interaction is almost 2 a.u. for the core molecular orbital corresponding to the 1s spinor of atomic silver, but that the effect on the outermost molecular orbital falls to 0.1 millihartree. For the valence molecular orbitals the relativistic corrections including Coulomb interactions is, however, 6 millihartree, indicating that the Dirac-Coulomb hamiltonian represents the most important relativistic effects on chemical properties, to which magnetic corrections are small.

If one considers relativistic correlation effects, the magnetic corrections on pair correlations involving core orbitals are the largest relativistic many-body effects, as we have considered in detail for the argon atom [26]. But from a chemical perspective, Breit interaction effects are localised in core orbitals, which implies that their impact on chemical properties such as bond lengths and vibrational frequencies will be comparatively small.

The localised nature of the Breit interaction is reflected in the results of Table 6 where it is revealed that the total Breit interaction correction to the electronic energy of AgCl is well-approximated by the sum of ionic contributions from Ag^+ and Cl^-. Since both Ag^+ and Cl^- are closed-shell ions, the Breit interaction enters into their electronic structures only in contributions of exchange type, since the direct interaction vanishes identically for closed-shell systems. The concentration of small-component density near nuclei effectively extinguishes all exchange contributions involving small components of either the Coulomb or exchange interactions except those involving amplitudes centred on a single nucleus. In order to generate a significant multi-centre matrix element of the Breit interaction, there would need to be a direct current-current interaction involving delocalised *unpaired* electrons. Long-range multi-centre Breit interaction corrections can be expected to be of importance in the electronic band structures of metals, and would be amplified for heavy elements. In an isolated molecule, however, one requires paramagnetic interactions involving two heavy centres before a direct matrix element of the Breit interaction can contribute significantly to its valence electronic structure.

TABLE 5. Orbital energies (in a.u.) for AgCl using a Hartree-Fock (HF), Dirac-Coulomb Hamiltonian (DHF) and Dirac-Coulomb-Breit Hamiltonian (DHFB).

MO	HF	DHF	DHFB
1	-913.86228	-943.12139	-941.23497
2	-134.90975	-142.02539	-141.85422
3	-125.21282	-131.53099	-131.23901
4	-125.21282	-125.09647	-124.89743
5	-125.21130	-125.09533	-124.89629
6	-104.68603	-105.10631	-105.03931
7	-25.94690	-27.41891	-27.39353
8	-21.97600	-23.20375	-23.15903
9	-21.97600	-22.03354	-22.00567
10	-21.97227	-22.03116	-22.00327
11	-14.70534	-14.65160	-14.63783
12	-14.70534	-14.64850	-14.63474
13	-14.70242	-14.41927	-14.41295
14	-14.70242	-14.41690	-14.41058
15	-14.70076	-14.41504	-14.40873
16	-10.41507	-10.49815	-10.49503
17	-7.87199	-7.92786	-7.92225
18	-7.86974	-7.86612	-7.86310
19	-7.86974	-7.86463	-7.86161
20	-4.03428	-4.31193	-4.30721
21	-2.71349	-2.91536	-2.90823
22	-2.70985	-2.72195	-2.71776
23	-2.70985	-2.71876	-2.71458
24	-0.95179	-0.96779	-0.96758
25	-0.58035	-0.57416	-0.57313
26	-0.57938	-0.56708	-0.56604
27	-0.57938	-0.55210	-0.55190
28	-0.57225	-0.54809	-0.54778
29	-0.57225	-0.54061	-0.54037
30	-0.34798	-0.35657	-0.35623
31	-0.34423	-0.35351	-0.35330
32	-0.34423	-0.35041	-0.35031

6.5. RELATIVISTIC MANY-BODY PERTURBATION THEORY

In Table 7 we present Dirac-Hartree-Fock and second-order many-body perturbation theory results for HCl, using uncontracted correlation consistent

TABLE 6. Total electronic energies of AgCl in a.u. E(HF) denotes the Hartree-Fock energy, E(DHF) is the Dirac-Hartree-Fock energy, E(DHF)+E(B) includes the first-order Breit interaction energy as a perturbation, and E(DHFB) includes the Breit interaction self-consistently.

E(HF)	E(DHF)	E(B)	E(DHF)+E(B)	E(DHFB)
-5657.00309	-5775.07031	3.67885	-5771.39146	-5771.39440

basis sets. In addition, equilibrium bond lengths and vibrational frequencies are calculated, and compared with results obtained by Visscher *et al.* [27]. The calculations were performed using our own direct implementation of many-body-perturbation theory, based on the non-relativistic formulation of White and Head-Gordon [28]. It is similar in general content to the approach adopted in the program DIRAC, except that our implementation benefits from the shorter integral list afforded by the choice of a G-spinor basis set.

TABLE 7. A comparison of vibrational frequencies and equilibrium bond lengths for $^1H^{35}Cl$ calculated with a variety of basis sets. The results labelled † are from Visscher *et al.*, 1996.

	cc-pVDZ(+)		cc-pVTZ(+)	
Method	d_e $(a.u.)$	ν (cm^{-1})	d_e $(a.u.)$	ν (cm^{-1})
DHF	2.411	3123	2.395	3135
DHF†	2.413	3124	2.394	3132
DHF+MBPT2	2.428	3030	2.406	3041
DHF+MBPT2†	2.434	3019	2.409	3041
Experiment	2.409	2991		

The agreement with the results in [27] is excellent for the cc-pVTZ(+) basis and good for the cc-pVDZ(+) basis. The label (+) denotes that basis set has been used in an *uncontracted* form, which accounts entirely for the apparent differences in the results (Visscher *et al.* adopted the original contraction scheme). We find much less variation in the cc-pVDZ(+) and cc-pVTZ(+) results as a consequence of the greater degree of variational freedom in an uncontracted basis set but the contraction plays only a minor role in any event.

6.6. RELATIVISTIC DENSITY FUNCTION THEORIES

In the formulation of Engel, Keller and Dreizler [29], the Feynman-gauge representation of the relativistic exchange-correlation energy, $E_{xc}[j^{\mu}]$, is a functional of the four-current, $j^{\mu} = (\varrho, \boldsymbol{j})$,

$$E_{xc}[j^{\mu}] = F[j^{\mu}] - T_s[j^{\mu}] - \frac{1}{2}\int\int \frac{j^{\nu}(\boldsymbol{x})j_{\nu}(\boldsymbol{y})}{|\boldsymbol{x}-\boldsymbol{y}|} d^3x\, d^3y$$

In the absence of external magnetic fields and with the neglect of radiative corrections requiring renormalization, Dirac-Kohn-Sham equations are obtained

$$\left\{c\boldsymbol{\alpha}\cdot\boldsymbol{p} + \beta c^2 + v_{ext}(\boldsymbol{r}) + v_c(\boldsymbol{r}) + v_{xc}(\boldsymbol{r})\right\}\psi_k(\boldsymbol{r}) = E_k\,\psi_k(\boldsymbol{r})$$

with eigenvalue, E_k. The external field potential, $v_{ext}(\boldsymbol{r})$, and the Coulomb potential between electrons, $v_c(\boldsymbol{r})$, are classical scalar quantities. The cost of calculating matrix elements of $v_c(\boldsymbol{r})$ may be reduced to an effectively non-relativistic level using $\boldsymbol{J}$-matrix methods because $\varrho(\boldsymbol{r})$ is a scalar, not a spinor.

The precise form of the exchange-correlation potential, $v_{xc}(\boldsymbol{r})$, is unknown, but it is defined by the functional derivative

$$v_{xc}(\boldsymbol{r}) = \frac{\delta E_{xc}(\varrho)}{\delta\varrho}$$

We assume published non-relativistic charge-density functionals of $(\varrho, |\boldsymbol{\nabla}\varrho|)$ in the first instance. However, we note that we wish also to investigate charge-current functionals which include $\boldsymbol{j}$.

Multicentre integrals over functionals of $\varrho(\boldsymbol{r})$ and $|\boldsymbol{\nabla}\varrho(\boldsymbol{r})|$ are evaluated using methods described by Becke [30]. The molecule is divided into nuclear-centred cells, and Voronoi polyhedra constructed which extinguish the cusp in $\varrho(\boldsymbol{r})$ at all other nuclei. The contribution from each cell is evaluated by numerical quadrature in a spherical polar coordinate system. We employ a radial mapping

$$r = \xi\frac{(1+x)}{(1-x)}, \quad -1 \le x \le 1$$

where ξ is chosen according to an empirical prescription linked to the atomic radius. The integration over x is with N_r-point Gauss-Legendre or Gauss-Chebyshev quadrature. The angular integrals are performed using tabulated angular weights, $\{w_i\}$ and abscissae, $\{\theta_i, \varphi_i\}$ in the Lebedev solid angle formulae. The solid angle integral centred at each nucleus is performed according to

$$\int f(\theta,\varphi)\, d\Omega = 4\pi\sum_{i=1}^{N_L} w_i f(\theta_i, \varphi_i)$$

where $N_L = \{50, 86, 110, 146, 194, 302, 590, 770\}$. In order to demonstrate that this approach is successful, we evaluate

$$\begin{aligned} F_1 &= \int \varrho(\boldsymbol{r})\, d\boldsymbol{r} \\ F_2 &= \int \varrho(\boldsymbol{r})^{4/3}\, d\boldsymbol{r} \\ F_3 &= \int \varrho(\boldsymbol{r})^{4/3} \frac{x^2}{(1 + 0.004\, x^2)}\, d\boldsymbol{r} \end{aligned}$$

where $x = |\boldsymbol{\nabla} \varrho|/\varrho^{4/3}$.

TABLE 8. Sample calculations of F_i using a Dirac-Hartree-Fock wavefunction for CO. The atomic basis set for both C and O was 13s8p. The exact value of F_1 is 14.0. No use of axial symmetry has been made in the angular integrations. The number of nucleus-centred quadrature points is indicated as (N_r, N_L).

	(32, 50)	(48, 50)	(48, 110)	(64, 194)
F_1	14.0001485	14.0001877	13.9999835	13.9999978
F_2	16.2992512	16.2993715	16.2992262	16.2992489
F_3	329.327409	329.356448	329.320408	329.289413

The essential conclusion conveyed by the results in Table 8 is that we can adapt quantum chemical DFT methods to relativistic DFT with little change. The reason for this is that the construction of densities eliminates the spinor structure in favour of scalar quantities at all stages of the calculation, except in the final diagonalisation of the matrix representations of the Dirac-Kohn-Sham equations. Our conclusion based on this observation is that the use of quasi-relativistic methods in density functional theories conveys little or no advantage, and one may just as well enjoy the convenience afforded by a four-component representation if a relativistic density functional theory is required. This point of view has already been adopted by Liu and his collaborators [31] in the construction of the Beijing Density Functional program, BDF, a development of the non-relativistic ADF program from Amsterdam. Compared to the development of efficient methods for the evaluation of multi-centre G-spinor integrals, the extension of our techniques to relativistic DFT has proved rather straightforward.

7. Conclusion

The distinguishing feature of our formulation of the relativistic electronic structure problem is the use of explicit representations of the charge-current densities in a G-spinor basis set. This approach hides most of the relativistic features of the problem in a few key numerical procedures, so that subsequent steps, such as the evaluation of electric and magnetic interaction matrix elements, and the calculation of many-body corrections, is almost identical to the practices of non-relativistic quantum chemistry. For a given set of Gaussian parameters, the G-spinor set which forms our primary computational basis offers the atomic basis of irreducible dimension, and the kinetic balance prescription transfers the advantages of a non-relativistic "family basis" set to the evaluation of relativistic integrals and matrix elements. A link has been established with non-relativistic methods, through the explicit use of the identity $(\boldsymbol{\sigma} \cdot \boldsymbol{p})(\boldsymbol{\sigma} \cdot \boldsymbol{p}) = p^2\boldsymbol{I}$, so that we can generate spin-orbitals directly in a G-spinor basis, and extract the exact non-relativistic limit of our calculations as a special case of our relativistic formulation.

In dealing with relativistic many-body problems, and with electric and magnetic properties, some appreciation of the role of how the negative-energy states are handled in relativistic QED seems to be essential if fundamental errors are to be avoided. Given a discrete basis of four-spinors, direct implementation of the rules of relativistic QED appears to be simpler than the incorporation of relativistic corrections to non-relativistic spin-orbitals and the prescriptions of non-relativistic QED. Of course there are always questions of personal taste and the computational efficiency of particular computer programs to consider, and others may favour the non-relativistic formulation for a combination of these reasons. But our choice is to adopt the Dirac equation, relativistic QED, four-spinors, the charge-current representations embodied in BERTHA, and all the simplifying features that go with them. Given the breadth of territory which one can cover when starting from this point of view, it would be misleading to suggest that this will always be the most efficient approach from a computational point of view, particularly when a non-relativistic spin-independent description suffices: our non-relativistic formulation, for example, makes no use of the separation of spatial and spin coordinates, but it proves to be convenient for our purposes. The heavy elements, however, bring with them both computational and physical complexity, and a four-component approach is appropriate in this case, both to simplify the treatment of many competing relativistic and electrodynamic effects, and to offer a greater insight into how they operate at the molecular level. We remain convinced that our vigorous defence of this position at the 1997 meeting of QSCP [3] was justified, and look forward to participating in future developments in relativistic quantum chemistry.

Acknowledgements

I wish to thank Ian Grant, Haakon Skaane, and Stephen Wilson for many years of fruitful and happy collaboration, Knut Fægri, for his unfailing patience in answering my often ill-conceived questions on many subjects, and Jon Lærdahl for insightful comments on the manuscript. I am grateful to the organisers of QSCP IV for their support, and for organising this stimulating meeting. Financial support is gratefully acknowledged from Prof. Frank Larkins, the Australian Research Council, and the British Engineering and Physical Science Research Council. I wish also to thank Professor Ken Ghiggino and the members of the School of Chemistry at the University of Melbourne for their hospitality.

References

1. R. P Feynman, Rev. Mod. Phys., **20**, 367 (1948); R. P. Feynman, *Phys. Rev.*, **76**, 769 (1949).
2. H. Kleinert, *Path integrals in quantum mechanics and polymer science*, 2nd edition, World Scientific, Singapore (1995).
3. H. M. Quiney, H. Skaane, I. P. Grant, *Adv. Quant. Chem.*, **32**, 1 (1998).
4. P. A. M. Dirac, *Proc. Royal Soc. (London) A*, **117**, 610 (1928).
5. H. M. Quiney and I. P. Grant, *J. Phys. B: At. Mol. Opt. Phys.*, **27**, L299 (1992).
6. H. A. Bethe and E. E. Salpeter, *Quantum mechanics of one- and two-electron systems*, Springer-Verlag, Göttingen, Berlin, Heidelberg (1957).
7. J. Sucher, *Physica Scripta*, **36**, 271 (1987).
8. R. P. Feynman, *Phys. Rev.*, **76**, 749 (1949).
9. W. H. Furry, *Phys. Rev.*, **81**, 115 (1951).
10. M. Gell-Mann, and F. Low, *Phys. Rev.*, **84**, 350 (1951).
11. B. Swirles, *Proc. Royal Soc. (London) A*, **152**, 625 (1935).
12. H. M. Quiney, I. P. Grant and S. Wilson, *J. Phys. B: At. Mol. Opt. Phys.*, (1985).
13. Y. Ishikawa, K. Koc, *this volume.*
14. G. E. Brown and D. G. Ravenhall, *Proc. Royal. Soc. (London) A*, **208**, 552-559 (1951).
15. L. N. Labzowsky, *Sov. Phys. JETP*, **32**, 94 (1971).
16. J. Sapirstein, *Physica Scripta*, **36** 801 (1987).
17. I. P. Grant and H. M. Quiney, *Adv. At. Mol. Phys.*, **23**, 37-86 (1988).
18. K. G. Dyall, I. P. Grant, S. Wilson, *J. Phys. B: At. Mol. Opt. Phys.*, **17**, L45(1984); K. G. Dyall, I. P. Grant, S. Wilson, *J. Phys. B: At. Mol. Opt. Phys.*, **17**, 493 (1984); K. G. Dyall, I. P. Grant, S. Wilson, *J. Phys. B: At. Mol. Opt. Phys.*, **17**, 1201 (1984).
19. V. R. Saunders, *Methods of computational molecular physics*, (eds. G. H. F. Diercksen and S. Wilson), p. 1, Reidel, Dordrecht (1983).
20. J. Almlöf, K. Faegri jr, and K. Korsell, *J. Comput. Chem.*, **3**, 385 (1982).
21. T. Saue, K. Fægri jr, T. Helgaker and O. Gropen, *Mol. Phys.*, **91**, 937 (1997).
22. L. Visscher, *Theor. Chem. Acc.*, **98**, 68 (1997).
23. M. Challacombe, E. Schwegler and J. Almlöf, *J. Chem. Phys.*, **104**, 4685 (1996); G. R. Ahmadi and J. Almlöf, *Chem. Phys. Lett.*, **246**, 364 (1996); C. A. White and M. Head-Gordon, *J. Chem. Phys.* , **104**, 2620 (1996).

24. K. G. Dyall, I. P. Grant, C. T. Johnson, and F. A. Parpia, *Comp. Phys. Commun.*, **55**, 425 (1989).
25. I. E. Macarthy and E. Weigold, *Rep. Prog. Phys.*, **54**, 789 (1991).
26. H. M. Quiney, I. P. Grant, and S. Wilson, *J. Phys. B: At. Mol. Opt. Phys.*, **23**, L271 (1990).
27. L. Visscher, J. Styszynski, and W. C. Nieuwpoort, *J. Chem. Phys.*, **105**, 1987 (1996).
28. C. A. White and M. Head-Gordon, *J. Chem. Phys.* , **105**, 5061 (1996).
29. E. Engel and R. M. Dreizler, *J. Comput. Chem.*, **20**, 31 (1999); E. Engel, S. Keller and R. M. Dreizler, *Phys. Rev.*, **A53**, 1367 (1996).
30. A. D. Becke, *J. Chem. Phys.*, **88**, 2547 (1988).
31. W. Liu, G. Hong, D. Dai, and M. Dolg, *Theor. Chem. Acc.*, **96**, 75 (1997).

VARIATIONAL PRINCIPLE IN THE DIRAC THEORY: SPURIOUS SOLUTIONS, UNEXPECTED EXTREMA AND OTHER TRAPS

MONIKA STANKE AND JACEK KARWOWSKI
Instytut Fizyki, Uniwersytet Mikołaja Kopernika
ul. Grudziądzka 5, 87-100 Toruń, Poland

Abstract. The dependence of the Rayleigh-Ritz Dirac energy of several states of one-electron atoms on nonlinear parameters is analyzed in detail. It is shown that, if the kinetic balance condition is not fulfilled by the trial functions, then the energy hypersurfaces in the space of the nonlinear parameters (the exponents of the basis functions) contain multiple extrema which may easily be taken for solutions of the corresponding variational minimax problems. Also the problem of spurious solutions is addressed. In particular we demonstrate that in different ranges of the nonlinear parameters the same eigenvalue of the Hamiltonian matrix may approximate different states of the atom. Close analogy between the Dirac and the corresponding Lévy-Leblond variational problems leads us to the conclusion that in relativistic variational calculations establishing correct relations between the components of the wavefunction is, in general, more important and more difficult to maintain than the eliminating of the influence of the negative-energy solutions on the bound-state wavefunctions.

1. Introduction

The Rayleigh-Ritz variational procedure, when applied to solving the Schrödinger eigenvalue problem, leads to rather simple and well established computational procedures. Its generalization to the case of the relativistic Dirac-Coulomb eigenvalue equation, although it has resulted in many successfull implementations (a complete set of references may be found in ref. [1]), is far from being trivial and there are still many questions to be addressed. The main source of the difficulty is the unboundedness from below of the Dirac Hamiltonian and the multi-component character of the wavefunction.

J. Maruani et al. (eds.), New Trends in Quantum Systems in Chemistry and Physics, Volume 1, 175–190.

A proper construction of the basis sets, so that they fulfil boundary conditions selected to make the variational space orthogonal to the negative continuum, gives probably the most powerful tool allowing to avoid instabilities and the variational collapse [2-11]. The importance of retaining certain relations between the components of the wavefunction has led to the so called kinetic balance condition [2-6]. Combining these two kinds of restrictions, Grant and his coworkers formulated a set of sufficient conditions to be fulfilled by a basis set in order to avoid the variational collapse [2-4].

The presence of spurious solutions in the algebraic Dirac-Coulomb problem stimulated a series of works aimed at their elimination [10-12]. As a consequence of these studies, a variational approach to the Dirac-Coulomb eigenvalue problem in which electron and positron states are simultaneously included and which is free of the spurious solutions and avoids variational collapse has been formulated [11].

The question how to control the behaviour of the variational energy with not too strictly constrained variational trial functions motivated the formulation of a number of minimax principles [13-17]. In our recent paper [18], hereafter referred to as I, we analyzed the behaviour of the variational Dirac energies of the ground-states of one-electron atoms as functions of the basis function nonlinear parameters. In particular, we studied the influence of the non-exact fulfilment of either the boundary conditions or the relationship between the large and small components of the trial functions on the results of the variational calculations. As a consequence, we have found several examples of violation of some of the minimax principles. As a side-effect of this study we have obtained an infinite number of variational solutions for which the variational energies are strictly degenerate with the exact ground-state energies but the corresponding wave-functions, though square-integrable, are completely wrong (they may even be orthogonal to the exact ones). We have also demonstrated that the same applies to all one-electron maximum angular momentum states ($1s_{1/2}$, $2p_{3/2}$, $3d_{5/2}$, $\ldots$).

In the present work we extend our previous study to excited states. In particular, we analyze the structure of the energy hypersurfaces in the space of the nonlinear parameters showing the existence of multiple extrema and spurious solutions. We demonstrate that, depending upon the choice of the nonlinear parameters, the same eigenvalue of the energy matrix may approximate energies of different states of the atom. Besides, we found a close analogy between the Dirac and the corresponding non-relativistic Lévy-Leblond [19] variational problems. This has led us to the conclusion that in relativistic variational calculations establishing the correct relations between the components of the wavefunction is, in general, more important (and also more difficult to maintain) than eliminating the influence of the negative-energy solutions on the bound-state wavefunctions. In this paper, we concentrate on discussing the basis sets which are being used mainly (but not only) in the context of studies on implementations of mini max principles [13], [20] and which do not fulfil the kinetic balance condition. One should expect that these kinds of trial functions may lead to some pathological behaviour of the corresponding variational problems. In particular, we do not discuss the basis sets which are known to be "safe" such as the one designed by the Oxford Group [2], [3] or the one fulfilling the condition of Goldman [11].

Atomic units are used in this paper. In some cases, in order to make equations more transparent, the electron mass m is written explicitly. The velocity of light is taken to be $c = 137.0359895$.

2. Formulation of the problem

Similarly as in I, we are concerned with the Dirac equation for an electron in the field of a stationary potential $V = -Z/r$:

$$\begin{pmatrix} -\frac{Z}{r} - E & c\left(\frac{\kappa}{r} - \frac{d}{dr}\right) \\ c\left(\frac{\kappa}{r} + \frac{d}{dr}\right) & -\frac{Z}{r} - 2mc^2 - E \end{pmatrix} \begin{pmatrix} \Phi_{\mathcal{D}}^L(r) \\ \Phi_{\mathcal{D}}^S(r) \end{pmatrix} = 0, \tag{1}$$

where $\kappa = \pm(j + \frac{1}{2})$ for $\ell = j \pm \frac{1}{2}$ is the Dirac angular momentum quantum number, $\Phi_{\mathcal{D}}^L$ and $\Phi_{\mathcal{D}}^S$ are the radial parts of the large and small components of the wavefunction and E is the energy relative to mc^2. We compare variational solutions of the Dirac equation with the these of the Schrödinger equation in the Lévy-Leblond form [19]:

$$\begin{pmatrix} -\frac{Z}{r} - E & c\left(\frac{\kappa}{r} - \frac{d}{dr}\right) \\ c\left(\frac{\kappa}{r} + \frac{d}{dr}\right) & -2mc^2 \end{pmatrix} \begin{pmatrix} \Phi_{\mathcal{L}}^L(r) \\ \Phi_{\mathcal{L}}^S(r) \end{pmatrix} = 0. \tag{2}$$

We define the Rayleigh quotient as

$$K[\Psi] = \frac{\langle\Psi|H|\Psi\rangle}{\langle\Psi|\Psi\rangle}, \tag{3}$$

where H stands for either the Dirac or Lévy-Leblond Hamiltonian,

$$\Psi = \begin{pmatrix} \Psi^L \\ \Psi^S \end{pmatrix}, \tag{4}$$

and

$$\Psi^L = \frac{1}{r} A^L(\Omega)\Phi^L(r) \text{ and } \Psi^S = \frac{1}{r} A^S(\Omega)\Phi^S(r) \tag{5}$$

are, respectively, large and small components of the wavefunction. The angular and spin parts of the trial functions, $A(\Omega)$, are equal to the exact ones. The radial functions are selected in the form

$$\Phi^L = \sum_{k=0}^{N-1} C_k^L \phi_k^L, \quad \Phi^S = \sum_{k=0}^{N-1} C_k^S \phi_k^S, \tag{6}$$

where C_k^L, C_k^S are variational parameters and ϕ^L, ϕ^S are the basis functions. The basis functions are taken as

$$\phi_k^L(r) = r^{k+g} e^{-\alpha r^t}, \quad \phi_k^S(r) = r^{k+g} e^{-\beta r^t} \tag{7}$$

where α and β are the nonlinear parameters, while g and t define the type of the radial basis set. In this study three kinds of the basis sets have been used:

– Drake and Goldman [7], if $g = \gamma$ and $t = 1$, where

$$\gamma = \sqrt{\kappa^2 - \frac{Z^2}{c^2}}, \tag{8}$$

– Slater, if $g = 1$ and $t = 1$,
– Gaussian, if $g = 1$ and $t = 2$.

Let us note that even with $\alpha = \beta$ none of these basis sets fulfils the kinetic balance condition and, consequently, does not belong to the class defined by Grant as *physically acceptable* [4].

Variation of $K[\Psi]$ with respect to the linear parameters leads, in the case of the Dirac equation, to the algebraic $2N \times 2N$ eigenvalue problem:

$$\begin{pmatrix} \boldsymbol{H}_{LL} - E_D^v \boldsymbol{S}_{LL} & c\boldsymbol{H}_{LS} \\ c\boldsymbol{H}_{SL} & \boldsymbol{H}_{SS} - E_D^v \boldsymbol{S}_{SS} \end{pmatrix} \begin{pmatrix} \boldsymbol{C}_v^L \\ \boldsymbol{C}_v^S \end{pmatrix} = 0, \tag{9}$$

where

$$\begin{aligned} (\boldsymbol{H}_{LL})_{kk'} &= \langle \phi_k^L | -\frac{Z}{r} | \phi_{k'}^L \rangle, \\ (\boldsymbol{H}_{LS})_{kk'} &= \langle \phi_k^L | \frac{\kappa}{r} - \frac{d}{dr} | \phi_{k'}^S \rangle, \\ (\boldsymbol{H}_{SL})_{kk'} &= \langle \phi_k^S | \frac{\kappa}{r} + \frac{d}{dr} | \phi_{k'}^L \rangle, \\ (\boldsymbol{H}_{SS})_{kk'} &= \langle \phi_k^S | -\frac{Z}{r} - 2mc^2 | \phi_{k'}^S \rangle, \end{aligned}$$

$(\boldsymbol{S}_{LL})_{kk'} = \langle \phi_k^L | \phi_{k'}^L \rangle$, $(\boldsymbol{S}_{SS})_{kk'} = \langle \phi_k^S | \phi_{k'}^S \rangle$, E_D^v is the eigenvalue and

$$\boldsymbol{C}_v = \begin{pmatrix} \boldsymbol{C}_v^L \\ \boldsymbol{C}_v^S \end{pmatrix}$$

is the corresponding eigenvector. The highest N eigenvalues of the matrix are supposed to approximate the corresponding Dirac energies. However, as discussed by several authors [7],[8],[10-12],[21], in some cases the so called *spurious roots* (matrix eigenvalues which do not correspond to any of the eigenvalues of the Dirac Hamiltonian) may appear. The remaining N eigenvalues of the matrix are located by approximately $2mc^2$ below the highest N eigenvalues, i.e. in the negative continuum.

In the Lévy-Leblond case the following $N \times N$ matrix eigenvalue equation is obtained [2],[3]:

$$\left(\boldsymbol{H} - E^v_{\mathcal{L}} \boldsymbol{S}_{LL}\right) \boldsymbol{C}^L_v = 0, \tag{10}$$

where

$$\boldsymbol{H} = \boldsymbol{H}_{LL} + \frac{1}{2m} \boldsymbol{H}_{LS} \boldsymbol{S}^{-1}_{SS} \boldsymbol{H}_{SL}.$$

This set of equations is equivalent to the corresponding Schrödinger algebraic problem if (and only if) the basis set fulfils the kinetic balance condition [2],[3]. However, contrary to Eq. (9), Eq. (10) is derived from a pseudo-eigenvalue equation (2), for which the set of energies is bounded from below.

3. Results and discussion

The energy obtained by solving the algebraic Dirac (9) and Lévy-Leblond (10) equations depends upon the nonlinear parameters α and β and upon the type of the basis set (i.e. upon the values of N, g and t). In this section we present results of an analysis of this dependence for several $s_{1/2}$ and $p_{1/2}$ states of one-electron atoms obtained using basis sets with $N = 1, 2, 3$.

In the first subsection, the eigenvalues of the Dirac (9) and Lévy-Leblond (10) Hamiltonian matrices in N-function bases ($N = 1, 2$), ${}_N E^{n\ell j}_{\mathcal{D}}(\alpha, \beta, g, t)$ and ${}_N E^{n\ell j}_{\mathcal{L}}(\alpha, \beta, g, t)$ respectively, are plotted versus α and β in a number of figures. The plots are composed of a set of equi-energetic lines drawn in the (α, β) coordinate plane and are identified as ${}_N\mathcal{D}^{n\ell j}_{g,t}$ (Dirac) and ${}_N\mathcal{L}^{n\ell j}_{g,t}$ (Lévy-Leblond). Maxima, saddle points and minima in the energy surfaces are marked, respectively, as

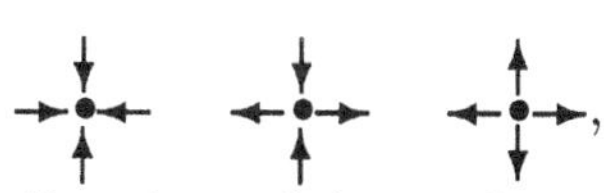

where the arrows show directions of the gradient.

In the second subsection, the highest eigenvalue (for $N = 1$) and two highest eigenvalues (for $N = 2, 3$) of the Hamiltonian matrix, corresponding to the extrema of the energy surfaces for $N = 1, 2$ are, collected in a table and their mutual relations are discussed.

3.1. GEOMETRICAL STRUCTURE OF THE ENERGY SURFACES

In the case of $N = 1$ the Dirac and Lévy-Leblond energy may both be expressed analytically as a function of α and β [2],[20]. The energy surface always contains exactly one extremum and this extremum is always a saddle point. For all the states described by nodeless wavefunctions, i.e. for the states with $\kappa = -n$, the trial functions which fulfil the exact boundary conditions for $r \to 0$ and for $r \to \infty$ (Drake and Goldman in the Dirac case and Slater in the Lévy-Leblond case) give the exact energy at the saddle point. A detailed analysis of the geometrical structure of the energy surfaces ${}_1\mathcal{D}_{g,t}^{1s_{1/2}}$ and ${}_1\mathcal{L}_{g,t}^{1s_{1/2}}$ for $(g,t) = (\gamma, 1)$, $(1,1)$, $(1,2)$ has been given in I. Qualitatively, the description of all other $\kappa = -n$ states is very similar.

The remaining (i.e. $|\kappa| < n$) states can never be described exactly with $N = 1$ in any of the basis sets used in this paper. If $N = 1$, the overall structure of the energy surfaces is, for these states, quite similar to that for $\kappa = -n$ ones, although there are some significant differences. In figure 1 the highest eigenvalue of $N = 1$ Hamiltonian matrix corresponding to the $2p_{1/2}$ (i.e. to $\kappa = 1$) state of a $Z = 90$ hydrogen-like atom is plotted versus α and β. As for $\kappa = -n$ (c.f. I), also in this case, the Dirac and the Lévy-Leblond surfaces are very similar [one has to remember that the Dirac case of $g = \gamma$ ($g = 1$) should be compared with the Lévy-Leblond case of $g = 1$ ($g = \gamma$)]. The root of the secular equation corresponding to the saddle point parameters (α_0, β_0), depending upon the case, approximates either the Dirac $2p_{1/2}$ or Schrödinger $n = 2$ energy. The only feature which may cause some surprise is that the approximation given by the saddle-point energy resulting from a trial function with an incorrect behaviour at the origin is, in some cases, significantly better than the one given by the function which for $r \to 0$ is asymptotically exact.

In figure 2 differences between either the $2p_{1/2}$ Dirac or $2p$ Schrödinger energy and the saddle point energies obtained using several different trial functions are plotted versus Z. As one can see, there is no simple rule which would link the type of the boundary conditions fulfilled by the basis functions and the accuracy of the resulting energy. In the case of a single basis function with the correct boundary conditions at the nucleus the Dirac saddle point is always above the exact energy (c.f [4], [10]), contrary to the case when the boundary conditions are not fulfilled. For the Slater-type basis the Dirac saddle point may be located above (small Z) as well as below (large Z) the exact energy. For the Gaussian trial functions the saddle point is always below the exact energy. The behaviour of the Lévy-Leblond solutions is quite similar. Let us note, that the Dirac and Lévy-Leblond energies behave in a similar way when either in both cases the trial

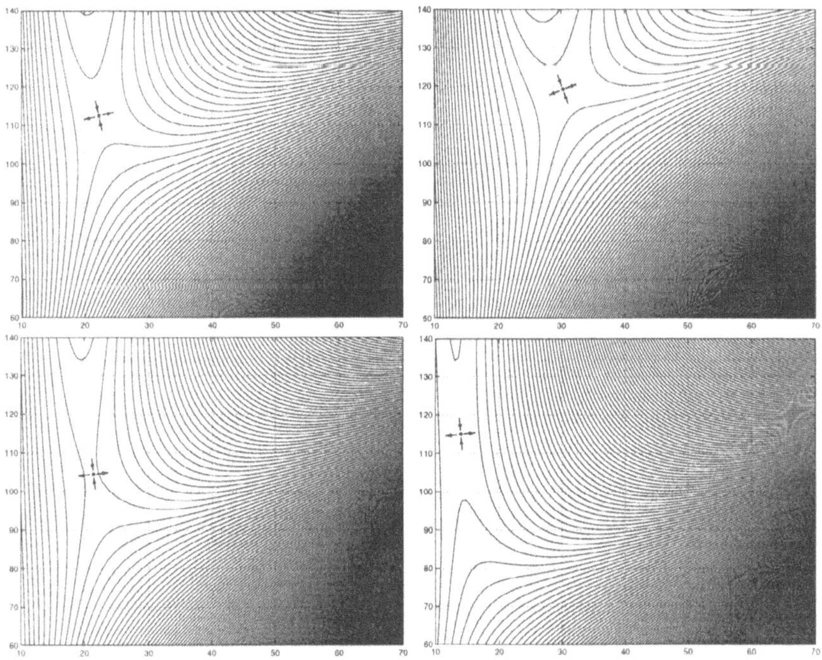

Figure 1. Plots ${}_1\mathcal{D}^{2p_{1/2}}_{\gamma,1}$ (upper-left), ${}_1\mathcal{D}^{2p_{1/2}}_{1,1}$ (upper-right) ${}_1\mathcal{L}^{2p_{1/2}}_{1,1}$ (lower-left), and ${}_1\mathcal{L}^{2p_{1/2}}_{\gamma,1}$ (lower-right) for the case of a $Z = 90$ hydrogen-like atom in the (α, β) coordinate plane. The saddle point energies ${}_1E^{2p_{1/2}}_{\mathcal{D}}(21, 108, \gamma, 1) = -1125.2$ a.u. and ${}_1E^{2p_{1/2}}_{\mathcal{D}}(31, 122, 1, 1) = 1196.5$ a.u. are, respectively, by 67.1 a.u higher and by 4.2 a.u lower than the exact Dirac $2p_{1/2}$ energy. In the Lévy-Leblond case, ${}_1E^{2p_{1/2}}_{\mathcal{L}}(21, 105, 1, 1) = -944.8$ a.u. and ${}_1E^{2p_{1/2}}_{\mathcal{L}}(15, 115, \gamma, 1) = -816.0$ a.u. are, respectively, by 67.7 a.u and by 196.5 a.u higher than the exact $n = 2$ Schrödinger energy.

function at the origin is exact ($g = \gamma$ for Dirac, $g = 1$ for Lévy-Leblond) or in both cases g is larger than the exact value ($g = 1$ for Dirac, $g = 2$ for Lévy-Leblond). Then, in this case we have another example of when the "variational collapse" is related to incorrect boundary conditions and to wrong relations between the components of the trial function rather than to the unboundedness from below of the Dirac Hamiltonian. Another instructive feature of this example is showing that in this case the correct upper-bound solution may give a poorer approximation to the energy than a solution without any bound properties.

The structure of the energy surfaces becomes considerably more complex when the basis set is enlarged. The two highest eigenvalues of the Hamiltonian matrices of $s_{1/2}$ and $p_{1/2}$ states of the one-electron $Z = 90$

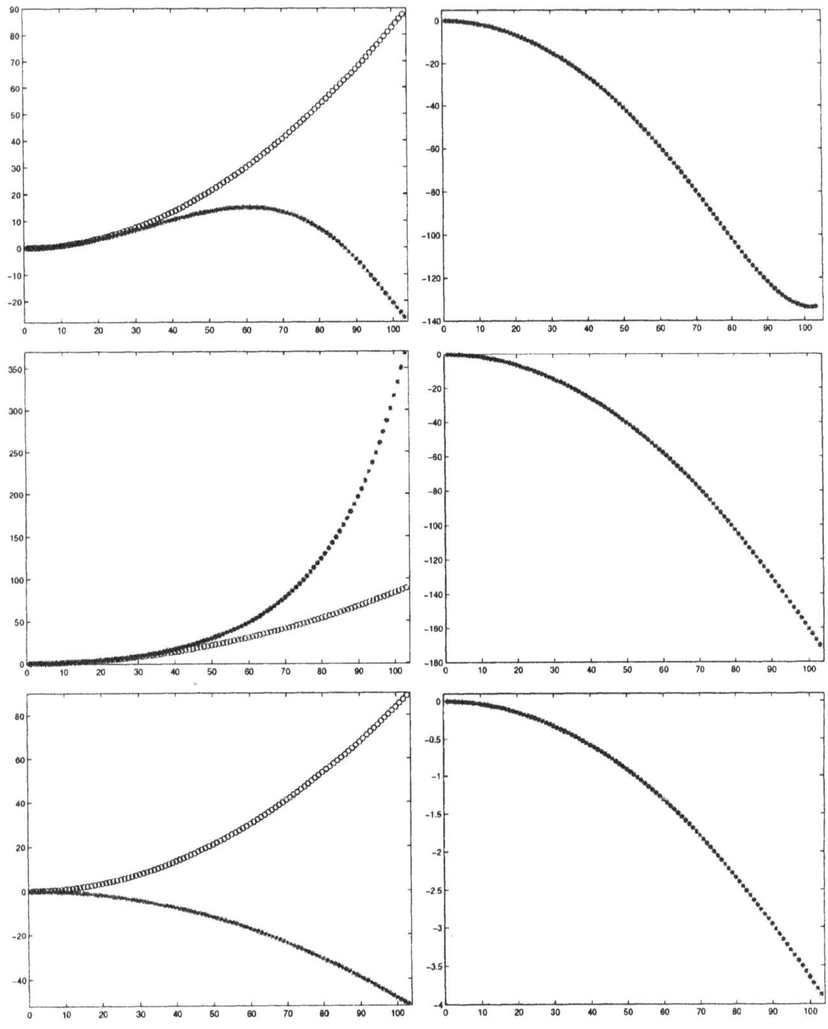

Figure 2. In the first-row subfigures differences between the exact Dirac $2p_{1/2}$ energies and the saddle point values are plotted versus Z. The cases of $g = \gamma$, $t = 1$ (Goldman and Drake) and $g = 1$, $t = 1$ (Slater) are displayed in the left subfigure (open circles and asterisks, respectively). The case of $g = 1$, $t = 2$ (Gaussian) functions is shown in the right subfigure. In the second-row and in the third-row subfigures similar differences between the exact Schrödinger $2p$ energies and the ones derived from the Lévy-Leblond equation are displayed. The open-circle curves correspond to $g = 1$, $t = 1$ basis functions. The asterisk curves in the second-row left subfigure correspond to $g = \gamma$, $t = 1$; in the second-row right – to $g = 1$, $t = 2$; in the third-row left – to $g = 2$, $t = 1$; in the third-row right – to $g = 2$, $t = 2$.

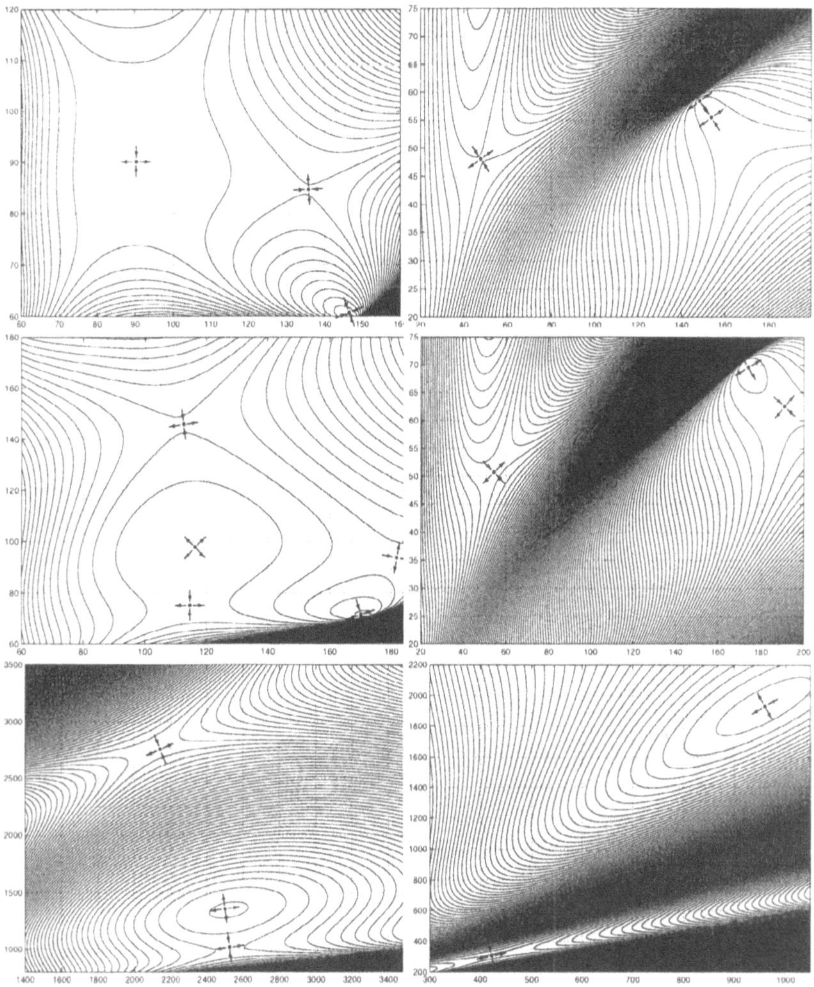

Figure 3. Plots ${}_2\mathcal{D}_{\gamma,1}^{1s_{1/2}}$ (upper left), ${}_2\mathcal{D}_{\gamma,1}^{2s_{1/2}}$ (upper right), ${}_2\mathcal{D}_{1,1}^{1s_{1/2}}$ (middle left), ${}_2\mathcal{D}_{1,1}^{2s_{1/2}}$ (middle right), ${}_2\mathcal{D}_{1,2}^{1s_{1/2}}$ (lower left) and ${}_2\mathcal{D}_{1,2}^{2s_{1/2}}$ (lower right) for a $Z = 90$ hydrogen-like atom in the (α, β) coordinate plane. The saddle points at $\alpha = \beta = 90$ in the upper-left figure and at $\alpha = \beta = 48.1$ in the upper-right figure correspond to the exact Dirac energies of, respectively, $1s_{1/2}$ and $2s_{1/2}$ states.

atom obtained with various $N = 2$ basis sets are plotted as functions of the non-linear parameters α and β in figures 3-5[1]. The most striking difference

[1]Let us note that the Dirac algebraic eigenvalue problem (9), for a given κ, is identical

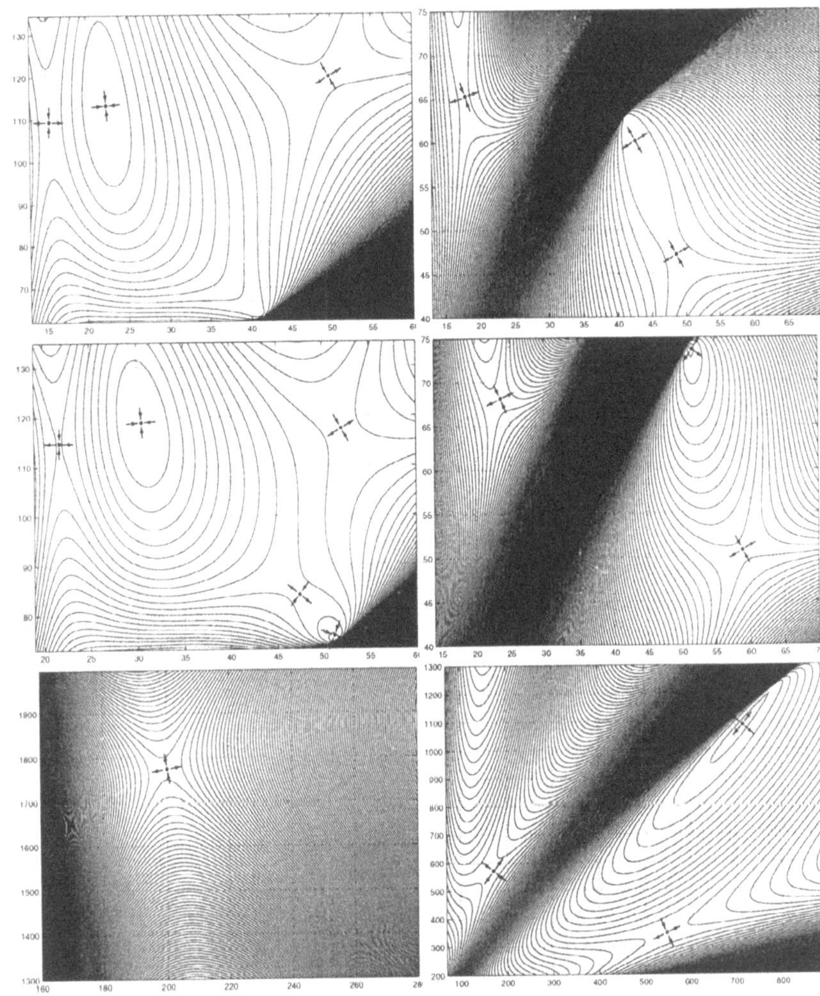

Figure 4. Plots ${}_2\mathcal{D}^{1p_{1/2}}_{\gamma,1}$ (upper left), ${}_2\mathcal{D}^{2p_{1/2}}_{\gamma,1}$ (upper right), ${}_2\mathcal{D}^{1p_{1/2}}_{1,1}$ (middle left), ${}_2\mathcal{D}^{2p_{1/2}}_{1,1}$ (middle right), ${}_2\mathcal{D}^{1p_{1/2}}_{1,2}$ (lower left) and ${}_2\mathcal{D}^{2p_{1/2}}_{1,2}$ (lower right) for a $Z = 90$ hydrogen-like atom in the (α, β) coordinate plane. The designation "$1p_{1/2}$" is used for the second highest ("spurious") root of the $N = 2$ eigenvalue problem. The saddle point at $\alpha = \beta = 48.1$ in the upper-right figure corresponds to the exact Dirac $2p_{1/2}$ energy.

when compared to the analogous plots for $N = 1$, is a multitude of all kinds of extrema spread over a large range of non-linear parameter values. The

to the one for $-\kappa$ as long as $\alpha = \beta$. However the energy surfaces in the (α, β) coordinate plane are, for these two states, different if $\alpha \neq \beta$

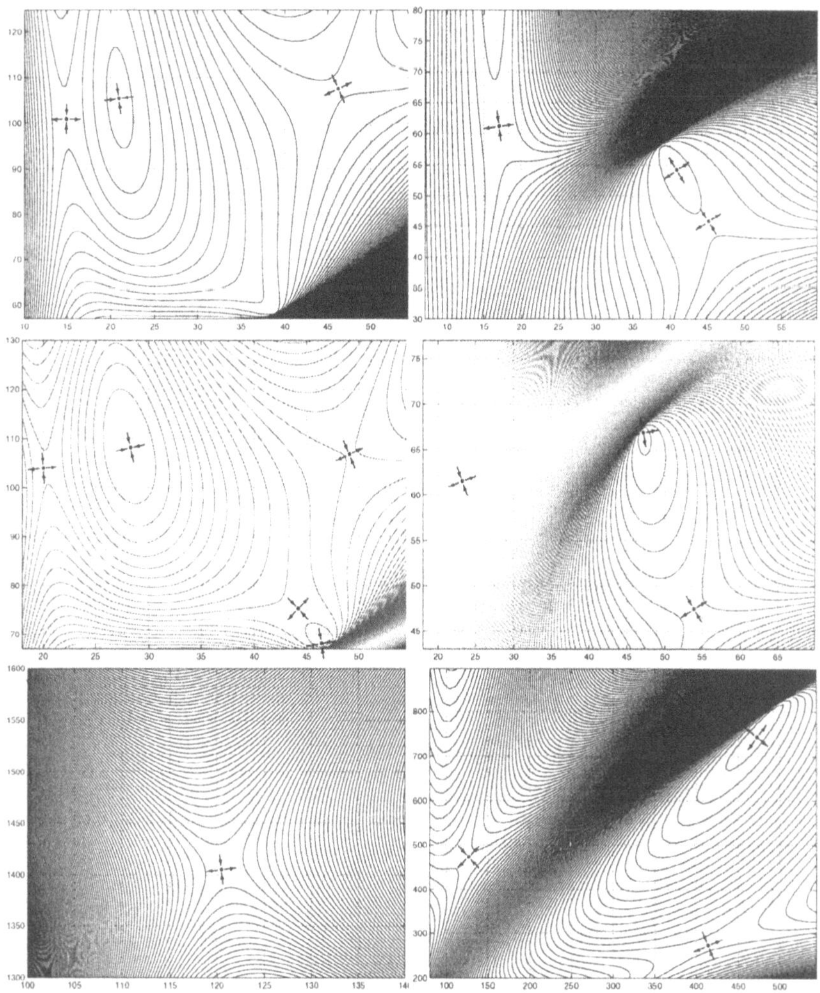

Figure 5. Plots ${}_2\mathcal{L}_{1,1}^{1p_{1/2}}$ (upper left), ${}_2\mathcal{L}_{1,1}^{2p_{1/2}}$ (upper right), ${}_2\mathcal{L}_{1.25,1}^{1p_{1/2}}$ (middle left), ${}_2\mathcal{L}_{1.25,1}^{2p_{1/2}}$ (middle right), ${}_2\mathcal{L}_{1,2}^{1p_{1/2}}$ (lower left) and ${}_2\mathcal{L}_{1,2}^{2p_{1/2}}$ (lower right) for a $Z = 90$ hydrogen-like atom in the (α, β) coordinate plane. The designation "$1p_{1/2}$" is used for the second highest ("spurious") root of the $N = 2$ eigenvalue problem. The saddle point at $\alpha = \beta = 45$ in the upper-right figure corresponds to the exact Schrödinger $2p$ energy.

energy surfaces corresponding to the Dirac ground state in the Drake and Goldman and in the Slater basis sets look entirely different. The extremum giving the exact Dirac $1s_{1/2}$ energy (the saddle point $\alpha = \beta = 90$ in the upper-left subfigure of fig 3) splits to three different saddle points when the Slater basis is used (middle-left subfigure of fig. 3). On the contrary, the energy surfaces for $2s_{1/2}$ are in these two basis sets nearly identical (upper-right and middle-right subfigures of fig. 3).

The close similarity between the Dirac and the Lévy-Leblond matrix eigenvalue problems is best seen when one compares the corresponding subfigures of fig. 4 (Dirac) and fig. 5 (Lévy-Leblond). The energy surfaces for these two equations remain nearly identical, independent of the kind of basis set as well as upon the range of the nonlinear parameters. In order to get the best analogy between the Dirac and the Lévy-Leblond basis sets, we have chosen the following correspondence: (1) the exact behaviour at the origin, i.e. $g = \gamma$ for Dirac and $g = 1$ for Lévy-Leblond (the first-row subfigures); (2) the lowest power r in the basis set is larger by ≈ 0.25 than the exact one, i.e. $g = 1$ for Dirac and $g = 1.25$ for Lévy-Leblond (the second-row subfigures); (3) the Gaussian functions with $g = 1$ in both cases (the third-row subfigures).
The plots corresponding to the Gaussian basis are, in general, very much different and they contain fewer extrema (there may be more extrema outside the range of the nonlinear parameters investigated in this work). Perhaps the only exception is the ground-state Gaussian surface ${}_2\mathcal{D}_{1,2}^{1s_{1/2}}$ which is similar to ${}_2\mathcal{D}_{1,1}^{1s_{1/2}}$ (fig. 3).

3.2. THE ENERGY VALUES

The highest eigenvalues of the Dirac Hamiltonian matrices for $\kappa = -1$ ($s_{1/2}$ states) and for $\kappa = 1$ ($p_{1/2}$ states) in the bases with $t = 1$ and $g = \gamma, 1$ are collected in table 1. In the case of $N = 1$ the highest eigenvalue, and in the cases of $N = 2$ and $N = 3$ two highest eigenvalues have been displayed. The calculations have been performed at all (α, β) points which correspond to extrema in either $ns_{1/2}$ or $np_{1/2}$ with $n = 1, 2$; $g = \gamma, 1$ and $N = 1, 2$. The energies may be compared with the exact Dirac values: -4618 ($1s_{1/2}$), -1192 ($2s_{1/2}$, $2p_{1/2}$), -512 ($3s_{1/2}$, $3p_{1/2}$). It is interesting to note that in the majority of cases in a given (α, β) point two extrema appear, usually (but not always) in the matrices corresponding to the values of N differing by 1.

In order to stress some similarities between the different energy surfaces, the energies in table 1 have been divided into 9 groups numbered by the index $q = 1, 2, \ldots, 9$. Among these groups, 7 are "complete" (they contain one extremum from each of the energy surfaces considered) and 2 are "incomplete". Here is a brief characterization of these groups:
$q = 1$: This group contains energies evaluated at the $N = 1$ saddle point values of (α, β). In all cases ${}_1E_1 = {}_2E_1$ with ${}_2E_1$ also corresponding to an extremum. The case of $\kappa = -1$, $g = \gamma$ gives the exact $1s_{1/2}$ solution and,

TABLE 1. Values of the nonlinear parameters corresponding to the extrema in the energy surfaces and the corresponding energy values for two lowest states of the $Z = 90$ hydrogen-like atom obtained using Drake and Goldman ($g = \gamma$) and Slater ($g = 1$) basis functions. In the consecutive columns the case number q, the identification of the state (κ) and of the basis functions (g), the coordinates (α, β) of the extremum and the Hamiltonian matrix eigenvalues (${}_N E_m$) are given. The designation ${}_N E_m$ means the m-th eigenvalue of the matrix corresponding to the N-element basis set. The energy corresponding to an extremum is given in boldface and the subscript (s, n or x) identifies the kind of extremum (saddle, minimum or maximum, respectively)

q	κ	g	α	β	${}_1E_1$	${}_2E_1$	${}_2E_2$	${}_3E_1$	${}_3E_2$
1	−1	γ	90	90	**−4618**$_s$	**−4618**$_s$	123	**−4618**$_s$	−966
	−1	1	114	99	**−4527**$_s$	**−4527**$_s$	528	−4571	−568
	1	γ	22	112	**−1125**$_s$	**−1125**$_x$	−438	−1148	−587
	1	1	31	119	**−1197**$_s$	**−1197**$_x$	−485	−1219	−670
2	−1	γ	135	85	−4917	**−4611**$_s$	2013	−4617	−720
	−1	1	179	93	−5305	**−4461**$_s$	712	−4678	−1332
	1	γ	50	120	−922	**−1220**$_s$	2296	**−1220**$_s$	−387
	1	1	52	117	−1289	**−1225**$_s$	1005	**−1225**$_s$	−397
3	−1	γ	148	59	−7615	**−4567**$_n$	**−4567**$_x$	−6246	−4542
	−1	1	172	71	−7010	**−4396**$_x$	**−4396**$_n$	−6691	−4209
	1	γ	42	61	−2604	−1599	**−1198**$_n$	**−1198**$_x$	−732
	1	1	51	74	−2378	**−1215**$_x$	**−1215**$_n$	−1239	−500
4	−1	γ	156	55	−8566	−6783	**−4561**$_s$	−9355	**−4561**$_n$
	−1	1	189	62	−8986	−9310	**−4357**$_s$	−13611	**−4357**$_s$
	1	γ	48	48	−3804	−4430	**−1192**$_s$	−4585	**−1192**$_n$
	1	1	59	50	−4055	−5467	**−1169**$_s$	−6666	**−1169**$_n$
5	−1	γ	59	129	−4875	−4621	**−1643**$_n$	−4613	−1512
	−1	1	73	138	−4917	−4364	**−1749**$_n$	−4458	−1495
	1	γ	18	156	−1180	−1166	**−552**$_n$	−1107	**−552**$_s$
	1	1	25	191	−1367	−1272	**−681**$_n$	−1203	−680
6	−1	γ	56	191	−5394	−5063	**−1578**$_s$	−4677	**−1578**$_s$
	−1	1	66	205	−5293	−4838	**−1587**$_s$	−4481	**−1587**$_s$
	1	γ	18	181	−1235	−1196	**−552**$_s$	−1128	−549
	1	1	25	202	−1396	−1290	**−681**$_s$	−1215	**−681**$_s$
7	−1	γ	48	48	−3804	−4430	**−1192**$_s$	−4585	**−1192**$_s$
	−1	1	56	52	−3511	−4124	**−1166**$_s$	−4369	**−1166**$_n$
	1	γ	18	65	−1149	−1206	**−500**$_s$	−1172	**−500**$_s$
	1	1	23	68	−1196	−1257	**−539**$_s$	−254	**−539**$_x$
8	−1	1	114	76	−4806	**−4512**$_s$	−315	−4566	−953
	1	γ	15	109	−1057	**−1149**$_s$	−527	−1102	−547
	1	1	22	115	−1143	**−1215**$_s$	−631	−1195	−608
9	−1	1	112	146	−5168	**−4479**$_s$	−565	−4581	−921
	1	1	47	85	−1814	**−1222**$_s$	−206	**−1222**$_x$	−441

consequently, this energy value is also an extremum for all ${}_N E_1$. There are no spurious roots in this case. The second root (${}_N E_2$), with enlargement of the basis, seems to converge to the next Dirac eigenvalue.
$q = 2$: Similar to the case of $q = 1$, except that the nonlinear parameters correspond to an extremum of ${}_2E_1$.
$q = 3$: A very strange case. The extrema in ${}_2E_1$ and ${}_2E_2$ coincide and these two energies are degenerate. In this point the energy surfaces merge (a maximum in the lower surface coincides with a minimum in the upper surface). The energy values corresponding to $\kappa = -1$ and to $\kappa = 1$ are close, respectively, to the Dirac $1s_{1/2}$ and $2p_{1/2}$ energies, except that the Dirac energies do not exhibit this kind of degeneracy.
$q = 4$: Another troublesome case. Spurious roots appear for *both* $s_{1/2}$ and $p_{1/2}$ symmetries. The matrix eigenvalues corresponding to the extrema give rather good approximation to the Dirac energies (the one with $\kappa = 1$ and $g = \gamma$ gives the exact $2p_{1/2}$ energy) however the presence of the spurious roots creates the well known difficulty.
$q = 5 - q = 7$: Three cases interesting and similar to each other. The extrema correspond to the second roots (${}_2E_2$). Both first and second roots reasonably well approximate the Dirac eigenvalues and there are no spurious roots. The case $q = 7$, $\kappa = -1$, $g = \gamma$ gives the exact $2s_{1/2}$ Dirac energy.
The "incomplete" cases ($q = 8$ and $q = 9$) are, in a way, similar to the cases $q = 1$ and $q = 2$.

Another very interesting feature of the energy surfaces which may be deduced from a comparison of table 1 and figures 3-5 is their multi-level structure. As one can see, the same eigenvalue of the Hamiltonian matrix, depending upon the range of the nonlinear parameter values, may approximate different eigenvalues of the Hamiltonian operator. In all $n = 2$ surfaces each extremum is located on one of two "terraces" linked to each other by steep steps (the equienergetic lines are there so dense that they appear in the plots as black areas). A "terrace" corresponding to a given eigenvalue contains one or several extrema separated by rather shallow valleys.

The structure of the energy surfaces becomes more complicated for larger N and for an increasing number of nonlinear parameters. Therefore, a general classification scheme of the extrema and the formulation of criteria which would allow the identification of those extrema which are physically meaningful seems to be desirable.

4. Conclusions

- As it is known [2-4],[11], establishing correct relations between the components of the wavefunction leads to a numerically stable and "safe" variational procedure for a Dirac-Coulomb eigenvalue problem. Incorrect relations between the two components may produce quite unexpected results, as e.g. exact energies with completely wrong wavefunctions. This feature of the equations, when misunderstood, may lead to serious mistakes; however, when consciously used, may lead to new simple methods for the estimation of the energy values.

- Contrary to the case of the Schrödinger-equation-based variational principle, in variational calculations with two-component wavefunctions a "good" energy does not imply that the corresponding wavefunction is correct. In particular, a mysterious disappearing of spurious solutions when using the minimax principle [13] does not constitute a feature of this principle but is a consequence of a fortuitous selection of the extremum in the space of the nonlinear parameters for which such a root does not appear.
- In the case of two-component variational problems the energy surfaces as functions of nonlinear parameters contain a multitude of different kinds of extrema. Most of them reasonably well approximate the eigenvalues but not the wavefunctions. The number of these extrema grows up very fast with the number of the linear variational parameters. The selection of the correct extremum must be associated with an analysis of the resulting wavefunctions by at least checking its asymptotic behaviour (c.f. [10], [11]).
- The structure of the energy surface drastically changes when the number of the basis functions is enlarged. Therefore a frequently advocated strategy: "Choose the nonlinear parameters variationally for a simple problem and then enlarge the basis", may lead to erroneous results.

Acknowledgements

We are grateful to Prof. Ian P. Grant, FRS, for his most helpful comments. We also thank a referee for his remarks. This work has been supported by the Polish KBN under project No. 2 P03B 126 14.

References

1. P. Pyykkö, *Relativistic Theory of Atoms and Molecules: A Bibliography 1916-1985*, Lecture Notes in Chemistry, vol. 41, Springer, Berlin 1986; *A Bibliography 1986-1992* Lecture Notes in Chemistry, vol. 60, Springer, Berlin 1993; the most recent updates are available in the internet: http://www.csc.fi/lul/rtam/rtamquery.html
2. K. G. Dyall, I. P. Grant and S. Wilson, J. Phys. B: At. Mol. Phys. **17**, L45, 493, 1201 (1983).
3. I. P. Grant, J. Phys. B: At. Mol. Phys. **19**, 3187 (1986).
4. I. P. Grant, in *Atomic, Molecular & Optical Physics Handbook*, ed. G. W. F. Drake, Chapt. 22, p. 278. AIP Press, Woodbury, New York, USA, 1996.
5. W. Kutzelnigg, Int. J. Quantum Chem. **15**, 107 (1984).

6. L. Visscher, P. J. C. Aerts, O. Visser and W. C. Nieuwpoort, Int. J. Quantum Chem. **S25**, 131 (1991).
7. G. W. F. Drake and S. P. Goldman, Phys. Rev. A **23**, 2093 (1981).
8. W. E. Baylis and S. J. Peel, Phys. Rev. A **28**, 2552 (1983).
9. J. Wood, I. P. Grant and S. Wilson, J. Phys. B: At. Mol. Phys. **18**, 3027 (1985).
10. S. P. Goldman, Phys. Rev. A **31**, 3541 (1985).
11. S. P. Goldman, J. Phys. B: At. Mol. Opt. Phys. **25**, 629 (1992).
12. J. D. Talman, Phys. Rev. A **50**, 3525 (1994).
13. J. D. Talman, Phys. Rev. Letters **57**, 1091 (1986).
14. S. N. Datta and G. Deviah, Pramana, **30**, 387 (1988).
15. W. Kutzelnigg, Chem. Phys. **225**, 203 (1997).
16. M. Griesemer and H. Siedentop, J. London Math. Soc. **xx**, xxx (1999).
17. H. M. Quiney, H. Skaane and I. P. Grant, Advan. Quantum Chem. **32**, 1 (1998).
18. J. Karwowski, G. Pestka and M. Stanke, *Progress in Theoretical Chemistry* **xx**, xxx (1999).
19. J. M. Lévy-Leblond, Commun. Math. Phys. **6**, 286 (1967).
20. A. Kołakowska, J. Phys. B: At. Mol. Phys. **29**, 4515 (1996).
21. Z. Chen, G. Fonte and S. P. Goldman, Phys. Rev. A **50**, 3838 (1994).

RELATIVISTIC MULTIREFERENCE MANY-BODY PERTURBATION THEORY

MARIUS JONAS VILKAS, KONRAD KOC

AND

YASUYUKI ISHIKAWA

Department of Chemistry, University of Puerto Rico, P.O. Box 23346, San Juan, PR 00931-3346, USA

Abstract. Our recently developed relativistic multireference Møller-Plesset (MR-MP) perturbation theory has been applied, using Gaussian spinors, to low-lying even- and odd-parity states of a carbon isoelectronic sequence up to Z = 60. We have analysed quantitatively the way relativity alters asymptotic configuration interaction.

1. Introduction

Accurate calculations of heavy atoms, heavy-atom-containing molecules, and highly ionized ions must include relativistic, electron correlation and QED effects. To accurately account for relativistic and electron correlation effects, an intense effort in the last decade has been directed toward developing relativistic many-body theories. Among the relativistic many-body techniques developed recently are multiconfiguration Dirac-Fock (MC DF) self-consistent field (SCF) method [1, 2], single-reference relativistic many-body perturbation theory [3, 4, 5], relativistic coupled cluster theory [6, 7], and the relativistic configuration interaction method [8, 9].

For a large number of atomic and molecular systems, near-degeneracy in the valence spinors gives rise to a manifold of strongly interacting configurations, i.e., strong configuration mixing within a complex [10], and makes a multiconfigurational treatment mandatory. The classic examples in atomic physics are the near-degene- racy effects in ground-state beryllium [10, 11, 12, 13] and open-shell atoms with two or more open valence shells [10, 14, 15, 16, 17]. For reactive and excited-state energy surfaces

J. Maruani et al. (eds.), New Trends in Quantum Systems in Chemistry and Physics, Volume 1, 191–218.

of molecules, correlated methods based on single configuration SCF theory also fail to properly describe the separated fragments because of the near-degeneracy which follows the separation process [18].

In recent studies [19], we have developed a relativistic multireference Møller-Plesset (MR-MP) perturbation theory that combines the strengths of both MC DF SCF and many-body perturbation methods in application to a general class of quasidegenerate systems with multiple open valence shells. We have extended the single-reference relativistic many-body perturbation theory [5] to a relativistic MR-MP perturbation theory for systems with a manifold of strongly interacting configurations. The relativistic MR-MP perturbation theory for electron correlation is designed to treat a general class of openshell systems with two or more valence electrons that often exhibit quasidegeneracies. The essential feature of the theory is its treatment of the state-specific nondynamic correlation in zero order through quadratically convergent matrix MC DF SCF [20], and recovery of the remaining correlation, which is predominantly dynamic pair correlation, by second-order MR-MP perturbation theory.

In this study, relativistic MR-MP perturbation calculations are performed on low-lying J=0, 1, 2, 3 even and odd-parity states of carbon and carbon isoelectronic sequence up to Z=60 (Z is nuclear charge). We examine quantitatively the way relativity alters asymptotic configuration interaction (CI) [10]. It is inappropriate to use relativistic many-body perturbation theory based on a single-configuration reference state ($1s^2 2s^2 2p_{1/2}^2$ J=0) for low-Z members of the isoelectronic sequence due to strong mixing among several configuration states (i.e., quasidegeneracy). The results of relativistic multireference many-body perturbation calculations demonstrate that it can be applied equally successfully to low-Z members in which configuration mixing is strong through higher-Z members in which mixing tends to weaken due to asymptotic CI [10]. The implementation of relativistic MR-MP procedure, using G spinors (G for "Gaussian"), is reviewed in the next section. In Sec. 3, results of relativistic MR-MP calculations on carbon and carbonlike ions will be presented.

2. Theory

2.1. THE RELATIVISTIC NO-PAIR DIRAC-COULOMB-BREIT HAMILTONIAN

The effective N-electron Hamiltonian (in atomic units) for the development of our MC DF SCF and MR-MP algorithms is taken to be the relativistic "no-pair" Dirac-Coulomb (DC) Hamiltonian [21, 22],

$$H_{DC}^{+} = \sum_{i}^{N} h_D(i) + \mathcal{L}_+ \left(\sum_{i>j}^{N} \frac{1}{r_{ij}} \right) \mathcal{L}_+ . \tag{1}$$

$\mathcal{L}_+ = L_+(1)L_+(2)\ldots L_+(N)$, where $L_+(i)$ is the projection operator onto the space $D^{(+)}$ spanned by the positive-energy eigenfunctions of the matrix DF SCF equation [22]. $\mathcal{L}_+$ is the projection operator onto the positive-energy space $D^{(+)}$ spanned by the N-electron configuration-state functions (CSFs) constructed from the positive-energy eigenfunctions ($\in D^{(+)}$) of the matrix DF SCF. It takes into account the field-theoretic condition that the negative-energy states are filled and causes the projected DC Hamiltonian to have normalizable bound-state solutions. This approach is called the no-pair approximation [21] because the formulas of many-body perturbation theory includes only excitation to positive-energy virtual states from positive-energy bound states and virtual electron-positron pairs are not permitted in the intermediate states in many-body perturbation summation.

The eigenfunctions of the matrix DF SCF equation clearly separate into two discrete manifolds, $D^{(+)}$ and $D^{(-)}$, respectively, of positive-energy and negative-energy states. As a result, the positive-energy projection operators can be accommodated easily in many-body calculations. The formal conditions on the projection are automatically satisfied when only the positive-energy spinors ($\in D^{(+)}$) are employed. h_D is the Dirac one-electron Hamiltonian (in a.u.)

$$h_D(i) = c(\boldsymbol{\alpha}_i \cdot \mathbf{p}_i) + (\beta_i - 1)c^2 + V_{nuc}(r_i). \tag{2}$$

Here $\boldsymbol{\alpha}$ and β are the 4×4 Dirac vector and scalar matrices, respectively. $V_{nuc}(r)$ is the nuclear potential, which for each nucleus takes the form

$$Vnuc(r) = \begin{cases} -\frac{Z}{r}, r > R, \\ -\frac{Z}{2R}(3 - \frac{r^2}{R^2}), r \leq R. \end{cases} \tag{3}$$

The nuclei are modeled as spheres of uniform proton-charge distribution; Z is the nuclear charge. R is the radius of that nucleus and is related to the atomic mass, A , by R= $2.2677 \cdot 10^{-5} A^{1/3}$. Adding the frequency-independent Breit interaction,

$$B_{12} = -\frac{1}{2}[\boldsymbol{\alpha}_1 \cdot \boldsymbol{\alpha}_2 + (\boldsymbol{\alpha}_1 \cdot \mathbf{r}_{12})(\boldsymbol{\alpha}_2 \cdot \mathbf{r}_{12})/r_{12}^2]/r_{12}, \tag{4}$$

to the electron-electron Coulomb interaction, in Coulomb gauge, results in the Coulomb-Breit potential which is correct to order α^2 (α being the fine structure constant) [21]. Addition of the Breit term yields the no-pair Dirac-Coulomb-Breit (DCB) Hamiltonian [21, 22]

$$H^+_{DCB} = \sum_i^N h_D(i) + \mathcal{L}_+ \left(\sum_{i>j}^N \frac{1}{r_{ij}} + B_{ij} \right) \mathcal{L}_+, \tag{5}$$

which is covariant to first order and increases the accuracy of calculated fine-structure splittings and inner-electron binding energies.

The effective Hamiltonian approach is a viable one for atoms and molecules because it translates into mathematical formalism the idea that atoms and molecules are weakly-bound inhomogeneous systems in which pair production processes are absent. In atoms and molecules, particle numbers are conserved and one can treat particle number nonconserving terms as a small perturbation. Higher-order QED effects appear first in order α^3. To evaluate the higher-order QED effects one needs to go beyond the no-pair approximation [23, 24].

2.2. THE MATRIX MULTICONFIGURATION DIRAC-FOCK SCF METHOD

N-electron eigenfunctions of the no-pair DC Hamiltonian are approximated by a linear combination of M configuration-state functions, $\{\Phi_I^{(+)}(\gamma_I \mathcal{J}\pi)$ $;I = 1, 2, \ldots, M\}$, constructed from positive-energy eigenfunctions of the matrix DF SCF equation. The M configuration-state functions form a subspace $P^{(+)}$ of the positive-energy space $\mathcal{D}^{(+)}$

$$\psi_K(\gamma_K \mathcal{J}\pi) = \sum_I^M C_I^{\gamma_K \mathcal{J}\pi} \Phi_I^{(+)}(\gamma_I \mathcal{J}\pi). \tag{6}$$

Here the MC DF SCF wave function $\psi_K(\gamma_K \mathcal{J}\pi)$ is an eigenfunction of the angular momentum and parity operators with total angular momentum $\mathcal{J}$ and parity π. γ denotes a set of quantum numbers other than $\mathcal{J}$ and π necessary to specify the state uniquely. In the following, $C_I^{\gamma_K \mathcal{J}\pi}$ is abbreviated as $C_I^{\gamma_K}$. The total DC energy of the general MC DF state $|\psi_K(\gamma_K \mathcal{J}\pi)>$ can be expressed as

$$E^{MC}(\gamma_K \mathcal{J}\pi) =< \psi_K(\gamma_K \mathcal{J}\pi)|H^+_{DC}|\psi_K(\gamma_K \mathcal{J}\pi) >=$$

$$= \sum_{I,J=1}^{P(+)} C_I^{\gamma_K} C_J^{\gamma_K} < \Phi_I^{(+)}(\gamma_I \mathcal{J}\pi)|H^+_{DC}|\Phi_J^{(+)}(\gamma_J \mathcal{J}\pi) > . \tag{7}$$

Here it is assumed that $\psi_K(\gamma_K \mathcal{J}\pi)$ and $\Phi_I^{(+)}(\gamma_I \mathcal{J}\pi)$ are normalized.

The hermiticity of the Hamiltonian has been employed to reduce the number of terms in the summation, and the total energy can be expressed in terms of the unique elements of the one- and two-particle radial integrals,

$$E^{MC}(\gamma_K \mathcal{J} \pi) = \sum_{\alpha=1}^{N_t} t_\alpha I(a_\alpha b_\alpha) + \sum_{\beta=1}^{N_V} V_\beta R^{\nu_\beta}(a_\beta b_\beta, c_\beta d_\beta), \tag{8}$$

where N_t and N_V are the numbers of nonzero t_α and V_β coefficients. The short notation for the radial integrals has been used:

$$I(ab) = I(n_a \kappa_a n_b \kappa_b) =< \phi^{(0)}_{n_a \kappa_a}(r) | h_D(r) | \phi^{(0)}_{n_b \kappa_b}(r) > \tag{9}$$

$$R^\nu(ab, cd) = R^\nu(n_a \kappa_a n_b \kappa_b, n_c \kappa_c n_d \kappa_d) =$$

$$=< \phi^{(0)}_{n_a \kappa_a}(r_1) \phi^{(0)}_{n_b \kappa_b}(r_2) | \frac{r_<^\nu}{r_>^{\nu+1}} | \phi^{(0)}_{n_c \kappa_c}(r_1) \phi^{(0)}_{n_d \kappa_d}(r_2) > . \tag{10}$$

The generalized coefficients t_α and V_β are expressed in terms of nonzero angular coefficients t_α^{IJ} and V_β^{IJ}

$$t_\alpha = \sum_{\alpha'=1}^{N'_t} t_{\alpha'}^{IJ} \delta(\alpha, \alpha') \{2 - \delta_{IJ}\} C_I^{\gamma K} C_J^{\gamma K} \tag{11}$$

$$V_\beta = \sum_{\beta'=1}^{N'_V} V_{\beta'}^{IJ} \delta(\beta, \beta') \{2 - \delta_{IJ}\} C_I^{\gamma K} C_J^{\gamma K} \tag{12}$$

The angular coefficients t_α^{IJ} and V_β^{IJ} account for the symmetries of the radial integrals $I(a_\alpha b_\alpha)$ and $R^\nu(a_\beta b_\beta, c_\beta d_\beta)$, and the notations $\alpha = \{a_\alpha b_\alpha\}$ and $\beta = \{\nu_\beta, a_\beta b_\beta, c_\beta d_\beta\}$ have been used.

The following notations are used: the indices e and f denote occupied spinors, the indices p, q, r and s denote any of the occupied or virtual spinors (both positive and negative energy spinors), the indices I, J and K denote CI coefficients and the indices a, b, c, d, ν are reserved for the sets α and β describing unique radial integrals.

Given a trial orthonormal set of one-particle radial spinors $\{\phi^{(0)}_{n_p \kappa_p}(r)\}$ ($\in D^{(+)} \cup D^{(-)}$), the optimum occupied electronic radial spinors $\{\phi^{(+)}_{n_e \kappa_e}(r)\}$ ($\in D^{(+)}$) can be found by a unitary transformation $\mathbf{U} = 1 + \mathbf{T}$ via

$$\phi^{(+)}_{n_e \kappa_e}(r) = \frac{1}{r} \begin{pmatrix} P_{n_e \kappa_e}(r) \\ Q_{n_e \kappa_e}(r) \end{pmatrix} = \sum_{p \in D^{(+)} \cup D^{(-)}}^{2N_\kappa} \phi^{(0)}_{n_p \kappa_p}(r) U_{pe} =$$

$$= \sum_p^{2N_\kappa} \phi^{(0)}_{n_p \kappa_p}(r)(T_{pe} + \delta_{pe}). \tag{13}$$

Here, the summation extends over both N$_\kappa$ negative- and N$_\kappa$ positive-energy spinors. The summation involving negative-energy spinors may not be excluded [20, 25] because the negative-energy spinors form part of the complete set of eigenfunctions of MC DF mean-field equations and account for the polarization of the vacuum due to mean-field potential. $P_{n\kappa}(r)$ and $Q_{n\kappa}(r)$ are the large and small radial components and are expanded in N_κ G spinors, $\{\chi^L_{\kappa i}\}$ and $\{\chi^S_{\kappa i}\}$, that satisfy the boundary conditions associated with the finite nucleus,

$$P_{n\kappa}(r) = \sum_i^{N_\kappa} \chi^L_{\kappa i}\xi^L_{ki}, \qquad Q_{n\kappa}(r) = \sum_i^{N_\kappa} \chi^S_{\kappa i}\xi^S_{ki}. \tag{14}$$

Here $\{\xi^L_{ki}\}$ and $\{\xi^S_{ki}\}$ are linear variation coefficients.

In terms of the powers of the spinor variation parameters $\mathbf{T} = \{T_{pe}\}$, the energy $E^{MC}(\gamma_K \mathcal{J}\pi)$ in Eq. (8) is a fourth-order function of rotation matrix elements T_{pe}. Inserting the optimum spinor expression (13) into Eq. (8) and collecting the terms of the same power of $\mathbf{T}$, the energy expression to second-order $E^{(2)}(\mathbf{T})$ in T_{pe} is obtained

$$E^{(2)}(\mathbf{T}) = E^{(o)} + \Delta E^{(1)}(\mathbf{T}) + \Delta E^{(2)}(\mathbf{T}). \tag{15}$$

Variation of the approximate energy $E^{(2)}(\mathbf{T})$ with respect to parameters T_{pe} leads to the Newton-Raphson (NR) equations:

$$\frac{\partial E^{(2)}(\mathbf{T})}{\partial T_{pe}} = g^o_{pe} + \sum_{qf} h^{oo}_{pe,qf} T_{qf} = 0, \tag{16}$$

where the gradient with respect to T_{pe} is

$$g^o_{pe} = \frac{\partial \Delta E^{(1)}(\mathbf{T})}{\partial T_{pe}} = \sum_{\alpha=1}^{N_t} t_\alpha [I(pb_\alpha)\delta(e,a_\alpha) + I(a_\alpha p)\delta(e,b_\alpha)]$$

$$+\sum_{\beta=1}^{N_V} V_\beta \cdot \{R^{\nu_\beta}(pb_\beta, c_\beta d_\beta)\delta(e,a_\beta) + R^{\nu_\beta}(a_\beta p, c_\beta d_\beta)\delta(e,b_\beta)$$

$$+R^{\nu_\beta}(a_\beta b_\beta, pd_\beta)\delta(e,c_\beta) + R^{\nu_\beta}(a_\beta b_\beta, c_\beta p)\delta(e,d_\beta)\} \tag{17}$$

and the Hessian matrix with respect to T_{pe} and T_{qf} is

$$h^{oo}_{pe,qf} = \frac{\partial^2 \Delta E^{(2)}(\mathbf{T})}{\partial T_{pe}\partial T_{qf}} =$$

$$= \sum_{\alpha=1}^{N_t} t_\alpha [I(pq)\delta(e,a_\alpha)\delta(f,b_\alpha) + I(qp)\delta(e,b_\alpha)\delta(f,a_\alpha)] + \sum_{\beta=1}^{N_V} V_\beta$$

$$\times\{R^{\nu_\beta}(pq, c_\beta d_\beta)\delta(e, a_\beta)\delta(f, b_\beta) + R^{\nu_\beta}(pb_\beta, qd_\beta)\delta(e, a_\beta)\delta(f, c_\beta)$$
$$+R^{\nu_\beta}(pb_\beta, c_\beta q)\delta(e, a_\beta)\delta(f, d_\beta) + R^{\nu_\beta}(a_\beta p, qd_\beta)\delta(e, b_\beta)\delta(f, c_\beta)$$
$$+R^{\nu_\beta}(a_\beta p, c_\beta q)\delta(e, b_\beta)\delta(f, d_\beta) + R^{\nu_\beta}(a_\beta b_\beta, pq)\delta(e, c_\beta)\delta(f, d_\beta)$$
$$+R^{\nu_\beta}(qp, c_\beta d_\beta)\delta(f, a_\beta)\delta(e, b_\beta) + R^{\nu_\beta}(qb_\beta, pd_\beta)\delta(f, a_\beta)\delta(e, c_\beta)$$
$$+R^{\nu_\beta}(qb_\beta, c_\beta p)\delta(f, a_\beta)\delta(e, d_\beta) + R^{\nu_\beta}(a_\beta q, pd_\beta)\delta(f, b_\beta)\delta(e, c_\beta)$$
$$+R^{\nu_\beta}(a_\beta q, c_\beta p)\delta(f, b_\beta)\delta(e, d_\beta) + R^{\nu_\beta}(a_\beta b_\beta, qp)\delta(f, c_\beta)\delta(e, d_\beta)\}. \quad (18)$$

To account for the orthogonality constraints, terms involving Lagrange multipliers must be added to the energy functional:

$$W = E^{(2)}(\mathbf{T}) + \sum_{ef}^{N_w} \lambda_{ef}(\delta_{ef} - S_{ef}) \quad (19)$$

where $S_{ef} = \langle\ \phi^{(0)}_{n_e\kappa_e}(r) \mid \phi^{(0)}_{n_f\kappa_f}(r)\rangle = \delta_{ef}$ is the overlap between spinors e and f (an orthonormal trial set of radial spinors was assumed) and $\{\lambda_{ef}\}$ are the Lagrange multipliers.

The CI coefficients $C_I^{\gamma K}$ (Eq.6) are not constant, and variations over them must also be incorporated in the second-order energy. Consider two sets of CI coefficients - $\mathbf{C}^\gamma = \{C_I^\gamma\}$ (optimum) and $\mathbf{C}^{(0)\gamma} = \{C_I^{(0)\gamma}\}$(approximate). Using Taylor expansion of the energy and expansion of ΔC_I^γ in terms of the CI vectors $\{C_I^{(0)\gamma'}\}$

$$\Delta C_I^\gamma = C_I^\gamma - C_I^{(0)\gamma} = \sum_{\gamma'=1}^{N_{CSF}} A_{\gamma'} C_I^{(0)\gamma'} - C_I^{(0)\gamma} = \sum_{\gamma'=1}^{N_{CSF}} B_{\gamma'} C_I^{(0)\gamma'} \quad (20)$$

the second-order energy can be expressed in terms of ΔC_I^γ or $B_{\gamma'}$. Let us define

$$h^{co}_{\gamma',pe} = h^{oc}_{pe,\gamma'} = \sum_I^{N_{CSF}} [\sum_\alpha^{N_t} t^I_\alpha \{I(pb_\alpha)\delta(e, a_\alpha) + I(a_\alpha p)\delta(e, b_\alpha)\}$$

$$+\sum_{\beta=1}^{N_V} V^I_\beta \cdot \{R^{\nu_\beta}(pb_\beta, c_\beta d_\beta)\delta(e, a_\beta) + R^{\nu_\beta}(a_\beta p, c_\beta d_\beta)\delta(e, b_\beta)$$

$$+R^{\nu_\beta}(a_\beta b_\beta, pd_\beta)\delta(e, c_\beta) + R^{\nu_\beta}(a_\beta b_\beta, c_\beta p)\delta(e, d_\beta)\}]C_I^{(0)\gamma'} \quad (21)$$

with

$$t_\alpha^K = \frac{\partial t_\alpha}{\partial C_K^\gamma}|_{C^{(0)\gamma}} = \sum_{\alpha'=1}^{N_t} t_{\alpha'}^{IJ}\delta(\alpha,\alpha')\{C_J^{(0)\gamma}\delta(I,K) + C_I^{(0)\gamma}\delta(J,K)\} \quad (22)$$

$$V_\beta^K = \frac{\partial V_\beta}{\partial C_K^\gamma}|_{C^{(0)\gamma}} = \sum_{\beta'=1}^{N_V} V_{\beta'}^{IJ}\delta(\beta,\beta')\{C_J^{(0)\gamma}\delta(I,K) + C_I^{(0)\gamma}\delta(J,K)\} \quad (23)$$

and

$$g_{\gamma'}^c = 2E^{(0)\gamma}\delta_{\gamma'\gamma} \quad (24)$$

$$h_{\gamma',\gamma''}^{cc} = 2E^{\gamma'}\delta_{\gamma'\gamma''}. \quad (25)$$

If we add $B_{\gamma'}$ to our set of variational parameters, we obtain the second-order Newton-Raphson equations

$$\begin{pmatrix} g_{pe}^o \\ g_{\gamma'}^c \end{pmatrix} + \sum_{qf\gamma''} \begin{pmatrix} h_{pe,qf}^{oo} & h_{pe,\gamma''}^{oc} \\ h_{\gamma',qf}^{co} & h_{\gamma',\gamma''}^{cc} \end{pmatrix} \begin{pmatrix} T_{qf} \\ B_{\gamma''} \end{pmatrix} = \begin{pmatrix} 0 \\ 0 \end{pmatrix}. \quad (26)$$

As with the spinor orthogonality constraints, the normalization condition $\sum_{\gamma'} A_{\gamma'}A_{\gamma'} = 1$ of the CI vectors must be incorporated:

$$W = E^{(2)}(\mathbf{T}, \Delta\mathbf{C}^\gamma) + \sum_{ef}\lambda_{ef}(\delta_{ef} - S_{ef}) + \Lambda(1 - \sum_{\gamma'} A_{\gamma'}A_{\gamma'}). \quad (27)$$

For the ground electronic state, the Hessian matrix possesses N_κ positive and N_κ negative eigenvalues corresponding to a minimum and a maximum, respectively, in the space of large and small component parameters. Therefore, the energy functional is minimized with respect to spinor rotations between the occupied electronic spinors and the positive energy virtual spinors. The functional is maximized with respect to spinor rotations between the occupied electronic spinors and the negative energy spinors. By maximazing the vacuum charge-current density polarization contribution, the MC DF mean-field potential defines its dressed vacuum.

The quadratically convergent NR algorithm for relativistic MC DF SCF calculations has been discussed in detail in previous work [20]. To remove the arbitrariness of the MC SCF spinors and density weighting, the canonical SCF spinors are transformed into natural spinors $\{\omega_{n_p\kappa_p}^{(+)}\}$ for subsequent perturbation calculations [26]. The key to successful implementation of the subsequent MR-MP perturbation theory calculations is rapid convergence of our quadratically convergent matrix MC DF SCF method [20] for a general class of MC DF wave functions for openshell quasidegenerate systems.

2.3. RELATIVISTIC MULTIREFERENCE MANY-BODY PERTURBATION THEORY

The no-pair DC Hamiltonian H_{DC}^{+} is partitioned into an unperturbed Hamiltonian and a perturbation term following Møller and Plesset [27],

$$H_{DC}^{+} = H_0 + V, \tag{28}$$

where the unperturbed model Hamiltonian H_0 is a sum of "average" DF operators F_{av}

$$H_0 = \sum_{i}^{N} F_{av}(i), \qquad V = H_{DC}^{+} - \sum_{i}^{N} F_{av}(i). \tag{29}$$

Here, the one-body operator F_{av} diagonal in $\{\omega_{n_p\kappa_p}^{(+)}\}$ may be defined by

$$\begin{aligned} F_{av} &= \sum_{p\in D(+)} |\omega_{n_p\kappa_p}^{(+)} >< \omega_{n_p\kappa_p}^{(+)}| f_{av} |\omega_{n_p\kappa_p}^{(+)} >< \omega_{n_p\kappa_p}^{(+)}| \\ &= \sum_{p\in D(+)} |\omega_{n_p\kappa_p}^{(+)} > \varepsilon_p^{+} < \omega_{n_p\kappa_p}^{(+)}|, \end{aligned} \tag{30}$$

where

$$\varepsilon_p^{+} =< \omega_{n_p\kappa_p}^{(+)}| f_{av} |\omega_{n_p\kappa_p}^{(+)} >, \qquad f_{av} = h_D + \sum_{p}^{occ} \widetilde{n}_p \left(J_p - K_p\right). \tag{31}$$

The generalized fractional occupation $\widetilde{n}_p$ is related to diagonal matrix elements of the first-order reduced density matrix constructed in natural spinors by

$$\widetilde{n}_p = D_{pp} = \sum_{I}^{P(+)} C_I^{\gamma\kappa} C_I^{\gamma\kappa} n_{n_p\kappa_p}[I], \tag{32}$$

where $n_{n_p\kappa_p}[I]$ is the occupation number of the $n_p\kappa_p$ shell in the CSF $\Phi_I(\gamma_I \mathcal{J}\pi)$. J_p and K_p are the usual Coulomb and exchange operators constructed in natural spinors.

The unperturbed Hamiltonian H_0 may be given in second quantized form,

$$H_0 = \sum_{p \in D(+)} \left\{ a_p^+ a_p \right\} \varepsilon_p^+, \tag{33}$$

where $\left\{ a_p^+ a_p \right\}$ is a normal product of creation and annihilation operators, a_p^+ and a_p, respectively. The zero-order Hamiltonian, H_0, is arbitrary but should be chosen as close to the full Hamiltonian H_{DC}^+ as possible so that the perturbation series converges rapidly in low order. The zero-order Hamiltonian is usually chosen to be a sum of effective one-electron operators (Møller-Plesset partitioning [27]). For closed-shell systems, the best results have been obtained with Møller-Plesset partitioning, i.e. the sum of closed-shell Fock operators as H_0. An effective one-body operator for general MC DF SCF closely related to the closed-shell Fock operator is the "average" DF operator F_{av} , a relativistic generalization of a nonrelativistic average Fock operator [26]. The theory provides a hierarchy of well-defined algorithms that allow one to calculate relativistic correlation corrections in non-iterative steps and, in low order, yields a large fraction of the dynamical correlation. In this form of partitioning, perturbation corrections describe relativistic electron correlation, including cross contributions between relativistic and correlation effects.

Many-electron wave functions correct to α^2 may be expanded in a set of CSFs that spans the entire N-electron positive-energy space $D^{(+)}$, $\{\Phi_I^{(+)}(\gamma_I \mathcal{J}\pi)\}$, constructed in terms of Dirac one-electron spinors. Individual CSFs are eigenfunctions of the total angular momentum and parity operators and are linear combinations of antisymmetrized products of positive-energy spinors ($\in D^{(+)}$). The one-electron spinors are mutually orthogonal so the CSFs $\{\Phi_I^{(+)}(\gamma_I \mathcal{J}\pi)\}$ are mutually orthogonal. The unperturbed Hamiltonian is diagonal in this space:

$$H_0 = \sum_{I}^{D(+)} |\Phi_I^{(+)}(\gamma_I \mathcal{J}\pi) > E_I^{CSF} < \Phi_I^{(+)}(\gamma_I \mathcal{J}\pi)|, \tag{34}$$

so that

$$H_0|\Phi_I^{(+)}(\gamma_I \mathcal{J}\pi) >= E_I^{CSF}|\Phi_I^{(+)}(\gamma_I \mathcal{J}\pi) > (I = 1, 2, \ldots). \tag{35}$$

Since the zero-order Hamiltonian is defined as a sum of one-electron operators F_{av} (Eq. 29), E_I^{CSF} is a sum of the products of one-electron energies defined by ε_q^+ and an occupation number $n_{n_q\kappa_q}[I]$ of the $n_q\kappa_q$ shell in the CSF $\Phi_I^{(+)}(\gamma_I \mathcal{J}\pi)$:

$$E_I^{CSF} = \sum_q^{D(+)} \varepsilon_q^+ n_{n_q\kappa_q}[I]. \tag{36}$$

The subset, $\{\Phi_I^{(+)}(\gamma_I\mathcal{J}\pi); I = 1, 2, \ldots, M\}$, with which we expand the MC DF SCF function $\psi_K(\gamma_K\mathcal{J}\pi)$ (Eq. 6) also defines an active subspace $P^{(+)}$ spanned by $\psi_K(\gamma_K\mathcal{J}\pi)$ and its M-1 orthogonal complements, $\{\psi_K(\gamma_K\mathcal{J}\pi); K = 1, 2, \ldots, M\}$. The matrix of H_{DC}^+ in this subspace is diagonal

$$< \psi_K(\gamma_K\mathcal{J}\pi)|H_{DC}^+|\psi_L(\gamma_L\mathcal{J}\pi) >= \delta_{KL}E^{MC}(\gamma_K\mathcal{J}\pi), \tag{37}$$

where

$$E_K^{(0)} =< \psi_K(\gamma_K\mathcal{J}\pi)|H_0|\psi_K(\gamma_K\mathcal{J}\pi) >= \sum_I^M C_I^{\gamma_K} C_I^{\gamma_K} E_I^{CSF} = \sum_p^{occ} \varepsilon_p^+ \tilde{n}_p$$

and

$$E_K^{(1)} =< \psi_K(\gamma_K\mathcal{J}\pi)|V|\psi_K(\gamma_K\mathcal{J}\pi) > .$$

The residual space in the positive-energy subspace is $Q^{(+)} = D^{(+)} - P^{(+)}$, which is spanned by CSFs $\{\Phi_I^{(+)}(\gamma_I\mathcal{J}\pi); I = M+1, M+2, \ldots\}$.

Application of Rayleigh-Schrödinger perturbation theory provides order-by-order expressions of the perturbation series for the state approximated by $|\psi_K(\gamma_K\mathcal{J}\pi) >$,

$$E_K(\gamma_K\mathcal{J}\pi) = E^{MC}(\gamma_K\mathcal{J}\pi) + E_K^{(2)} + E_K^{(3)} + \ldots, \tag{38}$$

where

$$E_K^{(2)} =< \psi_K(\gamma_K\mathcal{J}\pi)|V\mathcal{R}V|\psi_K(\gamma_K\mathcal{J}\pi) > \tag{39}$$

and

$$E_K^{(3)} =< \psi_K(\gamma_K\mathcal{J}\pi)|V\mathcal{R}(H_0 - E_K^{(1)})\mathcal{R}V|\psi_K(\gamma_K\mathcal{J}\pi) > . \tag{40}$$

Here, $\mathcal{R}$ is the resolvent operator,

$$\mathcal{R} = \frac{\mathcal{Q}^{(+)}}{E_K^{(0)} - H_0}, \mathcal{Q}^{(+)} = \sum_I^{Q(+)} |\Phi_I^{(+)}(\gamma_I\mathcal{J}\pi) >< \Phi_I^{(+)}(\gamma_I\mathcal{J}\pi)|. \tag{41}$$

The projection operator $\mathcal{Q}^{(+)}$ projects onto the subspace $Q^{(+)}$ spanned by CSFs $\{\Phi_I^{(+)}(\gamma_I\mathcal{J}\pi); I = M+1, M+2, \ldots\}$. Using the spectral resolution of the resolvent operator acting on $V|\Phi_I^{(+)}(\gamma_I \mathcal{J}\pi)>$, the second-order correction may be expressed as

$$E_K^{(2)} = \sum_{IJ} C_I^{\gamma K} C_J^{\gamma K} < \Phi_I^{(+)}(\gamma_I \mathcal{J}\pi)|V\mathcal{R}V|\Phi_J^{(+)}(\gamma_J \mathcal{J}\pi)> =$$

$$= \sum_{L=M+1}^{Q(+)} \sum_{I,J=1}^{P(+)} C_I^{\gamma K} C_J^{\gamma K} \frac{< \Phi_I^{(+)}(\gamma_I \mathcal{J}\pi)|V|\Phi_L^{(+)}(\gamma_L \mathcal{J}\pi)>}{E_J^{CSF} - E_L^{CSF}} \times$$

$$\times < \Phi_L^{(+)}(\gamma_L \mathcal{J}\pi)|V|\Phi_J^{(+)}(\gamma_J \mathcal{J}\pi)> . \tag{42}$$

In this form, all perturbation corrections beyond first order describe relativistic electron correlation for the state approximated by the MC DF SCF wavefunction $|\psi_K(\gamma_K \mathcal{J}\pi)>$. When the effective electron-electron interaction is approximated by the instantaneous Coulomb interaction $\frac{1}{r_{ij}}$, relativistic electron correlation is termed Dirac-Coulomb (DC) correlation [5]. Inclusion of the frequency-independent Breit interaction in the effective electron-electron interaction yields the no-pair DCB Hamiltonian (Eq. 5), and relativistic electron correlation arising from the DCB Hamiltonian is the DCB correlation [5].

Summations over the CSFs in Eqs. 34 through 42 are restricted to CSFs ($\in D^{(+)}$) constructed from the positive-energy branch ($D^{(+)}$) of the spinors, effectively incorporating into the computational scheme the "no-pair" projection operator $\mathcal{L}_+$ contained in the DC and DCB Hamiltonians. Further, the CSFs $\Phi_L^{(+)}(\gamma_L \mathcal{J}\pi)$ ($\in Q^{(+)}$) generated by excitations higher than double, relative to the reference CSFs $\Phi_I^{(+)}(\gamma_I \mathcal{J}\pi)$ ($\in P^{(+)}$), do not contribute to the second- and third-order because for them $< \Phi_I^{(+)}(\gamma_I \mathcal{J}\pi)|V|\Phi_L^{(+)}(\gamma_L \mathcal{J}\pi)> = 0$, $< \Phi_I^{(+)}(\gamma_I \mathcal{J}\pi)|H_{DC}^{+}|\Phi_L^{(+)}(\gamma_L \mathcal{J}\pi)> = 0$.

Neglecting interactions with the filled negative-energy sea, i.e. neglecting virtual electron-positron pairs in summing the MBPT diagrams, we have a straightforward extension of nonrelativistic MBPT. Negative energy states ($\in D^{(+)}$), as part of the complete set of states, do play a role in higher-order QED corrections. Studies have appeared which go beyond the "no-pair" approximation where negative-energy states are needed to evaluate the higher-order QED effects [28, 23, 29, 24, 30, 31]. Contributions from the negative energy states due to creation of virtual electron-positron pairs are of the order α^3 [28, 23, 29, 24, 4, 30, 31], and estimations of the radiative corrections are necessary in order to achieve spectroscopic accuracy for higher Z. In the present study, the lowest-order radiative corrections were estimated for each state to achieve better accuracy.

2.4. COMPUTATIONAL METHOD

The large radial component is expanded in a set of Gaussian-type functions (GTF) [32]

$$\chi^L_{\kappa i}(r) = A^L_{\kappa i} r^{n[\kappa]} \exp(-\zeta_{\kappa i} r^2) \tag{43}$$

with $n[\kappa] = -\kappa$ for $\kappa < 0$, and $n[\kappa] = \kappa + 1$ for $\kappa > 0$. $A^L_{\kappa i}$ is the normalization constant. The small component basis set, $\{\chi^S_{\kappa i}(r)\}$ is constructed to satisfy the boundary condition associated with the finite nucleus with a uniform proton charge distribution [32]. With the finite nucleus, GTFs of integer power of r are especially appropriate basis functions because the finite nuclear boundary results in a solution which is Gaussian at the origin [32]. Basis functions which satisfy the nuclear boundary conditions are also automatically kinetically balanced. Imposition of the boundary conditions results in particularly simple forms with spherical G spinors [32].

For all the systems studied, even-tempered basis sets [33] of Gaussian-type were used for MC DF SCF. In basis sets of even-tempered Gaussians, the exponents, $\{\zeta_{\kappa i}\}$ are given in terms of the parameters, α and β, according to the geometric series

$$\zeta_{\kappa i} = \alpha\beta^{i-1}; i = 1, 2, \ldots, N_\kappa. \tag{44}$$

In MC DF SCF calculations on carbonlike species, the parameters α and β are optimized until a minimum in the DF total energy is found. The radial functions that possess a different κ quantum number but the same quantum number ℓ are expanded in the same set of basis functions (e.g., the radial functions of $p_{1/2}$ and $p_{3/2}$ symmetries are expanded in the same set of p-type radial Gaussian-type functions). The nuclei were again modeled as spheres of uniform proton charge in every calculation. The nuclear model has been discussed in detail in Ref. [32].

Virtual spinors used in the MR-MP perturbation calculations were generated in the field of the nucleus and all electrons (V^N potential) by employing the "average" DF operator F_{av} (Eq. 30). The order of the partial-wave expansion, L_{max}, the highest angular momentum of the spinors included in the virtual space, is $L_{max} = 7$ throughout this study. All-electron MR-MP perturbation calculations including the frequency-independent Breit interaction in the first and second orders of perturbation theory are based on the no-pair Dirac-Coulomb-Breit Hamiltonian, H^+_{DCB}. The speed of light was taken to be 137.0359895 a.u. Radiative corrections, or the Lamb shifts, were estimated for each state by evaluating the electron self-energy and vacuum polarization following an approximation scheme discussed by Indelicato, Gorceix, and Desclaux [31]. The code described in Refs. [31, 34] was adapted to our basis set expansion calculations for this purpose. In this scheme [34], the screening of the self energy is estimated by employing the charge density of a spinor integrated to a short distance from the origin, typically 0.3 Compton wavelength. The ratio of the integral computed with an MC DF SCF spinor and that obtained by using the corresponding hydrogenic spinor is used to scale the self-energy correction for a bare nuclear charge computed by Mohr [28]. The effect on the term energy splittings of mass polarization and reduced mass are non-negligible. In the present study, however, we neglect these effects.

3. Results and Discussion

With four valence electrons, ground and low-lying excited states of carbon and carbonlike ions exhibit the near degeneracy characteristic of a manifold of strongly interacting configurations within the $n = 2$ complex [10, 35]. The carbon isoelectronic sequence provides an extreme example of how relativity alters asymptotic CI. For neutral carbon, the ground state is nominally $2s^22p^2\ ^3P_0$, but it is strongly mixed with the $2s^22p^2\ ^1S_0$, $2p^4\ ^3P_0$ and $2p^4$ $^1S_0$ configuration states. At higher Z, however, the $2p^4$ configurations have much higher energy than the $2s^22p^2$, and their mixing becomes weak. In other words, relativity alters the strong CI such that correlation configurations, which are significant in low-Z ions, become negligible in high-Z ions where relativistic effects are significant [10]. In the present study, we employ recently developed relativistic MR-MP perturbation theory code [19] based on expansions in G spinors to examine the effects of strong configuration mixing on electron correlation for low-Z through high-Z carbonlike ions. We first give a detailed account of MC DF SCF and MR-MP calculations applied to carbonlike ions with Z=10, 20, and 30.

Table 1 displays the computed MC DF SCF energies, E_{MCDF}, of the lowest J=0 ($^3P_0$), J=1 ($^3P_1$), and J=2 ($^3P_2$) even-parity states of Ne^{+4} (Z=10), Ca^{+14} (Z=20), and Zn^{+24} (Z=30) along with the configuration mixing coefficients. The MC DF SCF is a complete active space SCF within the $n = 2$ complex. The second column of the Table indicates the number of complete active space CSFs, N_{CSF}, arising from the $n = 2$ complex. The MC DF energy, E_{MCDF}, is given in the third column. In the MC DF SCF calculations, the 1s spinor was kept doubly occupied, and the remaining 4 electrons were treated as active electrons in generating complete active space CSFs within the $n = 2$ shells. MC DF SCF calculations on the carbonlike ions with Z=10, 20, and 30 were performed to obtain a single set of spinors for all the J=0, 1, and 2 fine-structure states by optimizing the J-averaged MC energies:

TABLE 1. MC DF SCF energies and configuration mixing coefficients for the lowest $J = 0,1,2$ even-parity states of carbonlike ions. Notations $2p_{1/2} = 2p_-$ and $2p_{3/2} = 2p_+$ were used.

State	N_{CSF}	E_{MCDF}
	Ne^{+4} (Z=10)	
$J=0$ ($^3P_0$)	4	-120.725767
$J=1$ ($^3P_1$)	2	-120.723760
$J=2$ ($^3P_2$)	4	-120.720011
	Ca^{+14} (Z=20)	
$J=0$ ($^3P_0$)	4	-540.771797
$J=1$ ($^3P_1$)	2	-540.687998
$J=2$ ($^3P_2$)	4	-540.596945
	Zn^{+24} (Z=30)	
$J=0$ ($^3P_0$)	4	-1270.949137
$J=1$ ($^3P_1$)	2	-1270.207544
$J=2$ ($^3P_2$)	4	-1269.907970

State	$C_{2s^2 2p_-^2}$	$C_{2s^2 2p_- 2p_+}$	$C_{2s^2 2p_+^2}$	$C_{2p_-^2 2p_+^2}$	$C_{2p_- 2p_+^3}$	$C_{2p_+^4}$
			Ne^{+4} (Z=10)			
$J=0$ ($^3P_0$)	0.81980		-0.55836	0.07665		-0.10142
$J=1$ ($^3P_1$)		0.99187			0.12729	
$J=2$ ($^3P_2$)		0.58662	0.79974	0.10313	0.07530	
			Ca^{+14} (Z=20)			
$J=0$ ($^3P_0$)	0.89796		-0.42628	0.08755		-0.06540
$J=1$ ($^3P_1$)		0.99424			0.10718	
$J=2$ ($^3P_2$)		0.74262	0.66060	0.07547	0.08019	
			Zn^{+24} (Z=30)			
$J=0$ ($^3P_0$)	0.96989		-0.22200	0.09617		-0.02803
$J=1$ ($^3P_1$)		0.99617			0.08749	
$J=2$ ($^3P_2$)		0.95158	0.29411	0.03138	0.08373	

$$E^{MC}_{J-ave}(\gamma_K \mathcal{J}\pi) = \sum_{J=0,1,2} (2\mathcal{J}+1)E^{MC}(\gamma_K \mathcal{J}\pi) / \sum_{\mathcal{J}'}(2\mathcal{J}'+1) \qquad (45)$$

instead of performing state-specific MC DF SCF calculations on each fine-structure state. The approach is especially effective for computing small fine-structure splittings (i.e., near degeneracy among the $2p_{1/2}$ and $2p_{3/2}$ spinors).

The magnitude of the configuration mixing coefficients is a measure of configuration interaction. The electronic configurations, $2s^2 2p^2_{1/2}$, $2s^2 2p^2_{3/2}$,

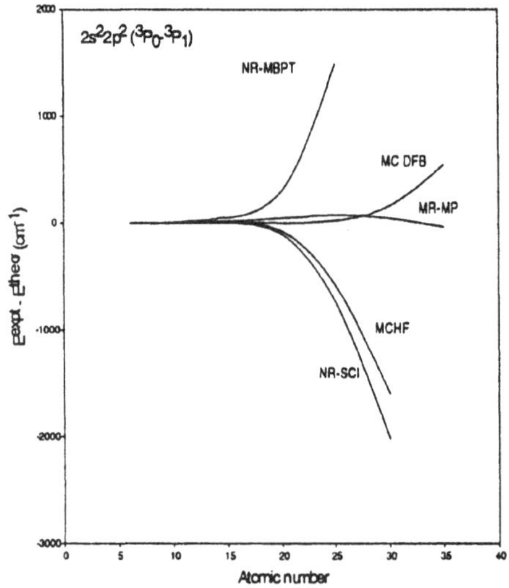

Figure 1. Deviations of the theoretical $2s^22p^2$ fine-structure separations, $^3P_0$ - $^3P_1$, from the experimental data (in cm^{-1}) as functions of Z.

TABLE 2. Contributions to the total energies (in a.u.) of $2s^22p^2\ ^3P_J$ of some carbonlike ions.

Z	E_{MCDF}	$B^{(1)}$	$E^{(2)}_{DC}$	$B^{(2)}$	LS	E_{tot}
			$2s^22p^2\ ^3P_0$			
10	-120.725767	0.015746	-0.152927	-0.001729	0.010857	-120.864677
20	-540.771797	0.149773	-0.166455	-0.007769	0.126073	-540.796248
30	-1270.949137	0.547108	-0.170089	-0.018812	0.502891	-1270.590929
			$2s^22p^2\ ^3P_1$			
10	-120.723760	0.015592	-0.152935	-0.001725	0.010861	-120.862828
20	-540.687998	0.145945	-0.166952	-0.007721	0.126306	-540.716726
30	-1270.207544	0.523372	-0.172102	-0.018250	0.504875	-1269.874524
			$2s^22p^2\ ^3P_2$			
10	-120.720011	0.014985	-0.152948	-0.001722	0.010869	-120.859697
20	-540.596945	0.138962	-0.167700	-0.007694	0.126450	-540.633378
30	-1269.907970	0.503745	-0.177124	-0.018307	0.505025	-1269.599655

$2p^2_{1/2}2p^2_{3/2}$, and $2p^4_{3/2}$ give rise to four J=0 even-parity states (two $^3P_0$ states and two $^1S_0$ states), and they do interact strongly in low-Z species. The $2p_{1/2}$ and $2p_{3/2}$ spinors are nearly degenerate in low-Z Ne^{+4} ion because rel-

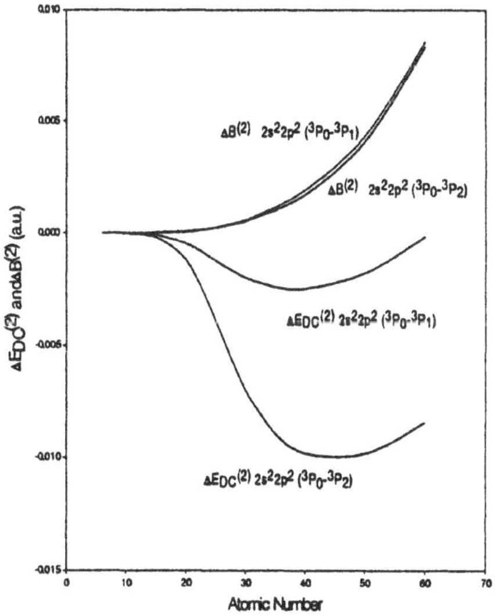

Figure 2. Contributions of the second-order Dirac-Coulomb, $\Delta E_{DC}^{(2)}$, and second-order Breit, $\Delta B^{(2)}$, correlation energies to the $2s^2 2p^2$ fine-structure separations, $^3P_0$ -$^3P_{1,2}$, as functions of Z (in a.u.).

ativistic effects are small, and the CSFs arising from $2s^2 2p_{1/2}^2$ and $2s^2 2p_{3/2}^2$ configurations are nearly degenerate, and there is a strong configuration interaction between them. Four-configuration MC DF SCF calculations yield the configuration mixing coefficients, 0.81980, -0.55836, 0.07665 and -0.10142, respectively, for the lowest J=0 ($^3P_0$) state of Ne^{+4}, with coefficients nearly equal in magnitude for the two CSFs arising from the $2s^2 2p_{1/2}^2$ and $2s^2 2p_{3/2}^2$. As Z increases, relativity lifts the near degeneracy and significantly weakens the configuration interaction between the two CSFs because it induces a large separation between the $2p_{1/2}$ and $2p_{3/2}$ spinor energies and simultaneously a smaller separation between the $2s_{1/2}$and $2p_{1/2}$ spinor energies (the $2s_{1/2}$ and $2p_{1/2}$ spinor energies become asymptotically degenerate in the hydrogenic limit). Table 1 displays just such a trend as the nuclear charge increases. Four-configuration MC DF SCF on the J=0 state of Zn^{+24} yields the configuration mixing coefficients, 0.96989 and -0.22200, respectively, for the two CSFs arising from the $2s^2 2p_{1/2}^2$ and $2s^2 2p_{3/2}^2$. The configuration interaction between the two CSFs for Zn^{+24} (Z=30) is reduced dramatically by relativity, making $2s^2 2p_{1/2}^2$ the dominant configuration. The $n = 2$ complex gives rise to two CSFs for the J=1, even-parity state. These CSFs come from the electronic configurations $2s^2 2p_{1/2} 2p_{3/2}$

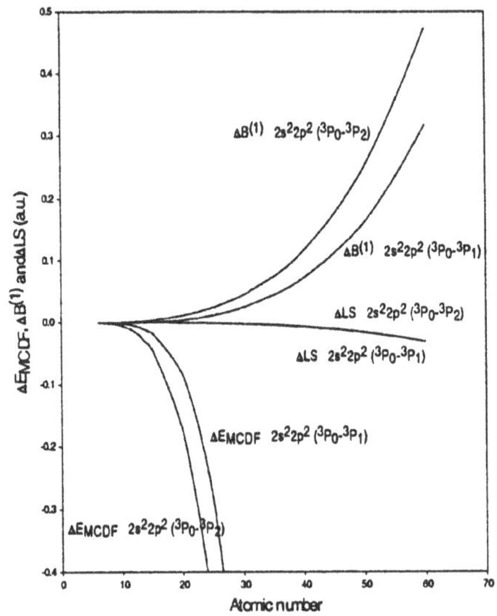

Figure 3. Contributions of the Lamb shift, ΔLS, and first-order Breit, $\Delta B^{(1)}$, corrections to the $2s^2 2p^2$ fine-structure separations, $^3P_0$ -$^3P_{1,2}$, as functions of Z (in a.u.).

and $2p_{1/2}2p^3_{3/2}$, the former being the dominant confguration with a mixing coefficient 0.99187. Thus the J=1 state does not exhibit near-degeneracy in the low- to high-Z series.

Within the $n = 2$ complex, the electronic configurations, $2s^2 2p_{1/2}\, 2p_{3/2}$, $2s^2 2p^2_{3/2}$, $2p^2_{1/2}2p^2_{3/2}$, and $2p_{1/2}2p^3_{3/2}$, give rise to J=2, even-parity CSFs (two $^3P_2$ and two $^1D_2$ states), and these interact strongly. Four-configuration MC DF SCF calculations on the ground $^3P_2$ state, including the four J=2 CSFs of Ne^{+4}, yield configuration mixing coefficients of 0.58662, 0.79974, 0.10313, and 0.07530, respectively, for the $2s^2 2p_{1/2}2p_{3/2}$, $2s^2 2p^2_{3/2}$, $2p^2_{1/2}2p^2_{3/2}$, and $2p_{1/2}2p^3_{3/2}$ CSFs, indicating near degeneracy, while the configuration mixing coefficients become 0.95158, 0.29411, 0.03138 and 0.08373 for the heavier Zn^{+24} with the configuration $2s^2 2p_{1/2}2p_{3/2}$ being more dominant. Again, as Z increases, relativity causes a large separation of the $2p_{1/2}$ and $2p_{3/2}$ spinor energies and weakens the configuration interaction between the $2s^2 2p_{1/2}2p_{3/2}$ and $2s^2 2p^2_{3/2}$ CSF.

The bulk of the experimentally determined fine-structure term energies are reproduced by the MC DF calculations within the $n = 2$ complex. In Ne^{+4}, the lowest $^3P_1$ (J=1) and $^3P_2$ (J=2) state energies computed in two- and four-configuration MC DF SCF calculations are, respectively, 440 cm^{-1} and 1263 cm^{-1} above the ground $^3P_0$ (J=2) state computed in four-

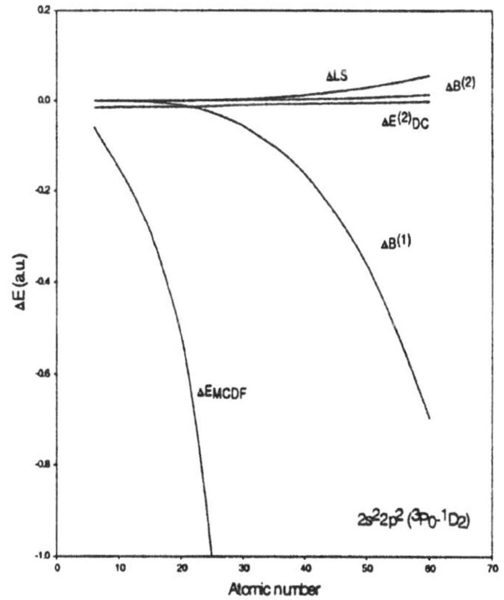

Figure 4. Contributions from each order of perturbation theory to the term energy $2s^2 2p^2\ ^3P_0$ - $2s^2 2p^2\ ^1D_2$ (in a.u.).

configuration MC DF SCF, while experimental values are, respectively, 413 cm^{-1} and 1111 cm^{-1} [36, 37]. For Zn^{+24}, the lowest $^3P_0$ (J=0) and $^3P_1$ (J=1) state energies computed in two- and four-configuration MC DF SCF calculations are, respectively, 162761 cm^{-1} and 228510 cm^{-1} above the ground $^3P_2$ (J=2) state computed by four-configuration MC DF SCF, while the corresponding experimental values are, respectively, 157717 cm^{-1} and 218009 cm^{-1} [36, 37]. The residual discrepancy is primarily due to dynamic correlation and radiative corrections unaccounted for in the MC DF SCF calculations.

State-specific MR-MP calculations were carried out on the lowest ^3P_J (J=0, 1, 2) states as well as eleven low-lying excited states of carbon and sixteen carbonlike ions with Z=7, 8, 9, 10, 11, 12, 13, 14, 15, 20, 25, 30, 35, 40, 50, 60. The fourteen lowest states consist of two J=0 ($^3P_0$ and $^1S_0$), one J=1 ($^3P_1$), and two J=2 ($^3P_2$ and $^1D_2$) even-parity states arising from the $2s^2 2p^2$ configuration, one J=0 ($^1S_0$), two J=2 ($^3P_2$ and $^1D_2$) even-parity states arising from the $2p^4$, and two J=2 ($^5S_2^o$ and $^1D_2^o$), three J=1 ($^3D_1^o$, $^3S_1^o$, and $^1P_1^o$), and one J=0 ($^3P_0^o$) odd-parity states arising from the $2s2p^3$. Critically evaluated experimental data are available for these ions up to Z=35 [36, 37]. Table 2 displays the computed MR-MP second-order DC energies, $E_{DC}^{(2)}$, first- and second-order Breit interaction energies, $B^{(1)}$ and $B^{(2)}$, and the Lamb shift correction, LS, of the lowest J=0 ($^3P_0$), J=1

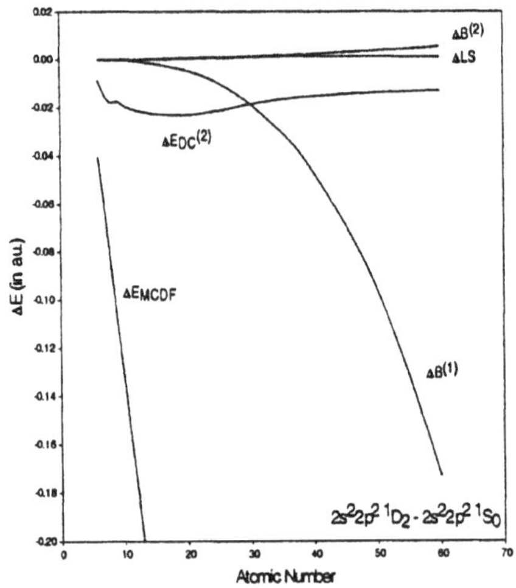

Figure 5. Contributions from each order of perturbation theory to the term energy $2s^2 2p^2\ ^1D_2$ - $2s^2 2p^2\ ^1S_0$ (in a.u.).

($^3P_1$), and J=2 ($^3P_2$) even-parity states of Ne^{+4} (Z=10), Ca^{+14} (Z=20), and Zn^{+24} (Z=30). The radiative corrections, or the electron self -energy and vacuum polarization, were estimated by employing the method described by Indelicato and Kim [31, 34]. The total energies E_{tot}, the sum of the MC DF SCF energies, and correlation and radiative corrections, are given in the last column of the Table.

In Table 3, a detailed comparison of theoretical and experimental data is made on the term energies (cm^{-1}) of the thirteen low-lying even- and odd-parity states of carbonlike ions with Z=10, 20 and 30, relative to the ground J=0 ($2s^2 2p^2\ ^3P_0$) state. Experimental term energy separations [36, 37] are reproduced in the second column for comparison. Theoretical term energy separations of the low-lying excited states, given in the third column of the Table, were computed by subtracting the total energy of the ground J=0 ($2s^2 2p^2\ ^3P_0$) state from those of the excited levels. The last two columns (denoted NR-SCI and NR-MBPT) contain the term energy separations obtained in previous nonrelativistic correlated calculations for comparison. In the NR-MBPT calculations [38], multireference second-order perturbation theory was employed to account for electron correlation for the ions with 10≤Z≤30. In the NR-SCI calculations [39], combination of perturbation theory and simplified configuration interaction (SCI) was employed to account for electron correlation in some carbonlike ions. Relativistic cor-

TABLE 3. Comparison of the term energy separations obtained by second-order MR-MP calculations on Ne^{+4} (Z=10), Ca^{+14} (Z=20) and Zn^{+24} (Z=30) with experiment and previous work. Deviations of the theoretical results from experimental data are given in parentheses. Energies (in cm^{-1}) are given relative to the ground $2s^22p^2\ ^3P_0$ state.

Level	Experiment	MR-MP	NR-SCI	NR-MBPT
		Ne^{+4}		
$2s^22p^2\ ^3P_1$	413	405(8)	404(9)	395(18)
$^1D_2$	30291	29967(324)	30578(-278)	29721(570)
$^1S_0$	63915	63120(795)	64113(-198)	63310(605)
$2s2p^3\ ^5S_2^o$	88402	88105(297)	87844(558)	87712(689)
$^3D_1^o$	175926	174580(1346)	176095(-169)	172842(3084)
$^3P_0^o$	208188	207500(688)	209266(-1078)	205788(2400)
$^3S_1^o$	279372	278415(957)	280252(-880)	279397(-25)
$^1D_2^o$	270555	269736(819)	271079(-524)	268705(1850)
$^1P_1^o$	303815	302688(1127)	306157(-2342)	301656(2159)
$2p^4\ ^3P_2$	412678	413632(-954)	416615(-3937)	406866(5812)
$^1D_2$	436582	437974(-1392)	440505(-3923)	427945(8637)
$^1S_0$	500475	501399(-924)	507417(-6942)	493168(7307)
		Ca^{+14}		
$2s^22p^2\ ^3P_1$	17555	17504(51)	17658(-103)	17111(444)
$^1D_2$	108595	108448(147)	108326(269)	107860(735)
$^1S_0$	197648	197291(357)	197429(219)	197132(516)
$2s2p^3\ ^5S_2^o$	275894	275769(125)	276006(-112)	276516(-622)
$^3D_1^o$	497583	497057(526)	498304(-721)	495662(1921)
$^3P_0^o$	581695	581419(276)	583354(-1659)	580113(1582)
$^3S_1^o$	728889	728426(463)	731009(-2120)	727455(1434)
$^1D_2^o$	729693	729364(329)	731271(-1578)	728002(1691)
$^1P_1^o$	814361	813982(379)	816978(-2617)	812348(2013)
$2p^4\ ^3P_2$	1107558	1107847(-289)	1111030(-3472)	1105028(2530)
$^1D_2$	1195159	1195674(-515)	1198527(-3368)	1191802(3357)
$^1S_0$	1350868	1351286(-418)	1356144(-5276)	1347181(3687)
		Zn^{+24}		
$2s^22p^2\ ^3P_1$	157717	157667(50)	159737(-2020)	152482(5235)
$^1D_2$	429317	429120(197)	423597(5720)	419114(10203)
$^1S_0$	582277	582146(131)	576384(5893)	570101(12176)
$2s2p^3\ ^5S_2^o$	704942	703815(1127)	701636(3306)	
$^3D_1^o$	1026512	1026282(230)	1021590(4922)	
$^3P_0^o$	1238573	1237903(670)	1231683(6890)	
$^3S_1^o$	1428104	1428266(-162)	1419147(8957)	
$^1D_2^o$	1514472	1515328(-856)	1504649(9823)	
$^1P_1^o$	1696908	1696764(144)	1683925(12983)	
$2p^4\ ^3P_2$	2108474	2109188(-714)	2105001(3472)	2091429(-17045)
$^1D_2$	2382294	2382120(174)	2373526(8768)	2358965(22329)
$^1S_0$	2701912	2701760(152)	2686041(-15871)	2669546(32366)

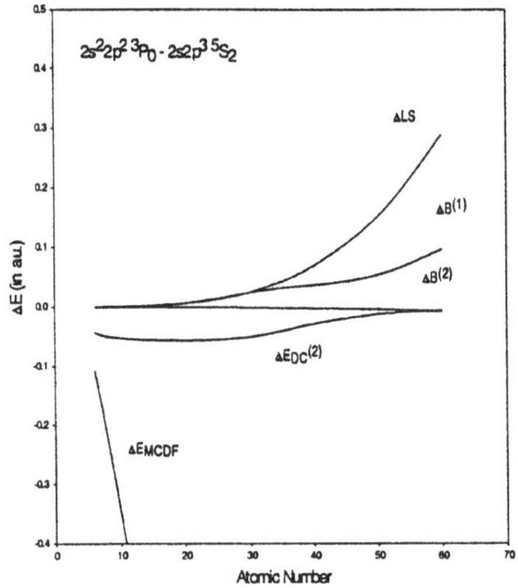

Figure 6. Contributions from each order of perturbation theory to the term energy $2s^22p^2\,^3P_0$ - $2s2p^3\,^5S_2$ (in a.u.).

rections were included in the Breit-Pauli approximation. The deviations from experiment of the calculated nonrelativistic term energy separations are nearly the same as those obtained by our relativistic MR-MP for the low-Z Ne^{+4}, but increase in magnitude rapidly as Z increases, manifesting the inadequacy of the Breit-Pauli approximation. At low-Z, both nonrelativistic and relativistic multireference perturbation theories disagree with experiment by 1-2%. The accuracy of both the nonrelativistic and relativistic calculations for the low-Z ions is limited by the approximate treatment of electron correlation.

Figure 1 illustrates the differences $E^{\exp}(^3P_0 - {}^3P_1) - E^{theor}(^3P_0 - {}^3P_1)$ between theoretical and experimental fine-structure energy separations (in cm^{-1}), $2s^22p^2\,^3P_0 - 2s^22p^2\,^3P_1$, as functions of the atomic number Z. The deviations from experiment of the fine-structure separations computed by NR-MBPT [38], NR-SCI [39], and nonrelativistic multiconfigurational Hartree-Fock (MCHF) [40], are also given to illustrate the sharp increases as Z increases. Both NR-MBPT and NR-SCI, as well as MCHF, start to show significant deviations from experiment for Z>20. The failure to reproduce experimental fine-structure separations may be attributed to the absense of fully relativistic treatment including QED corrections. The fine-structure separations computed by MC DF SCF plus first-order Breit interaction correction (denoted MC DFB in Fig. 1) starts to deviate significantly beyond

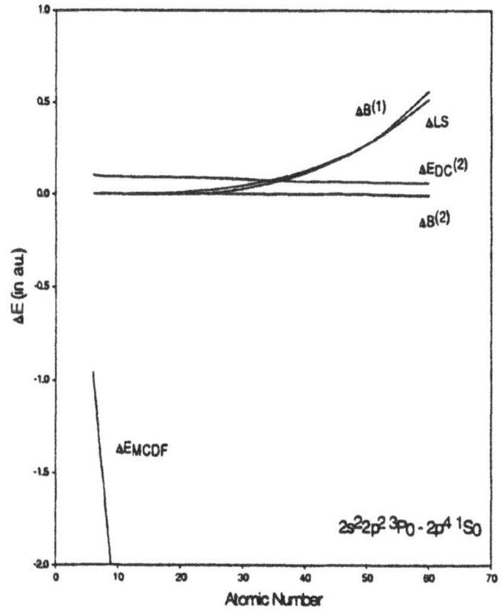

Figure 7. Contributions from each order of perturbation theory to the term energy $2s^2 2p^2\ ^3P_0$ - $2p^4\ ^1S_0$ (in a.u.).

Z~30, necessitating dynamic correlation and radiative corrections. Figure 1 illustrates that relativistic MR-MP calculations (curve labeled MR-MP in the Figure), which include the Breit interaction in the effective electron-electron interaction, as well as the Lamb shifts, result in significant corrections and yield close agreement between the calculated and experimental fine-structure energy separations throughout the 6$\leq$ Z$\leq$ 35 series.

Figures 2 and 3 illustrate the contributions from each order of perturbation theory to fine-structure energy separations, $2s^2 2p^2\ ^3P_0 - 2s^2 2p^2\ ^3P_1$ and $2s^2 2p^2\ ^3P_0 - 2s^2 2p^2\ ^3P_2$ as functions of Z. These contributions were computed by subtracting the energy of the ground J=0 ($^3P_0$) even-parity state from those of the J=1, 2 fine-structure states in each order of perturbation theory displayed in Table 2. Besides the zero-order contribution, $\Delta E_{MCDF} = E_{MCDF}(^3P_0) - E_{MCDF}(^3P_{2,1})$, the dominant contribution to the difference in the fine-structure separations is due to the first-order Breit interaction $\Delta B^{(1)} = B^{(1)}(^3P_0) - B^{(1)}(^3P_{2,1})$, which is why the MC DF SCF plus first-order Breit interaction correction predicts the fine-structure separations accurately up to Z~30. For higher Z, however, the contributions from dynamic correlation and QED corrections, $\Delta E^{(2)}_{DC}$, $\Delta B^{(2)}$, and ΔLS, also become important. None of the higher-order relativistic effects are fully accounted for by the nonrelativistic correlated methods, which account for relativistic effects solely by employing Breit-Pauli Hamiltonian, and the dif-

TABLE 4. Energies (in cm^{-1}) of $2s^22p^2\ ^3P_2$, $^1D_2$ and $^1S_0$ even-parity states of carbon and carbonlike ions relative to the ground $2s^22p^2\ ^3P_0$ state.

Z	$2s^22p^2\ ^3P_2$		$2s^22p^2\ ^1D_2$		$2s^22p^2\ ^1S_0$	
	E^{theor}	E^{exp}	E^{theor}	E^{exp}	E^{theor}	E^{exp}
60	4481354		8948917		9326278	
50	2003517		3977884		4272264	
40	756050		1484339		1704088	
35	423840	423460	827134	828075	1012750	1012339
30	218027	218009	429120	429317	582146	582277
25	98698	98804	212235	212320	332904	333240
20	35828	35917	108448	108595	197291	197648
15	8989	9031	59506	59681	119531	119960
14	6378	6415	52734	52927	107319	107792
13	4388	4419	46516	46729	95718	96243
12	2907	2933	40716	40957	84573	85163
11	1839	1858	35232	35506	73750	74423
10	1095	1111	29967	30291	63120	63915
9	603	615	24847	25236	52509	53538
8	301	306	19747	20274	42458	43186
7	126	130	14791	15316	31776	32689
6	40	43	9452	10194	20181	21648

ference between the calculated and experimental fine-structure separations tends to diverge as Z increases.

In Tables 4, 5, and 6, a detailed comparison of theoretical and experimental data is made on the term energies (cm^{-1}) of the low-lying even- and odd-parity states of carbon and carbonlike ions with Z=6-60, given relative to the ground J=0 ($2s^22p^2\ ^3P_0$) state. Theoretical term energy separations, E^{theor}, of the low-lying excited states were computed by subtracting the total energy of the ground J=0 ($2s^22p^2\ ^3P_0$) state from those of the excited levels. Experimental term energy separations E^{exp} [36, 37, 41] are reproduced in an adjacent column for comparison. Experimental data are not available for ions with Z=40, 50, 60.

We see that the theoretical term energy separations differ from experiment by approximately 7% near the low-Z end and by less than 0.1% at Z=35. Although the deviation between theory and experiment increases to the level of a few percent near the low-Z end, it is consistently below 1% in the range $12 \leq Z \leq 35$, quite good agreement.

Figs. 4 and 5, respectively, illustrate the contributions from each order

TABLE 5. Energies (cm^{-1}) of $2s2p^3$ odd-parity states of carbon and carbonlike ions relative to the ground $2s^22p^2\ ^3P_0$ state.

Z	$2s2p^3\ ^5S_2^o$ E^{theor}	E^{exp}	$2s2p^3\ ^3D_1^o$ E^{theor}	E^{exp}	$2s2p^3\ ^3P_0^o$ E^{theor}	E^{exp}
60	6101989		6653593		10946547	
50	3221623		3675602		5533006	
40	1601310		1978685		2649402	
35	1083711	1087299	1434085	1433398	1811238	1813193
30	703815	704942	1026282	1026512	1237903	1238573
25	442954	443199	722782	723180	849998	850271
20	275769	275894	497057	497583	585335	581695
15	167507	167731	322632	323379	379716	379931
14	150075	150320	291618	292440	343612	344066
13	133573	133837	261466	262377	308648	309142
12	117834	118115	232007	233034	274421	274969
11	102719	103010	203097	204262	240765	241370
10	88105	88402	174580	175926	207500	208188
9	73887	74194	146297	147898	174456	175257
8	60026	60325	118081	120059	141422	142397
7	46360	46785	89336	92252	106718	109224
6	33149	33735	60113	64090	73956	75255
Z	$2s2p^3\ ^3S_1^o$ E^{theor}	E^{exp}	$2s2p^3\ ^1D_2^o$ E^{theor}	E^{exp}	$2s2p^3\ ^1P_1^o$ E^{theor}	E^{exp}
60	11194685		15280025		15807972	
50	5751916		7434040		7851610	
40	2849826		3367950		3676112	
35	2007318	2006021	2242457	2239520	2490026	2490473
30	1428266	1428104	1515328	1514472	1696764	1696908
25	1025230	1025564	1048663	1048774	1172479	1172639
20	728426	728889	729364	729693	813982	814361
15	489927	490551	484193	484700	541281	541981
14	446264	446941	439840	440401	491998	492774
13	403462	404197	396428	397045	443731	444589
12	361335	362138	353736	354419	396236	397181
11	319715	320588	311573	312321	349296	350329
10	278415	279372	269736	270555	302688	303815
9	237240	238294	228017	228900	256180	257384
8	195954	197086	186153	187052	209465	210462
7	153979	155127	143382	144187	160890	149188
6	110408	105799	99425		113252	119878

TABLE 6. Energies (cm^{-1}) of $2p^4\ ^3P_2$, $^1D_2$ and $^1S_0$ even-parity states of carbon and carbonlike ions relative to the ground $2s^22p^2\ ^3P_0$ state.

Z	$2p^4\ ^3P_2$		$2p^4\ ^1D_2$		$2p^4\ ^1S_0$	
	E^{theor}	E^{exp}	E^{theor}	E^{exp}	E^{theor}	E^{exp}
60	13038645		17613316		22253257	
50	7174541		9253656		11371979	
40	3889244		4707889		5557770	
35	2863439	2860683	3345145	3346758	3862279	3865945
30	2109188	2108474	2382120	2382294	2701760	2701912
25	1545930	1545721	1698701	1698342	1913393	1912928
20	1107847	1107558	1195674	1195159	1351286	1350868
15	742967	742540	793827	793113	902806	902338
14	675219	674750	720299	719510	820221	819717
13	608611	608090	648211	647335	739058	738525
12	542941	542338	577292	576295	659058	658445
11	478008	477272	507285	506128	579918	579190
10	413632	412678	437974	436582	501399	500475
9	349647	348325	369165	367400	423261	421980
8	285937	283759	300742	298292	345268	343305
7	222249	220290	232483		266915	
6	158586		164559		187074	

of perturbation theory to the term energy separations, $2s^22p^2\ ^3P_0 - 2s^22p^2\ ^1D_2$ and $2s^22p^2\ ^1D_2 - 2s^22p^2\ ^1S_0$ as functions of Z. The largest contribution to the term energy separations is due to the zero-order contribution, ΔE_{MCDF}, of course. Near the low-Z end, the contribution from $\Delta E^{(2)}_{DC}$, $\Delta B^{(2)}$ and ΔLS are small and nearly constant. As Z increases, however, the contribution from the first-order Breit interaction $\Delta B^{(1)}$ grows rapidly.

The contributions from each order of perturbation theory to the term energy separations, $2s^22p^2\ ^3P_0 - 2s2p^3\ ^5S_2$ and $2s^22p^2\ ^3P_0 - 2p^4\ ^1S_0$, are given, respectively, in Figs. 6 and 7. In the independent particle model, the ground $^3P_0$ state arises nominally from the electronic configuration $2s^22p^2$, whereas the $^5S_2$ and $^1S_0$ states arise from $2s2p^3$ and $2p^4$, respectively. As the Lamb shift corrections tend to differ noticeably for states arising from different electronic configurations, the contribution to the term energy separations from the difference in the Lamb shift, ΔLS, grows as rapidly as $\Delta B^{(1)}$ as Z increases while the contributions $\Delta E^{(2)}_{DC}$ and $\Delta B^{(2)}$ are small and nearly constant.

References

1. Desclaux, J.P.: 1973, *At. Data Nucl. Data Tables* **12**, 311; Desclaux, J.P.: 1975, *Computer Phys Comm* **9**, 31; Desclaux, J.P.: 1985, In *"Atomic Theory Workshop on Relativistic and QED Effects in Heavy Atoms"*, Ed. H. P. Kelly and Y-K Kim, AIP Conf. Proc. New York **136**, 162.
2. Grant, I.P., McKenzie, B.J., Norrington, P.H., Mayers, F.F. and Pyper, N.C.: 1980, *Comput. Phys. Commun.* **21**, 207; Parpia, F.A., Froese Fischer, C. and Grant, I.P.: 1996, *Comput. Phys. Commun.* **94**, 249.
3. Johnson, W.R., Idress, M. and Sapirstein, J.: 1987, *Phys. Rev. A* **35**, 3218.
4. Quiney, H.M., Grant, I.P. and Wilson, S.: 1990, *J. Phys. B* **23**, L271.
5. Ishikawa, Y. and Koc, K.: 1994, *Phys. Rev A* **50**, 4733.
6. Lindroth, E. and Salomonson, S.: 1990, *Phys. Rev. A* **41**, 4659.
7. Eliav, E., Kaldor, U. and Ishikawa, Y.: 1994, *Phys. Rev. A* **49**, 1724.
8. Beck, D.R.: 1988, *Phys. Rev. A* **37**, 1847.
9. Chen, M.H., Cheng, K.T. and Johnson, W.R.: 1993, *Phys. Rev. A* **47**, 3692.
10. Weiss, A.W. and Kim, Y.-K.: 1995, *Phys. Rev. A* **51**, 4487.
11. Liu, Z.W., and Kelly, H.P.: 1991, *Phys. Rev. A* **43**, 3305.
12. Morrison, J.C. and Froese Fischer, C.: 1987, *Phys. Rev. A* **35**, 2429.
13. Salomonson, S., Lindgren, I. and Martenssson, A.-M.: 1980, *Physica Scripta* **21**, 351; Lindroth, E. and Martensson-Pendrill, A.-M.: 1996, *Phys. Rev. A* **53**, 3151.
14. Avgoustoglou, E., Johnson, W.R., Plante, D.R., Sapirstein, J., Sheinerman, S. and Blundell, S. A.: 1992, *Phys. Rev. A* **46**, 5478; Avgoustoglou, E., Johnson, W. R., Liu, Z. W. and Sapirstein, J.: 1995, *Phys. Rev. A* **51**, 1196.
15. Avgoustoglou, E. and Beck, D.R.: 1998, *Phys Rev A* **57**, 4286.
16. Dzuba, V.A., Flambaum, V.V. and Kozlov, M. G.: 1996, *Phys. Rev. A* **54**, 3948; Dzuba, V.A. and Johnson, W.R.: 1998, *Phys. Rev. A* **57**, 2459.
17. Beck, D.R. and Cai, Z.-Y.: 1989, *Phys. Rev. A* **40**, 1657; Beck, D.R.: 1992, *Phys. Rev. A* **45**, 1399; Beck, D.R.: 1997, *Phys. Rev. A* **56**, 2428.
18. Werner, H.J.: 1987, *Adv. Chem. Phys.* **69**, 1.
19. Vilkas, M.J., Koc, K. and Ishikawa, Y., 1998, *Chem. Phys. Letters* **296**, 68; Vilkas, M. J., Ishikawa, Y. and Koc, K.: 1999, *Phys. Rev. A* **60**, 2808.
20. Vilkas, M.J., Ishikawa, Y. and Koc, K.: 1998, *Phys. Rev. E* **58**, 5096; Vilkas, M.J., Ishikawa, Y. and Koc, K.: 1998, *Chem. Phys. Letters* **280**, 167; Vilkas, M.J., Ishikawa, Y. and Koc, K.: 1998: *Int. J. Quantum Chem.* **70**, 813.
21. Sucher, J.: 1980, *Phys. Rev. A* **22**, 348.
22. Mittleman, M.H.: 1981, *Phys. Rev. A* **24**, 1167.
23. Blundell, S.A., Mohr, P.J., Johnson, W.R. and Sapirstein, J.: 1993, *Phys. Rev. A* **48**, 2615.
24. Labzowsky, L., Karasiev, K., Lindgren, I., Persson, H. and Salomonson, S.: 1993, *Physica Scripta* **T46**, 150; Lindgren, I., Persson, H., Salomonson, S., Karasiev, V., Labzowsky, L., Mitrushenkov, A. and Tokman, M.: 1993, *J. Phys. B* **26**, L503; Lindgren, I., Persson, H., Salomonson, S. and Labzowsky, L.: 1995, *Phys. Rev. A* **51**, 1167.
25. Quiney, H.M.: 1999, *This volume.*
26. Hirao, K.: 1992, *Chem. Phys. Letters* **190**, 374; Hirao, K.: 1993, *Chem. Phys. Letters* **201**, 59; Nakano, H.: 1993, *J. Chem. Phys.* **99**, 7983.
27. Møller, C. and Plesset, M. S.: 1934, *Phys. Rev.* **46**, 618.
28. Mohr, P.J.: 1992, *Phys. Rev. A* **46**, 4421; Mohr, P.J. and Kim, Y.-K.: 1992, *Phys. Rev. A* **45**, 2727; Mohr, P.J., Plunien, G. and Soff, G.: 1998, *Phys. Rep.* **293**, 227.
29. Sapirstein, J.: 1987, *Physica Scripta* **36**, 801; Sapirstein, J.: 1988, *Nucl. Instrum. Meth. B* **31**, 70; Mallampalli, S. and Sapirstein, J.: 1996, *Phys. Rev. A* **54**, 2714; Mallampalli, S. and Sapirstein, J.: 1998, *Phys. Rev. A* **57**, 1548.
30. Drake, G.W.F.: 1982, *Adv. At. Mol. Phys.* **18**, 399.
31. Indelicato, P., Gorceix, O. and Desclaux, J.P.: 1987, *J. Phys. B* **20**, 651.

32. Ishikawa, Y., Quiney, H.M. and Malli, G.L.: 1991, *Phys. Rev. A* **43**, 3270; Koc, K. and Ishikawa, Y.: 1994, *Phys. Rev. A* **49**, 794; Ishikawa, Y., Koc, K. and Schwarz, W.H.E.: 1997, *Chem. Phys.* **225**, 239.
33. Schmidt, M.W. and Ruedenberg, K.: 1979, *J. Chem. Phys.* **71**, 3951.
34. Kim, Y.-K.: 1990, *in Atomic Processes in Plasmas, AIP Conf. Proc. No. 206*, 19.
35. Koc, K., Ishikawa, Y., Kagawa, T. and Kim, Y.-K.: 1996, *Chem. Phys. Letters* **263**, 338.
36. Edlén, B.: 1982, *Physica Scripta* **26**, 71.
37. Edlén, B.: 1985, *Physica Scripta* **31**, 345.
38. Vilkas, M.J., Gaigalas, G. and Merkelis, G.: 1992, *Lithuanian J. Phys.* **31**, 84; Vilkas, M.J., Martinson, I., Merkelis, G., Gaigalas, G. and Kisielius, R.: 1996, *Physica Scripta* **54**, 281.
39. Bogdanovich, P., Gaigalas, G., Momkauskaite, A. and Rudzikas, Z.: 1997, *Physica Scripta* **56**, 231.
40. Froese Fischer, Ch. and Saha, H.P.: 1985, *Physica Scripta* **32**, 181.
41. Fuhr, J.R., Martin, W.C., Musgrove, A., Sugar, J. and Wiese, W.L.: 1997, NIST Atomic Spectroscopic Database, available at http://physics.nist.gov/PhysRefData/contents.html (December 1997).

RELATIVISTIC VALENCE BOND THEORY AND ITS APPLICATION TO METASTABLE XE_2

S. KOTOCHIGOVA AND E. TIESINGA
National Institute of Standards and Technology,
Gaithersburg, MD 20899, USA

AND

I. TUPITSYN
Physics Department, St. Petersburg University,
198904 St. Petersburg, Russia

Abstract. We present a new version of the relativistic configuration interaction valence bond (RCIVB) method. It is designed to perform an *ab initio* all-electron relativistic electronic structure calculation for diatomic molecules. A nonorthogonal basis set is constructed from numerical Dirac-Fock atomic orbitals as well as relativistic Sturmian functions. A symmetric reexpansion of atomic orbitals from one atomic center to another is introduced to simplify the calculation of many-center integrals. The electronic structure of the metastable $(5p^5 6s + 5p^5 6s)$ Xenon molecule is calculated and the influence of different configurations on the formation of the molecule is analyzed.

1. Introduction

Recent years have seen a significant increase of interest in the *ab initio* valence bond (VB) approach [1, 2]. Early interest in this method started with Heitler and London [3] who were the first to introduce the valence bond theory in the late twenties. The basic idea of the method is to construct the molecular wave function from atomic orbitals localized at the different atomic centers. Consequently the covalent bonding between atoms is described in terms of an "exchange effect" initiated by the overlap of orbitals of participating atoms. At large internuclear separations R the molecular wave function has a pure atomic form that appropriately describes

J. Maruani et al. (eds.), New Trends in Quantum Systems in Chemistry and Physics, Volume 1, 219–242.

the molecular dissociation limit. At short internuclear separations orbitals around different centers have considerable overlap or non-orthogonality which leads to a large exchange effect and creates a bond. Therefore it is crucial for the method to use a formalism incorporating the non-orthogonality of the basis wave functions. Application of the VB theory for many-electron molecules only became practical with the advent of large computers.

An important development for the valence bond theory was the use of "ionic structures", in which the participating atoms obtain positive and negative charges. Coulson and Fischer [4] showed that ionic structures included into the calculation lead to deformation of the atomic wave functions and hence to more realistic description of molecular formation.

Development of the VB theory was furhter directed towards optimization of the shapes of orbitals in order to introduce correlation effects. These models for electronic structure of a molecule were proposed by Goddard [5] and Gerratt [6]. The distinctive features of the model are that it takes into account the different ways of coupling the electron spins together to give total electron spin S and that the atomic orbitals are optimized by applying a direct minimization procedure. These ideas formed the basis of the spin-coupled theory [7] which was applied to the electronic structure of several diatomic molecules [8, 9] whose spectroscopic constants showed a 90 - 95 % agreement with observed values. Its concepts have been used to develop sophisticated spin-coupled approaches [10, 11, 2, 12, 13, 14, 15] that take into account a considerable amount of the chemically significant electron correlation effects. Another development [5] lead to the so-called generalized valence bond method[16, 17, 18], which combines features of self-consistent field and spin-coupled approaches.

In a recent development of the method we incorporated numerical Hartree-Fock (HF) non-orthogonal atomic orbitals [19, 20, 21, 22] to construct the molecular wave function. Unlike many applications of the VB theory where analytic atomic orbitals of Gaussian type are used as basis functions, this way of constructing molecular wave functions avoids the need for large basis sets. The analytic functions do not display a correct behavior at the nuclei and in the asymptotic region which thus require a large number of Gaussians. In our basis it seems sufficient to use a single HF orbital for each inner shell and a few additional excited orbitals to describe valence electrons. This is true because the atomic HF orbitals already form a good representation of the atom. They have the right number of nodes and are orthogonal with respect to other HF orbitals localized at the same center.

More crucial for the success of Refs. [19, 20, 21, 22] is the implementation of a full configuration interaction (CI) procedure based on non-orthogonal basis functions. Each molecular configuration is constructed from atomic configurations, covalent or ionic. In turn each atomic configuration is described by Slater determinants constructed from the numerical HF orbitals.

In the current paper we introduce a relativistic development in valence bond theory. Much of the structure of Refs. [19, 20, 21, 22] is retained in this relativistic version of VB theory with the primary difference being that the basis functions are now constructed from four-component atomic Dirac spinors. These functions are numerically obtained by solving the integro-differential Dirac-Fock (DF) equations for atoms. In addition to using DF functions to optimize our basis we use configurations with fractional occupation. These configurations are constructed using relativistic version of the Hyper-Hartree-Fock method [23]. In this approach every atomic configuration is described by a density matrix of mixed states corresponding to the configuration average. In the relativistic calculations we average over all relativistic configurations which are created from the same non-relativistic configuration. This averaged configuration has the property that it is the solution of non-relativistic HF equations when the speed of light goes to infinity in the relativistic DF equations.

In a CI expansion, excited configurations help to describe the formation of the molecule. Describing these excited configurations on the basis of Dirac-Fock functions has proven to be ineffective because the DF orbitals increase rapidly in size with increasing general quantum number n. Moreover, a complete set of these functions contains, in addition to discrete functions, continuum functions which are clearly impractical computationally.

To improve the characteristics of the excited states we instead use a set of Sturmian functions. The idea to use Sturmian functions as virtual states to model correlation effects in an atom was first developed by Sherstuk and Pavinsky [24, 25]. It was shown that these functions form a complete set of discrete functions with similar asymptotic behavior and orbital size as the occupied valence orbitals. These characteristics make Sturmian basis functions very efficient in describing correlation effects (See section 2).

The one- and two-electron integrals are calculated using a modified Löwdin's reexpansion procedure [26]. The modification concerns the fact that Löwdin's method is not very efficient when reexpanding strongly localized orbitals. Instead the integration region is divided in two and in each region the slowly varying part of the basis function centered in the other region is reexpanded. This symmetric reexpansion has much faster convergence characteristics than the method proposed in Ref. [26].

Our approach is, in principle, an all-electron calculation, which, for instance, allows us to evaluate the electronic densities at the nuclear sites and to calculate hyperfine structure constants. Hence, the dynamics of all electrons in a molecule is accounted for. However, often this is not necessary. Deep lying orbitals do not take part in the molecular formation.

Therefore we introduce core and valence orbitals, where core electrons will not participate in the CI.

The method is designed to calculate the electronic potential surfaces and other electronic properties of dimers composed of atoms with any nuclear charge Z, any number of electrons, and any level of excitation. In this report, it is applied here to obtaining the electronic potentials for two interacting Xe atoms. This method is very suitable for studying collisional problems, because it naturally provides the physically realistic description of interacting atoms at large internuclear separations.

There are two reasons for calculating the metastable Xe potentials. Firstly, it has been proposed that metastable noble gases such as xenon might be good candidates for Bose-Einstein Condensation [27]. Secondly, the analysis of recent experiment [28] which studied the real-time dynamics of ultra-cold collisions with metastable and double excited xenon atoms, requires a theoretical treatment of the potentials over a wide range of internuclear separations. This study is a first attempt to obtain *ab initio* relativistic electronic potentials of metastable Xe atoms. Previous semi-empirical investigation of metastable xenon gas has focused on the long-range interactions between two distant atoms [27]. However in many situations the intermediate and short internuclear separations are essential for a complete understanding of cold collisions. For instance, the ability to hold metastable atoms with a reasonable density in a magnetic or optical trap depends on the full shape of the adiabatic potential curves. The asymptotic description of the potential curves is not sufficient and an *ab initio* calculation is required. Notice that the ability to hold metastable Xe not only depends on the adiabatic potentials but also on the "width" of the potentials. This width describes Penning and associative ionization of colliding metastable Xe. This paper does not address these ionization issues.

2. Atomic basis functions for occupied and unoccupied orbitals

The numerical atomic Dirac-Fock wave functions that describe occupied molecular orbitals were obtained by solving integro-differential DF equations for the self-consistent field of the configuration average. The equations for the self-consistent field were derived by applying the Hartree-Fock method to the eigenvalue and eigenfunction problem of the relativistic energy operator

$$\hat{H} = \sum_i \hat{h}_i + \sum_{i \neq j} \hat{v}_{ij}, \tag{1}$$

for each atom, where $\hat{h}_i = c(\vec{\alpha}_i \vec{p}_i) + \beta_i c^2 - Z/r_i$ is the Dirac operator for an electron in the field of a nucleus of charge Z and $\hat{v}_{ij} = 1/|r_i - r_j|$ is the Coulomb operator for the electron-electron interaction; α and β are the

Dirac matrices and c is the speed of light. We express all equations using atomic units where $\hbar = m_e = 1$. One atomic unit of length is 1 a.u. = 0.0529177 nm.

In the central-field approximation the single-particle wave function is a four-component Dirac spinor

$$\psi_{n\kappa\mu}(\vec{r}) = \frac{1}{r}\begin{pmatrix} P_{n\kappa}(r) & \chi_{\kappa\mu}(\theta,\vartheta) \\ iQ_{n\kappa}(r) & \chi_{-\kappa\mu}(\theta,\vartheta) \end{pmatrix}, \tag{2}$$

where $\vec{r}$ is the electron coordinate, $P_{n\kappa}(r)$ and $Q_{n\kappa}(r)$ are the large and small components of the wave function, respectively; $\chi_{\kappa\mu}(\theta,\vartheta)$ is the spin-orbit wave function, corresponding to an eigenvalue $\kappa = \ell(\ell+1) - (j+1/2)^2$, where ℓ and j are the orbital and total angular momentum quantum numbers and μ is the projection quantum number of j.

The antisymmetric N-electron atomic wave functions are obtained in Slater determinant form $det^A = det(\psi_1(\vec{r_1}), ..., \psi_N(\vec{r_N}))/\sqrt{N!}$ where the functions ψ_i are of the form given in Eq. (2) and superscript A labels the atom. To construct a many-electron wave function which belongs to the configuration

$$K = (n_1 l_1 j_1)^{q_1} ... (n_A l_A j_A)^{q_A}$$

we select determinants in which q_1 one-electron functions belong to shell $n_1 l_1 j_1$, and so on. Notice that for convenience we use the notation $n\ell j$ instead of $n\kappa$ to label the orbital $\psi_{n\kappa\mu}$. Both notations uniquely define a relativistic orbital. Note also that every non-relativistic configuration has split into several relativistic configurations.

Atomic configurations with fractional occupation are constructed for the orbitals in open shells. Fractional occupation is introduced via a configuration average. The average of Dirac-Fock orbitals is taken over all states of each relativistic configuration and all relativistic configurations belonging to the same non-relativistic configuration. This averaging ensures convergence of Dirac-Fock solutions to the non-relativistic Hartree-Fock solutions when the speed of light is infinity.

The unoccupied or virtual orbitals are described by Sturmian functions. These Sturmian orbitals are obtained by solving integro-differential Dirac-Fock-Sturm equations. These equations can be derived from the DF equations for occupied valence orbitals. The Coulomb interaction between electron and nucleus in these equations is multiplied with a factor λ. Furthermore, the one-electron energy is held constant at the energy of a valence electron. Solving these equations for eigenvalues $\lambda > 1$ we obtain a complete set of eigenfunctions. The complete set of these Sturmian functions is discrete and each orbital has approximately the same radius and the same asymptotic behavior as the corresponding valence orbital. These Sturmian orbitals help to construct excited configurations and lead to a compact and rapidly converging CI procedure.

3. Molecular wave function

The total molecular wave function Ψ_{AB} for a N-electron two-atomic molecule AB is introduced as a linear combination of molecular Slater determinants det_α^{AB},

$$\Psi_{AB} = \sum_\alpha C_\alpha det_\alpha^{AB} \tag{3}$$

where every molecular determinant is the antisymmetrized product of two atomic determinants

$$det_\alpha^{AB} = \hat{A}(det_\alpha^A \cdot det_\alpha^B) \tag{4}$$

The coefficients C_α in (3) are obtained by solving a generalized eigenvalue matrix problem described by the equation

$$\hat{H}_{AB}\vec{C} = E\hat{S}_{AB}\vec{C}, \tag{5}$$

where $\hat{H}_{AB}$ is the Hamiltonian matrix of atoms A and B and their mutual Coulomb interactions. The nonorthogonality matrix $\hat{S}_{AB}$, which is a scalar product of Slater determinants, is given by

$$(\hat{S}_{AB})_{\alpha\beta} = < det_\alpha^{AB} \mid det_\beta^{AB} > = (D_{\alpha\alpha}D_{\beta\beta})^{-1/2} D_{\alpha\beta}, \tag{6}$$

where $D_{\alpha\beta} = det| < \alpha_i|\beta_j > |$ is the determinant of the matrix of overlap integrals
$S_{i,j}^{\alpha\beta} = < \alpha_i|\beta_j >$ between orbitals α_i and β_j belonging to Slater determinants det_α^{AB} and det_β^{AB}, respectively. The α_i (β_j) stand for atomic orbitals $\psi_{n\kappa}(\vec{r})$ centered at either nucleus A or B.

It is convenient to define one- and two-particle density transition matrices. The one-particle density matrix is given by

$$\rho_1^{\alpha,\beta}(\vec{r},\vec{r}') = (D_{\alpha\alpha}D_{\beta\beta})^{-1/2} D_{\alpha\beta} \sum_{i,j}^{N} (S^{-1})_{i,j}^{\alpha,\beta} \psi_i(\vec{r})\psi_j^*(\vec{r}'), \tag{7}$$

where $\psi_i(\vec{r})$ and $\psi_j^*(\vec{r}')$ are atomic orbitals and the two-particle density matrix is

$$\rho_2^{\alpha,\beta}(\vec{r_1},\vec{r_2}|\vec{r_1}',\vec{r_2}') = (D_{\alpha\alpha}D_{\beta\beta})^{-1/2} \sum_{i\neq k}^{N}\sum_{j\neq l}^{N} D_{i,j,k,l}^{\alpha,\beta} \cdot \psi_i(\vec{r_1})\psi_j^*(\vec{r_1}')\psi_k(\vec{r_2})\psi_l^*(\vec{r_2}') \tag{8}$$

where

$$D^{\alpha\beta}_{i,j,k,l} = D_{\alpha\beta}\varepsilon_{i,k}\varepsilon_{j,l}[(S^{-1})^{\alpha,\beta}_{i,j}(S^{-1})^{\alpha,\beta}_{k,l} - (S^{-1})^{\alpha,\beta}_{i,l}(S^{-1})^{\alpha,\beta}_{k,j}]$$

$$\varepsilon_{i,k} = \begin{cases} 1 & i < k \\ -1 & i > k \end{cases} . \tag{9}$$

Introducing the Hamiltonian $\hat{H}_{AB}$ through one- and two-electron terms and the Coulomb repulsion u_{AB} between the nuclei,

$$\hat{H}_{AB} = \sum_{i=1}^{N} \hat{h}_i + \sum_{i\neq j}^{N} \hat{v}_{ij} + u_{AB} , \tag{10}$$

we describe one-electron matrix elements using Eq. (7) in the form[1]

$$< det^{AB}_{\alpha} | \sum_{i=1}^{N} \hat{h}_i | det^{AB}_{\beta} > = (D_{\alpha\alpha}D_{\beta\beta})^{-1/2} D_{\alpha\beta} \sum_{i,j=1}^{N} (S^{-1})^{\alpha,\beta}_{i,j} \cdot < \alpha_i | \hat{h} | \beta_j > , \tag{11}$$

where

$$\hat{h} = c(\vec{\alpha}\vec{p}) + \beta c^2 - \frac{Z_A}{|\vec{r} - \vec{R}_A|} - \frac{Z_B}{|\vec{r} - \vec{R}_B|} , \tag{12}$$

and where R_A, and R_B are the nuclear coordinates. See Section 4 for more details.

Two-electron matrix elements are obtained using Eq. (8) in the form [1]

$$< det^{AB}_{\alpha} | \sum_{i\neq j}^{N} \hat{v}_{ij} | det^{AB}_{\beta} > = (D_{\alpha\alpha}D_{\beta\beta})^{-1/2} \sum_{i,k=1}^{N} \sum_{j,l=1}^{N} D^{\alpha\beta}_{i,j,k,l} \cdot < \alpha_i, \alpha_j | \frac{1}{r_{12}} | \beta_k, \beta_l > , \tag{13}$$

where the evaluation of the matrix element $< \alpha_i, \alpha_j | 1/r_{12} | \beta_k, \beta_l >$ will be discussed in Sections 5 and 6.

4. Wave function reexpansion

Two-center integrals are calculated using a symmetrical reexpansion procedure when a product of two wave functions localized at different centers appears in the integrands. The reexpansion procedure is based on techniques proposed by Löwdin [26]. Assume that the atomic nuclei are situated at position A and B respectively (see Fig. 1). The coordinates can be related to an arbitrary origin O and the z-axis is directed along the internuclear axis AB. The nuclear coordinates are $\vec{R}_{\rm A}$ and $\vec{R}_{\rm B}$ and the electrons coordinates are r_i with $i = 1, 2, \ldots$. The following geometrical relations exist for any of the electrons

$$\vec{R} = \vec{R}_{\rm B} - \vec{R}_{\rm A}; \quad \vec{r}_{\rm iA} = \vec{r}_i - \vec{R}_{\rm A}; \quad \vec{r}_{\rm iB} = \vec{r}_i - \vec{R}_{\rm B}$$
$$\vec{r}_{\rm iA} - \vec{r}_{\rm iB} = \vec{R}_{\rm B} - \vec{R}_{\rm A} = \vec{R}. \tag{14}$$

In the central-field approximation the atomic orbitals from centers A and B are Dirac spinors in Eq. (2) which can be written in the form

$$\psi_a(\vec{r}) = \frac{f_a(|\vec{r}|)}{|\vec{r}|}\chi_{\ell_a,j_a\mu_a}(\vec{r}), \qquad \psi_b(\vec{r}) = \frac{f_b(|\vec{r}|)}{|\vec{r}|}\chi_{\ell_b,j_b\mu_b}(\vec{r}), \tag{15}$$

where $f(r)$ describes the large P or small Q component of the radial one-electron wave function. The spin-orbit wave function $\chi_{\ell,j,\mu} = \chi_{\kappa\mu}$ in terms of the spherical harmonics $Y_{\ell,m}(\vec{r})$ is

$$\chi_{\ell,j,\mu}(\vec{r}) = \sum_{m,\sigma=\pm\frac{1}{2}} C^{j\mu}_{\ell m,\frac{1}{2}\sigma} Y_{\ell,m}(\vec{r}) \cdot \Phi_\sigma, \tag{16}$$

where $C^{j\mu}_{\ell m,\frac{1}{2}\sigma}$ are Clebsch-Gordan coefficients and Φ_σ is a spin function.

To accelerate convergence of the reexpansion we modified the standard Löwdin reexpansion procedure by dividing the range of integration into two exclusive regions V_A and V_B where we assume that the region V_A contains atom A and the region V_B contains atom B. We only apply the reexpansion procedure to the "tails" of wave functions occurring in a given region. To describe this we introduce the step-wise functions:

$$\Theta_A(\vec{r}) = \begin{cases} 1 & \vec{r} \in V_A \\ 0 & \vec{r} \in V_B \end{cases} \quad \text{and} \quad \Theta_{\rm B}(\vec{r}) = \begin{cases} 0 & \vec{r} \in V_A \\ 1 & \vec{r} \in V_B \end{cases} \tag{17}$$

The reexpansion of a wave function centered at B onto center A has the form

$$\Theta_A(\vec{r})\frac{f_b(|\vec{r}_B|)}{|\vec{r}_B|}\chi_{\ell_b,j_b\mu_b}(\vec{r}_B) = \frac{1}{r_A}\sum_{\ell,j}\zeta_b(A,\ell,j,\mu_b|r_A)\chi_{\ell j\mu_b}(\vec{r}_A), \tag{18}$$

where

$$\zeta_b(A,\ell,j,\mu_b|r_A) = \zeta_b(A,\kappa_b,\mu_b|r_A) = \sum_{m,\sigma} C^{j_b\mu_b}_{\ell_b m\frac{1}{2}\sigma} C^{j\mu_b}_{\ell m,\frac{1}{2}\sigma} \frac{K_{\ell m}K_{\ell_b m}}{R} \cdot$$
$$\int_{max(|r_A|,|R-r_A|)}^{|R+r_A|} dr_B f_b(r_B) P^{|m|}_{\ell_b}(|\frac{r_B^2 + R^2 - r_A^2}{2r_B R}|) \cdot$$
$$P^{|m|}_{\ell}(|\frac{r_A^2 + R^2 - r_B^2}{2r_B R}|), \tag{19}$$

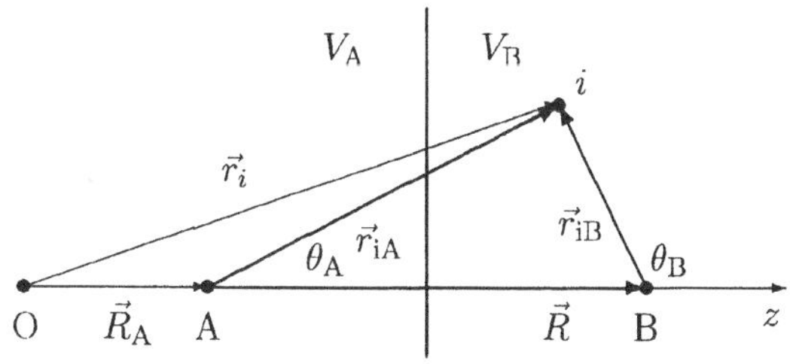

Figure 1. Definition of nuclear and electron coordinates. The nuclei are located at A and B and the electron is located at i. The integration regions V_A and V_B are also shown.

where $P_\ell^{|m|}$ are the standard associated Legendre polynomials, $\sigma = m\text{-}\mu$ and

$$K_{\ell m} = \sqrt{\frac{(2\ell+1)}{2}\frac{(\ell-|m|)!}{(\ell+|m|)!}}.$$

Using symmetrical reexpansion the product of two wave functions, calculated on the different centers can be described as

$$\frac{f_a(r_A)}{r_A}\chi_{\ell_a,j_a\mu_a}(\vec{r}_A)\frac{f_b(r_B)}{r_B}\chi_{\ell_b,j_b\mu_b}(\vec{r}_B) = \frac{f_a(r_A)}{r_A}\chi_{\ell_a,j_a\mu_a}(\vec{r}_A)\cdot$$
$$\frac{1}{r_A}\sum_{\ell,j}\zeta_b(A,\ell,j,\mu_a|r_A)\chi_{\ell,j\mu_b}(\vec{r}_A) + \frac{f_b(r_B)}{r_B}\chi_{\ell_b,j_b\mu_b}(\vec{r}_B)\cdot$$
$$\frac{1}{r_B}\sum_{\ell,j}\zeta_a(B,\ell,j,\mu_b|r_B)\chi_{\ell,j\mu_a}(\vec{r}_B). \tag{20}$$

Expression (20) is used to calculate the one- and two-electron two-center integrals. As an example we present the final expression for the overlap integral between orbitals α_i and β_j from $\det_\alpha^{AB}$ and $\det_\beta^{AB}$, respectively,

$$S_{i,j}^{\alpha\beta} = \int_0^\infty dr\{P_a(r)\zeta_{P_b}(A,\kappa_a,\mu_a|r_A) + P_b(r)\zeta_{P_a}(B,\kappa_b,\mu_b|r_B)\} +$$
$$\int_0^\infty dr\{Q_a(r)\zeta_{Q_b}(A,-\kappa_a,\mu_a|r_A) + Q_b(r)\zeta_{Q_a}(B,-\kappa_b,\mu_b|r_B)\}. \tag{21}$$

5. Coulomb-type two-center integrals

In this section we will find expressions for the Coulomb integrals, in a form convenient for computation, using the reexpanded atomic orbitals. We have extended the notation for the electron coordinates in the following way:

$$\vec{r}_{12} = \vec{r}_1 - \vec{r}_2$$

where the r_i and R_α are defined with respect to an arbitrary origin.

Following Eq. (13) the matrix elements for the Coulomb interaction are

$$\gamma_{ac,bd} = \int d\vec{r}_1 \int d\vec{r}_2 \frac{\rho_{a,c}(\vec{r}_{1A})\rho_{b,d}(\vec{r}_{2B})}{|\vec{r}_1 - \vec{r}_2|}, \tag{22}$$

where $\rho_{a,c}(\vec{r}_{1A}) = \psi_a^*(\vec{r}_{1A})\psi_c(\vec{r}_{1A})$ and $\rho_{b,d}(\vec{r}_{2B}) = \psi_b^*(\vec{r}_{2B})\psi_d(\vec{r}_{2B})$ are densities of four-component electronic wave functions centered at A and B, respectively. Since the Coulomb potential for an electron density ρ is defined as

$$U(\vec{r}) = \int d\vec{r}' \frac{\rho(\vec{r}')}{|\vec{r} - \vec{r}'|}, \tag{23}$$

the Coulomb matrix elements Eq.(22) can be written as

$$\gamma_{ac,bd} = \int d\vec{r}_2\, U_{ac}(\vec{r}_{2A})\rho_{bd}(\vec{r}_{2B})\,. \tag{24}$$

Now we expand the densities $\rho(\vec{r})$ and Coulomb potentials $U(\vec{r})$ that are created by these densities in terms of the spherical harmonics. For $\rho_{a,c}(\vec{r})$ we have

$$\rho_{ac}(\vec{r}) - \sum_{\ell,m} \sqrt{\frac{(\ell+1)}{4\pi}} g^\ell(j_a\mu_a, j_c, \mu_c) Y_{\ell,m}(\vec{r})\rho_{ac}(r), \tag{25}$$

where $g^\ell(j\mu, j', \mu')$ are relativistic Gaunt coefficients:

$$g^\ell(j\mu, j', \mu') = \frac{\sqrt{(2j+1)(2j'+1)}}{2\ell+1}(-1)^{\mu'+\frac{1}{2}} C^{\ell,\mu-\mu'}_{j\mu,j'-\mu'} C^{\ell,0}_{j-\frac{1}{2},j'\frac{1}{2}}, \tag{26}$$

and $\rho_{ac}(r)$ is the radial electronic density:

$$\rho_{ac}(r) = P_a(r)P_c(r) + Q_a(r)Q_c(r). \tag{27}$$

A similar expansion can be written for the density ρ_{bd} near center B. The potential $U_{ac}(\vec{r})$ can be introduced as

$$U_{ac}(\vec{r}) = \frac{1}{r} \sum_{k=|j_a-j_c|,\mu}^{j_a+j_c} v^k_{ac}(r) \cdot Y_{k,\mu}(\vec{r}), \tag{28}$$

where

$$v_{ac}^{k}(r) = \sqrt{\frac{4\pi}{2k+1}} g^{k}(j_a\mu_a, j_c\mu_c) \cdot \int dr' \rho_{ac}^{k}(r') \frac{(r_<)^k}{(r_>)^{k+1}} \cdot \tag{29}$$

A similar expression can be obtained for U_{bd}.

The integration region of $\vec{r}$ in Eq. (22) is divided in two half planes V_A and V_B and a surface S between the planes. Furthermore using the Laplace equation $\Delta U(\vec{r}) = -4\pi\rho(\vec{r})$ and applying Green's theorem, we rewrite Eq. (24) as

$$\gamma_{ac,bd} = \gamma_{ac,bd}^{(V_A)} + \gamma_{ac,bd}^{(V_B)} + \gamma_{ac,bd}^{(S)}, \tag{30}$$

where

$$\begin{aligned} \gamma_{ac,bd}^{(V_A)} &= \int_{V_A} d\vec{r}\, U_{ac}(\vec{r}_A)\rho_{bd}(\vec{r}_B), \\ \gamma_{ac,bd}^{(V_B)} &= \int_{V_B} d\vec{r}\, U_{bd}(\vec{r}_B)\rho_{ac}(\vec{r}_A), \\ \gamma_{ac,bd}^{(S)} &= -\frac{1}{4\pi}\int_S dS \left[U_{ac}(\vec{r}_A)\frac{\partial}{\partial z}U_{bd}(\vec{r}_B) - U_{bd}(\vec{r}_B)\frac{\partial}{\partial z}U_{ac}(\vec{r}_A)\right]. \end{aligned} \tag{31}$$

Using Eq. (28) and the symmetrical reexpansion of Eq. (18) for the density $\rho_{bd}(\vec{r}_B)$ around center A and for the density $\rho_{ac}(\vec{r}_A)$ around center B we obtain for the volume integrals

$$\begin{aligned} \gamma_{ac,bd}^{(V_A)} &= \sum_{k}^{\ell_a+\ell_c} \sum_{k'}^{\ell_b+\ell_d} \int dr v_{ac}^{k}(r) \zeta_{bd}(A, k', \mu_d - \mu_b | r_A), \\ \gamma_{ac,bd}^{(V_B)} &= \sum_{k}^{\ell_b+\ell_d} \sum_{k'}^{\ell_a+\ell_c} \int dr v_{bd}^{k}(r) \zeta_{ac}(B, k', \mu_a - \mu_c | r_B). \end{aligned} \tag{32}$$

For the surface (S) integrals we have

$$\begin{aligned} \gamma_{ac,bd}^{(S)} = \sum_{k}^{\ell_a+\ell_c} \sum_{k'}^{\ell_b+\ell_d} \Bigg(&\frac{1}{R} v_{ac}^{k}(\frac{R}{2}) v_{bd}^{k'}(\frac{R}{2}) \delta_{\mu,0} - \\ 4\,\pi &\int_{R/2}^{\infty} dr \frac{1}{r} v_{ac}^{k}(\frac{R}{2}) v_{bd}^{k'}(\frac{R}{2}) \\ &\frac{\partial [Y_{k,\mu}(R/2r, 0) Y_{k',\mu}^{*}(R/2r, 0)]}{\partial R} \Bigg), \end{aligned} \tag{33}$$

where $\mu = \mu_a - \mu_c = \mu_d - \mu_b$.

6. Exchange-type integrals

The exchange type interaction matrix element is

$$h_{ac,bd} = \int d\vec{r}_1 \int d\vec{r}_2 \cdot \psi_a^*(\vec{r}_{1A}) \cdot \psi_b(\vec{r}_{1B}) \frac{1}{r_{12}} \psi_c(\vec{r}_{2A}) \cdot \psi_d^*(\vec{r}_{2B}) \tag{34}$$

where $\psi(\vec{r})$ is defined by Eq.(2), a,c denote one-electron functions centered at A, and b,d denote functions centered at B.

We can separate the relativistic exchange type integral into a part for the large component P and a part for the small component Q:

$$h_{ac,bd} = h_{ac,bd}^P + h_{ac,bd}^Q \tag{35}$$

Let us consider the integral $h_{ac,bd}^P$ for the large component. Equivalent expressions for the small component integrals can be derived. For the product of functions of different centers we use the two-center expansion formula (20). The spatial integrations for the exchange type integral is divided in four parts via

$$\int d\vec{r}_1 \int d\vec{r}_2 \;=\; \left(\int_{V_A} d\vec{r}_1 + \int_{V_B} d\vec{r}_1\right)\left(\int_{V_A} d\vec{r}_2 + \int_{V_B} d\vec{r}_2\right),$$

where the half planes $V_{\mathrm{A,B}}$ are defined in Fig. 1. Hence the exchange type integral has the four contributions

$$h_{ac,bd}^P = h_{ac,bd}^{P,(AA)} + h_{ac,bd}^{P,(BB)} + h_{ac,bd}^{P,(AB)} + h_{ac,bd}^{P,(BA)}, \tag{36}$$

where the first two terms of Eq. (36) correspond to integrals where the wave functions of electrons 1 and 2 are in overlapping region and the last two terms of Eq. (36) are related to integrals where the wave functions are in non-overlapping region. Using an expansion of the Coulomb interaction $1/r_{12}$ around center A in terms of spherical harmonics

$$\frac{1}{r_{12}} = 4\pi \sum_{k=0}^{\infty} \frac{1}{2k+1} u_k(r_{1A}, r_{2A}) \sum_{\mu} Y_{k\mu}(\vec{r}_{1A}) Y_{k\mu}^*(\vec{r}_{2A}),$$

where $u_k(r_1, r_2) = r_<^k / r_>^{k+1}$, we have for terms with overlapping wave functions

$$h_{ac,bd}^{P,(AA)} = \sum_{jl} \sum_{j'l'} \sum_{k} g^k(j_a\mu_a, j\mu_c) g^k(j_d\mu_d, j'\mu_b) \cdot$$
$$\int_0^{\infty} dr_2 P_d(r_2)\; \zeta_b(A, l', j', \mu_b | r_2) \cdot$$
$$\int_0^{\infty} dr_1 P_a(r_1)\; \zeta_c(A, l, j, \mu_c | r_1)\; u_k(r_1, r_2). \tag{37}$$

The integral $h_{ac,bd}^{P,(BB)}$ is solved similarly. The integrals $h_{ac,bd}^{P,(AB)}$ and $h_{ac,bd}^{P,(BA)}$ are evaluated using techniques similar to those in Section 5.

The main complexity in evaluating exchange type integrals is that it includes infinite sums. These expansions however converge very fast due to the symmetrical reexpansion procedure. It is sufficient to use 8 to 10 terms in expansion (37) to obtain exchange type integrals with an accuracy of 10^{-6}.

7. Metastable Xe_2

We now use the RCIVB method, described in above sections, to calculate the electronic potentials of metastable Xe_2. Xe atom has 54 electrons and in the calculation we define the 48 electrons in the closed shells $1s^2 2s^2 2p^6 ... 5s^2$ of each Xe atom as the core. An R-dependent all-electron core potential is calculated exactly and included in the Hamiltonian. The outer $5p^5$ and $6s$ orbitals of metastable Xe are valence orbitals. Furthermore we use the Sturmian $6p, 5d, 7s$ virtual or unoccupied orbitals to enhance the correlation. Various covalent and ionic configurations are constructed by distributing electrons from the $5p^5 6s$ configuration over the relativistic $5p_{1/2}^2$, $5p_{3/2}^3$, $6s_{1/2}$, $6p_{1/2}$, $6p_{3/2}$, $5d_{3/2}$, $5d_{5/2}$, and $7s_{1/2}$ orbitals. In total there are 97 relativistic configurations in our CI expansion. This particular choice of configurations which ignores the $5p_{1/2}^1 5p_{3/2}^4 nl$ configurations, takes into account the majority of the correlation and restricts the number of one and two electron integrals to a manageable number. In total we construct 1991 molecular determinants from these configurations. The list of configurations is given in Tables 1 and 7. For clarity the molecular configurations with atomic configurations interchanged are not listed in Tables 1 and 7 but they are included in our calculation. This ensures that the gerade ("g")/ungerade ("u") symmetry of molecule is satisfied.

The configurations are divided into several groups on the basis of their role in the formation of the molecule. The first line in the table describes the core orbitals. The metastable Xe $5p^5 6s + 5p^5 6s$ is the leading valence configuration.

The next group of configurations include configurations where one Xe atom is in the metastable state and the other Xe atom is in a higher excited state that has the same parity as the metastable atom. These configurations serve to obtain a correct description of the metastable atom and consequently a correct dissociation limit for the molecule. The group of configurations with "opposite parity" include two atomic configurations of opposite parity and contribute to the formation of a molecule at short and intermediate internuclear separation.

Two groups of configurations contribute to the long-range polarization

TABLE 1. List of molecular configurations used in the calculations

Atom A		Atom B	Description
$1s^2_{1/2}...5s^2_{1/2}$	+	$1s^2_{1/2}...5s^2_{1/2}$	core
$5p^2_{1/2}5p^3_{3/2}6s_{1/2}$	+	$5p^2_{1/2}5p^3_{3/2}6s_{1/2}$	metastable
$5p^2_{1/2}5p^3_{3/2}6s_{1/2}$	+	$5p^2_{1/2}5p^3_{3/2}5d_{5/2}$	same parity
$5p^2_{1/2}5p^3_{3/2}6s_{1/2}$	+	$5p^2_{1/2}5p^3_{3/2}5d_{3/2}$	
$5p^2_{1/2}5p^3_{3/2}6s_{1/2}$	+	$5p^2_{1/2}5p^3_{3/2}7s_{1/2}$	
$5p^2_{1/2}5p^3_{3/2}5d_{3/2}$	+	$5p^2_{1/2}5p^3_{3/2}5d_{3/2}$	
$5p^2_{1/2}5p^3_{3/2}5d_{5/2}$	+	$5p^2_{1/2}5p^3_{3/2}5d_{5/2}$	other neutral
$5p^2_{1/2}5p^3_{3/2}5d_{3/2}$	+	$5p^2_{1/2}5p^3_{3/2}5d_{5/2}$	
$5p^2_{1/2}5p^3_{3/2}5d_{3/2}$	+	$5p^2_{1/2}5p^3_{3/2}7s_{1/2}$	
$5p^2_{1/2}5p^3_{3/2}5d_{5/2}$	+	$5p^2_{1/2}5p^3_{3/2}7s_{1/2}$	
$5p^2_{1/2}5p^3_{3/2}7s_{1/2}$	+	$5p^2_{1/2}5p^3_{3/2}7s_{1/2}$	
$5p^2_{1/2}5p^2_{3/2}6s^2_{1/2}$	+	$5p^2_{1/2}5p^4_{3/2}$	
$5p^2_{1/2}5p^3_{3/2}6s_{1/2}$	+	$5p^2_{1/2}5p^4_{3/2}$	
$5p^2_{1/2}5p^3_{3/2}5d_{3/2}$	+	$5p^2_{1/2}5p^4_{3/2}$	
$5p^2_{1/2}5p^3_{3/2}5d_{5/2}$	+	$5p^2_{1/2}5p^4_{3/2}$	
$5p^2_{1/2}5p^3_{3/2}6p_{1/2}$	+	$5p^2_{1/2}5p^4_{3/2}$	
$5p^2_{1/2}5p^3_{3/2}6p_{3/2}$	+	$5p^2_{1/2}5p^4_{3/2}$	
$5p^2_{1/2}5p^3_{3/2}7s_{1/2}$	+	$5p^2_{1/2}5p^4_{3/2}$	
$5p^2_{1/2}5p^4_{3/2}$	+	$5p^2_{1/2}5p^4_{3/2}$	
$5p^2_{1/2}5p^3_{3/2}6p_{1/2}$	+	$5p^2_{1/2}5p^3_{3/2}7s_{1/2}$	
$5p^2_{1/2}5p^3_{3/2}6p_{3/2}$	+	$5p^2_{1/2}5p^3_{3/2}7s_{1/2}$	
$5p^2_{1/2}5p^3_{3/2}6s_{1/2}$	+	$5p^2_{1/2}5p^3_{3/2}6p_{1/2}$	opposite parity
$5p^2_{1/2}5p^3_{3/2}6s_{1/2}$	+	$5p^2_{1/2}5p^3_{3/2}6p_{3/2}$	
$5p^2_{1/2}5p^3_{3/2}6p_{1/2}$	+	$5p^2_{1/2}5p^3_{3/2}6p_{1/2}$	C_6
$5p^2_{1/2}5p^3_{3/2}6p_{3/2}$	+	$5p^2_{1/2}5p^3_{3/2}6p_{3/2}$	
$5p^2_{1/2}5p^3_{3/2}6p_{1/2}$	+	$5p^2_{1/2}5p^3_{3/2}6p_{3/2}$	
$5p^2_{1/2}5p^3_{3/2}5d_{3/2}$	+	$5p^2_{1/2}5p^3_{3/2}6p_{3/2}$	C_8
$5p^2_{1/2}5p^3_{3/2}5d_{3/2}$	+	$5p^2_{1/2}5p^3_{3/2}6p_{1/2}$	
$5p^2_{1/2}5p^3_{3/2}5d_{5/2}$	+	$5p^2_{1/2}5p^3_{3/2}6p_{1/2}$	
$5p^2_{1/2}5p^3_{3/2}5d_{5/2}$	+	$5p^2_{1/2}5p^3_{3/2}6p_{3/2}$	
$5p^2_{1/2}5p^3_{3/2}$	+	$5p^2_{1/2}5p^3_{3/2}6s_{1/2}5d_{3/2}$	ionic
$5p^2_{1/2}5p^3_{3/2}$	+	$5p^2_{1/2}5p^3_{3/2}5d_{3/2}5d_{5/2}$	
$5p^2_{1/2}5p^3_{3/2}$	+	$5p^2_{1/2}5p^3_{3/2}6s_{1/2}5d_{5/2}$	
$5p^2_{1/2}5p^3_{3/2}$	+	$5p^2_{1/2}5p^3_{3/2}6s_{1/2}6p_{1/2}$	
$5p^2_{1/2}5p^3_{3/2}$	+	$5p^2_{1/2}5p^3_{3/2}5d_{3/2}6p_{1/2}$	
$5p^2_{1/2}5p^3_{3/2}$	+	$5p^2_{1/2}5p^3_{3/2}5d_{5/2}6p_{3/2}$	
$5p^2_{1/2}5p^3_{3/2}$	+	$5p^2_{1/2}5p^3_{3/2}6s_{1/2}6p_{3/2}$	
$5p^2_{1/2}5p^3_{3/2}$	+	$5p^2_{1/2}5p^3_{3/2}6p_{1/2}6p_{3/2}$	
$5p^2_{1/2}5p^3_{3/2}$	+	$5p^2_{1/2}5p^3_{3/2}5d_{5/2}6p_{1/2}$	
$5p^2_{1/2}5p^3_{3/2}$	+	$5p^2_{1/2}5p^3_{3/2}5d_{3/2}6p_{3/2}$	

TABLE 2. List of molecular configurations. Continued

Atom A		Atom B	Description
$5p_{1/2}^2 5p_{3/2}^3$	+	$5p_{1/2}^2 5p_{3/2}^3 6s_{1/2}^2$	ionic
$5p_{1/2}^2 5p_{3/2}^3$	+	$5p_{1/2}^2 5p_{3/2}^3 5d_{3/2}^2$	
$5p_{1/2}^2 5p_{3/2}^3$	+	$5p_{1/2}^2 5p_{3/2}^3 5d_{5/2}^2$	
$5p_{1/2}^2 5p_{3/2}^3$	+	$5p_{1/2}^2 5p_{3/2}^3 6p_{1/2}^2$	
$5p_{1/2}^2 5p_{3/2}^3$	+	$5p_{1/2}^2 5p_{3/2}^3 6p_{3/2}^2$	
$5p_{1/2}^2 5p_{3/2}^3$	+	$5p_{1/2}^2 5p_{3/2}^3 7s_{1/2}^2$	
$5p_{1/2}^2 5p_{3/2}^3$	+	$5p_{1/2}^2 5p_{3/2}^3 6s_{1/2} 7s_{1/2}$	
$5p_{1/2}^2 5p_{3/2}^3$	+	$5p_{1/2}^2 5p_{3/2}^4 6p_{1/2}$	
$5p_{1/2}^2 5p_{3/2}^3$	+	$5p_{1/2}^2 5p_{3/2}^4 6s_{1/2}$	
$5p_{1/2}^2 5p_{3/2}^3$	+	$5p_{1/2}^2 5p_{3/2}^4 7s_{1/2}$	
$5p_{1/2}^2 5p_{3/2}^3$	+	$5p_{1/2}^2 5p_{3/2}^4 5d_{3/2}$	
$5p_{1/2}^2 5p_{3/2}^3$	+	$5p_{1/2}^2 5p_{3/2}^4 6p_{3/2}$	
$5p_{1/2}^2 5p_{3/2}^3$	+	$5p_{1/2}^2 5p_{3/2}^4 5d_{5/2}$	

interactions between metastable atoms. These are the "C_6" and "C_8" configurations, which give rise to the induced dipole-dipole $-C_6/R^6$ and to the induced dipole-quadrapole $-C_8/R^8$ interactions, respectively. The next group are the so called "ionic" configurations, where participating atoms have positive and negative charges. These configurations contribute at short internuclear separation. The last group of configurations presents so called "other neutral" configurations and include configurations which do not play a significant role in the molecular formation. Some of these configurations contribute to the atomic structure.

The use of physically realistic configurations in the CI ensures that molecular quantities such as dissociation energies, exchange and Coulomb type interactions and dipole-dipole long-range interactions are introduced correctly. Later we demonstrate our ability to analyze the influence of each group of configurations on the formation of the molecule by plotting the configuration weights as function of internuclear separation. The configuration weights are sensitive indicators of the molecular wave function.

The configuration weights for each state of the seven groups are determined by diagonilizing the matrix H_{AB}. Since there are many eigenstates which have a lower energy than the metastable potentials we applied direct diagonalization instead of a Davidson iterative procedure.

For the present calculation we focus on the lowest metastable potentials of Xe_2 which are accessible in experiments with ultra-cold metastable Xe [28]. These potentials dissociate to the three limits ($5p^5 6s + 5p^5 6s$) $[3/2]_2 + [3/2]_2$, $[3/2]_2 + [3/2]_1$ and $[3/2]_1 + [3/2]_1$ and have a projection of their total electron angular momentum along the internuclear axis $\Omega = 0, 1, 2, 3,$

4. The atomic notation is explained in Ref. [29]. The metastable potentials are identified from a comparison with experimental atomic energies [29], the degeneracy at the dissociation limit for a given Ω and the CI weights.

The potentials are used to obtain molecular constants of metastable Xe_2. The equilibrium distance, R_e, rotational, ω_e, and vibrational, B_e, frequencies at R_e, and the dissociation energy, D_e, are shown in Table 3. Potentials in Table 3 are ordered with increasing total energy. The second column numbers the potentials. The first column of Table 3 indicates the symmetry labels of the potentials. In relativistic notation these correspond to the projection quantum number Ω and "g" and "u" symmetry. Table 4 describes the molecular dissociation limits of the potentials shown in Table 3.

As an example of the structure of the metastable potentials, we show in Fig. 2 the potentials with Ω=0, that dissociate to the lowest metastable atomic $5p^5 6s$ levels. The potentials create three groups of curves with R_e between 10 a.u. and 13.3 a.u., 12.5 a.u. and 14.1 a.u., and 12.4 a.u. and 13.7 a.u.. The curves do not exhibit strong avoided crossings. For $\Omega = 0$ there are eight "g" potentials and six "u" potentials. The splittings between the three dissociation limits are related to the $J = 1$ and $J = 2$ fine-structure splitting of the $5p^5 6s$ [3/2] term. These splittings are in a good agreement (7%) with experimental atomic data [29] where the experimental difference between the two metastable atomic levels $[3/2]_2$ and $[3/2]_1$ is $\Delta E_{exp.} =$ 977.6 cm^{-1} which is compared to the energy distance between the two lowest molecular dissociation limits in Fig. 2. The energy difference between the first and the third limit in Fig. 2 agrees to the same level of accuracy with twice the atomic energy difference $\Delta E_{exp.}$.

Along with the potential energy surfaces we evaluated the CI weights of the potentials. These weights are the square of the CI amplitudes of Eq. (3). Figures 3 and 4 show the summed weights of the groups of configurations defined in Tables 1 and 7 for the g and u lowest $\Omega = 3$ curves as function of internuclear separation. The weight of each group is obtained as the sum of all CI weights of determinants that belong to the configurations of this group. The two $\Omega = 3$ potentials are quantitively different. From Table 3 we see that the lowest curve is much deeper and has an equilibrium distance at approximately 10 a.u. The second potential has an equilibrium distance of 13.5 a.u. For both figures the metastable configuration has the largest CI weight. Figs. 3 and 4 also show there exist two distinctive regions of internuclear separations where the curves are qualitatively different. For $R \leq$ 14 a.u. there is very "strong" configuration interaction. Many configurations contribute to the formation of the molecule. Except for the metastable configuration the CI weights of different groups have similar values.

This situation changes for larger internuclear separations. The CI weights

TABLE 3. Molecular constants of $\mathbf{Xe_2}$ potentials dissociating to $(5p^5 6s + 5p^5 6s)$ states (The energy equivalent 1 cm^{-1} is 29.9792458 GHz.)

$\Omega_{g/u}$	Index	R_e (a.u.)	ω_e (cm^{-1})	B_e (cm^{-1})	D_e (cm^{-1})
0_g	1	10.0	18.3	0.0046	1650
0_u	2	10.0	19.0	0.0046	1636
0_u	3	10.4	23.4	0.0043	1533
0_g	4	10.4	24.0	0.0042	1517
0_g	5	13.3	7.6	0.0026	314
0_u	6	12.5	12.8	0.0029	856
0_g	7	12.6	11.0	0.0029	579
0_u	8	13.2	8.6	0.0026	366
0_g	9	13.5	7.3	0.0025	296
0_u	10	13.5	7.3	0.0025	295
0_g	11	14.1	6.7	0.0023	239
0_g	12	12.4	6.6	0.0030	649
0_u	13	12.6	6.7	0.0029	566
0_g	14	13.7	10.7	0.0024	348
1_u	1	10.4	18.8	0.0042	1606
1_g	2	10.2	20.6	0.0044	1576
1_u	3	10.3	22.0	0.0042	1525
1_g	4	13.3	7.1	0.0026	306
1_u	5	12.5	12.8	0.0029	855
1_g	6	12.6	11.2	0.0029	581
1_u	7	13.1	8.8	0.0027	365
1_g	8	13.6	6.7	0.0025	278
1_u	9	13.6	7.5	0.0025	278
1_g	10	14.1	6.6	0.0023	235
1_u	11	13.0	10.5	0.0027	514
1_g	12	13.4	7.1	0.0026	381
2_g	1	10.2	20.4	0.0044	1596
2_u	2	10.2	20.6	0.0044	1515
2_g	3	13.4	7.2	0.0025	284
2_u	4	12.4	13.0	0.0030	851
2_g	5	12.6	11.6	0.0029	585
2_g	6	13.7	6.7	0.0024	254
2_u	7	14.0	6.5	0.0023	227
2_g	8	13.1	7.0	0.0027	412
3_u	1	9.93	20.41	0.0046	1521
3_g	2	13.5	6.85	0.0025	255
3_u	3	12.3	13.44	0.0030	844
3_g	4	13.6	6.81	0.0025	258
4_g	1	13.6	6.43	0.0025	228

TABLE 4. Dissociation limits of $\mathbf{Xe_2}$ potentials from Table 3

Ω	Indices from Table 3	Dissociation limit
0	1 - 5	$[3/2]_2 + [3/2]_2$
	6 - 11	$[3/2]_2 + [3/2]_1$
	12 - 14	$[3/2]_1 + [3/2]_1$
1	1 - 4	$[3/2]_2 + [3/2]_2$
	5 - 10	$[3/2]_2 + [3/2]_1$
	11, 12	$[3/2]_1 + [3/2]_1$
2	1-3	$[3/2]_2 + [3/2]_2$
	4-7	$[3/2]_2 + [3/2]_1$
	8	$[3/2]_1 + [3/2]_1$
3	1,2	$[3/2]_2 + [3/2]_2$
	3,4	$[3/2]_2 + [3/2]_1$
4	1	$[3/2]_2 + [3/2]_2$

have a more smooth and predictable behaviour as function of R. For example, the "opposite" parity configurations show an exponential behaviour consistent with the fact that these configurations can only contribute when the atomic wave function overlap. The "C_6" and "C_8" weights have a $1/R^6$ and $1/R^8$ behaviour, respectively. Notice also the "other neutral" group includes configurations which contribute to the atomic structure, and the CI weight of this group has a finite value at large internuclear separations.

8. Relativistic effects

An important check on any relativistic theory is its convergence to a non-relativistic limit when the speed of light goes to infinity. In fact, the difference between the relativistic and the "$c \rightarrow \infty$" potentials shows the value of relativistic effects embedded in the Dirac equations including the spin-orbit interaction. The knowledge of these relativistic effects for core and valence electrons can be used in semiempirical and effective potential approaches to construct realistic potentials.

To provide this data we performed three calculations: relativistic, applying the RCIVB method, non-relativistic, using the relativistic code with

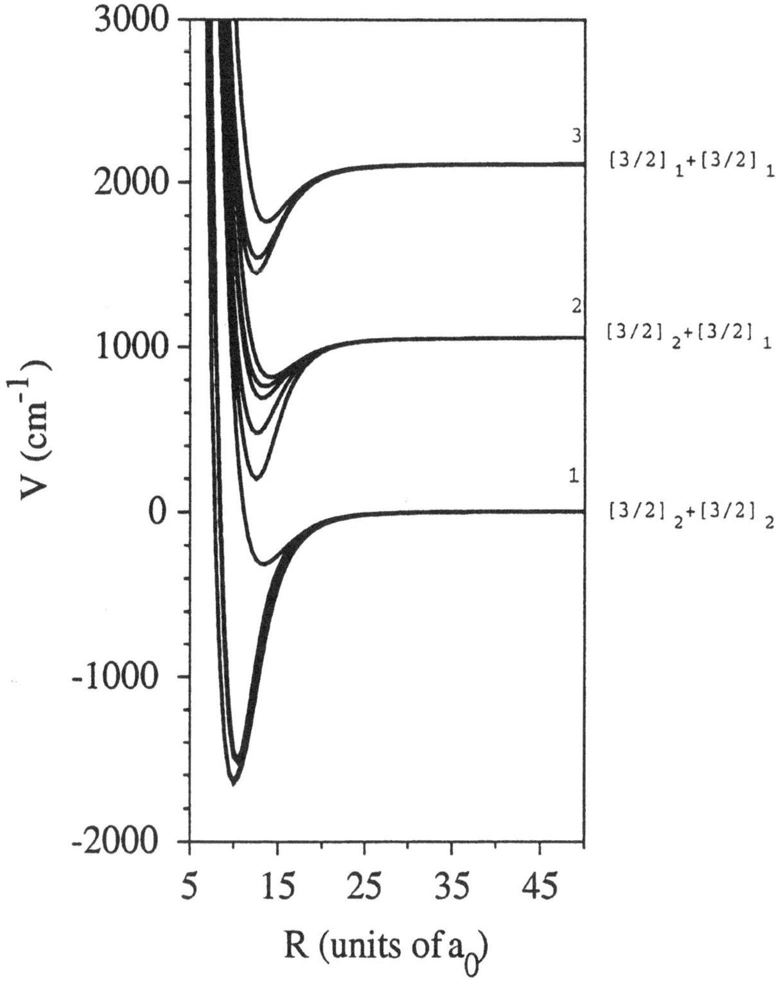

Figure 2. The $\Omega = 0$ metastable $5p^5 6s + 5p^5 6s$ Xe potentials as a function of internuclear separation.

the speed of light set to 10^4 x c, and a non-relativistic based on a CIVB method [22] using Hartree-Fock orbitals. We find that the total relativistic core energy is 430.06 a.u. (1 a.u.= 4.359743×10^{-18} J) lower than the total non-relativistic core energy. The valence energies are different as well. Fig. 5 presents the valence energies of the two lowest $\Omega = 3$ potentials,

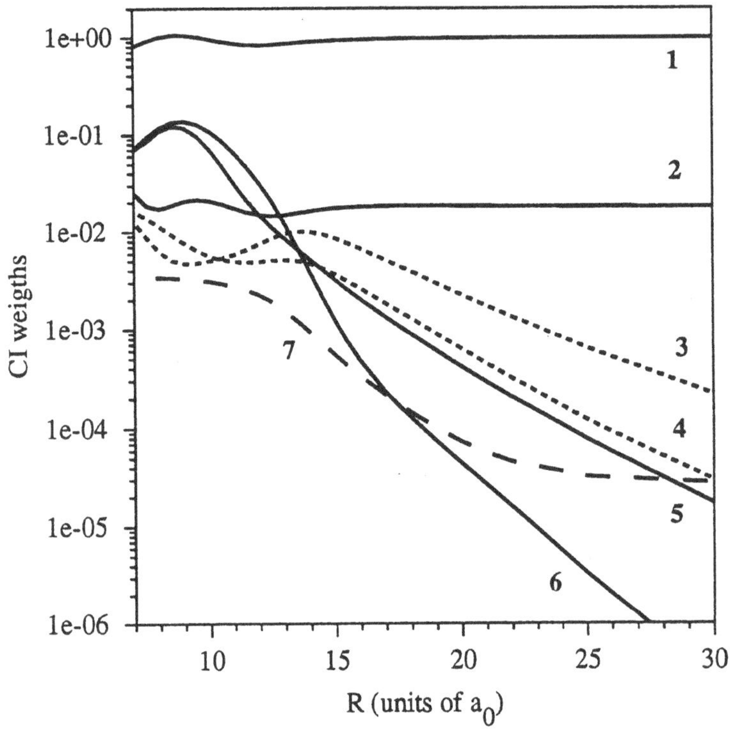

Figure 3. The configuration interaction weights of the lowest level of $\Omega=3$ as a function of internuclear separation. The six lines correspond with 1) the metastable configuration, 2) configurations with the same parity as a metastable, 3) "C_6" configurations, 4) "C_8" configurations", 5) "ionic" configurations, 6) configurations with "opposite parity", 7) "other neutral" configurations.

calculated in the three different approximations: relativistic, "c$\rightarrow\infty$", and non-relativistic as function of a internuclear separation. Comparing these results shows that when c $\rightarrow\infty$ the relativistic code converges to the non-relativistic limit (see the top four solid and dotted curves in Fig. 5). We explain the small difference between the two results by the fact that our CI expansion does not include configurations with $5p^1_{1/2}\ 5p^4_{3/2}$ shells in our calculation in order to avoid unacceptably large matrices. This means that we do not provide a complete set of relativistic configurations belonging to non-relativistic configurations with a $5p^5$ shell, which is the necessary condition for a complete convergence.

Moreover, comparison of relativistic and non-relativistic curves in Fig. 5 shows that the contraction of relativistic orbitals leads to a shorter equi-

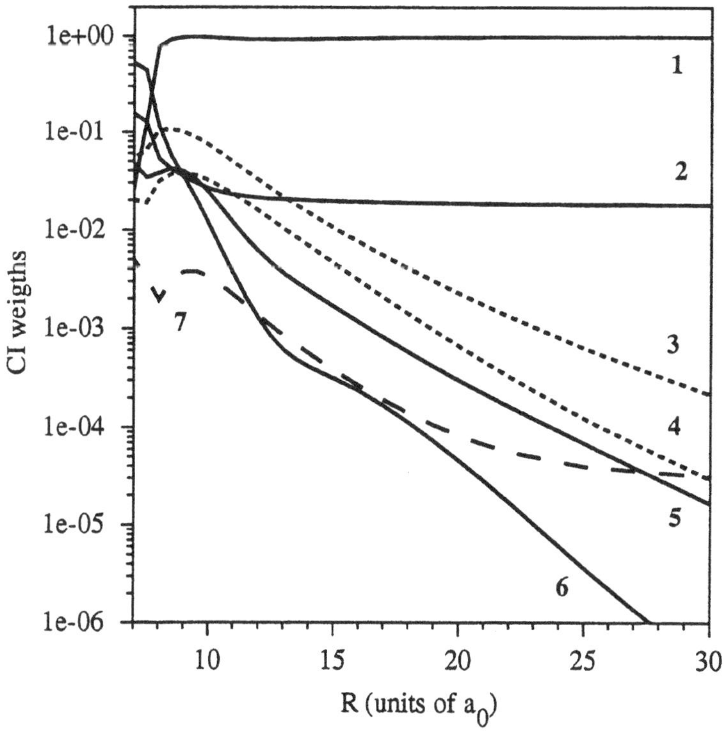

Figure 4. The configuration interaction weights of the second level of $\Omega=3$ as a function of internuclear separation. The lines have the same interpretation as those of Fig. 3.

librium distance. For example, for the lowest potential R_e is smaller by 0.21 a.u.. For the same potential the non-relativistic dissociation energy is bigger by 230 cm^{-1}.

9. Conclusions

We presented a new version of the *ab initio* relativistic configuration interaction valence bond method. The method uses nonorthogonal four-component Dirac-Fock orbitals and relativistic Sturmian wave functions. To further optimize these functions we introduced fractional occupation of the orbitals and density matrices of mixed states that correspond to the configuration average. The physically realistic atomic orbitals in our model lead to a compact description of a molecule and to an efficient configuration interaction procedure.

For a long time it has been thought that the use of numerical atomic

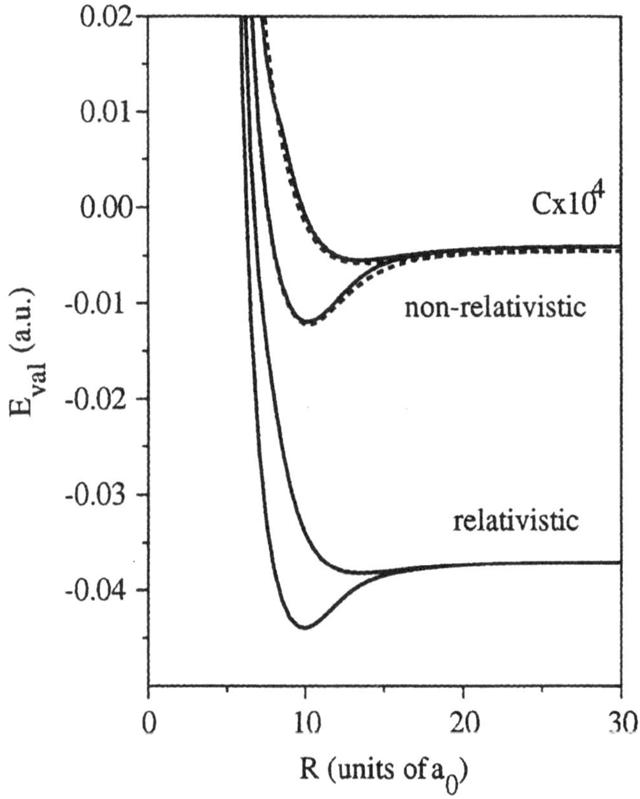

Figure 5. The valence energy calculated in relativistic (bottom two solid lines), "c→ ∞" (upper two solid lines), and non-relativistic (dotted lines) approximations for the two lowest $\Omega = 3$ potentials of metastable Xe_2.

wave functions in a molecular basis leads to a complex calculation of the many-center integrals. Here we develop a symmetric reexpansion procedure which ensures a fast convergence of the reexpansion.

We apply our method to a calculation of the electronic potentials of metastable Xe_2 dimer which is relevant for ultra-cold collisions. This application demonstrates the ability of the valence bond theory to provide a physically realistic description of interacting atoms.

We investigated the reliability of the model by comparing the electronic structure near the molecular dissociation limits with the corresponding energy structure of metastable Xe atom which is known experimentally. The structures are found to be in good agreement. The difference between relativistic and non-relativistic total energies of metastable Xe_2 is large and highlights the importance of a non-perturbative relativistic approach. Relativistic effects on the shape of the potentials show a contraction of bond lengths and a slight decreasing of the binding energies.

References

1. McWeeny, R. (1992). *Methods of Molecular Quantum Mechanics.* Academic Press, London.
2. Klein, D. J. and Trinajstic, N. (1990). *Valence bond theory and chemical structures.* Elsevier, Amsterdam.
3. Heitler, W. and London, F. (1927). Wechselwirkung neutraler Atome and homoopolare Bindung nach der Quantenmechanik. *Z. Physik*, 44:455-472.
4. Coulson, C. A. and Fischer, I. H. (1949). Notes on the Molecular Orbital Treatment of the Hydrogen Molecule. *Phil. Mag.*, 40:386-393.
5. Goddard, W. A. (1967). Improved Quantum Theory of Many-Electron Systems. I. Construction of Eigen functions of $\hat{S}^2$ which satisfy Pauli's Principle. II. The Basic Method. *Phys. Rev.*, 157:73-80 and 81-93.
6. Gerratt, J. and Lipscomb, W. N. (1968). Spin-coupled wave functions for atoms and molecules. *Proc. natn. Acad. Sci. U.S.A.*, 59:332-335.
7. Gerratt, J. (1971). General theory of spin-coupled wave functions for atoms and molecules. *Adv. atom. mol. Phys.*, 7:141-221.
8. Wilson, S. and Gerratt, J. (1975). Calculation of potential energy curves for the ground state of the hydrogen molecule. *Molec. Phys.*, 30:777-787.
9. Pyper, N. C. and Gerratt, J. (1977). Spin-coupled theory of molecular wavefunctions: applications to the structure and properties of $LiH(X^1\Sigma^+)$, $BH(X^1\Sigma^+)$, $Li_2(X^1\Sigma_g^+)$ and $HF(X^1\Sigma^+)$. *Proc. R. Soc. Lond. A*, 355:407-439.
10. Cooper, D. L., Gerratt, J., and Raimondi, M. (1987). Modern valence bond theory. *Adv. Chem. Phys.*, 69:319-397.
11. Cooper, D. L., Gerratt, J., and Raimondi, M. (1988). Spin-coupled valence bond theory. *Int. Rev. Phys. Chem.*, 7:59-80.
12. Cooper, D. L., Gerratt, J., and Raimondi, M. (1990). The spin-coupled valence bond description of benzenoid aromatic-molecules. *Top. Curr. Chem.*, 153:41-50. In *Advances in the Theory of Benzenoid Hydrocarbons*, I. Gutman, S. J. Cyvin, Editors. Berlin , New York , Springer-Verlag.
13. Cooper, D. L., Gerratt, J., and Raimondi, M. (1990). The spin-coupled approach to electronic structure. *Mol. Simulation*, 4:293-312.
14. Cooper, D. L., Gerratt, J., and Raimondi, M. (1991). Applications of spin-coupled valence bond theory. *Chem. Rev.*, 91:929-964.
15. Gerratt, J., Cooper, D. L., Karadakov, P. B. and Raimondi, M. (1997). Modern valence bond theory. *Chem. Soc. Revs.*, 26:87-100.
16. Langlois, J., Muller, R., Coley, T., Goddard III, W., Ringnalda, M., Won, Y., and Friesner, R. (1990). Pseudospectral generalized valence-bond calculations: application to methylene, ethylene, and silylene. *J. Chem. Phys.*, 92:7488-7497.
17. Miller, R., Langlois, J., Ringnalda, M., Friesner, R., and Goddard III, W. (1994). A generalized direct inversion in the iterative subspace approach for generalized valence bond wave functions. *J. Chem. Phys.*, 100:1226-1235.
18. Tannor, D., Marten, B., Murphy, R., Friesner, R., Sitkoff, D., Nicholls, A., Ringnalda, M., Goddard III, W., and Honig, B. (1994). Accurate first principles calculation of molecular charge distributions and solvation energies from ab initio quantum

mechanics and continuum dielectric theory. *J. Am. Chem. Soc.*, 116:11875-11882.
19. Kotochigova, S. and Tupitsyn, I. (1995). Electronic structure of molecules by the numerical generalized-valence-bond wave functions. *Int. J. Quant. Chem.*, 29:307-312.
20. Kotochigova, S. and Tupitsyn, I. (1998). Accurate ab initio calculation of molecular constants. *J. Res. Natl. Inst. Stand. Technol.*, 103:201-204.
21. Kotochigova, S. and Tupitsyn, I. (1998). Hyperfine structure constants for diatomic molecules. *J. Res. Natl. Inst. Stand. Technol.*, 103:205-207.
22. Kotochigova, S., Tiesinga, E., and Tupitsyn, I. (1999). Non-relativistic *ab initio* calculation of the interaction potentials between metastable Ne atoms. *Phys. Rev. A*, submitted for publication.
23. Slater, J. C. (1974). *The self-consistent field for molecules and solids* vol. 4, N.Y., McGraw-Hill.
24. Pavinsky, P. P. and Sherstuk, A. I. (1974). *Problems in theoretical physics*, Leningrad, vol. 1.
25. Sherstuk, A. I. (1975). Possibility of excluding a continuous spectrum from the calculation of electronic states of a molecular hydrogen ion. The Sturm expansion of the Green's function in two-center problems. *Opt. Spectrosc.*, 38:601-602.
26. Löwdin, P. O. (1956). On the historical development of the valence bond method and the non-orthogonality problem. *Advances in Physics*, 5:1-14.
27. Doery, M. R., Vredenbregt, E. J. D., Op de Beek, S. S., Beijerinck, H. C. W., and Verhaar, B. J. (1998). Limit on suppression of ionization in metastable neon traps due to long-range anisotropy. *Phys. Rev. A*, 58:3673-3682.
28. Orzel, C., Bergeson, S. D., Kulin, S., and Rolston, S. L. (1998). Time-resolved studies of ultracold ionizing collisions. *Phys. Rev. Lett.* 80:5093-5096.
29. Moore, C. E. (1952). *Atomic Energy Levels*, National Bureau Standards, Washington, D.C.

RELATIVISTIC QUANTUM CHEMISTRY OF SUPERHEAVY TRANSACTINIDE ELEMENTS

GULZARI L. MALLI
Department of Chemistry, Simon Fraser University
Burnaby, B.C., Canada V5A 1S6

Abstract. In this paper we report relativistic coupled-cluster calculations for tetrahedral $RfCl_4$. We assumed the Gaussian nuclear model implemented in MOLFDIR and our universal basis set was contracted with the atomic balance and kinetic energy constraint as implemented in MOLFDIR. In our most extensive calculation 24 electrons were correlated and 144 virtual spinors were included in the active space. The correlation and total energies obtained for $RfCl_4$ in this calculation were -0.2001 and -40538.3441 hartrees, respectively. Extensive calculations that correlate 30-40 electrons and include much larger active spinor spaces are in progress.

1. Introduction

Recently the chemistry and physics of man-made superheavy transactinide elements with atomic number $Z > 103$ have been vigorously investigated [1-26], both experimentally and theoretically. It is well-established [21-28] that Dirac's relativistic quantum mechanics (RQM) is mandatory for a proper understanding of electronic structure of atoms and molecules of heavy actinides and superheavy transactinide elements. Furthermore, it is well recognized [17,21,24-27] that the relativistic effects are expected to be so large in transactinide chemistry (TAC) that the dynamics of even the valence electrons of atoms of the transactinide elements (TAE) would be significantly affected, so that extrapolation of the chemical properties of these heaviest elements from their lighter homologs may be untrustworthy.

Moreover, due to both the direct and indirect relativistic effects, the electron configuration of the superheavy Lawrencium (Z=103) and some of the translawrencium elements may turn out to be different from what would be expected from their position in the periodic table. Needless to say, the consequences of the pronounced relativistic effects would be clearly manifested in the chemistry and physics of the superheavy elements (SHE). Therefore, in order to study the atomic and molecular systems of the superheavy elements, it is appropriate to

J. Maruani et al. (eds.), New Trends in Quantum Systems in Chemistry and Physics, Volume 1, 243–255.

develop generalization of Dirac's relativistic equation for an electron to many-electron atomic and molecular systems. Although the relativistic Hamiltonian for a many-electron system cannot be written down exactly, the principal effects of relativity can be described reasonably well by the so-called relativistic or Dirac-Fock Hamiltonian, which consists of a sum of one-electron Dirac Hamiltonians (H_D) plus the electron-electron Coulomb interaction amongst electrons.

The relativistic or DF-SCF theory for molecular systems was developed by Malli and Oreg [29]. It has been used extensively in the investigation of the relativistic effects for molecules of the heavy actinide as well as superheavy transactinide elements [21-28]. Due to very large effects of relativity in systems of SHE, there is an obvious need for the experimentalists in this area of research to rely on theoretical ab initio fully relativistic calculations which may help and even guide them in their research in the translawrencium chemistry (TLC). We have reported *ab initio all-electron fully relativistic* Dirac-Fock-Coulomb (DFC) and Dirac-Fock-Breit (DFB) calculations for the molecules of the actinides, viz.: thorium tetrafluoride [22], uranium hexafluoride [23] and the tetrachloride of the superheavy element rutherfordium [27]. In addition, we have presented the dissociation energy calculated from *the* first *ab initio all-electron* Dirac-Fock-Coulomb (DFC), Dirac-Fock-Breit (DFB) and non-relativistic (NR) Hartree-Fock (HF) wavefunctions for these systems.

Recently [26] we have discussed the effects of relativity on bonding, dissociation energy, bond length, electronic structure, gap between the highest occupied and lowest unoccupied molecular spinors (HOMS and HUMS, respectively), the Mulliken population analysis for a large number of molecules of the superheavy transactinide elements Rf Db Sg, ekaplatinum E110, ekagold E111, ekamercury E112, etc.

I shall discuss here the results of our *first* ab initio *fully* relativistic (four-component) coupled-cluster (RCC) calculations for $RfCl_4$, a polyatomic of the superheavy transactinide, and I hope that the calculations based upon our *molecular spinor self-consistent field* (MS-SCF) theory [29], which include *simultaneously relativistic* and *correlation* effects, may lead to an understanding of the electronic structure and bonding in these novel, very short-lived species of superheavy transactinide elements.

2. Dirac-Fock-Breit treatment for molecules of superheavy transactinide elements

We present here only a brief outline of the Dirac-Fock-Breit (DFB) formalism for molecules and refer the reader to more extensive accounts for details [21-23, 26-30].

The approximate relativistic Dirac-Fock-Coulomb Hamiltonian (H_{DC}) for an N-electron molecular system containing n nuclei, under the Born-Oppenheimer approximation (omitting the nuclear repulsion terms which are constant for a given molecular configuration) can be written (in atomic units):

$$H_{DC} = \sum_{i=1}^{N} H_D(i) + \sum_{i<j} \frac{1}{r_{ij}} \quad (1)$$

In Eq. (1), the $H_D(i)$ consists of the Dirac's kinetic energy operator, mass energy and nuclear attraction of the i-th electron and has the well-known expression, viz.:

$$H_D(i) = c\alpha_i \cdot p_i + (\beta_i - 1)c^2 + V_{nuc} \quad (2)$$

where Dirac's matrix operators α and β have the usual 4x4 matrix representation.

The rest-mass energy of an electron has been subtracted in Eq. (2), in order to get its binding energy, and the potential V_{nuc} due to n *finite nuclei* of the molecular system is taken as the sum of their nuclear potentials: $V_{nuc} = \sum_n V_n$ (3),

and for molecular systems involving heavy atoms (with $Z > 70$) a finite nuclear model is invariably used. We shall use the Gaussian nuclear model [30] in which a single Gaussian function is used for each nuclear charge distribution. The advantage of using this nuclear model in basis-set calculations on polyatomics with finite nuclei is that all multicentre integrals can be calculated analytically, with a Gaussian basis set, in a straightforward way. It should be pointed out, however, that for heavy atoms the total relativistic Dirac-Fock electronic energy obtained with the Gaussian nuclear model is, in general, *higher* than that obtained with either the spherical-ball, point nucleus or the Fermi nuclear model. The overall higher total atomic energy obtained with the Gaussian nuclear model is mostly reflected in the higher energies of the lowest angular momenta *atomic spinors* (AS), viz.: $1s_{1/2}$, $2s_{1/2}$ and $2p_{1/2}$, which are closest to the nucleus; however, the energies of all other atomic spinors of an atom are calculated to be almost identical using the various finite nuclear models.

The instantaneous Coulomb repulsion between the electrons is treated *non-relativistically* in the Dirac-Coulomb Hamiltonian and the magnetic and retardation corrections to it are generally included perturbationally as discussed later.

The N-electron wavefunction Φ for the closed-shell molecular system is taken as a single Slater determinant (SD), also called an antisymmetrized product (AP) of one-electron 4-component molecular spinors (MS) [29], viz.:

$$\Phi = (N!)^{-1/2} \,|\, \phi_1(1)\, \phi_2(2)\, \phi_3(3) \,. \;.\; .\; \phi_N(N) \,|. \qquad (4)$$

The molecular spinors (MS) ϕ_i are generally taken to form an orthonormal set and can be constructed so as to transform like the extra or *additional irreducible representations* (EIR) of the double symmetry group of the molecule under investigation [29]. The energy expectation value E can then be written as:

$$E = \langle \Phi | H | \Phi \rangle / \langle \Phi | \Phi \rangle \qquad (5)$$

The molecular spinors ϕ_i are expressed in terms of the large and small components

$$\phi_i^X = \sum_{q=1}^{n} C_{iq}^X \chi_q^X, \quad X = L \text{ or } S.. \qquad (6)$$

The χ_q^L and ϕ_i can be symmetry adapted; however, we shall ignore the double group symmetry labels. for the molecular spinors. The basis spinors χ_q^X will be constrained to obey the kinetic balance relation, viz.:

$$\chi_q^S = (\sigma \cdot p)\, \chi_q^L. \qquad (7)$$

Then following Malli and Oreg [29] the Dirac-Hartree-Fock- Roothaan (DHFR) or relativistic Hartree-Fock-Roothaan (RHFR) SCF equations for closed-shell molecules can be written as:

$$Fc_i \;\; = \varepsilon_i S c_i \qquad (8)$$

where F is the Dirac-Fock matrix operator, ε_i is the orbital energy of molecular spinor (MS) ϕ_i, and S is the overlap matrix. All matrix elements occurring in DHFR-SCF calculations for polyatomics in general can be expressed [29] in terms of the types of matrix elements that arise in nonrelativistic (NR) Hartree-Fock-Roothaan (HFR) SCF calculations for polyatomic molecules, and well-developed techniques for the evaluation of these matrix elements, using Gaussian type functions, have been in existence for decades.

The electron-electron Coulomb interaction is treated nonrelativistically, as mentioned earlier in this section, in H_{DC}. The Breit interaction [31] consisting of the magnetic and retardation terms was proposed to remedy partially this defect of H_{DC} and the addition of Breit interaction (B_{ij}) to H_{DC} leads to the Dirac-Coulomb-Breit (DCB) Hamiltonian H_{DCB}, which has the form:

$$H_{DCB} = H_{DC} + \sum_{i<j} B_{ij}. \tag{9}$$

The B_{ij} in Eq. (9) is usually written as:

$$B_{ij} = -\frac{1}{2}\{(\alpha_i \cdot \alpha_j) r_{ij}^{-1} + (\alpha_i \cdot r_{ij})(\alpha_j \cdot r_{ij})\, r_{ij}^{-3}\} \tag{10}$$

Twice the first term in Eq. (10), called the magnetic or Gaunt interaction, is the dominant part of the Breit interaction; the retardation term is about 10% of the Gaunt interaction, and in general the contribution of the Breit interaction is fairly marginal compared to the Coulomb interaction term. The use of H_{DCB} as a starting point for variational molecular calculations leads to the Dirac-Fock-Breit (DFB) SCF equations. The Dirac-Fock-Breit matrix operator occurring in the DFB-SCF equations involves the matrix elements of the magnetic and retardation interactions. The expressions of the matrix elements of the magnetic or Gaunt interaction are given in Malli and Oreg [29]. We have recently included the Gaunt interaction perturbationally, with H_{DC} as the unperturbed Hamiltonian, in relativistic calculations on ThF_4 [22], UF_6 [23], $RfCl_4$ [27]. The retardation term, however, has not been included in relativistic molecular calculations for heavy and superheavy elements.

3. Universal Gaussian basis set for relativistic coupled-cluster calculations on superheavy transactinide elements

Recently we have developed [32-34] a generator-coordinate Dirac-Fock method (GCDF) for closed and open-shell systems, and we have reported an accurate relativistic universal Gaussian basis set (RUGBS) for atomic systems up to E113 (ekathallium, with Z=113). We have used our RUGBS in all of our Dirac-Fock (DF) SCF and Dirac-Fock-Breit (DFB) calculations on numerous molecules of the transactindes with the MOLFDIR package [30], assuming the Gaussian nuclear model. The RUGBS was contracted using the general contraction scheme along with the atomic balance procedure as implemented in the MOLFDIR code [30]. The RUGBS for the small (S) component of atoms used in our calculations were obtained from the RUGBS of the corresponding large (L) component, such that the L and S components of each atom satisfy the so-called kinetic balance condition as implemented in MOLFDIR [30].

The exponents of the L component of the RUGBS used in our calculations are given in our earlier papers, and the DF-SCF total energies obtained with our contracted relativistic RUGBS for various atoms (using the Gaussian nuclear model) up to Sg (Z=106) are in excellent agreement with those obtained with the numerical finite difference scheme [35]. It should be pointed out that the total DF-SCF average of configuration energy for the transactinides, obtained by Desclaux [35] using numerical finite difference methodology with the spherical ball nuclear model is about 7-10 hartree (1 hartree=27.211 eV) *lower* than that obtained by us using the MOLFDIR [30] code, which, however, as mentioned above, uses the Gaussian nuclear model. This difference is due mostly to the different nuclear models used in these two calculations, and the energy difference, as expected, is reflected mostly in the energies of the 1s and 2s atomic spinors of the transactinide under investigation.

These differences in total energy as well as in the energies of the innermost atomic spinors should be kept in mind while comparing the total relativistic DFC as well as non-relativistic HF atomic energies calculated with the basis-set expansion method using different nuclear models, especially for systems involving the heaviest transactinide atoms. However, for the light atoms: O, F, Cl, etc., the difference in nuclear model is not very significant and the total DF-SCF energies for such atoms reported by Desclaux [35] are in excellent agreement with those obtained by us using the MOLFDIR code [30].

The contracted RUGBS for all the atoms of a molecule under study is used in all of our calculations. However, the contracted RUGBS for the L components only, were used in the nonrelativistic (NR) limit Hartree-Fock (HF) calculations, which were also performed with the MOLFDIR [30] package using the Gaussian nuclear model. The total NR HF energies for point nucleus Rf ($6d^2$ average) and other transactinides reported in Desclaux [35], however, are in general lower by about 3–4 hartrees than that reported here as expected, since it is well-known that point nucleus calculations yield *lower* total energy for an atom than that obtained from the corresponding finite nuclear model calculation, in accord with our results reported here.

4. Dirac-Fock-Breit calculations for molecules of the superheavy transactinide elements

We have investigated a large number of molecular systems of superheavy transactinide elements and a partial list of the systems that have been investigated includes the following: RfBr, RfCl, $ZrCl_4$, $HfCl_4$, RfF_4, $RfCl_4$, $RfBr_4$, DbCl, DbO, DbBr, $NbCl_5$, TaCl5, $DbCl_5$, $NbOCl_3$, $TaOCl_3$, $DbOCl_3$, $NbBr_5$, TaBr5, $DbBr_5$, SgBr, SgCl, SgO, $SgCl_6$, WCl_6, $MoCl_6$, SgF_6, WF_6, MoF_6, $SgBr_6$, SgO_2Cl_2, $SgOCl_4$, WO2Cl2, $WOCl_4$, BhBr, BhCl, $BhCl_6$, $ReCl_6$, $BhOCl_4$, HsO_4, $HsCl_6$, OsO_4, $OsCl_6$, E110Cl_6, E110F_6, E111H, E111Cl, E111Br, E112Cl_2,

$E112Cl_4$, $E112F_2$, $E112F_4$, HgF_4, $E114Cl_2$, $E114Cl_4$, E117Cl, $E118F_2$, $E118F_6$ ....

We have discussed in our earlier papers [21-23,25-28] thc salient results of our *ab initio* all-electron fully relativistic Dirac-Fock (Breit) SCF calculations for numerous molecules of the actinide and superheavy transactinides listed above; and we refer the reader to our papers for further details. We next discuss the relativistic coupled-cluster (RCC) treatment for molecules of the superheavy transactinide elements and present here the results of the *first* RCC calculations [36] for a polyatomic involving the superheavy element Rutherfordium, viz.: $RfCl_4$.

5. Relativistic coupled-cluster methodology

The coupled-cluster (CC) method [37] has emerged as a very powerful tool for calculating the *correlation effects* in atomic and molecular systems, as it includes electron correlation to high order and is *size extensive*, a property of particular importance for heavy systems, for which relativistic effects are also very significant. In order to treat both the relativistic and electron correlation effects *simultaneously*, the relativistic coupled-cluster (RCC) has been developed [38,39] by interfacing the relativistic Dirac-Fock (Breit) SCF theory with the CC method, and results have been reported for a number of atoms and the hydrides CdH and SnH_4 [40].

Very recently, we have performed extensive RCC calculations on molecules of heavy and superheavy elements: gold hydride (AuH), thorium tetrafluoride (ThF_4) [28] and $RfCl_4$, the tetrachloride of the superheavy element rutherfordium [36] The RCC method and its recent applications to atomic and molecular systems [28,36,38-40] are summarized below.

Although the relativistic many-body Hamiltonian for atomic and molecular systems cannot be expressed in closed potential form; nonetheless the nonrelativistic many-body formalism can be extended to the relativistic domain by using the formalism based on effective potentials and derived with arbitrary accuracy from quantum electrodynamics (QED) as described by Lindgren [37,41]. The transition from the nonrelativistic to the fully relativistic case requires two major modifications: (1) *two-component* Pauli-Schrödinger *spinorbitals* are replaced by *four-component* Dirac *spinors*, (2) instantaneous electron-electron Coulomb interactions are supplanted by the irreducible multiphoton interactions with the radiative and renormalization counter terms.

The starting point for the RCC method (with single and double excitations) which includes relativistic and electron correlation effects simultaneously to high order for molecules, is the projected Dirac-Coulomb (or Dirac-Coulomb-Breit) Hamiltonian [42,43]:

$$\mathrm{H}+ = H_0 + \mathrm{V}, \tag{11}$$

where

$$H_0 = \sum_i \Lambda_i^+ H_D(i) \Lambda_i^+ \tag{12}$$

$$H_D(i) = c\alpha_i \cdot p_i + c^2(\beta_i - 1) + V_{nuc}(i) + U(i) \tag{13}$$

and

$$\mathrm{V} = \sum_{i<j} \Lambda_i^+ \Lambda_j^+ (V_{eff})_{ij} \Lambda_j^+ \Lambda_i^+ - \sum_i \Lambda_i^+ U(i) \Lambda_i^+ \tag{14}$$

An arbitrary potential U is included in the unperturbed Hamiltonian H_0 and subtracted from the perturbation V, and this potential is chosen to approximate the effect of electron-electron interaction and it may be the Dirac-Fock self-consistent field potential. The $\Lambda+$ is a product of projection operators onto the positive energy states of the Dirac Hamiltonian H_D, and because of the projection operators, the Hamiltonian $H+$ has normalizable bound state solutions. This approximation is known as the no-virtual-pair approximation since virtual electron positron pairs are not allowed in intermediate states. The form of the effective potential V_{eff} depends upon the gauge used and, in the particular Coulomb gauge (in atomic units, correct to the second order in the fine structure constant α), it has the form:

$$V_{eff} = \frac{1}{r_{12}} + B_{12} + O(\alpha^3), \tag{15}$$

where B_{12} is the frequency-independent Breit interaction that is defined in Eq. (10) above.

The no-pair Dirac-Coulomb-Breit Hamiltonian H+ may be rewritten [42,43], in the second-quantized form, in terms of normal-ordered products of the spinor operators $\{r^+s\}$ and $\{r^+s^+ut\}$:

$$H = H_+ - \langle 0|H_+|0\rangle = \sum_{rs} f_{rs}\{r^+s\} + \frac{1}{4}\sum_{rstu}\langle rs\|tu\rangle\{r^+s^+ut\}, \tag{16}$$

where f_{rs} and $\langle rs\|tu\rangle$ are elements of one-electron Dirac-Fock and antisymmetrized two-electron Coulomb-Breit interaction matrices over Dirac four-component spinors, respectively.

The effect of the projection operators Λ+ is now taken over by the normal ordering, denoted by the curly braces in the equation above, which requires annihilation operators to be moved to the right of the creation operators as if all anticommutation relations vanish.

The Fermi level is set at the top of the highest occupied positive energy state, and the negative energy state is ignored. The no-pair approximation leads to natural and straightforward extension of the nonrelativistic coupled-cluster theory.

The multireference valence-universal Fock space coupled-cluster approach, however, defines and calculates an effective Hamiltonian in a low-dimensional model (or P) space, with eigenvalues approximating some desirable eigenvalues of the physical Hamiltonian. According to Lindgren's formulation of the open shell CC method [44], the effective Hamiltonian has the form

$$H_{eff} = PH\,\Omega P \tag{17}$$

where Ω is the normal-ordered wave operator:

$$\Omega = \{\exp(S)\} \tag{18}$$

The excitation operator S is defined in the Fock-space coupled-cluster approach with respect to a closed-shell reference determinant. In addition to the traditional decomposition into terms with different total (l) number of excited electrons, S is partitioned according to the number of valence holes (m) and valence particles (n) to be excited with respect to the reference determinant:

$$S = \sum_{n\geq 0}\sum_{m\geq 0}\sum_{l\geq m+n} S_l^{(n,m)} \tag{19}$$

The upper indices in the excitation amplitudes reflect the partitioning of the Fock space into sectors, which correspond to the different numbers of electrons in the physical system. This partitioning allows for partial decoupling of the open shell CC equations, since the equations in each sector do not involve excitation amplitudes from higher sectors. The eigenvalues of the effective Hamiltonian given above in a sector directly yield the correlated energies in that sector with respect to the correlated (0,0) reference state. The transition energies may be ionization potentials, electron affinities or excitation energies, according to the presence of valence holes and / or valence particles.

The lower index l in Eq. (19) is truncated at $l = 2$. The resulting coupled cluster with single and double excitations (CCSD) scheme involves the fully self-consistent, iterative calculation of all one- and two-body virtual excitation amplitudes, and sums all diagrams with these excitations to infinite order. Negative energy states are excluded from the Q space, and the diagrammatic summations in the CC equations are carried out only within the subspace of the positive energy branch of the Dirac-Fock spectrum.

The implementation of the 4-component matrix Dirac-Fock and relativistic CC calculations is done by expansion of atomic or molecular spinors in basis sets of Gaussian 4-component spinors. Kinetic and atomic balance conditions are imposed on the basis of avoiding variational collapse. The four-component method involves generating the orbitals or spinors by Dirac-Fock calculations, followed by applying the coupled-cluster scheme at the singles-and-doubles (CCSD) level. The DF functions and matrix elements are calculated using the MOLFDIR package [30].

The coupled-cluster stage is more complicated in the four-component case than in the non-relativistic or two-component cases, due to the appearance of complex orbitals or complex spin-orbit integrals. Explicit complex algebra is avoided, but the necessary real algebra is heavier than in the one-component coupled-cluster case. The full double-group symmetry is used at the Dirac-Fock level, while only Abelian subgroups are considered in the RCC code.

6. Relativistic coupled-cluster calculations for molecules of superheavy elements: $RfCl_4$

Recently we have reported [27] *ab initio* all-electron fully relativistic molecular spinor Dirac-Fock (DF) self-consistent field (SCF) and nonrelativistic limit Hartree-Fock (HF) SCF calculations at four Rf-Cl bond distances for the ground state of tetrahedral (T_d) rutherfordium tetrachloride ($RfCl_4$). The dominant magnetic part (also called the Gaunt interaction) of the Breit interaction correction for RfCl4 was calculated perturbationally to be 66.85 hartree (1 hartree=27.211 eV). We have investigated [27] the effects of relativity on the bond length, dissociation energy, volatility, etc. for $RfCl_4$. However, both the *relativistic* (four-component) and *electron correlation* effects have not been included so far for any polyatomic involving a superheavy transactinide element. The $RfCl_4$ is a prototype of a polyatomic of a superheavy element and is a good candidate for the investigation of both the relativistic and electron correlation effects (which may not be additive) *simultaneously* via the relativistic coupled-cluster method.

We report here the *first* RCC calculations [36] for tetrahedral $RfCl_4$, at the Rf-Cl bond length of 2.385 Angstrom, optimized at the DF level by us [27]. It should be mentioned that in our calculations of electron correlation energy the CCSD approximation is used. Dirac–Fock–Breit, Dirac–Fock and nonrelativistic limit HF calculations were performed with the MOLFDIR package [30] as described in our earlier paper [27] at four distances assuming a tetrahedral geometry for $RfCl_4$. We assumed the Gaussian nuclear model implemented in MOLFDIR [30] and our universal Gaussian basis set (UGBS) [32-34]. The UGBS was contracted with the atomic balance and kinetic energy constraint as implemented in MOLFDIR to make the calculations feasible with the computer facilities available to us. The contracted basis set for the large (L) and small component (S) of Rf is [12s 15p 12d 7f (L) | 17s 22p 22d 15f 9g (S)], while for Cl, the correlation consistent basis set of Dunning given in MOLFDIR [30] was used, viz., Cl: [4s 4p 1d | 3s 5p 4d 1f (S)]. We realize that much larger contracted basis may be necessary for accurate CCSD calculations; however as is well known the number of integrals increases as $\sim N^4$, where N is the number of basis functions. Hence a judicious choice has to be made as to the size of the contracted basis sets to be used in CCSD calculations, which require hundreds of hours of CPU on medium size computers and about 30 gigabyte of disk space (except where direct SCF approach is used) for a medium size molecule like $RfCl_4$.

In our preliminary CCSD calculations, 24 electrons were correlated in 4 calculations differing in the number of active virtual spinors. The molecular CCSD calculations for closed-shell ground state of tetrahedral $RfCl_4$ were performed with the MOLFDIR code [30], while the calculations for the Rf and Cl atoms with open-shell ground states were performed with the Fock-space coupled-cluster (FSCC) code [38]. In our most extensive calculation, 24 electrons were correlated and 144 virtual spinors with energies up to 1.5 hartree were included in the active space. The correlation and total energies obtained for $RfCl_4$ in this calculation are –0.2001 and –40538.3441 hartrees, respectively. Extensive calculations that correlate 30-40 electrons and include much larger active spinor spaces are in progress.

In conclusion, very extensive relativistic coupled-cluster calculations would be mandatory to take into account *simultaneously* the effects of relativity and of electron correlation, which are very significant for superheavy systems with a large number of electrons such as those involving transactinide elements. We have just made a start in this area of research and the future holds many challenges and promises.

Acknowledgments

I am most grateful to Prof. Jean Maruani for his kind invitation to lecture at the Fourth European Workshop on Quantum Systems in Chemistry and Physics held at INJEP, Marly-le-Roi, France, in April 1999. I sincerely thank the organizers, especially Prof. Jean Maruani and Dr. Steve Wilson, for their warm hospitality and generous financial assistance. Most of the calculations reported here were performed on the supercomputer of the National Scientific Energy Research

Computing Center (NSERC) of the Lawrence Berkeley National Laboratory (LBNL). This superb supercomputing facility, sponsored and financed by the US Department of Energy (DOE), was indispensable for our enormous calculations. I gratefully acknowledge the award of a supercomputing grant by DOE, and thank Dr. Jonathan Carter, consultant at NERSC, for his valuable advice and help. Finally I would like to acknowledge my research coworkers whose names appear in the publications cited below.

References

1. Y.V. Lobanov, V.I. Kuznetsov, V.A. Druin, V.P. Perelygin, K.A. Gavrilov, S.P. Tret'yakova, V.M. Plotko, G.N. Flerov and Y.T. Oganessian, *Phys. Lett.* **13**, 73 (1964).
2. I. Zvara, Y.T. Chuburkov, R. Tsaletka, T.S. Zvarova, M.R. Shalaevskii and B.V. Shilov, *Sov. Atomic Energy* **21**, 709 (1966).
3. A. Ghiorso, M. Nurmia, J. Harris, K. Eskola and P. Eskola, *Phys. Rev. Lett.* **22**, 1317 (1969).
4. C.E. Bemis, Jr., R.J. Silva, D.C. Hensley, O.L. Keller, Jr., J.R. Tarrant, L.D. Hunt, P.F. Dither, R.L. Hahn and C.D.Goodman, *Phys. Rev. Lett.* **31**, 647 (1973).
5. A. Ghiorso, M. Nurmia, K. Eskola, J. Harris and P. Eskola, *Phys. Rev. Lett.* **22**, 1498 (1970).
6. G.T. Seaborg and W.D. Loveland, *The Elements Beyond Uranium*, John Wiley, New York, 1990.
7. Proceedings of the Robert A. Welch Foundation, *Conference on Chemical Research XXXIV, Fifty Years with Transuranium Elements*, Houston, Texas, October 1990.
8. H.W. Gaeggeler, *J. Radioanal. Nucl. Chem.*, Articles, **183**, 261 (1994).
9. A. Turler, *Radiochim. Acta* **72**, 7 (1996).
10. X. Schadel, *Radiochim. Acta* **70/71**, 207 (1995).
11. B. Wierczinski, J. Alstad, K. Eberhardt, B. Eichler, H. Gaeggeler, G. Herrmann, D. Jost, A. Nahler, M. Pense-Maskow, A.V.R. Reddy, G. Skarnemark, N. Trautman and A. Turler, *Radiochim. Acta* **69**, 77 (1995).
12. S.N. Timokhin, A.B. Yakushev, H. Xu, V.P. Perelygin and I. Zvara, *J.Radioanal.Nucl.Chem.Lett.* **212**, 31 (1996).
13. A.B. Yaushev, S.N. Timokhin, M.V. Vedeneev, X. Honggui and I. Zvara, *J.Radioanal. Nucl. Chem.*, Articles, **205**, 63 (1996).
14. S. Hofmann, *Rep.Progr.Phys.* **61**, 639 (1998).
15. V. Ninov, K.E. Gregorich, W. Loveland, A. Ghiorso, D.C. Hoffman, D.M. Lee, H. Nitsche, W.J. Swiatecki, U.W. Kirbach, C.A. Laue, J.L. Adams, J.B. Patin, D.A. Shaughnessy, D.A. Strellis and P.A. Wilk, *Phys.Rev.Lett.* **83**, 1104 (1999).
16. *Chem. Eng. News*, June 14, 1999, page 6.
17. I.P. Grant and N.C. Pyper, *Nature* **265**, 715 (1977).
18. G.L. Malli (ed.), *Relativistic Effects in Atoms, Molecules and Solids*, NATO ASI Series B: Physics, vol. **87** (Aug. 1981, Vancouver, BC, Canada), Plenum Press, New York, 1983.
19. M.A.K. Lodhi, *Superheavy Elements*, Pergamon Press, Oxford, 1973.
20. K. Kumar, *Superheavy Elements*, IOP Publishing, Bristol, UK, 1989.
21. G.L. Malli, in *Relativistic and Electron Correlation Effects in Molecules and Solids*, NATO ASI Series B: Physics, vol. **318** (Aug. 1992, Vancouver, BC, Canada), Plenum Press, New York, pp. 1-16 (1994).
22. G.L. Malli and J. Styszynski, *J.Chem.Phys.* **101**, 10736 (1994).
23. G.L. Malli and J. Styszynski, *J.Chem. Phys.* **104**, 1012 (1996).
24. V.G. Pershina, *Chem.Rev.* **96**, 1977 (1996).

25. G.L. Malli: "Ab-Initio All-Electron Fully Relativistic Dirac-Fock-Breit Calculations for Compounds of the Heaviest Elements: the Transactinides Rutherfordium through EkaAstatine Z=117", invited lecture at *Fourth Workshop on Physics and Chemistry of the Heaviest Elements*, Göteborg, Sweden, June 1997.
26. G.L. Malli, in *Transactinide Elements*: Proceedings of the 41st Welch Conference on Chemical Research, October 1997, Welch Foundation, Houston, Texas, 1997, pp. 197-228.
27. G.L. Malli and J. Styszynski, *J.Chem.Phys.* **109**, 4448 (1998).
28. G.L. Malli, J. Styszynski, E. Eliav, U. Kaldor and L. Visscher, "Relativistic Fock-Space Coupled-Cluster Calculations for Molecules with Heavy Elements: ThF_4 and AuH", Poster presented at the Research Conference on Relativistic Effects in Heavy-Element Chemistry and Physics, European Research Conferences, April 1999, Acquafredda di Maratea, Italy.
29. G. Malli and J. Oreg, *J.Chem.Phys.* **63**, 830 (1975).
30. L. Visscher, O. Visser, P.J.C. Aerts, H. Merenga and W.C. Nieuwpoort, Comp. Phys. Comm. 81, 120 (1994).
31. G. Breit, *Phys.Rev.* **36**,553 (1930); *ibid.* **39**, 616 (1932).
32. G.L. Malli, A.B.F. da Silva and Y. Ishikawa, *Phys.Rev.*A **47**, 143 (1993).
33. G.L. Malli, A.B.F. da Silva and Y. Ishikawa, *J.Chem.Phys.* **101**, 6829 (1994).
34. G.L. Malli and Y. Ishikawa, *J.Chem.Phys.* **109**, 8759 (1998) .
35. J.-P. Desclaux, *Atomic Data & Nucl. Data Tables* **12**, 311 (1973).
36. G.L. Malli, Relativistic coupled-cluster calculations on $RfCl_4$ (unpublished results).
37. (a) I. Lindgren, in *Lecture Notes in Chemistry*, Vol. **52**, *Many-Body Methods in Quantum Chemistry*, U. Kaldor (ed.), Springer, Berlin, 1989, p. 293.
(b) J. Paldus, in *Relativistic and Electron Correlation Effects in Molecules and Solids*; NATO-ASI Ser. B, vol. **318**, G.L. Malli (ed.), Plenum Press, New York, 1994, p. 207.
(c) *Theoret. Chim. Acta* **80**, issues 2-6 (1991).
38. E. Eliav and U. Kaldor, *Chem.Phys.Lett.* **248**, 405 (1996).
39. L. Visscher, K.G. Dyall and T.J. Lee, *J.Chem.Phys.* **105**, 8769(1996).
40. E. Eliav, U. Kaldor and B.A. Hess, *J.Chem.Phys.* **108**, 3409(1998).
41. I. Lindgren, *Nucl. Instr. Meth. Phys. Res.* B **31**, 102 (1998).
42. J. Sucher, *Phys.Rev.* A **22**, 348 (1980); *ibid.*, *Phys. Scr.* **36**, 271 (1987).
43. W. Buchmuller and K. Dietz, *Z. Phys.* C**5**, 45 (1980).
44. I. Lindgren and J. Morrison, *Atomic Many-Body Theory*, Springer-Verlag, Berlin, 1986.

Part IV

Valence Theory

THE NATURE OF BINDING IN HRgY COMPOUNDS (Rg = Ar, Kr, Xe; Y = F, Cl) BASED ON THE TOPOLOGICAL ANALYSIS OF THE ELECTRON LOCALISATION FUNCTION (ELF)

S. BERSKI [a)], B. SILVI [b)], J. LUNDELL [c)], S. NOURY [b)] AND Z. LATAJKA [a)]

[a)] *Faculty of Chemistry, University of Wroclaw, F. Joliot-Curie 14, 50-383 Wroclaw, Poland*
[b)] *Laboratoire de Chimie Théorique (UMR CNRS 7616), UPMC, 4 place Jussieu, 75252 Paris Cedex 05, France*
[c)] *Laboratory of Physical Chemistry, A.I. Virtasen Aukio 1, P.O. Box 55, University of Helsinki, 00014 Helsinki, Finland*

Abstract. For the first time the nature of binding in HrgY compounds (Rg = Ar, Kr, Xe; Y = F, Cl) is elucidated on the ground of the topological analysis of the electron localisation function (ELF). The binding between rare gas (Rg) and halogen (Y) is classified as an unshared-electron interaction type due to the lack of a bonding attractor between the C(Rg) and C(Y) atomic cores. The partial charge transfer from rare gas to halogen ranges between 0.6 and 0.7 e, in agreement with values from Bader's Atoms-in-Molecules decomposition technique and Mulliken's Population Analysis. The contribution of the V(Rg) basin to the delocalisation of the V(Y) basin is by 10-20% larger than that from V(Y) to V(Rg), as revealed by the contribution to the population variance parameter, and this effect corresponds to the direction of charge transfer. A set of valence-bond structures is proposed. The largest contribution (55-75%) comes from the HRg^+Y^- ionic limit. The minimum of the electron localisation function between halogen and rare-gas atoms ranges between 0.25 and 0.35.

1. Introduction

The inertness of the rare gases (Rg) toward chemical bond formation has been attributed to the «stable octet» outer electron structure typical for the group 18 elements. In 1962, the first rare-gas compound $Xe^+[PtF_6]^-$ was observed by Bartlett [1]. Today, the chemistry of Xe comprises compounds with Xe-Y (Y = N, C, F, O, Cl, Br, I and S) bonds [2]. The chemistry of krypton is mainly confined to krypton difluoride (KrF_2) and its derivatives. No stable neutral ground state molecules are known for the lighter rare-gas atoms: helium, neon and argon.

J. Maruani et al. (eds.), New Trends in Quantum Systems in Chemistry and Physics, Volume 1, 259–279.

Recently, Pettersson et al. [3,4] reported and characterized by means of matrix-isolation infrared spectroscopy several novel rare-gas compounds including HXeCl, HXeBr, HXeI, HKrCl and XeH_2. These molecules have in common an Rg-H fragment, which is bound to an atom with a negative partial charge. Calculations at the MP2 level [2-4] suggested that these HRgY molecules in their neutral ground state are linear. The Xe-H distance vary between 1.66 and 1.86 Å, which is much less than deduced from the Xe-H pair potential van der Waals minimum (3.8 Å [5]). For Kr-H bonds the calculated bond distances are close to 1.45 Å [2]. According to the Mulliken Population Analysis the HRgY molecules are charge transfer species. Most of the positive partial charge is centered on the rare-gas atoms, while the other fragments possess the counterbalancing negative partial charges, the hydrogen in a lower extent than, for example, the halogen in the halogen containing HRgY species.

In this paper the topological analysis of the Electron Localisation Function is adopted to study on the nature of binding in HRgY compounds (Rg = Ar, Kr, Xe; Y = F, Cl). The mean electron populations, the atomic populations obtained from the Mulliken's population analysis and Bader's atoms-in-molecules decomposition technique are compared.

2. Computational details

Molecular properties of the HRgY molecules were studied in the framework of density functional theory using the gradient-corrected correlation functional by Lee, Yang and Parr [6], combined with the Becke3 exchange functional [7,8]. The GAUSSIAN 94 [9] package of computer codes was used for ab initio calculations. The B3LYP-calculated wave functions were used as an input for the topological analysis of the Electron Localization Function (ELF), as implemented in TopMoD [10]. Bader's Atoms-in-Molecules (AIMs) decomposition technique for the electron density was performed using the program implemented in GAUSSIAN 94.

The all-electron, split-valence 6-311++G(d,p) basis set was used for all calculations on HRgY (Rg = Ar, Kr; Y = F, Cl) species. To study the effect of the basis set enlargement on the ELF analysis, the larger basis sets 6-311++G (2d,2p) and 6-311++G (3df,3pd) including several polarization functions on each atom was tested. For the Xe-compounds the Stuttgart effective core potentials [11], denoted as SECP, was used. This basis set includes 8 valence electrons in the valence space of Xe. The standard Pople-type 6-311++G(d,p) basis set was used for lighter atoms in the calculations of the HXeY species. To test the pseudopotential effect additional calculations on HXeY were performed with all electron basis set of Huzinaga (Huz)[12]. The optimized structures of the HRgY species, at various levels of theory, are collected in Table 1.

TABLE 1. Ab-initio calculated properties of HRgY species at different computational levels

Molecule	Method	r(Rg-H) / Å	r(Rg-Y) / Å
HArF	UMP2/6-31G(d,p)	1.393	1.941
HArCl	UMP2/6-31G(d,p)	1.394	2.501
HKrF	UMP2/4333/433/4 (Kr) /6-31G(d,p) (F) /6-311G(d,p) (H)	1.526	2.038
HKrCl	UMP2/4333/433/4 (Kr) /533/5111 (Cl) /6-311G(d,p) (H)	1.534	2.687
HXeF	UMP2 /43333/4333/43 (Xe) /6-31G(d,p) (Cl) / 6-311G(d,p) (H)	1.669	2.139
HXeCl	UMP2 /43333/4333/43 (Xe) /533/511 (Cl) / 6-311G(d,p) (H)	1.674	2.852

3. Topological analysis of the Electron Localisation Function (ELF)

Current analysis of the bonding in molecules and crystals are performed by projection techniques on atomic basis functions (Mulliken, Löwdin, Mayer, Natural Population Analysis). This yields scattered results for the same system, which depend upon the basis set quality [13]. An alternative approach is provided by the topological interpretation of the gradient field of a local, well defined function such as the theory of Atoms in Molecules [14] by Bader, which is based on the analysis of the electron density. The topological analysis of the Becke and Edgecombe [15] electron localisation function has been recently proposed. This takes advantage of the ELF as a measure of the Pauli repulsion as shown recently by Silvi and Savin [16]. The ELF function is defined as:

$$\eta(\mathbf{r}) = \frac{1}{1+\left[\dfrac{D_\sigma(\mathbf{r})}{D_\sigma^0(\mathbf{r})}\right]^2}$$

and for a single determinental wave function built from Hartree-Fock or Kohn-

Sham orbitals φ_i,

$$D_\sigma(\mathbf{r}) = \frac{1}{2}\sum_i \left|\nabla\varphi_i(\mathbf{r})\right|^2 - \frac{1}{8}\frac{\left|\nabla\rho(\mathbf{r})\right|^2}{\rho(\mathbf{r})}$$

and

$$D_\sigma^0(\mathbf{r}) = C_F\rho(\mathbf{r})^{5/3}.$$

$D_\sigma(\mathbf{r})$ has the physical meaning of the excess local kinetic energy density due to Pauli repulsion [17] and $D^0{}_\sigma(\mathbf{r})$ is the Thomas-Fermi kinetic energy density which can be regarded as a «renormalization» factor. The C_F is the Fermi constant with a value of 2.871 a. u.. The range of values of the ELF function is $0 \leq \eta \leq 1$. Where electrons are alone, or form pairs of antiparallel spins, the Pauli principle has little influence on their behavior and the excess local kinetic energy has a low value, whereas at the boundaries between such regions the probability of finding parallel spin electrons close together is rather high and the excess local kinetic energy has a large value.

The gradient field of ELF provides a partition of the molecular space into basins of attractors having a clear chemical meaning. A gradient dynamical system is a field of vectors $\mathbf{X} = \nabla\eta(\mathbf{r})$ defined on a manifold (M). By integrating over all vectors one can build trajectories which start at their α-limit and end at their ω-limit. The ω-limits are always sets of critical points at which $\nabla\eta(\mathbf{r}) = 0$. The α-limits are not always critical points (asymptotic behavior). Critical points that are only the ω-limit of trajectories are called attractors whereas the set of points defining the trajectories ending at a given attractor is the basin of the attractor. The boundary between two basins is called separatrix. In the case of the gradient field of the ELF function, the attractors may be single points, circles or spheres, according to the symmetry of the system, since ELF transforms as the totally symmetric representation of the system point group. Note that, actually, cases "almost" circle or "almost" sphere attractors occur in regions of space dominated by a local cylindrical or spherical electron-nucleus potential. There are two kinds of basins, at the one hand are the core basins which encompass nuclei (with Z>2) and on the other hand are valence basins, which form the outermost shell of electron density. These latter are characterized by their synaptic order which is the number of core basins with which the given valence basin share a common boundary. The nomenclature of the valence basins has been given in ref 18. It is important to note that this description of the chemical bond allows to adopt a point of view which is complementary of that of valence. In the standard valence picture a bond is considered as a link joining one atom to another. Here we have the number of cores and a given piece of glue (the valence basin) by means of which core basins are stuck on.

The classification of chemical bonds from the ELF analysis was based on the presence of point, ring and spherical attractors. From chemical point of view there are three types of attractors (i.e. local maxima of the electron density): core, bonding which is located between the core attractors and non-bonding. Following Bader, there are basically two kinds of bonding interaction: shared-electron interaction and closed-shell interaction. In the topological theory of the ELF gradient field the shared electron interaction is characterized by occurrence of a di or polysynaptic basin which is missing in the closed-shell/unshared interaction. Covalent, dative and metallic bonds are subclasses of the shared electron interaction whereas ionic, hydrogen, electrostatic and van der Waals belong to the other class. The case of hydrogen is special because it has no core attractor and the above classification criteria can not be applied. Thus, the topological analysis maps the basic chemical ideas onto a rigorous mathematical model and therefore removes the ambiguities and uncertainties in the standard definitions of bonds. Furthermore, this new topological analysis is in accord with the general ideas of chemical bonding.

The population of a basin is defined as the integral of the one-electron density over the basin:

$$\overline{N}_i = \int_{\Omega_i} \rho(\mathbf{r})d\mathbf{r}$$

The quantum mechanical uncertainity of the basin population (the standard deviation $\sigma(\overline{N}_i)$) can be calculated from the population variance ($\sigma^2(\overline{N}_i)$). The variance is expressed in terms of the diagonal elements of the first ($\rho(\mathbf{r})$) and second order ($\pi(\mathbf{r}_1,\mathbf{r}_2)$) density matrices as

$$\sigma^2(\overline{N};\Omega) = \sum_{j\neq i}\left(\overline{N}_i(\Omega_i).\overline{N}_j(\Omega_j) - \overline{N}_{ij}(\Omega_i,\Omega_j)\right)$$

where r_i denotes the space and spin coordinates of the electron labeled i.

As remarked by Noury et al. [19], the variance $\sigma^2(\overline{N}_i)$ can be written as the sum of contributions arising from the other basins:

$$\sigma^2(\overline{N};\Omega) = \sum_{j\neq i}\left(\overline{N}_i(\Omega_i).\overline{N}_j(\Omega_j) - \overline{N}_{ij}(\Omega_i,\Omega_j)\right)$$

in which $\overline{N}_{ij}$ is the actual number of pairs between the Ω_i and Ω_j basins:

$$\overline{N}_{ij} = \int_{\Omega_i}\int_{\Omega_j} \pi(\mathbf{r}_1,\mathbf{r}_2)d\mathbf{r}_1 d\mathbf{r}_2 \,.$$

The variance of the basin population has been recognized long time ago by Bader and Stephens [20] as a measure of the delocalization within the framework of Daudel's loge theory [21]. Therefore, for the sake of usefulness, the relative fluctuations parameter [20] has been introduced:

$$\lambda = \frac{\sigma^2(\bar{N}_i)}{\bar{N}_i}.$$

4. Results and Discussion

Figure 1 presents the reduction of localisation domains of the HArF molecule. For small value of ELF one can observe a large basin encompassing whole molecule, which is splitted at ELF of 0.17 into a single basin over halogen V(F) and second basin surrounding the ArH moiety. The second bifurcation occurs at ELF of 0.78 giving rise to the non-bonding attractor basin of argon V(Ar) and large domain over hydrogen V(Ar,H). Figure 1c illustrates the topology of attractor basins localised in the HArF molecule and this picture corresponds to all HRgY systems investigated with the all-electron basis set. In the case of the pseudopotential approximation used a core attractor of a rare gas is missing. The schematic representation of attractors associated with localisation basins is shown in Figure 2.

There is one core attractor of halogen C(Y) for Y = F, Cl and one core attractor of rare-gas atom C(Rg) for Rg = Ar, Kr, Xe. When the effective core potential approximation is used the core attractor of Xe is missing. The valence electrons are reflected by monosynaptic attractors V(Rg) and V(Y) corresponding to the non-bonding electron density. Due to the axial symmetry of all molecules - the point group is $C_{\infty v}$ - the V(Rg) and V(Y) attractors are of a circular shape. In the case of hydrogen the topology of ELF yields one protonated, disynaptic attractor V(Rg, H), which is of a point type. The values of the basin population ($\bar{N}$), the relative fluctuation (λ) and the percentage contribution to the variance ($\sigma^2(\bar{N})$) of the various HRgY spe-cies are collected in Tables 2 to 4.

According to Silvi and Savin [16], the interaction between two atoms is of covalent, dative or metallic type if at least one bonding attractor is localised between two core attractors. The bonding attractor means the polysynaptic attractor, which basin has a common separatrix with two or more core basins. In the case of the HRgY molecules the topology of ELF yields only one such attractor localised between C(Rg) and C(Y). However, its basin possesses only one common separatrix with C(Rg), hence it can be interpreted as a monosynaptic basin of the non-bonding electron density of the rare gas. Therefore, this attractor is not a bonding one and the interaction between rare-gas and halogen is classified as unshared-electron interaction. Additionally, it is not due to pairing of electrons and is presumably of the electrostatic origin.

The nature of the interaction between rare-gas and halogen is also elucidated using Bader's Atoms-in-Molecules decomposition technique [14], which divides the continuos electron distribution into non-overlapping atomic basins. Table 5 contains the properties of the charge density at critical points (CP) localised for the Rg-H and X-Rg bonds. The binding between rare-gas and halogen atoms has properties typical of ionic or closed-shell interactions - low ρ and positive $\nabla^2\rho$, whereas the bonding between rare-gas and hydrogen atoms has features typical of a covalent interaction. Therefore, the interpretation of the nature of binding in the Rg-Y bonds, based on the topological analysis of ELF, is supported by AIMs method.

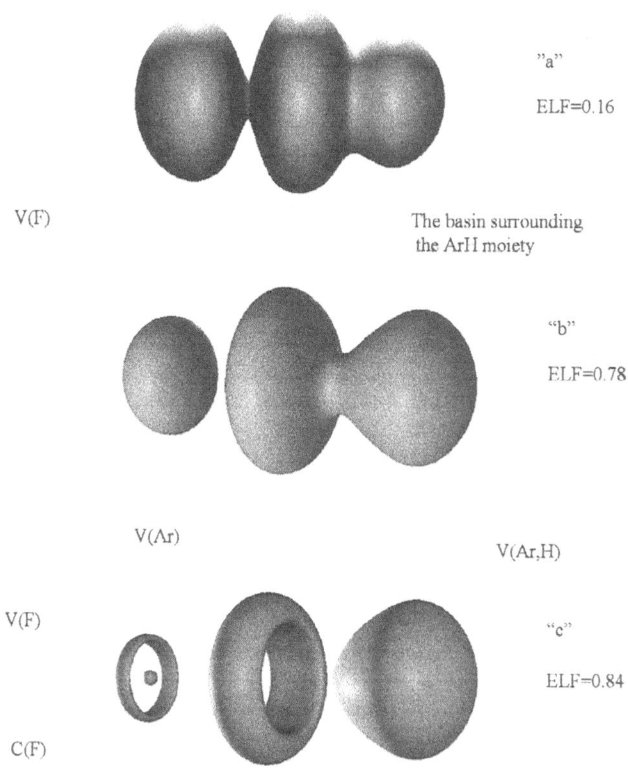

Figure 1. The reduction of localisation basins in HArF. At low values of ELF ("a") there are three localisation basins: two core (not visible in the picture) and one valence encompassing the whole molecule. The bifurcation at ELF=0.17 ("b") splits the common valence basin into the fluorine atomic one and the basin encapsulating the ArH moiety. A further bifurcation occurs at ELF=0.79 ("c"), giving rise to non-bonding attractor basins for fluorine V(F) and argon V(Ar) as well as a large basin over hydrogen V(Ar,H).

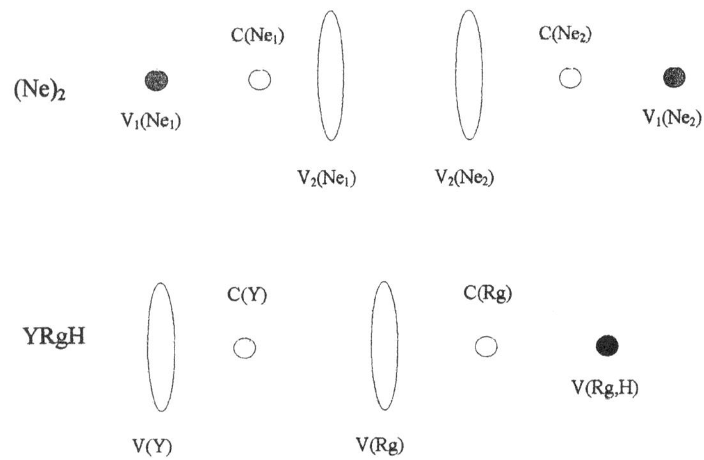

Figure 2. The comparison of attractors localised in $(Ne)_2$ and HRgY (Rg = Ar, Kr, Xe; Y = F, Cl) systems according to the topological analysis of the Electron Localisation Function. Note the different locations of the circular attractors of the non-bonding electron density of halogen V(Y) in HRgY and of neon $V_2(Ne_1)$ in $(Ne)_2$ in respect to the position of V(Rg) and $V_2(Ne_2)$. For the sake of simplicity cores of Ne as well as rare-gas and halogen atoms are placed at the same distance. The core, bonding and non-bonding point attractors are represented by white, black and grey dots, respectively, valence attractors by solid lines.

TABLE 2. The mean electron population (in e.) of the HRgY compounds obtained by the topological analysis of the ELF

Molecule	Basis set	C(Rg)	C(Y)	V(Rg)	V(Y)	V(Rg,H)
HArF	6-311++G(d,p)	10.07	2.16	6.61	7.48	1.69
HArCl	6-311++G(d,p)	10.06	10.07	6.73	7.51	1.61
HKrF	6-311++G(d,p)	27.81	2.14	6.64	7.51	1.89
	6-311++G(2d,2p)	27.76	2.20	6.97	7.46	1.55
	6-311++G(3df,3pd)	27.81	2.16	6.54	7.51	1.98
HKrCl	6-311++G(d,p)	27.76	10.08	7.09	7.51	1.49
	6-311++G(2d,2p)	27.76	10.08	7.06	7.54	1.50
	6-311++G(3df,3pd)	27.81	10.07	6.63	7.56	1.91
HXeF	Huz/6-311G(d,p)	45.81	2.18	6.84	7.48	1.65
	SECP/6-311G(d,p)	-	2.16	6.66	7.49	1.66
HXeCl	Huz/6-311G(d,p)	45.79	10.08	6.93	7.54	1.60
	SECP/6-311G(d,p)	-	10.07	6.77	7.54	1.59

TABLE 3. The relative fluctuation λ of electron density within basins localised in the HRgY compounds obtained by the topological analysis of the ELF

Molecule	Basis set	C(Rg)	C(Y)	V(Rg)	V(Y)	V(Rg,H)
HArF	6-311++G(d,p)	0.06	0.18	0.21	0.10	0.53
HArCl	6-311++G(d,p)	0.06	0.06	0.20	0.13	0.57
HKrF	6-311++G(d,p)	0.05	0.18	0.29	0.10	0.50
	6-311++G(2d,2p)	0.05	0.18	0.28	0.11	0.48
	6-311++G(3df,3pd)	0.05	0.18	0.30	0.10	0.49
HKrCl	6-311++G(d,p)	0.05	0.06	0.27	0.13	0.51
	6-311++G(2d,2p)	0.05	0.06	0.27	0.12	0.51
	6-311++G(3df,3pd)	0.05	0.06	0.28	0.12	0.52
HXeF	Huz/6-311G(d,p)	0.03	0.18	0.31	0.11	0.45
	SECP/6-311G(d,p)	-	0.17	0.17	0.10	0.40
HXeCl	Huz/6-311G(d,p)	0.03	0.06	0.30	0.13	0.49
	SECP/6-311G(d,p)	-	0.06	0.13	0.14	0.44

Figure 2 shows a comparison between the topology of attractors in the exemplary non-bonding system $(Ne)_2$ previously studied by Silvi and Savin [16] and the HRgY compounds. The neon atoms were placed at a distance of 1.0Å and calculations were carried out at the B3LYP / 6-311++G (2d,2p) computational level. One can notice that in $(Ne)_2$ the monosynaptic attractors of Ne non-bonding electron densities are of point ($V_1(Ne_1)$, $V_1(Ne_2)$) and circular shapes ($V_2(Ne_1)$, $V_2(Ne_2)$) and the latter ones are localised between two C(Ne) attractors with a «face to face» orientation. Presumably an accumulation of the electron density (ρ) at very small distance results in a large spatial deformation of ρ, which is well reflected by circular attractors. Figure 3 presents the reduction of the localisation basins found in $(Ne)_2$. For ELF=0.5 (Figure 3b) there are two, well separated basins associated with the non-bonding electron density of the Ne atoms, which is maximally delocalised. At higher values of ELF both basins undergo bifurcations yielding two valence basins: $V_2(Ne)$ of circular attractors associated with «outer» electron density and $V_1(Ne)$ of the point attractors associated with the «inner» electron density. The mean electron populations computed for the $V_2(Ne_1)$ and $V_2(Ne_2)$ basins yield large values of 7.39e in comparison to rather small $\overline{N}$ found for a $V_1(Ne_1)$ or $V_1(Ne_2)$ of 0.29e. An inspection of the relative fluctuation parameter reveals that the electron density in the $V_1(Ne_1)$ and $V_1(Ne_2)$ basins is almost entirely delocalised, with λ equal to 0.89, while for the $V_2(Ne_1)$ and $V_2(Ne_2)$ basins a small delocalisation is predicted with a value for λ of 0.14.

TABLE 4. The contribution to the variance σ^2 obtained by the ELF. Percentage contributions lower than 10% are omitted

	Contribution analysis [%]
HArF	
C(F)	95% V(F)
C(Ar)	17% V(Ar,H), 81% V(Ar)
V(Ar)	19% V(F), 35% C(Ar), 45% V(Ar,H)
V(F)	15% V(Ar,H), 33% V(Ar), 47% C(F)
V(Ar,H)	12% C(Ar), 13% V(F), 71% V(Ar)
HArCl	
C(Cl)	97% V(Cl)
C(Ar)	16% V(Ar,H), 82% V(Ar)
V(Ar)	17% V(Cl), 37% C(Ar), 44% V(Ar,H)
V(Cl)	14% V(Ar,H), 24% V(Ar), 57% C(Cl)
V(Ar,H)	11% C(Ar), 16% V(Cl), 68% V(Ar)
HKrF	
C(F)	96% V(F)
C(Kr)	16% V(Kr,H), 82% V(Kr)
V(Kr)	15% V(F), 31% V(Kr,H), 54% C(Kr)
V(F)	13% V(Kr,H), 36% V(Kr), 47% C(F)
V(Kr,H)	11% V(F), 22% C(Kr), 64% V(Kr)
HKrCl	
C(Cl)	97% V(Cl)
C(Kr)	90% V(Kr)
V(Kr)	13% V(Cl), 27% V(Kr,H), 60% C(Kr)
V(Cl)	14% V(Kr,H), 26% V(Kr), 58% C(Cl)
V(Kr,H)	15% C(Kr), 18% V(Cl), 66% V(Kr)
HXeF	
C(F)	96% V(F)
C(Xe)	89% V(Xe)
V(Xe)	12% V(F), 23% V(Xe,H), 65% C(Xe)
V(F)	17% V(Xe,H), 32% V(Xe), 48% C(F)
V(Xe,H)	18% V(F), 18% C(Xe), 63% V(Xe)
HXeCl	
C(Cl)	97% V(Cl)
C(Xe)	10% V(Xe,H), 89% V(Xe)
V(Xe)	11% V(Cl), 23% V(Xe,H), 66% C(Xe)
V(Cl)	16% V(Xe,H), 24% V(Xe), 58% C(Cl)
V(Xe,H)	19% C(Xe), 19% V(Cl), 61% V(Xe)

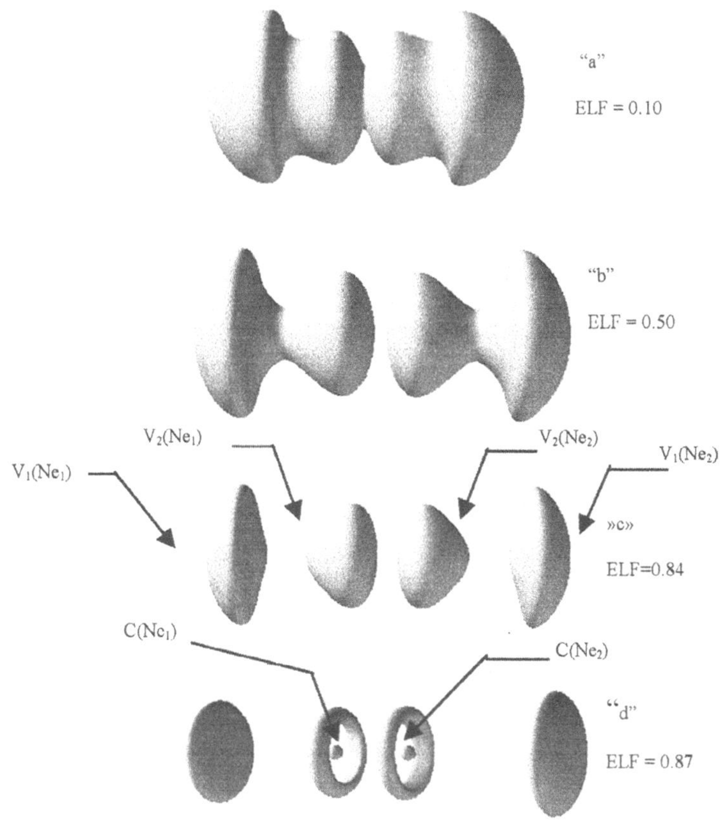

Figure 3. The reduction of localisation basins in the repulsive neon dimer, $(Ne)_2$. At low values of ELF ("a") there is one valence localisation basin surrounding the whole molecule and two core domains of neon - $C(Ne_1)$ and $C(Ne_2)$ (not visible on the picture). The bifurcation at ELF=0.31 splits the common valence basin into two well separated basins encapsulating Ne_1 and Ne_2 atoms ("b"). A further bifurcation occurs at ELF=0.60, giving rise to two valence non-bonding basins, $V_1(Ne)$ and $V_2(Ne)$, localised for each neon atom ("c"). The circular shape of the $V_2(Ne_1)$ and $V_2(Ne_2)$ basins is well represented in "d".

In the HRgY compounds the V(Y) attractor is separated from V(Rg) by presence of the core attractor C(Y), and the binding in the HRgY species differ essentially from the situation found in the (repulsive) neon dimer. The different position of V(Y) in respect to V(Rg) suggest that possible accumulations of the electron density, described by approximate (-Y+ -Rg +) H polarisation scheme, yields an additional stabilisation by dipole-dipole interaction.

TABLE 5. Selected bond critical properties [a] of molecules obtained by the topological analysis of the charge density (AIMs)

	ρ[e]	$\nabla^2\rho$	r_1[Å]	r_2[Å]
HArF - B3LYP/6-311++G(d,p)				
Ar - H	0.190	-0.348	1.898	0.713
F - Ar	0.098	+0.398	1.754	1.914
HArCl - B3LYP/6-311++G(d,p)				
Ar - H	0.183	-0.320	1.911	0.702
Cl - Ar	0.048	+0.136	2.428	2.299
HKrF - B3LYP/6-311++G(d,p)				
Kr - H	0.166	-0.251	1.976	0.888
F - Kr	0.092	+0.307	1.828	2.024
HKrF - B3LYP/6-311++G(2d,2p)				
Kr - H	0.164	-0.245	1.974	0.888
F - Kr	0.092	+0.312	1.832	2.019
HKrCl - B3LYP/6-311++G(d,p)				
Kr - H	0.159	-0.232	2.009	0.871
Cl - Kr	0.043	+0.083	2.564	2.513
HKrCl - B3LYP/6-311++G(2d,2p)				
Kr - H	0.158	-0.223	2.004	0.874
Cl - Kr	0.044	+ 0.086	2.582	2.495
HXeF - B3LYP/Huz/6-311G(d,p)				
Xe - H	0.133	-0.006	1.982	1.153
F - Xe	0.083	+0.280	1.883	2.160
HXeCl- B3LYP/Huz/6-311G(d,p)				
Xe - H	0.133	-0.062	2.022	1.121
Cl - Xe	0.035	+0.089	2.632	2.758

a) ρ is the value of the charge density at the critical point, $\nabla^2\rho$ is the second derivative of the charge density there, r_1 and r_2 are the approximate distances from the critical point to attractor A and attractor B in the bond A-B.

The tree-diagrams of $(Ne)_2$ and HArF presented in Fig. 4 illustrate the reduction of localisation basins caused by an increase of value of the electron localisation function. One can notice that values associated with bifurcations are different in both systems. Furthermore, basins of the valence electron density V(Ne) in the neon dimer are split into V_1(Ne) and V_2(Ne), in contrast to HArF, where only one basin associated with the non-bonding electron density of fluor V(F) is observed.

The populations of the C(F) and C(Ar) core basins with $\overline{N}$ of 2.16e and 10.07e, respectively, are slightly larger than the formal values predicted on the basis of the $2n^2$ formula, where n is principal quantum number. As it was shown previously by Noury et al. [19], it can be associated with frequent visits of valence electrons into a core region. For the HKrY species the mean electron population of C(Kr) is about 0.2e less than the formal value of 28e due to the [Kr] = $[Ar3d^{10}]4s^24p^6$ electron configuration. Presumably it may be explained by the non-negligible contribution of the 3d core electrons to the valence shell. The analysis of C(Rg) and C(Y) using the larger 6-311++G (2d,2p) and 6-311++G (3df,3pd) basis sets reveal only small (0.05e) differences in $\bar{N}$ compared to the value using the 6-311++G (d,p) basis set. In the case of HXeY molecules the population of the xenon core C(Xe) presents an effect similar to that observed for C(Kr). The mean electron population is about 0.2e less than the formal value of 46e expected on the basis of an [Xe] = $[Kr4d^{10}]5s^25p^6$ electron configuration. We may assume an explanation analogous to that proposed for the Kr core, due to a contribution of *4d* electrons to the valence shell.

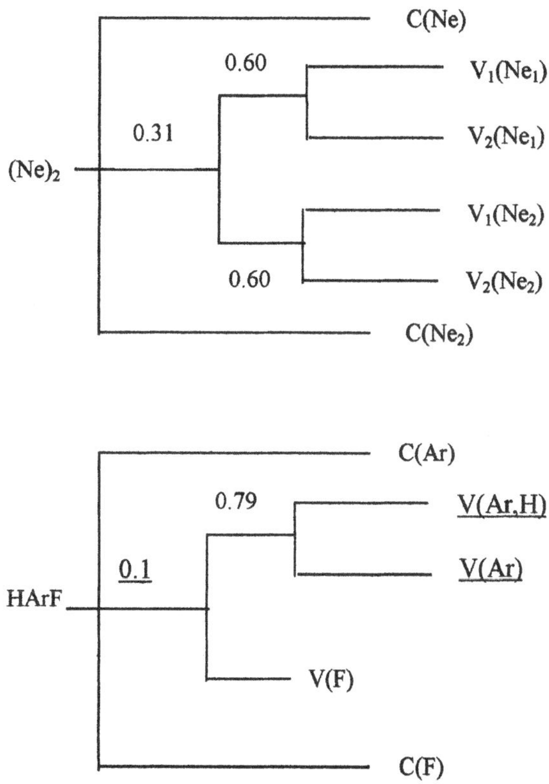

Figure 4. Comparison of tree-diagrams presenting a reduction of localisation basins in the $(Ne)_2$ and HArF systems. The values of the ELF correspond to catastrophes when a basin undergoes a bifurcation.

The hydrogen atom is described by a protonated, disynaptic attractor V(Rg,H) with the mean electron populations of basin equal to 1.69e and 1.61e for HArF and HArCl, respectively. The computations performed for HKrF anf HKrCl with the same basis set [6-311++G(d,p)] give values of 1.89e and 1.49e, respectively, which suggests that relatively large difference (about 0.4e) is associated with the exchange of F by Cl. Adding one set of polarization functions on HKrF induces lowering of $\bar{N}$ from 1.89e to 1.55e, whereas for HKrCl no changes are observed. Using the highly polarized 6-311++G(3df,3pd) basis set rises the values to 1.98 and 1.91e for HKrF and HKrCl, respectively. It is obvious that the population of the V(Rg,H) basin depends on the quality of an adopted basis set, and this effect should be carefully investigated in the future. Furthermore, it is possible that a large alteration of the V(Rg,H) basin set is related to the phenomenon of a contribution of the 3d electrons to the valence shell. For HXeF and HXeCl, the mean electron population of the V(Xe,H) basin (the ECP approximation), 1.66e and 1.59e respectively, does not differ essentially from the $\bar{N}$ values found for HArF and HArCl.

The mean electron population of V(Rg) basins in HArF and HArCl are 6.61e and 6.73e, respectively. For HKrF and HKrCl the calculations at the 6-311++G(d,p) basis set estimate values of 6.64e and 7.09e, and for HXeF and HXeCl values of 6.66e and 6.77e, respectively. Thus, when fluorine is replaced with Cl in the HRgY species the population of V(Rg) rises by slightly (0.1-0.4e) for all the chlorine-containing compounds. No big changes between various basis sets are seen, even though the 6-311++G(2d,2p) basis estimate slightly larger values than the two other basis sets. The mean electron population of the V(Rg) basin, which deviates between 6.6e and 7.1e depending on the considered system and the quality of the basis set, suggests that, respective to the isolated atom consisting of 8 valence electrons, the valence shell of the rare gas is essentially missing one electron.

Because the V(Rg,H) basins have an electron population less than 2.0e the missing electron density is transferred to regions of the non-bonding electron density of halogen V(Y). In fact, the computed $\bar{N}$ of V(Y) basins are between 7.46e and 7.56e. The compounds with Cl atoms show a slight increase of the mean electron population V(Y) compared to the fluorine compounds being 0.05e for HArF and 0.06e for HXeF and this effect can be explained by larger electron affinity of Cl than F.

The use of the pseudopotential approximation for xenon reveals nominal changes for C(Y), V(Y) and V(Rg,H) basins in comparison to calculations with all-electron basis set. However, the mean electron population of V(Rg) is diminished by 0.18e and 0.16e for HXeF and HXeCl, respectively. The relative fluctuations λ are presented in Table 3 for all studied HRgY molecules. According to the topological analysis, there are only minor fluctuations of the electron density within the core basins C(Ar), C(Kr) and C(Xe), ranging between 0.03e

(Xe) and 0.06e (Ar). Furthermore, the electron density of the core basins is mainly exchanged with the non-bonding electron density V(Rg), about 81-90%, and the V(Rg,H) basin, about 17-10% (Table 4). Comparable small λ values of 0.06 are obtained for C(Cl) but for C(F) values almost three times larger (0.18) are found. As was predicted for the C(Rg) basins the electron density is mainly exchanged with non-bonding electron densities V(Y). Comparison of λ computed for the monosynaptic, valence basins V(Rg) and V(Y) shows slightly larger values for rare gases than for halogens, and the relative fluctuations range between 0.13 and 0.30 for V(Rg) and 0.10 and 0.14 for V(Y). The largest relative fluctuations are observed for the V(Rg,H) basins, which range between 0.45 and 0.57. The pseudopotential effect in the HXeY cases reveals a λ of V(Rg,H) decreased by about 0.05.

The percentage contributions to the variance (σ^2) for the Ar, Kr and Xe-containing molecules are shown in Table 4. Results of the contribution analysis reveal some differences in delocalisation of the electron density for the HArY and HKrY compounds. The largest contribution to delocalisation of the V(Ar) basin (about 45%) comes from V(Ar,H). However, for the krypton-containing molecules the largest exchange of the V(Kr) electron density appears with the C(Kr) basin, being 54% in HKrF and 60% in HKrCl. The observed difference can be understood by a large contribution of the *3d* core electrons to the valence shell for the Kr-atom. Additionally, both basins of the non-bonding electron densities V(Rg) and V(Y) presents mutual delocalisation of the electron density, i.e. V(F) $\Leftrightarrow$ V(Ar), V(Cl) $\Leftrightarrow$V(Ar) in HArF and HArCl and V(F) $\Leftrightarrow$V(Kr), V(Cl) $\Leftrightarrow$V(Kr) in HKrF and HKrCl. The percentage contributions of the V(Kr) basin to the delocalisation of V(Cl) or V(F), and of V(Ar) either to V(Cl) or V(F) are about 20% and 10%, respectively, larger than observed in opposite way, i.e. the contribution of V(Y) to the delocalisation of V(Rg). The analyzes performed for HXeF and HXeCl support this picture, thus the V(Xe) and V(F) or V(Cl) basins mutually delocalise the electron density. The percentage contribution of V(Xe) to the delocalisation of V(F) is 20% larger than contribution of V(F) to V(Xe). Therefore, this effect can be associated with the observed result that the rare-gas atoms donate the electron density to the halogen atoms.

The large depopulation of the V(Rg) basins and the increased electron population of V(Y) suggest that the formation of the HRgY molecules is associated with a transfer of the electron density. The corresponding sums of mean electron populations for basins V(Rg)+V(Rg,H)+C(Rg) and V(Y)+C(Y) result in polarisations of the molecules: $[HAr]^{+0.6}F^{-0.6}$, $[HAr]^{+0.6}Cl^{-0.6}$, $[HKr]^{+0.7}F^{-0.7}$, $[HKr]^{+0.6}Cl^{-0.6}$, $[HXe]^{+0.7}F^{-0.7}$, $[HXe]^{+0.6}Cl^{-0.6}$ computed with the 6-311++G(d,p) basis sets (HArY and HKrY) and the ECP approximation (HXeY). The saturation of basis set to 6-311++G(2d,2p) or 6-311++G(3df,3pd) yields $[HKr]^{+0.7}F^{-0.7}$and $[HKr]^{+0.7}Cl^{-0.7}$. The magnitude of the partial charge transfer (Δq) is rather stable along the series of studies molecules and the difference between F and Cl isomers appears only for HKrY and HXeY compounds. The value of Δq increases by 0.1e in cases of HKrF and HXeF what can be explained by larger electronegativity of fluorine, which stronger polarises the valence shell of the rare-gas atom. On the contrary, the argon-containing species indicate the Ar-atom to be a «hard» rare-gas with a very small static polarisation and only nominal charge transfer is predicted by the ELF method. We must emphasize that generally the results achieved by the topological analysis of the ELF correspond to the values found on the basis of the Mulliken Population Analysis. For HArF and HArCl the electron density is transferred to the halogen and Δq is equal to 0.571e and 0.573e, respectively. For HKrF(Cl) the value of Δq rises to 0.652e (0.665e) and for HXeF(Cl) to 0.707e (0.704e). On this basis we may suggest that a relatively simple MPA analysis yields a reasonable description of charge transfer effects in the HRgY systems.

A satisfactory discussion of the charge transfer effects as investigated above on the basis of the topological analysis of ELF and the Mulliken Population Analysis may be complemented by the topological analysis of the electron density done in line of Bader [14]. In Table 6 there are collected the integrated atomic populations computed for all studied molecules. The Ar and Kr containing species possess the net positive charge on hydrogen being in range from +0.233e (HArCl) to +0.085e (HKrF) obtained at the B3LYP / 6-311++G (d,p) level. Interestingly, going to HXeF and HXeCl the hydrogen atom becomes negatively charged with net atomic charges equaled to -0.252e and -0.126e, respectively. The comparison of the atomic populations computed for the rare-gas and halogen (Y) atoms presents that in all molecules Rg is positively and Y negatively charged. Assuming the charge-transfer formula of the $[HRg]^{+\delta} Y^{-\delta}$ type the computations based on the integrated atomic populations [B3LYP / 6-311++G (d,p)] results in the respective polarisation schemes: $[HAr]^{+0.64}F^{-0.64}$, $[HAr]^{+0.56}Cl^{-0.56}$, $[HKr]^{+0.66}F^{-0.66}$, $[HKr]^{+0.57}Cl^{-0.57}$. For xenon containing systems the $[HXe]^{+0.71}F^{-0.71}$ and $[HXe]^{+0.63}Cl^{-0.63}$ formulas are predicted. The adoption of the 6-311++G (2d,2p) basis set for HKrF and HKrCl leads to a Δq of 0.672 and 0.589e, respectively. The achieved picture of charge-transfer effects is similar to that predicted on the basis of topological analysis of the ELF, which presented a transfer of about 0.6e-0.7e from the rare gas to the halogen.

As surveyed by Coulson [22] the nature of binding in xenon compounds having an even number of fluorine atoms (XeF_2, XeF_4, XeF_6, XeF_8) may be described by a valence-bond resonance model. For XeF_2 a resonance between the ionic limits $F^-(XeF)^+ \leftrightarrow (FXe)^+F^-$ yielded an approximate charge distribution of $F^{-0.5}$-Xe^{+1}-$F^{-0.5}$. Adopting this idea for the binding in HRgY compounds allows us to identify three valence-bond structures resonating together:

$$\text{(I) } H^+ \, Rg \, Y^- \quad \leftrightarrow \quad \text{(II) } H\text{-}Rg^+ \, Y^- \quad \leftrightarrow \quad \text{(III) } H\text{-}Rg^- \, Y^+$$

All mesomeric structures illustrate the possible localisation of the positive charge, i.e. on the naked proton in structure I and on the rare gas or halogen in structures II and III, respectively. A simple calculation using the data obtained from the topological analysis of ELF yields the approximate percentage weights for the three mesomeric structures of the various HRgY molecules:

HArF: I - 16%, II - 66%, III - 18%;
HArCl: I - 19%, II - 60%, III - 21%;
HKrF: I - 7%, II - 75%, III - 18%;
HKrCl: I - 25%, II - 56%, III - 19%;
HXeF: I - 17%, II - 59%, III - 25%;
HXeCl: I - 20%, II - 61%, III - 19%.

TABLE 6. Integrated atomic populations computed on the basis of the topological analysis of the charge density (AIMs)

Molecule	Basis set	H	Rg	Y
HArF	6-311++G(d,p)	0.781	17.581	9.638
HArCl	6-311++G(d,p)	0.768	17.677	17.556
HKrF	6-311++G(d,p)	0.915	35.421	9.664
	6-311++G(2d,2p)	0.918	35.410	9.672
HKrCl	6-311++G(d,p)	0.878	35.548	17.574
	6-311++G(2d,2p)	0.879	35.532	17.589
HXeF	Huz/6-311G(d,p)	1.252	53.037	9.711
HXeCl	Huz/6-311G(d,p)	1.126	53.245	17.629

It appears that the largest approximate weight is on the H-Rg^+F^- form, and the positive charge is mainly localised on the rare-gas atom. Furthermore, this implies that the H-Rg bond is mostly covalent. Structures I and III possess weights which range between 7% (HKrF) and 25% (HKrCl) for structure I and between 18% (HArF, HKrF) and 25% (HXeF) for structure III. Comparison with HarY shows that H-Rg^+F^- structures possess larger weights than H-Rg^+Cl^- species. As pointed out by Coulson [22], the F atom has the advantage over chlorine as a ligand. The electron affinities are nearly equal but F is smaller, so that the electrostatic energy for the creation of charges Rg^+Y^- will be considerably larger for larger halogens like chlorine. The Cl-containing HRgCl molecules have larger approximate weights in the ionic limit H^+ Rg Cl^-. In conclusion, the H-Rg bond is more ionic in the HRgCl compounds than in the HRgF isomers. This, however, results also in a less ionic Rg-Y bond for the larger halogen species compared to the fluorine ones.

From the topological analysis of ELF it is obvious that bonding between rare-gas and halogen atoms is of unshared-electron type. Furthermore, additional stability may be gained due to electrostatic effects and the predicted electron density transfer supports this. Comparison of the total electron density (ρ) and the ELF along the axis of HArF is presented in Figure 5. The shell structure of atomic cores is reflected by one maximum for F corresponding to K-shell and two maxima of K and L core shells in the case of Ar. The inspection of ELF between the V(F) and V(Ar) maxima reveals one minimum with the ELF value about 0.25. The largest values of 0.35 were found for HXeF and HXeCl.

5. Conclusions

The following conclusions can be drawn from the present study.

1. The topological analysis of the electron localisation function adopted for HRgY compounds (Rg = Ar, Kr, Xe; Y = F, Cl) reveals two core attractors C(Rg) and C(Y) associated with the electron density of the atomic cores, two monosynaptic, valence attractors V(Rg) and V(Y) of non-bonding electron density for Rg and Y, and one protonated disynaptic attractor of hydrogen V(Rg,H).

2. There are no bonding, disynaptic attractors between rare-gas C(Rg) and halogen C(Y) cores. Thus, the binding is classified as an unshared-electron interaction type, and is presumably of electrostatic origin.

3. Comparison between an exemplary repulsive system $(Ne)_2$ and the neutral HRgY compounds presents a different localisation of the $V_2(Ne_2)$ and V(Rg) attractors with respect to the position of the second attractor of the non-bonding electron density, i.e. $V_2(Ne_1)$ and V(Y).

4. The mean electron population of the C(Kr) basin is about 0.2e less than the formal value of 28e predicted on the basis of the $[Ar3d^{10}]4s^2 4p^6$ electron configuration. This is presumably caused by participation of underlying the *3d* core electrons to the valence shell. This effect is missing for Ar core basins, where the 3s and 3p orbitals form the valence space.

5. The mean electron populations of the V(Rg,H) basins are similar for HArY and HXeY molecules and vary between 1.60e and 1.70e. For the HKrY compounds, $\bar{N}$ depends on the adopted basis set and varies between about 1.5e and 2.0e, which may be associated with saturation of the valence space when multiple sets of polarisation functions are used and with the contributing *3d* core electrons to V(Kr).

6. There is partial charge transfer from the rare-gas to the halogen atom, about 0.6e for HArF, HArCl, HKrCl and HxeCl, and slightly larger (0.7e) for HKrF and HXeF. Generally the achieved values correspond to those computed by the Mulliken Population Analysis, and the rather simple MPA method gives reasonable estimates of charge-transfer effects in the HRgY systems.

7. The observed distribution of electron density from rare gas to halogen is also reflected by the contribution to the population variance parameter, which represents the mutual exchange of the electron density of the V(Y) and V(Rg) basins. The variance is about 20% larger from V(Rg) to V(Y) than from V(Y) to V(Rg).

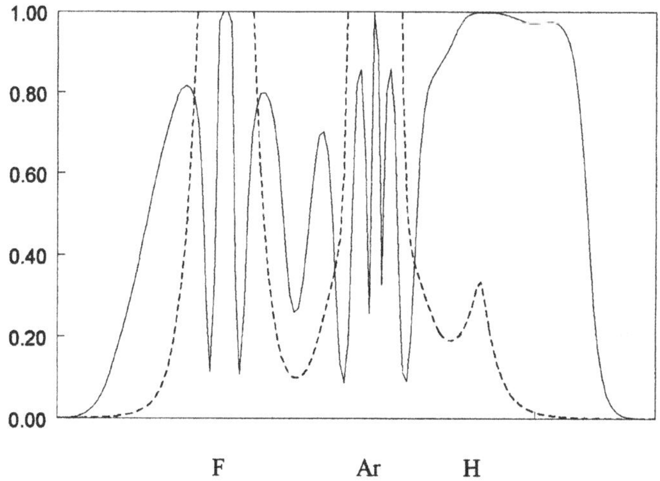

Figure 5. Comparison of the total electron density (dashed line) and electron localisation function ELF (solid line) along the axis of HarF, computed at the B3LYP / 6-311++G (d,p) level. The total electron density, for the sake of simplicity, was cut at the value of 1.0.

8. Among the proposed set of three resonating valence-bond structures, $H^{+}Rg\ F^{-}$, $H\text{-}Rg^{+}\ Y^{-}$ and $H\text{-}Rg^{-}\ Y^{+}$, the largest approximate weight is obtained for the $H\text{-}Rg^{+}\ Y^{-}$ ionic limit. Therefore, the positive charge is mainly localised on the rare-gas atom.

9. The analysis of the ELF along the molecular axis carried for HArF revealed a minimum localised between maxima corresponding to rare-gas (Rg) and halogen non-bonding electron densities with an ELF value of 0.25.

Acknowledgments

S.B. and Z.L. gratefully acknowledge the Wroclaw Supercomputer Center and Poznan Supercomputer Center for providing the computer time and the facilities where some of the calculations were performed. This work was partly supported by the Polish Committee for Scientific Research under grant nb 3.T09A.064.15. The CSC-Center for Scientific Computing (Espoo, Finland) is thanked for the computer mainframe time used during this study. J.L. acknowledges financial support from the Finnish Cultural Foundation and the Academy of Finland. Dr Alexis Markovits is warmly thanked for his assistance in putting the manuscript in final form.

References

1. Bartlett, N. (1962) Xenon hexafluoroplatinate (v) $Xe^{+}[PtF_6]^{-}$, *Proc.Chem.Soc.*, 218.
2. Pettersson, M., Lundell, J., and Räsänen, M. (1999) New rare-gas-containing neutral molecules, *Eur. J. Inorg. Chem.*, 729-737.
3. Pettersson, M., Lundell, J., and Räsänen, M. (1995) Neutral rare-gas-containing charge-transfer molecules in solid matrices. I. HXeCl, HXeBr, HXeI and HKrCl in Kr and Xe, *J.Chem.Phys.* **102**, 6423-6431.
4. Pettersson, M., Lundell, J., and Räsänen, M. (1995) Neutral rare-gas containing charge-transfer molecules in solid matrices. II. HXeH, HXeD and DXeD in Xe, *J. Chem. Phys.* **103**, 205-211.
5. Bickes Jr., R.W., Lantzsch, B., Toennies, J.P., and Walaschewski, K. (1973): Scattering experiments with fast hydrogen atoms: velocity dependence of the integral elastic cross section with rare gases in the energy range 0.001-1.00 eV, *Farad.Discuss.Chem. Soc.* **55**, 167-178.
6. Lee, C., Yang, W., and Parr, R.G. (1988) Development of the Colle-Salvetti correlation-energy formula into a functional of the electron density. *Phys.Rev.B 37* 785-789.
7. Becke, A.D. (1993) A new mixing of Hartree-Fock and local density-functional theories, *J.Chem.Phys.* **98**, 1372-1377.
8. Becke, A.D. (1993) Density-functional thermochemistry. III. The role of exact exchange, *J.Chem.Phys.* **98**, 5648-5662.
9. Frisch, M.J., Trucks, G.W., Schlegel, H.B., Gill, P.M.W., Johnson, B.G., Robb, M.A., Cheeseman, J.R., Keith, T.A., Petersson, G.A., Montgomery, J.A., Raghavachari, K., Al-Laham, M.A., Zakrzewski, V.G., Ortiz, J.V., Foresman, J.B., Cioslowski, J., Stefanov, B., Nanayakkara, A., Challacombe, M., Peng, C.Y., Ayala, P.Y., Chen, W., Wong, M.W., Andres, J.L., Replogle, E.S., Gomperts, R., Martin, R.L., Fox, D.J., Binkley, J.S., Defrees, D.J., Baker, J., Stewart, J.P., Head-Gordon, M., Gonzalez, C., and Pople, J.A.: GAUSSIAN 94, Gaussian, Inc., Pittsburgh, PA, 1995.
10. Noury, S., Krodikis, X., Fuster, F., and Silvi, B.: TopMoD package, 1997.
11. http://indy2.theochem.uni-stuttgart.de/pseudopotentiale/
12. Huzinaga, S.: *Gaussian Basis Sets for Molecular Calculations*, Elsevier, Amsterdam, 1984.
13. Bachrach, S.M. (1994) Population analysis and electron densities from Quantum Mechanics, in K.B. Lipkowski and D.B. Boyd (eds.) *Reviews in Computational Chemistry*; VCH, New York, pp. 171-227.
14. Bader, R.F.W. (1990) *Atoms in Molecules: a Quantum Theory*, Oxford University Press.
15. Becke, A.D., and Edgecombe, K.E. (1990) A simple measure of electron localization in atomic and molecular systems, *J.Chem.Phys.* **92**, 5397-5403.
16. Silvi, B., and Savin, A. (1994) Classification of chemical bonds based on topological analysis of electron localization function, *Nature* **371**, 683-686.
17. Savin, A., Jepsen, A.O., Flad, J., Andersen, O.K., Preuss, H., and von Schnering, H.G. (1992) Electron localization in the solid-state structures of the elements: the diamond structure, *Angew.Chem. Int.Ed.Engl.* **31**, 187-188.
18. Savin, A., Silvi, B., and Colonna, F. (1996) Topological analysis of the electron localization function applied to delocalized bonds, *Can.J.Chem.* **74**, 1088-1096.

19. Noury, S., Colonna, F., Savin, A., and Silvi, B. (1998) Analysis of delocalization in the topological theory of the chemical bond, *J.Mol.Struct.* **450**, 59-68.
20. Bader, R.F.W., and Stephens, M.E. (1974) Fluctuation and correlation of electrons in molecular systems, *Chem.Phys.Lett.* **26**, 445-449.
21. Daudel, R.: *Quantum Theory of The Chemical Bond*, Reidel, Dordrecht, 1974.
22. Coulson, J. (1964) The nature of bonding in xenon fluoride and related molecules, *J.Chem.Soc.*, 1442-1454.

SYMMETRY-SEPARATED ($\sigma+\pi$) *VS* BENT-BOND (Ω) MODELS OF FIRST-ROW TRANSITION-METAL METHYLENE CATIONS

FRANÇOIS OGLIARO[§], STEPHEN D. LOADES[‡], DAVID L. COOPER[*]
Department of Chemistry, University of Liverpool, P.O. Box 147, Liverpool L69 7ZD, United Kingdom

AND PETER B. KARADAKOV
Department of Chemistry, University of Surrey, Guildford, Surrey GU2 5XH, United Kingdom

Abstract. This is the first of two papers providing a modern valence bond explanation, within the framework of spin-coupled theory, of the general tendencies and characteristics of chemical bonding in the monocationic molecular systems $MCH_2,^+$ (M=Sc–Co). The present paper concentrates on a general study of two alternative representations of the metal-ligand double bond, namely the $\sigma+\pi$ and bent (Ω) bond descriptions. The close equivalence of these two models is established both within the valence and core parts of the wavefunction. Our results show that the spin degrees of freedom influence the preference for a bonding model to a much greater extent than do different choices for the core orbitals or variations in orbital flexibility. The presence of nonbonding electrons on the metal is found to reduce the size of the spin space for the Ω bond wavefunction relative to its $\sigma+\pi$ alternative; this favours energetically the $\sigma+\pi$ construction. In contrast, systems deprived of nonbonding electrons, such as $ScCH^+,_2$, $TiCH^{2+},_2$ and $VCH^{3+},_2$, prefer the bent bond model. Extension of the active space is found to advantage the $\sigma+\pi$ representation as a consequence of an escalation of the difference between the spin flexibilities of the two models, and this can lead to an inversion of the hierarchy of the two descriptions (as in the case of $ScCH_2,^+$). We find that modest variations of the molecular geometry do not modify the main conclusion of this survey: The classical $\sigma+\pi$ bond model offers the more appropriate description of the metal-methylene interaction. We discuss also the triplet character of the metal-carbon π bond, which is found to stem from the presence of unpaired nonbonding electrons on the metal and can be viewed as the result of a compromise between the preservation of some metal d–d exchange energy and the formation of a strong purely singlet-coupled covalent bond.

[§] *Present address: The Hebrew University of Jerusalem, Givat Ram Campus, 91904 Jerusalem, Israel*

[‡] *Present address: Uniqema, P.O. Box 2, 2800 AA Gouda, The Netherlands*

[*] Author to whom correspondence should be addressed

J. Maruani et al. (eds.), New Trends in Quantum Systems in Chemistry and Physics, Volume 1, 281–311.

1. Introduction

Over the past decade, the first-row transition metal methylene species have been the subject of an increasing number of experimental studies [1], mainly because they are considered to act as possible intermediates in many important catalytic reactions. The main aim of the experimental work has been to obtain reliable thermodynamic data for the metal-ligand bond strength, which is required in order to provide an accurate assessment of the energetics of elementary reaction steps involving bond-making and bond-breaking processes. In parallel, the moderate sizes of these molecules have made them an attractive target for the theoretical chemistry community [2–6]. Although the bonding mode in the first-row transition metal methylene molecules is relatively simple, it is known to be very difficult to treat at low levels of theory. As pointed out by Carter and Goddard [3], the Hartree-Fock (HF) method leads to very poor descriptions of these systems because of the weakness of the π interaction. Thus, MCH_2^+ systems define a challenging research area which is very suitable for testing the capabilities of various quantum chemical methods. There is little evidence about how valence bond (VB) based approaches would perform in this area - only very few VB theoretical studies on transition metal compounds have been published so far. Until now, our own studies have been concerned exclusively with relatively small hydride and/or hydrogen compounds [7–10]. A number of questions of chemical interest concerning the MCH_2^+ complexes remain to be answered. For instance, a complete rationalization and explanation of the observed bonding strength and dipole moment variation trends still does not exist. The consecutive filling of nonbonding metal orbitals within the series and the importance of the alternative description in terms of Lewis resonance structures have not been analyzed in sufficient detail. Our experience indicates that, in most cases, questions of this type can be answered in a straightforward manner by the results of *ab initio* spin-coupled (SC) calculations, which utilize a highly visual and yet sufficiently quantitative modern variant of VB theory. These questions will be addressed fully in the second part of this work [11], which focuses on the chemical characteristics of the MCH_2^+ (M=Sc–Co) complexes.

In this first part, we begin with a discussion of the chemical relevance of the two alternative bonding models capable of describing the metal-methylene interaction. Within a VB-style formalism, a double bond can be represented using either the classical separated bond scheme, or as a pair of bent bonds [12,13], as shown below for the case of a homonuclear double bond:

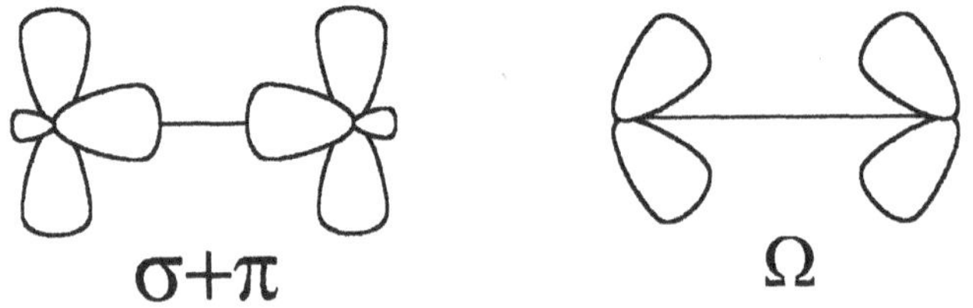

According to the variational principle, the better model is expected to be the one incorporating the more spin and/or orbital variational degrees of freedom. Preliminary SC calculations performed by Loades [10] suggest that, most probably, there is no single description suitable for all $MCH_2,^+$ (M=Sc–Co) complexes. Loades found that the Ω bond wavefunction of $ScCH_2,^+$ is lower in energy than its σ+π alternative although, on the contrary, $TiCH_2,^+$ opts clearly for the classical σ+π scheme [10]. Recently, we reinvestigated this problem using fully-variational SC calculations allowing core relaxation. The new results, which were partially presented elsewhere [13], are in very good agreement with Loades' conclusion that a universal lowest-energy bonding model does not exist. For inorganic complexes, these results also indicate that spin flexibility may be more important than orbital flexibility. Here we give a complete account of this work which has been extended to include the $MCH_2,^+$ (M=Sc–Co) series. The features of the two types of SC wavefunctions for multiple bonds are discussed in detail. Their core components are treated with special attention. In particular, fully-optimized sets of core orbitals are compared with cores taken from analogous '*N* in *N*' complete active space self-consistent field (CASSCF) calculations [14]. The importance of this comparison follows from the fact that for other systems [15], CASSCF doubly-occupied orbitals have been found to provide core descriptions for frozen-core SC calculations, which are almost as accurate as their fully-optimized counterparts. Next, we present a thorough explanation of the contraction of the spin space occurring in the bent bond solution. The increase of the degree of this contraction in parallel with the increase of the number of SC orbitals is discussed in detail, together with its influence on the energy difference (Δ_E) between the two bonding models. Additional insights into this issue are introduced by the parallel between the magnitude of the energy separation and the extent to which the wavefunction makes use of any available additional spin flexibility. In order to be able to make confident conclusions, we also carried out several calculations on species such as $TiCH^{2+},_2$, the hypothetical $VCH^{3+},_2$, and on excited states of $TiCH_2,^+$ and $VCH_2,^+$. For the same purpose, we then examined the influence of small distortions of the molecular geometry on Δ_E. In order to establish the relative merits of the two rival bonding models, the SC wavefunctions were compared to their more elaborate CASSCF counterparts. In our survey, the CASSCF wavefunctions provide the criteria for the extent to which the SC approach recovers nondynamical correlation energy, defined here as the energy separation between the CASSCF and HF energies.

Another point of interest is that the metal-ligand π bond of $ScCH_2,^+$ has been observed to have a considerable open-shell character. Using the modified coupled-pair functional (MCPF) formalism, It has been found [6] that the occupation numbers of two natural orbitals representing the π interaction are unexpectedly similar, and close to unity. This suggests that the π bond may contain a certain amount of triplet character. We address this question and provide a simple SC explanation for the origin of this triplet character and the reasons for its increase on passing from $ScCH_2,^+$ to $MnCH_2,^+$.

2. Spin-coupled model

This work is based on the SC approximation to the solution of the nonrelativistic Schrödinger equation [16]. The SC (or *full*-GVB [17]) wavefunction incorporates a significant amount of nondynamical correlation energy and has the distinct advantage of employing a single-configuration ansatz that is easy to handle and to interpret. It incorporates just one product of singly-occupied nonorthogonal active orbitals ϕ_μ:

$$\Psi_{SC} = \mathcal{A}(\Psi^{core}\, \phi_1\phi_2 \ldots \phi_N\, \Theta^N_{,SM}) \tag{1}$$

The core space Ψ^{core} is usually expressed as the product of doubly-occupied orbitals (*i.e.* $\psi_1^2\psi_2^2 \ldots \psi_n^2$) and a perfect-pairing (PP) core spin function Θ^{core}

$$\Theta^{core} = \sqrt{½}(\alpha_1\beta_2 - \alpha_2\beta_1)\sqrt{½}(\alpha_3\beta_4 - \alpha_4\beta_3) \ldots \sqrt{½}(\alpha_{2n-1}\beta_{2n} - \alpha_{2n}\beta_{2n-1}) \tag{2}$$

The spin function for the N active electrons $\Theta^N_{,SM}$ is expressed as a linear combination of simultaneous $\hat{S}^2$ and $\hat{S}_z$ spin eigenfunctions $\Theta^N_{k,SM}$ with quantum numbers S and M,

$$\Theta^N_{SM} = \sum_{k=1}^{f_S^N} C_{Sk}\Theta^N_{SM;k} \tag{3}$$

where the summation over the index k runs over all linearly independent spin eigenfunctions of this type. The expression for the spin space dimension $f^N_{,S}$, which is well-known from the literature [18], depends only on N and S. The construction of complete sets of spin eigenfunctions (spin bases) has received much attention, since it can be done in a number of different ways highlighting different features of the way in which the individual electron spins are coupled together to achieve the overall value of S. The more common construction methods are those due to Kotani [18], Rumer [19] and Serber [20]. The last one of these has often been able to provide the most detailed information about the symmetry properties of the total wavefunction.

In the Kotani basis, starting with the spin function for the first electron, the spin eigenfunctions are built up by successive addition, one by one, of the spins of the remaining electrons. The overall spin values at each stage of the construction process provide a convenient way of labelling the individual spin terms. For example the sequence (½)0(½) for a singlet system of 4 electrons (S=0) indicates that the spins of the first two electrons are singlet coupled and addition of the third electron results in a 3-electron doublet. The required final spin is achieved with the fourth electron and it is not necessary to indicate its value, as it is the same for all 4-electron singlet spin eigenfunctions. The Serber basis spin basis is built up in a similar fashion, but this time the main construction elements are provided by all two-electron singlet and triplet spin eigenfunctions. The first electron pair is used as starting point, and the remaining pairs are added one by one (plus an eventual final unpaired electron, if N is odd). In order to define a Serber spin eigenfunction, it is necessary to indicate the spin of each pair and the overall spin at each stage of the construction process. For example, the sequence (01)1(½) for a five-electron doublet shows that the first and the second pairs, a singlet and a triplet, respectively, are coupled to an overall triplet, while addition of the fifth electron leads to the required total spin of ½. The Rumer spin basis is generated by choosing $f^N_{,S}$ linearly independent spin eigenfunctions in which N–2S electrons form singlet pairs, and the remaining 2S electrons are assigned spin α. A convenient notation for a spin function of this type is provided by the list of its singlet pairs. For example for a five-electron doublet system the label (1–4,2–3) corresponds to a product of two two-electron spin singlet pairs and an α spin function for the last electron, 5. When the spin bases are ordered in the 'standard' way [21], the first spin function in the Rumer spin basis coincides with the last spin functions in the Kotani and Serber bases. This spin function, common to these three bases, is of the PP type.

The antisymmetrized product of the spatial part of the SC wavefunction with a particular spin eigenfunction $\Theta^N_{k,SM}$ defines a VB-type structure. It is possible to evaluate the contribution of each VB structure to the total SC wavefunction directly. However, in most cases it is sufficient to analyze instead the composition of the active-space spin function (Eq. (3)). The occupation number W_k of each spin function $\Theta^N_{k,SM}$ within $\Theta^N_{,SM}$ can be evaluated using one of several different schemes [22–25]. One of them, due to Chirgwin and Coulson [22], defines these occupation numbers as

$$W_k = C_{Sk} \sum_{l=1}^{f_S^N} C_{Sl} \langle \Theta_k | \Theta_l \rangle \tag{4}$$

In the Kotani and Serber bases, the spin eigenfunctions are orthogonal by construction, and their weights within $\Theta^N_{,SM}$ are given simply by the squares of the corresponding spin-coupling coefficients C_{Sk} (*i.e.* $W_k = C^2_{,Sk}$).

An alternative way of interpreting the optimal spin-coupling pattern is to examine the values of $\langle \hat{s}_i \cdot \hat{s}_j \rangle$ calculated over the active-space spin function [26]. These spin correlation matrix elements depend on the coupling of the electron spins associated with orbitals ϕ_i and ϕ_j: the limiting cases are $-\frac{3}{4}$ and $\frac{1}{4}$ for pure singlet and pure triplet couplings, respectively, whereas zero values indicate completely uncoupled spins. Values of $\langle (\hat{s}_i + \hat{s}_j)^2 \rangle$ provide the same type of information, but this time the special values are 2 for a triplet coupled pair, $^3/_2$ for two uncoupled electrons and zero for a singlet coupled pair [27]. The two spin-correlation scales are linked by the simple relation:

$$\langle (\hat{s}_i + \hat{s}_j)^2 \rangle = {}^3/_2 + 2\,\langle \hat{s}_i \cdot \hat{s}_j \rangle \tag{5}$$

which can be generalized to groups of electrons. In the case of two groups G_A and G_B of n_A and n_B electrons, respectively, we obtain the expression

$$\left\langle \left(\sum_{i \in G_A \cup G_B}^{n_A+n_B} \hat{s}_i^2 \right) \right\rangle = \sum_{i \in G_A}^{n_A} \sum_{j \in G_A}^{n_A} \langle \hat{s}_i \cdot \hat{s}_j \rangle + \sum_{i \in G_B}^{n_B} \sum_{j \in G_B}^{n_B} \langle \hat{s}_i \cdot \hat{s}_j \rangle + 2 \sum_{i \in G_A}^{n_A} \sum_{j \in G_B}^{n_B} \langle \hat{s}_i \cdot \hat{s}_j \rangle \tag{6}$$

in which the last term on the right-hand side accounts for the coupling between the two groups. When divided by the number of electron pairs, this term gives the average coupling between two electrons belonging to G_A and to G_B, respectively, $\langle \hat{s}(G_A) \cdot \hat{s}(G_B) \rangle$.

Typically, the active SC orbitals $\{\phi_\mu\}$ and the core orbitals $\{\psi_\nu\}$ are expressed as linear combinations of atom-centred functions $\{\chi_i\}$ as

$$\phi_\mu = \sum_{i=1}^{m} c_{\mu i} \chi_i \tag{7a}$$

and

$$\psi_\nu = \sum_{i=1}^{m} c_{\nu i} \chi_i \tag{7b}$$

A fully variational *ab initio* SC calculation would involve simultaneous optimization of the three sets of coefficients $\{C_{sk}\}$, $\{c_{\mu i}\}$, $\{c_{\nu i}\}$ which appear in the Eqs. (3) and (7). In contrast to classical VB theory, the SC formalism imposes no preconceptions on the shapes of the orbitals and on the degree of their localization. Thus, SC orbitals represent a unique outcome of the use of the variational principle to optimize the SC wavefunction of Eq. (1).

The fact that, in most cases, the optimized SC orbitals turn out to be well-localized about different atomic centres makes the analysis of spatial symmetry of the SC wavefunction different from that for molecular orbital wavefunctions built from delocalized orbitals. If Ψ_{SC} is nondegenerate, the effect of a symmetry operation R on the SC wavefunction can be expressed as

$$R\,\Psi_{SC} = \chi_R\,\Psi_{SC} \tag{8}$$

where χ_R is the character of R in the irreducible representation characterizing the spatial symmetry of Ψ_{SC}. As a rule, the core space Ψ^{core} transforms according to the totally symmetric irreducible representation of the point group of the molecule, so that we do not need to consider it in the spatial symmetry analysis. Let us assume that if any of the SC orbitals belong to definite irreducible representations of the point group of the molecule, then all of these irreducible representations are one-dimensional. The the right-hand side of Eq. (**Erreur! Source du renvoi introuvable.**) can be expanded as

$$R\,\Psi_{SC} = R\,\mathcal{A}(\phi_1\phi_2\ldots\phi_N\,\Theta^N{}_{,SM}) = (-1)^{N_R}\,\mathcal{A}(P^R\,\phi_1\phi_2\ldots\phi_N\,\Theta^N{}_{,SM}) \tag{9}$$

where P^R denotes the permutation of (some of) the SC orbitals caused by R and N_R stands for the number of SC orbitals that change sign under R. If P_R is a permutation of the electron coordinates, defined analogously to P^R,

$$P_R\,(-1)^{N_R}\,\mathcal{A}(P^R\,\phi_1\phi_2\ldots\phi_N\,\Theta^N{}_{,SM}) = (-1)^{N_R}\,\mathcal{A}(\phi_1\phi_2\ldots\phi_N\,P_R\,\Theta^N{}_{,SM}) = \varepsilon_{P^R}\,\chi_R\,\Psi_{SC} \tag{10}$$

where ε_{P^R} denotes the parity of P^R. As a direct consequence of Eq. (**Erreur! Source du renvoi introuvable.**), the active space spin function $\Theta^N{}_{,SM}$ has to satisfy the symmetry requirement

$$\Theta^N{}_{,SM} = \varepsilon_{P^R}\,\chi_R\,(-1)^{N_R}\,P_R\,\Theta^N{}_{,SM} \tag{11}$$

which can effectively decrease the number of free variational spin-coupling coefficients C_{sk} [see Eq. (3)].

There are several possibilities for further refinement of the SC wavefunction. By analogy with the CASSCF formalism, a multiconfigurational SC wavefunction can be built by adding ionic configurations to the covalent structures. Such an MCSC formalism is capable to recover all nondynamical correlation energy included in the CASSCF wavefunction, provided all possible ionic structures are explicitly included (a small difference may arise in the case when the core and SC orbitals come from a preliminary single-configuration SC calculation and are not re-optimized). If the goal is to include further correlation effects, especially of the dynamical type, it is also possible to augment the SC wavefunction with excited structures [15,28–30]. For high accuracy, it is usual to follow a SC or MCSC optimization with nonorthogonal CI calculation (SCVB approach [31]), which can produce results comparable to those obtained with CASSCF-CI calculations and is particularly suitable for treating excited states [8*a*].

3. Details of the calculations

The ions $ScCH_2^+$ and $TiCH_2^+$ were found by Ricca and Bauschlicher to adopt C_s symmetry [4], but the C_{2v} forms were calculated to be less stable by only 2 kcal/mol at the B3LYP level of theory. This small difference allows us to assume, in order to facilitate comparisons between results, that all the species we have chosen to study have C_{2v} symmetry (according to the standard spectroscopic orientation, all atoms lie in the $\hat{\sigma}_{yz}$ plane and the metal-carbon bond is aligned with the $\vec{z}$ axis), the CH bond lengths are fixed at 2.078 bohr and the HCH bond angles at 125°. The metal-carbon bond distances we used (in bohr, Sc–C=3.729, Ti–C=3.646, V–C=3.502, Cr–C=3.407, Mn–C=3.471, Fe–C= 3.434, Co–C=3.386), were taken from C_{2v}-symmetry ground state geometry optimizations [5] performed in the MCPF formalism ($^1A_1$ for $ScCH_2^+$, $^2A_1$ for $TiCH_2^+$, $^3B_2$ for VCH_2^+, $^4B_1$ for $CrCH_2^+$, $^5B_1$ for $MnCH_2^+$, $^4B_1$ for $FeCH_2^+$, $^3A_2$ for $CoCH_2^+$). The true ground state of the vanadium species is most probably $^3B_1$, rather than $^3B_2$. However, the energy separation between the two states computed at the B3LYP level of theory is very small (~0.5 kcal/mol), and this inversion in the ordering originates from the zero-point correction [4]. Here we consider the $^3B_2$ state of VCH_2^+ in order to have a simple sequential filling of the nonbonding orbitals throughout the MCH_2^+ series [11]. The calculations on di- and tri-cation species $TiCH_2^{2+}$ and VCH_2^{3+} were performed at the equilibrium geometries of the corresponding monocations. The same holds for the various excited states treated in this survey.

For consistency, we employed a high quality Gaussian-type basis set very similar to the one used in Ref. [5] to optimize the molecular geometries. It includes polarization functions which are known to be required for the proper description of bent bond solutions: As a result of the fact that the Ω bond wavefunction incorporates a larger number of orbital degrees of freedom than its σ+π counterpart, use of a poor-quality basis set could constitute a bias in favour of the σ+π construction. For metal atoms we employed an [8s4p3d] contraction of the (14s9p5d) primitive set of Wachters [32], supplemented by the set of polarization functions (3f)/[2f] developed by Bauschlicher *et al.* [33], and leading to a final basis set of the form (14s9p5d3f)/[8s4p3d2f]. The classical correlation-consistent Dunning [34] valence triple-ζ (cc-pVTZ) basis set, which corresponds to (10s5p2d1f)/[4s3p2d1f] for C and (5s2p1d)/[3s2p1d] for H, was used for light atoms. Only the pure spherical harmonic components of the basis functions were utilized.

The *minimal* SC and CASSCF active spaces involve the σ and π metal-carbon electron pairs (*i.e.* two a_1 orbitals and two b_1 orbitals), augmented if necessary with orbitals holding nonbonding electrons on the metal atom. From $TiCH_2^+$ to $MnCH_2^+$, the additional electrons are placed in nonbonding a_1 (d_δ=$d_{x^2-y^2}$), b_2 (d_π=d_{yz}), a_2 (d_δ=d_{xy}), a_1 ($s+\lambda d_\sigma$ =$s+\lambda d_{2z^2-x^2-y^2}$) metal orbitals, respectively, designated by ℓ_1, ℓ_2, ℓ_3, ℓ_4. For $FeCH_2^+$ the ℓ_1 orbital becomes doubly occupied, while in $CoCH_2^+$ the extra electron is in the ℓ_2 orbital, which makes it necessary to add one a_1 and one b_2 orbital in order to perform an '*N* in *N*' CASSCF calculation. Additional calculations were carried out with the *standard* active space, which includes all six valence electrons of CH_2. In this case, the active space is extended with the two a_1 and two b_2 orbitals required to describe the two CH bonding pairs.

Two types of symmetry adaptation were used in the SC calculations. The σ/π separation was achieved simply by insisting that certain orbitals should be of pure σ (a_1) and pure π (b_1) symmetry, respectively. In the Ω bond wavefunctions, the SC orbitals were constrained to transform into one another under appropriate operations of the C_{2v} point group: there are only two symmetry-unique SC orbitals describing the metal-methylene bond, the other two orbitals represent symmetry partners of the first two with respect to the molecular plane $\hat{\sigma}_{yz}$. The reflection of a pair of SC orbitals describing one CH bond through the $\hat{\sigma}_{xz}$ plane bisecting the HCH angle produces the pair of orbitals associated with the second CH bond. The orbitals occupied by unpaired electrons are constrained to belong to certain C_{2v} irreducible representations (see above). Except where stated explicitly, the spin analysis is performed on SC wavefunctions in which the orbitals are ordered so as to maximize the PP contribution. For example, the orbital ordering within the wavefunctions for $MnCH_2^+$ exploiting the *standard* active space is assumed to be:

$$(\phi^1{}_{C}\phi^1{}_{H}\phi^2{}_{C}\phi^2{}_{H})(\sigma_C\sigma_M\pi_C\pi_M)(\ell_1\ell_2\ell_3\ell_4) \text{ for } \Psi(\sigma+\pi)$$
$$(\phi^1{}_{C}\phi^1{}_{H}\phi^2{}_{C}\phi^2{}_{H})(\Omega^1{}_{C}\Omega^1{}_{M}\Omega^2{}_{C}\Omega^2{}_{M})\,(\ell_1\ell_2\ell_3\ell_4) \text{ for } \Psi(\Omega)$$

If there are no nonbonding electrons in the complex, or their number is less than 4, the corresponding orbitals are omitted from the product. The same holds for the orbitals describing CH bonds which are not present in the wavefunctions used in *minimal* active space calculations.

All calculations were carried out with the MOLPRO package [35], which incorporates a very efficient modern VB module [9,15,30]. Except where otherwise stated, the SC wavefunctions were determined through full simultaneous optimization of the core and valence subspaces. The transformation of the spin-coupling weights (W_k) and coefficients (C_{sk}) between the Rumer, Serber and Kotani spin bases was performed using the SPINS program [36].

4. Results and Discussion

4.1. BENT BOND *VS* SEPARATED BOND MODELS

A. Common features of the two models. At the SC level of theory, the metal-methylene double bond is described by four orbitals which closely resemble deformed hybrid atomic orbitals and have no contributions from the metal 4p valence shell. Throughout the MCH_2^+ (M=Sc–Co) series, these orbitals remain very similar qualitatively, regardless of the type of metal atom. For example, the orbitals for $CrCH_2^+$ shown in Fig. 1 bear strong resemblance to those for $TiCH_2^+$ presented elsewhere [13] (see also Ref. [11]). This confirms an earlier observation on the transferability of the SC model of metal-ligand interactions, made for MH and MH^+ (M = Sc–Cr, Y–Mo) systems [7,10]. Within the separated $\sigma+\pi$ bond framework the bonding orbitals, each of which is localized about a unique atomic site, form two pairs that are clearly associated with two-centre σ and π bonds, respectively. The axial interaction is described by a $s+\lambda d_\sigma$-like hybrid orbital on the metal atom σ_M, deformed towards the methylene fragment, and a sp^x-like hybrid orbital σ_C, centred on the carbon atom. The π bond pair consists of an almost pure d_π orbital π_M and a π_C orbital which is deformed in the direction of the metal atom, but remains distinctly $C(2p_\pi)$-based. The equivalent bond description uses two symmetry-related pairs. The two predominantly C-based orbitals Ω_C resemble deformed well-localized sp^x-hybrids, but their M-based partners Ω_M extend considerably towards the carbon atom. In both models the electron spins of the orbitals within bonding pairs are predominantly singlet-coupled, which suggests markedly covalent interactions [11]. The values of the correlation matrix elements $\langle \hat{s}_i \cdot \hat{s}_j \rangle$ range from −0.732 to −0.442 (see Table 1) which indicates that the triplet character of the metal-carbon bond does not exceed 20%.

The unpaired electrons, if any, occupy the almost exclusively metal orbitals ℓ_i. There are no obvious differences between nonbonding orbitals associated with the σ+π and Ω bond wavefunctions. The same observation holds for the orbitals ϕ_C and ϕ_H which describe the carbon-hydrogen interactions. The shapes of these orbitals are strongly reminiscent of the typical spin-coupled CH bond descriptions in various organic and inorganic systems [13]. The obvious differences in the shapes and in the degrees of localization of the four orbitals engaged in the metal-carbon double bonds within the σ+π and Ω bond schemes do not necessarily imply that the corresponding SC wavefunctions should be entirely dissimilar. In fact, any observable property depends on the overall wavefunction only, which suggests that a meaningful comparison of the σ+π and Ω bond models should analyze not only the orbitals directly involved in multiple bonds, but also the many-electron wavefunctions which incorporate these orbitals.

As a first step in the comparison between the overall SC wavefunctions corresponding to the σ+π and Ω bond schemes, we examined the impact of the core treatment on the energy characteristics of the two bonding models. Our objectives were (*i*) to establish that the core orbitals computed within the σ+π and Ω bond frameworks are almost completely equivalent, and (*ii*) to demonstrate the very close similarity between these core orbitals and those taken from a matching CASSCF calculation. The results of the SC calculations employing different choices of core orbitals are shown in Table 1 and suggest the following conclusions [the notation used to distinguish between different SC wavefunctions can be explained on the example of (Ω)/(σ+π) which indicates that the valence part of the SC wavefunction involves Ω bond orbitals, while the core orbitals come from a fully-variational σ+π SC calculation which, in turn, is denoted as (σ+π) / σ+π)]:

i) The core orbitals taken from fully-variational equivalent and separated bond SC calculations are very similar. They can be used interchangeably in order to discuss the energy separation between the two bonding models. Indeed, the wavefunction hierarchies established by the energies obtained using the same set of SC core orbitals [*i.e.* either the (Ω)/(Ω) and (σ+π)/(Ω) energies, or the (Ω)/(σ+π) and (σ+π)/(σ+π) energies] are in harmony with the fully-variational results [*i.e.* the (Ω)/(Ω) and (σ+π) / (σ+π) energies]. Naturally, the use of SC core orbitals calculated within the σ+π framework favours the separated bond model while, on the contrary, the Ω bond constructions are put to some advantage by core orbitals taken from bent bond calculations. The largest energy variations upon change of the set of core orbitals are observed for $ScCH_2^+$: frozen core orbitals originating from Ω bond and σ+π bond calculations produce Δ_E values of 0.89 and 3.27 mH, respectively, while fully variational calculations including core relaxation yield an energy separation of 1.90 mH.

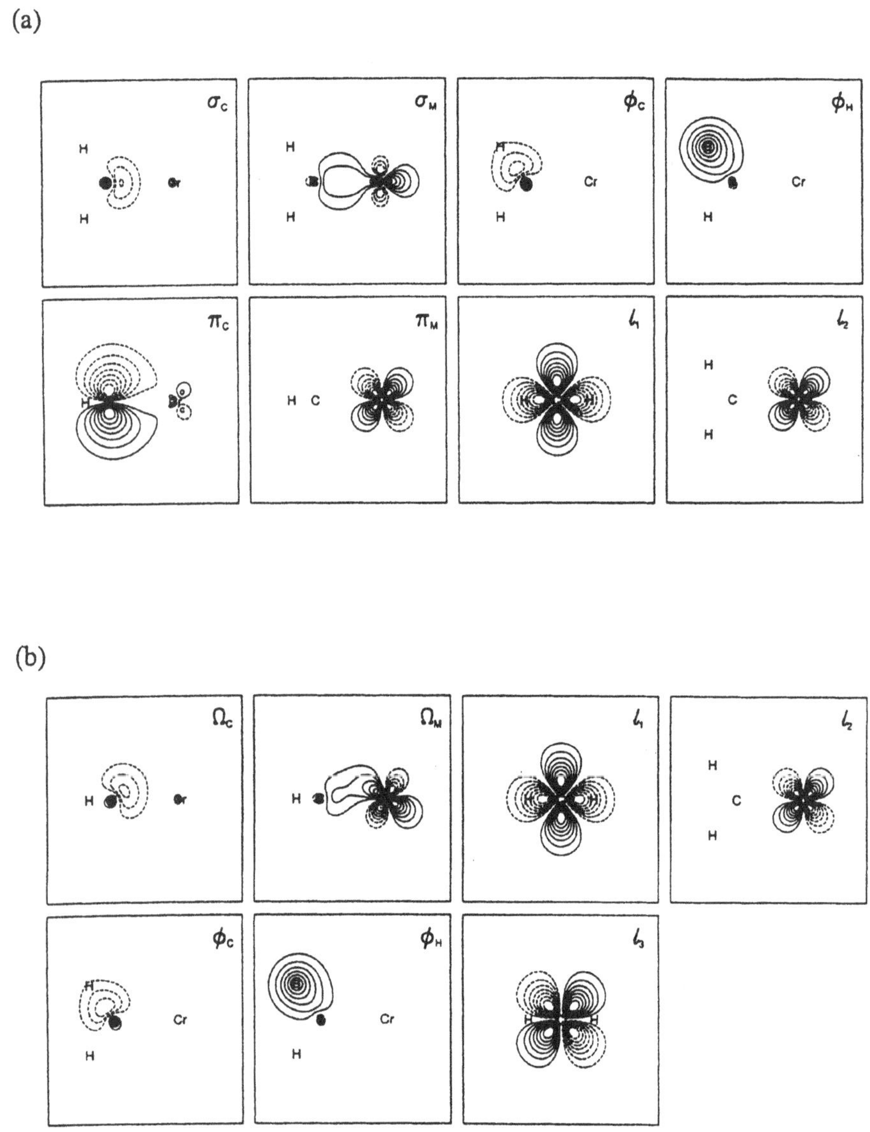

Figure 1. Fully-variational symmetry-unique SC orbitals for $CrCH_2^+$ calculated with a *standard* active space for (a) the σ+π model (b) the Ω model. Orbital ℓ_3 is shown only for the Ω model: it is practically indistinguishble between the two calculations.

ii) The use of identical frozen CASSCF core orbitals within the two SC wavefunctions as a replacement for the corresponding fully-variational SC core orbitals does not lead to significant changes in the energy separations between the two bonding models. It does not affect the ordering of the σ+π and Ω bond solutions in any of the MCH_2,$^+$ systems listed in Table 1 [compare the (σ+π)/(CAS) and (Ω)/(CAS) energies with their (Ω)/(Ω) and (σ+π)/(σ+π) counterparts]. Nor does the use of CASSCF core orbitals favour one of the two bonding models in a systematic fashion. For example, in the case of $CrCH_2$,$^+$ the CASSCF core creates a better environment for the bent bond solution (Δ_E decreases from 1.59 to 1.54 mH), whereas in the case of $MnCH_2$,$^+$ its use has an exactly opposite effect (Δ_E increases from 3.20 to 3.29 mH). The biggest change is observed for VCH_2,$^+$: the energy separation increases from 1.81 to 2.29 mH as a result of switching to CASSCF core orbitals.

The similarity between the three types of core orbitals (taken from fully-variational Ω bond and σ+π SC wavefunctions, and from a CASSCF calculation within an analogous active space) increases as we move from left to right in Table 1. This follows directly from the observation that the changes in the energy separation between the two multiple bond models resulting from different choices of the core orbitals become smaller in comparison to its magnitude. For example, in the case of $MnCH_2$,$^+$ the Δ_E variation caused by different core treatments is just 0.1 mH, which is negligible given the fact that Δ_E is close to 3.3 mH.

The second step in the comparison between the overall SC wavefunctions corresponding to the σ+π and Ω bond schemes focuses on their valence (or active) components. One standard way of examining these valence components as a whole, without going down to individual orbitals, is to perform Mulliken population analyses on the corresponding active spaces. The gross active-space populations of the basis functions on the metal atoms are given in Table 1. We observe a classical mixing of $3d^n4s^1$ and $3d^{n+1}$ asymptotes which will be discussed in detail in the forthcoming paper [11]. Clearly, in both SC solutions the metal atoms participate in the M–C double bonds with very much the same selections of basis functions: The largest differences between populations derived from σ+π and Ω bond SC wavefunctions do not exceed 0.03 electrons (see the s orbital populations of $ScCH_2$,$^+$ and $TiCH_2$,$^+$). The SC and CASSCF values are in very good agreement, as well. This shows that not only the core components, but also the valence parts of the two types of SC wavefunctions are remarkably similar. As a result, the σ+π and Ω bond constructions can be considered as two ways of representing essentially one and the same wavefunction. Moreover, both SC wavefunctions are found to resemble very closely their multiconfigurational CASSCF counterpart. In this way, the metal-methylene systems studied in this paper provide further examples of the close analogy between SC and CASSCF wavefunctions which represents a cornerstone of the CASVB formalism [9,37].

TABLE 1. Summary of main SC and CAS(SCF) results for the MCH_2^+ (M=Sc–Mn) series[a]

		$ScCH^+_{,2}$	$TiCH^+_{,2}$	$VCH^+_{,2}$	$CrCH^+_{,2}$	$MnCH^+_{,2}$
Relative energies, expressed in mH (and percentages)	(HF)/(HF)[b]	0(0)	0(0)	0(0)	0(0)	0(0)
	(CAS)/(CAS)	85.31 (100)	100.79(100)	113.74(100)	136.20(100)	147.33(100)
	$(\sigma+\pi)$/(CAS)	71.98 (84.4)	89.45 (88.7)	102.32(89.9)	124.20(91.2)	141.46(96.0)
	(Ω)/(CAS)	73.75 (86.4)	88.14 (87.5)	100.03(88.2)	122.66(90.0)	138.17(93.8)
	$(\sigma+\pi)/(\sigma+\pi)$	72.41 (84.9)	89.82 (89.1)	102.65(90.2)	124.50(91.4)	141.49(96.0)
	$(\Omega)/(\sigma+\pi)$	73.30 (85.9)	88.41 (87.7)	100.76(88.6)	122.91(90.2)	138.13(93.8)
	$(\sigma+\pi)/(\Omega)$	71.04 (83.3)	89.77 (89.1)	100.25(90.3)	124.44(91.3)	141.42(96.0)
	$(\Omega)/(\Omega)$	74.31 (87.1)	88.45 (87.7)	100.84(88.7)	122.91(90.4)	138.29(93.8)
	$(\sigma+\pi)/(\sigma+\pi)$ PP[c]	71.30 (83.6)	82.33 (77.9)	85.98 (75.6)	91.14 (66.7)	90.38 (61.3)
	$(\Omega)/(\Omega)$ PP[c]	74.22 (87.0)	87.37 (79.3)	87.91 (77.3)	93.70 (68.9)	99.72 (67.6)
Perfect-pairing weight[d]	$W^{PP}_{,\sigma+\pi}$	98.2	88.8	80.0	68.9	59.7
	$W^{PP}_{,\Omega}$	81.9	88.1	79.8	66.0	50.5
Mulliken population analysis	(CAS)/(CAS)	$s^{0.15}d^{1.38}$	$s^{0.15}d^{2.41}$	$s^{0.12}d^{3.45}$	$s^{0.10}d^{4.53}$	$s^{0.51}d^{5.10}$
	$(\sigma+\pi)/(\sigma+\pi)$	$s^{0.16}d^{1.34}$	$s^{0.18}d^{2.34}$	$s^{0.15}d^{3.39}$	$s^{0.12}d^{4.47}$	$s^{0.50}d^{5.08}$
	$(\Omega)/(\Omega)$	$s^{0.13}d^{1.34}$	$s^{0.15}d^{2.36}$	$s^{0.13}d^{3.41}$	$s^{0.10}d^{4.49}$	$s^{0.51}d^{5.10}$
Spin correlation elements $\langle \hat{s}_i \cdot \hat{s}_j \rangle$	σ_C and σ_M	–0.732	–0.722	–0.711	–0.669	–0.618
	π_C and π_M	–0.732	–0.644	–0.565	–0.477	–0.422
	Ω_C and Ω_M	–0.569	–0.672	–0.634	–0.550	–0.441

[a] *Minimal* active space calculations
[b] This notation indicates: (valence treatment)/(core treatment)
[c] Calculations carried out within the perfect-pairing restriction
[d] Within the Serber basis

B. Is there a preference for a particular bonding scheme? It was first suggested in Ref. [38], on the examples of SC wavefunctions for ethene and ethyne, that the preference for one or other of the bonding schemes comes as the result of a rather delicate balance between the variational freedom associated with the sets of spatial ($\{c_{\mu i}\}$, $\{c_{\nu i}\}$) and spin parameters (C_{Sk}) included in the SC wavefunction (see Eqs. (1), (3), and (6)). It can easily be demonstrated, using simple overlap considerations [13], that the Ω bond model always incorporates more orbital degrees of freedom than its σ+π counterpart. Assuming the predominant importance of orbital flexibility, the bent bond solution can be expected to be favoured energetically. Most of the earlier computational VB-style studies of multiple bonds have, indeed, reached the conclusion that the preferred multiple bond description is based on the bent bond model or, alternatively, that orbital degrees of freedom are more important than the degrees of freedom within the spin-coupling pattern. The results of our SC calculations on the MCH_2,$^+$ (M=Sc–Mn) series provide much evidence that contradicts this assumption, and seriously question the wide-spread belief [39–45] in the superiority of the bent bond model. The bent bond model provides a lower-energy wavefunction only in the case of $ScCH_2$,$^+$, whereas the σ+π construction proves to be energetically superior for all the remaining MCH_2,$^+$ systems (see Table 1). This gives us grounds to assume that the presence of nonbonding electrons favours the separated σ+π solution. Two further observations come to support this conjecture:

i) Although their monocationic predecessors opt for the σ+π bond model, di- and tricationic systems deprived of unpaired electrons, such as $TiCH^{2+}$,$_2$ and VCH^{3+},$_2$, are better described in the equivalent bond context: the corresponding Δ_E values are −11.0 and −13.8 mH, respectively.
ii) Within the PP approximation [which provides an efficient way of removing all coupling between the nonbonding electron(s) on the metal atom and the two bonding pairs], the equivalent bond solution is always lower in energy than its σ+π bond alternative (see Table 1). By comparing the percentages of correlation energy recovered, it can be seen that the PP approximation is a much more drastic restriction than is the use of fixed CASSCF instead of fully-variational SC core orbitals.

The first of these observations (*i*) shows in an unambiguous way that the preference for the separated bond model originates in these systems from the presence of nonbonding electrons on the metal atom. The second observation (*ii*) identifies the presence of non-PP spin-coupling patterns associated with the orbitals accommodating the nonbonding electron(s) and the orbitals involved in the metal-carbon bonds as the factor responsible for the lowering of the energy of the σ+π solution.

Within the MCH_2,$^+$ series, the number of spin degrees of freedom turns out to be more important than orbital flexibility. This can be shown by analyzing the relationship between the magnitude and the sign of Δ_E and the dimension of the spin space. For the 4-electron problem $ScCH^+$,$_2$ the size of the spin space remains the same in both wavefunctions and, due to the different orbital flexibilities, the energetic advantage is with the bent bond solution. The same argument applies also to the 4-electron systems $TiCH^{2+}$,$_2$ and VCH^{3+},$_2$, for which the computed Δ_E values are distinctly negative (see above). On the contrary, in the 5-electron treatment of $TiCH^+$,$_2$ ($f^5_{1/2}$=5), the numbers of independent spin parameters are different in the two bond models. This can be seen from Table 2, which reports the composition of the optimal active-space spin function in different spin bases. Not all of the five Serber or Rumer spin eigenfunctions contribute to the bent bond spin-coupling pattern, which can be explained using symmetry considerations. The five spin eigenfunctions making up the Rumer basis can be sketched as

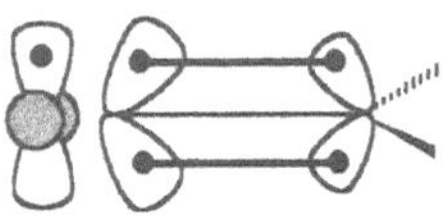

$^R\Theta^5_{1/2;1} = (1-2,3-4,5)$

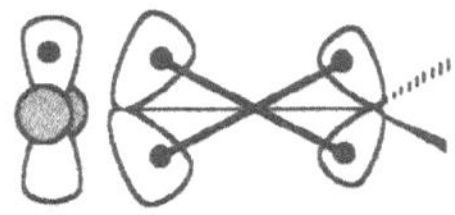

$^R\Theta^5_{1/2;2} = (1-4,2-3,5)$

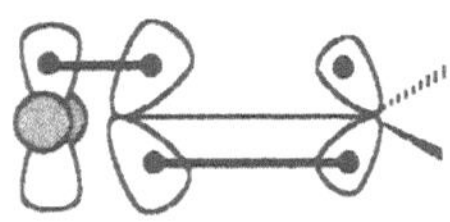

$^R\Theta^5_{1/2;3} = (1-2,4-5,3)$

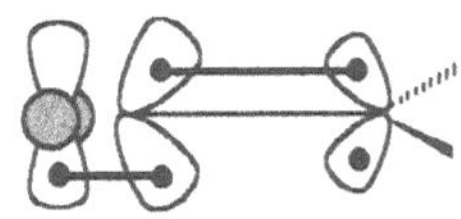

$^R\Theta^5_{1/2;5} = (2-5,3-4,1)$

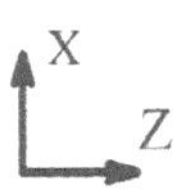

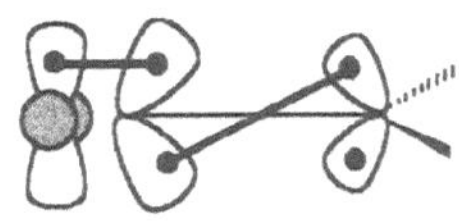

$^R\Theta^5_{1/2;4} = (2-3,4-5,1)$

TABLE 2. Chirgwin-Coulson weights of spin functions for the SC wavefunction of $TiCH_2^+$ ($f_{½,}^5=5$), expressed in the Kotani, Serber and Rumer spin bases[a,b]

k	Kotani $^K\Theta_{½,}^5{}_{;k}$	σ+π	Ω	Serber $^S\Theta_{½,}^5{}_{;k}$	σ+π	Ω	Rumer $^R\Theta_{½,}^5{}_{;k}$	σ+π	Ω
1	((½)1(3/2))	0.003	0.027	((10)1;(½))	0.006	0.041	(1–2,3–4,5)[c]	0.856	0.842
2	((½)1(½)1)	0.004	0.013	((11)1;(½))	0.001	0.000	(1–4,2–3,5)	0.036	0.049
3	((½)0(½)1)	0.084	0.041	((01)1;(½))	0.084	0.041	(1–2,4–5,3)	0.102	0.055
4	((½)1(½)0)	0.021	0.037	((11)0;(½))	0.021	0.037	(2–3,4–5,1)	0.001	0.000
5	((½)0(½)0)[c]	0.889	0.881	((00)0;(½))[c]	0.889	0.881	(2–5,3–4,1)	0.005	0.055

[a] *Minimal* active space calculations

[b] Orbitals are ordered as follows: ($\sigma_C\sigma_M\pi_C\pi_M\ell_1$) or ($\Omega^1{}_{,C}\Omega^1{}_{,M}\Omega^2{}_{,C}\Omega^2{}_{,M}\ell_1$)

[c] Perfect-pairing term

Spin functions $^R\Theta^5{}_{,½;1}$ and $^R\Theta^5{}_{,½;2}$ remain unchanged under the electron coordinate permutation $P_R = (1\ 3)(2\ 4)$ associated with the reflection $\sigma,\hat{}_{yz}$ and, as a consequence [see

Eq. (10) and the accompanying discussion], the corresponding spin-coupling coefficients represent independent variational parameters. Spin functions ${}^{R}\Theta^{5}_{,½;3}$ and ${}^{R}\Theta^{5}_{,½;5}$ are interchanged by the permutation (1 3)(2 4), and so the associated spin-coupling coefficients must be equal (see Table 2). It is straightforward to show that $(1\ 3)(2\ 4)^{R}\Theta^{5}_{,½;4} \equiv$ (4–1,2–5,3) may be expressed as the sum of all five spin terms (including ${}^{R}\Theta^{5}_{,½;4,}$); this can only be consistent with Eq. (10) if ${}^{R}\Theta^{5}_{,½;4}$ has no contribution to the active-space spin function. A similar contraction of the active spin space of the bent bond solution is observed within the Serber basis. The second spin function ${}^{S}\Theta^{5}_{,½;2}$ that corresponds to triplet couplings of the two bonding pairs of orbitals, which are then combined to yield an overall triplet (see Table 2), does not participate in the overall spin function for the Ω bond solution. Thus, the active-space spin function for the separated bond solution incorporates 4 independent variational parameters (*i.e.* five non-zero spin-coupling coefficients subjected to a normalization condition), instead of only two independent variational parameters (*i.e.* four non-zero spin-coupling coefficients, two of which must be equal) for the bent model. This leads to a difference $\Delta_s = -2$ between the numbers of spin degrees of freedom for the bent bond and σ+π models. In comparison with the scandium system, the preference for the bent bond representation observed in the titanium complex, despite its lower orbital flexibility, is a clear manifestation of the predominant significance of the spin degrees of variational freedom. The results obtained for the other systems provide further confirmation of this observation. Additionally, these results illustrate the evolution of the spin coupling pattern throughout the series.

TABLE 3. Spin-coupling coefficients (*C*) and Chirgwin-Coulson weights (*W*) of spin functions for the SC wavefunction of $VCH_2{}^{+}$ ($f_1{}^{6}$=9)[a,b,c]

	Function	$C(\sigma+\pi)$	$W(\sigma+\pi)$	$C(\Omega)$	$W(\Omega)$
${}^{S}\Theta_{1,;1}^{6}$	((11)2;1)	0.023	0.001	0.038	0.001
${}^{S}\Theta_{1,;2}^{6}$	((10)1;0)	−0.005	0.000	−0.003	0.000
${}^{S}\Theta_{1,;3}^{6}$	((11)1;0)	0.002	0.000	0.000	0.000
${}^{S}\Theta_{1,;4}^{6}$	((01)1;0)	−0.001	0.000	−0.003	0.000
${}^{S}\Theta_{1,;5}^{6}$	((10)1;1)	0.113	0.013	0.293	0.085
${}^{S}\Theta_{1,;6}^{6}$	((11)1;1)	−0.033	0.001	0.000	0.000
${}^{S}\Theta_{1,;7}^{6}$	((01)1;1)	0.400	0.160	0.293	0.085
${}^{S}\Theta_{1,;8}^{6}$	((11)0;1)	−0.158	0.025	−0.171	0.029
${}^{S}\Theta_{1,;9}^{6}$	((00)0;1)	0.894	0.800	0.893	0.797

[a] *Minimal* active space calculations
[b] Serber basis
[c] Orbitals are ordered as follows: $(\sigma_C\sigma_M\pi_C\pi_M\ell_1\ell_2)$ or $(\Omega^1{}_{,C}\Omega^1{}_{,M}\Omega^2{}_{,C}\Omega^2{}_{,M}\ell_1\ell_2)$

The composition of the active space spin-coupling patterns within the two SC wavefunctions for $VCH^{+}{}_{,2}$ ($f^{6}{}_{,1}$=9) in the Serber spin basis are given in Table 3. The spatial symmetry analyses of these spin-coupling patterns [see Eq. (10) and the accompanying discussion] indicate the presence of 8 independent spin variational parameters within the separated bond model, against just 4 within its bent bond rival. It is useful to mention that the two spin eigenfunctions which do not contribute to the active space spin-coupling pattern in the bent bond wavefunction, ${}^{S}\Theta^{6}{}_{,1;3} \equiv ((11)1;0)$ and ${}^{S}\Theta^{6}{}_{,1;6} \equiv ((11)1;1)$, are direct descendants of the spin eigenfunction which was found to be inactive in the bent bond wavefunction for $TiCH^{+}{}_{,2}$, ${}^{S}\Theta^{5}{}_{,½;2} \equiv ((11)1;(½))$. The contraction of the spin variational space that occurs when passing from the $\sigma+\pi$ to the Ω bond solution in $VCH^{+}{}_{,2}$, $\Delta_s = -4$, can be shown to arise from symmetry constraints analogous to those discussed in the case of $TiCH^{+}{}_{,2}$. Having this feature in mind, it is not surprising that the vanadium complex is found to opt for the $\sigma+\pi$ model (see Table 1). Similar contractions of the spin variational space are observed in the $CrCH^{+}{}_{,2}$ ($f^{7}{}_{,3/2}=14$) and $MnCH^{+}{}_{,2}$ ($f^{8}{}_{,2}=20$) complexes. The corresponding values of Δ_s, -6 and -8, respectively, suggest that these complexes should also prefer the separated bond description, which is confirmed by the computational results (see Table 1).

A careful examination of Tables 2 and 3 shows that several spin eigenfunctions, although allowed by symmetry, do not contribute much to the overall active space spin-coupling pattern. Examples are provided by ${}^{S}\Theta^{5}{}_{,½;2}$ in the $\sigma+\pi$ wavefunction for $TiCH^{+}{}_{,2}$ (see Table 2), and ${}^{S}\Theta^{6}{}_{,1;2}$ and ${}^{S}\Theta^{6}{}_{,1;4}$ in both wavefunctions for $VCH^{+}{}_{,2}$ (see Table 3). In the case of the $\sigma+\pi$ wavefunction for $TiCH^{+}{}_{,2}$, ${}^{S}\Theta^{5}{}_{,½;2} \equiv ((11)1;(½))$ provides unfavourable triplet couplings of the spins of the orbitals engaged in both the σ and π bonds. ${}^{S}\Theta^{6}{}_{,1;2}$ and ${}^{S}\Theta^{6}{}_{,1;4}$ for $VCH^{+}{}_{,2}$ are associated with a singlet coupling of the spins of the unpaired electrons in orbitals ℓ_1 and ℓ_2. The negligible weights of these spin eigenfunctions indicate that the metal centre prefers to keep these electrons triplet coupled so as to preserve a maximum of atomic d–d exchange energy. The existence of spin eigenfunctions with very small weights suggests that it would be fairer to compare the numbers of spin degrees of freedom in the $\sigma+\pi$ and bent bond models by taking into account only the spin eigenfunctions that have non-negligible contributions to the overall active space spin-coupling pattern. Thus, although for $TiCH^{+}{}_{,2}$ and $VCH^{+}{}_{,2}$ Δ_s takes values of -2 and -4, respectively, the $\sigma+\pi$ wavefunctions for these complexes have effectively just one and two, respectively, significant spin degrees of freedom more than their bent bond rivals. The corresponding 'effective' spin space contractions for $CrCH^{+}{}_{,2}$ and $MnCH^{+}{}_{,2}$ are -4 and -6, respectively. The spin eigenfunctions with negligible contributions to the active spin space in these complexes can be shown to be descendants of ${}^{S}\Theta^{6}{}_{,1;2}$ and ${}^{S}\Theta^{6}{}_{,1;4}$ for $VCH^{+}{}_{,2}$.

Our analysis of the active space spin functions in the $ScCH^{+}{}_{,2}$–$MnCH^{+}{}_{,2}$ series indicates a progressive increase of the difference between the numbers of symmetry-allowed spin degrees of freedom within the $\sigma+\pi$ and bent bond constructions, as well as of the difference between the numbers of spin eigenfunctions that have non-negligible contributions to the corresponding overall spin-coupling patterns. There are also clearly expressed inheritance patterns observed throughout the series of complexes which relate the structures of the symmetry-forbidden spin eigenfunctions, as well as of the spin eigenfunctions with very low weights.

The next step is to establish to what extent the difference between the spin degrees of freedom within the wavefunctions in the two bonding models can influence the energetic preference for one of them. In $ScCH^{+}{}_{,2}$, where the two bonding models have the same numbers of spin degrees of freedom ($\Delta_s = 0$), the preferred bond model, as we have seen already, is the bent bond one ($\Delta_E = -1.19$ mH). In $TiCH^+{,}_2$, the unpaired electron occupies the nonbonding orbital ℓ_1. This leads to a spin space contraction when passing from the $\sigma+\pi$ to the bent bond solution ($\Delta_s = -2$), which puts the former to an advantage ($\Delta_E = 1.37$ mH). In $VCH^+{,}_2$, the placement of the second unpaired electron in another nonbonding orbital, ℓ_2, leads to a decrease of Δ_s to -4. However, as we have shown above, this additional spin space contraction does not follow from the appearance of a new symmetry constraint, but rather from the propagation of a specific spin-coupling pattern that already exists in $TiCH^+{,}_2$. This suggests that the energy separation should not be expected to change too much in comparison to that found for the titanium species. Indeed, in $VCH^+{,}_2$ we observe only a very small increase of Δ_E to 1.81 mH. Similar arguments can be applied to $CrCH^+{,}_2$ ($\Delta_s = -6$), for which the energy separation ($\Delta_E = 1.61$ mH) is very close to those for $TiCH^+{,}_2$ and $VCH^+{,}_2$.

In $MnCH^+{,}_2$ ($\Delta_s = -6$) the fourth nonbonding electron occupies orbital ℓ_4 which is of a_1 symmetry. In a certain sense, this provides some extra orbital flexibility for describing the metal-methylene interaction: in $MnCH^+{,}_2$ the four bonding electrons have access to six active ($4a_1 + 2b_1$) orbitals instead of only five ($3a_1 + 2b_1$) or four ($2a_1 + 2b_1$) for the other complexes (see section 0). This should, in principle, lead to increased orbital flexibility and favour the bent bond model, since it is the representation that possesses the greater number of orbital degrees of freedom. However, on the contrary, we observe in $MnCH^+{,}_2$ a sharp increase of Δ_E to 3.21 mH, which obviously provides a strong piece of evidence for the predominance of spin variational freedom over orbital flexibility. In fact, extension of the active space does not necessarily have to favour the equivalent bond model in all situations. In some cases, the additional orbital flexibility may require additional spin degrees of freedom, so that the active space spin-coupling pattern could adjust itself better to the changes in the orbital shapes. One example in which an improvement in orbital flexibility favours the $\sigma+\pi$ solution is provided by the comparison between the SC descriptions of the ground state of $TiCH^+{,}_2$ ($^2A_1$) and the $^2A_2$ excited state obtained by placing the lone electron in the nonbonding orbital ℓ_2 (which is of a_2 symmetry). The dimension of the spin space is the same for these two states. In analogy with the ground state, the fourth Rumer spin eigenfunction $^R\Theta^5{}_{,\frac{1}{2};4}$ can also be shown to be symmetry-forbidden in the $^2A_2$ excited state. The difference between the numbers of independent spin variational parameters in the bent bond and $\sigma+\pi$ models for the $^2A_2$ excited state is the same as in the ground state, $\Delta_s = -2$, but the four bonding electrons have access to four active orbitals ($2a_1+2b_1$) in the $^2A_2$ state, against five ($3a_1+2b_1$) active orbitals in the ground state. However, the energy separation Δ_E between the two models for the $^2A_2$ excited state is just 0.64 mH, about two times smaller than that in the ground state ($\Delta_E = 1.37$ mH). Thus, the increased orbital flexibility in the ground state favours not the bent bond, but the $\sigma+\pi$ model in this case.

It is interesting to find out how the difference between the two bonding models would be influenced by the engagement of two nonbonding orbitals, which normally are unpaired, in a singlet nonbonding lone pair. For $VCH^+{,}_2$, it is possible to construct a $^1A_1$ excited state ($f^6{}_{,0}=5$) by placing the two unpaired electrons in two ℓ_1 δ-type orbitals. The difference between the numbers of independent spin variational parameters in the bent bond and $\sigma+\pi$ models for this excited state is $\Delta_s = -2$, which follows from an analysis similar to that for $VCH^+{,}_2$: the symmetry-forbidden terms in the bent bond solution,

${}^{S}\Theta^{6}{}_{,0;2} \equiv ((11)1;1)$ and ${}^{R}\Theta^{6}{}_{,0;4} \equiv (1\text{–}6,2\text{–}3,4\text{–}5)$ are direct descendants of ${}^{S}\Theta^{5}{}_{,½;2}$ and ${}^{R}\Theta^{5}{}_{,½;4}$ in Table 2. The energy separation between the two models increases from 1.81 mH in favour of the σ+π model in the ${}^{3}B_{2}$ ground state of $VCH^{+}{}_{,2}$ to 2.37 mH in the ${}^{1}A_{1}$ excited state. This increase is related to the fact that, in comparison to the ${}^{3}B_{2}$ ground state, the ${}^{1}A_{1}$ excited state makes one extra a_1 orbital accessible to the electrons involved in the metal-methylene bond.

Our results strongly suggest the existence of a direct relationship between the difference in the spin degrees of freedom incorporated in the two bonding models and the energetic preference for one of them. In most cases, just this difference, expressed by the parameter Δ_s, is sufficient to predict which bonding description will be favoured at the equilibrium geometry. In all examples studied in this paper, the decrease of the spin flexibility of the bent bond solution is due to the presence of nonbonding metal electrons which may be either singlet or triplet coupled. However, Δ_s on its own does not provide sufficient information for an estimate of the magnitude of the energy separation between the two models. Our calculations on excited states indicate that for an estimate of this type one should also take into account the actual use that the wavefunction makes of all available spin degrees of freedom. This observation is confirmed by the results presented in the following subsections.

C. Influence of extension of the active space. In our *standard* active space calculations, the four electrons involved in the two CH bonds which were previously placed in the core are now treated explicitly at the SC level of theory. Just as in the case of the *minimal* active space calculations, the PP approximation systematically favours the bent bond model over its σ+π bond alternative. The calculated values of Δ_E for the $MCH^{+}{}_{,2}$ (M=Sc–Cr) series are −2.69, −1.29, −1.78 and −2.39 mH, respectively. In fact, the opposite outcome would have been very surprising since, as we have mentioned earlier, when the non-PP spin-coupling patterns are neglected the bent bond solution immediately becomes the better model because of its greater orbital flexibility. According to the results of the full spin space calculations, the extension of the active space consistently benefits the σ+π bond model. In the case of $ScCH^{+}{}_{,2}$, the preference for the bent bond construction is reduced to only −0.30 mH, in comparison to −1.9 mH within the *minimal* active space. In systems, for which the *minimal* active space calculation indicates that the σ+π bond model is lower in energy, the *standard* active space calculation leads to a further increase of Δ_E. Within the *standard* active space calculations, the σ+π bond descriptions for the $MCH^{+}{}_{,2}$ (M=Ti–Cr) complexes are favoured by energy differences of 1.90, 2.21 and 1.96 mH, respectively, against just 1.37, 1.81 and 1.61 mH, respectively, within the *minimal* active space.

The explicit inclusion of the electrons from the CH bonds in the active space widens the gap between the spin flexibilities of the two bonding models by introducing more new symmetry constraints within the Ω bond wavefunction than within its σ+π counterpart. This is well-illustrated by the 8-electron ($f^{8}{}_{,0}$=14) active space for $ScCH^{+}{}_{,2}$. The individual contributions of the spin eigenfunctions making up the overall active space spin-coupling pattern are collected in Table 4. Three spin coupling modes (${}^{S}\Theta^{8}{}_{,0;3}$, ${}^{S}\Theta^{8}{}_{,0;6}$ and ${}^{S}\Theta^{8}{}_{,0;6}$) are symmetry-forbidden in the σ+π, as well as in the bent bond model. The reason for the appearance of these constraints is easier to comprehend when the CH bonding orbitals are taken in the order $\phi^{1}{}_{,C}\phi^{2}{}_{,C}\phi^{1}{}_{,H}\phi^{2}{}_{,H}$. Then all Serber spin eigenfunctions originating from the four-electron triplet patterns (10)1 and (01)1, irrespective of the bonding model and of the metal-carbon part of the wavefunction, are forbidden by symmetry because they change their signs when subjected to the electron coordinate permutation associated with the reflection in the $\sigma,\hat{}_{xz}$ plane [see Eq. (10) and the accompanying discussion].

Two additional spin eigenfunctions (${}^{S}\Theta^{8}_{,0;5}$ and ${}^{S}\Theta^{8}_{,0;7}$, see Table 4) are symmetry-forbidden only within the bent bond context. They both correspond to simultaneous triplet couplings of the two metal-carbon bonding pairs leading to an overall triplet. It is not difficult to establish that both ${}^{S}\Theta^{8}_{,0;5}$ and ${}^{S}\Theta^{8}_{,0;7}$ can be derived from the symmetry-forbidden spin eigenfunction ${}^{S}\Theta^{5}_{,½;2}$ present in the five-electron spin space for $TiCH^{+}_{,2}$ (see Table 2). Thus, for $ScCH^{+}_{,2}$ ($f^{8}_{,0}$=14) Δ_s decreases from 0 to −2 on passing from the *minimal* to the *standard* active space. Such additional symmetry constraints are also present in the *standard* active space wavefunctions for $TiCH^{+}_{,2}$ ($f^{9}_{,½}$=42), $VCH^{+}_{,2}$ ($f^{10}_{,1}$=90) and $CrCH^{+}_{,2}$ ($f^{11}_{,3/2}$=165), and can be rationalized using the inheritance patterns relating the structures of the symmetry-forbidden spin eigenfunctions discussed in the previous subsection. As a result, the Δ_s values for $TiCH^{+}_{,2}$, $VCH^{+}_{,2}$ and $CrCH^{+}_{,2}$ within the *standard* active space decrease to −7, −16 and −27, respectively, in comparison to −4, −6 and −8, respectively, within the *minimal* active space.

TABLE 4. Spin-coupling coefficients (C) and Chirgwin-Coulson weights (W) of spin functions for the SC wavefunction of $ScCH_{2},^{+}$ ($f_{0,}^{8}$=14)[a,b,c]

	Function	$C(\sigma+\pi)$	$W(\sigma+\pi)$	$C(\Omega)$	$W(\Omega)$
${}^{S}\Theta_{0,;1}^{8}$	((11)2;1)1;1)	0.042	0.002	0.042	0.002
${}^{S}\Theta_{0,;2}^{8}$	((10)1;0)1;1)	−0.107	0.012	−0.078	0.006
${}^{S}\Theta_{0,;3}^{8}$	((11)1;0)1;1)	0.000	0.000	0.000	0.000
${}^{S}\Theta_{0,;4}^{8}$	((01)1;0)1;1)	−0.107	0.012	−0.078	0.006
${}^{S}\Theta_{0,;5}^{8}$	((10)1;1)1;1)	−0.020	0.001	0.000	0.000
${}^{S}\Theta_{0,;6}^{8}$	((11)1;1)1;1)	0.000	0.000	0.000	0.000
${}^{S}\Theta_{0,;7}^{8}$	((01)1;1)1;1)	−0.020	0.001	0.000	0.000
${}^{S}\Theta_{0,;8}^{8}$	((11)0;1)1;1)	0.045	0.002	0.016	0.001
${}^{S}\Theta_{0,;9}^{8}$	((00)0;1)1;1)	−0.138	0.019	−0.167	0.028
${}^{S}\Theta_{0,;10}^{8}$	((10)1;1)0;0)	−0.123	0.015	−0.078	0.006
${}^{S}\Theta_{0,;11}^{8}$	((11)1;1)0;0)	0.000	0.000	0.000	0.000
${}^{S}\Theta_{0,;12}^{8}$	((01)1;1)0;0)	−0.123	0.015	−0.078	0.006
${}^{S}\Theta_{0,;13}^{8}$	((11)0;0)0;0)	0.127	0.016	−0.097	0.009
${}^{S}\Theta_{0,;14}^{8}$	((00)0;0)0;0)	−0.952	0.906	0.967	0.935

[a] *Standard* active space calculations
[b] Serber basis
[c] Orbitals are ordered as: $(\phi^{1}_{,C}\phi^{1}_{,H}\phi^{2}_{,C}\phi^{2}_{,H})(\sigma_{C}\sigma_{M}\pi_{C}\pi_{M})$ or $(\phi^{1}_{,C}\phi^{1}_{,H}\phi^{1}_{,H}\phi^{2}_{,H})(\Omega^{1}_{,C}\Omega^{1}_{,M}\Omega^{2}_{,C}\Omega^{2}_{,M})$

As can be shown with the use of simple overlap-based reasoning [13], the extension of the active space does not provide extra orbital degrees of freedom for the bent bond

As can be shown with the use of simple overlap-based reasoning [13], the extension of the active space does not provide extra orbital degrees of freedom for the bent bond solution. This suggests that the additional energy lowering observed for the separated bond model within the *standard* active space is not related to orbital flexibility but comes entirely from the increased differences between the numbers of spin degrees of freedom in the two models within the larger active space.

It is reasonable to inquire whether the trends observed upon passing from the *minimal* to the *standard* active space will be preserved upon further extension of the active space. This is particularly important in the case of ScCH^{+},$_{2}$, for which the energy separation between the two bonding models decreases sharply when the electrons within the CH bonds are transferred to the active space (see above). In order to address this question, we carried out additional calculations on the ScCH^{+},$_{2}$ using a 10-electron active space ($f^{10}{,}_{0}$=42) obtained by adding two core M(3p) electrons to the *standard* active space. This choice of the active space is, of course, arbitrary to some extent, but it can still furnish some useful qualitative indications. The calculations were carried out by means of the nonvariational version of the CASVB formalism, which provides a close approximation to the fully-variational SC solution. The nonvariational CASVB strategy described in Refs. 9 and 37 transforms the CASSCF wavefunction to a form dominated by a single covalent SC-like construction, which can be achieved using either overlap- or energy-based considerations. On passing from the 8-electron to the 10-electron calculation, Δ_s for ScCH^{+},$_{2}$ decreases from –2 to –8. In parallel with this, according to the energy-based CASVB calculations, Δ_E increases to 1.16 mH which clearly suggests that this further extension of the active space may establish a preference for the separated bond scheme. There are good reasons to expect that a fully-variational SC treatment would confirm this result and reverse the wavefunction hierarchy for ScCH^{+},$_{2}$. Thus, the preference for the Ω bond solution found with *minimal* active space calculations may turn out to be a computational artefact of rather limited importance.

D. Molecular distortions. The choice to work with the symmetric equilibrium molecular geometries only (C_{2v} point group for all molecules) can introduce an artificial bias in favour of one of the two bonding models. In fact, the bent bond scheme can be extremely sensitive to changes in the interatomic distances which is demonstrated by results presented in this subsection and in Ref. [10]. Moreover, the geometries used in our calculations were taken from theoretical structure optimizations [5], the accuracy of which may be limited: Recent B3LYP calculations [4] suggest that ScCH^{+},$_{2}$ may prefer to adopt a C_s geometry, in which the CH_2 group rotates to allow some back donation into empty Sc d orbitals of b_2 pseudo-symmetry. It is also necessary to establish whether small atomic displacements (for instance, due to molecular vibrations or solvent effects) could influence the description of bonding in our systems and modify the energy separation Δ_E to an appreciable extent. In order to address this points, we revisited the 5-electron treatment of TiCH^{+},$_{2}$ and considered five different types of geometry distortions, numbered from **I** to **V** as illustrated below. Except for distortion **V** (see below for an explanation), our results are summarized in the Fig. 2, in which the energy separation Δ_E and the weights of the PP terms W^{PP} are plotted against the parameters characterizing the distortions.

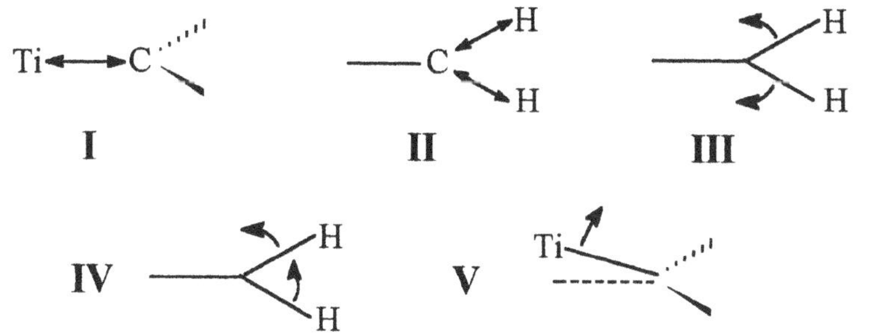

The totally symmetric (A_1) distortions, namely, the TiC stretch (**I**), the CH stretch (**II**), and the CH_2 valence bond angle bending (**III**) all preserve the C_{2v} symmetry. Motions **II** and **III** which do not involve the TiC bond directly have very little impact on the energy separation. The displacement of the two hydrogen atoms by 0.1 bohr away from the equilibrium CH distance modifies Δ_E by ~0.3 mH (see Fig. 2-II). The deviation of the CH_2 valence bond angle by ±10° from its equilibrium value leads to approximately the same variation of the energy separation (see Fig. 2-III). The modification of the TiC internuclear distance has a more pronounced effect on the value of Δ_E (see Fig. 2-I). The shortening of the TiC bond length is accompanied by a decrease of Δ_E. For very short interatomic separations the bent bond model even becomes the lower in energy. For example, when the equilibrium TiC bond length is reduced by 0.2 bohr, the Ω bond solution is placed ~1.4 mH below its $\sigma+\pi$ rival. In contrast to this, large interatomic separations favour the separated bond model which is demonstrated by the increase of Δ_E with the increase of the TiC interatomic distance. Fig. 2 shows that the spin-coupling pattern within the Ω bond model is more sensitive to small distortions than the spin-coupling pattern within the $\sigma+\pi$ bond scheme: Along distortion paths **I**, **II** and **III**, $W^{PP}{}_{,\Omega}$ changes much faster than $W^{PP}{}_{,\sigma+\pi}$.

The pairs of plots for distortions **I**, **II** and **III** indicate a well-defined correlation between the energy separation and the weight of the PP term, especially within the bent bond wavefunction. An increase of the weight of this term always favours the bent bond solution. At very short TiC bond distances the PP term dominates the active space spin-coupling pattern, the $\sigma+\pi$ model cannot make much use of its extra spin degrees of freedom, and the greater orbital flexibility of the bent bond construction makes it lower in energy. On the contrary, when the TiC bond is stretched, spin flexibility gains in significance, as the active space spin functions have to incorporate more significant contributions from non-PP spin eigenfunctions. Logically, this favours the model which offers a higher number of spin degrees of freedom, the $\sigma+\pi$ construction.

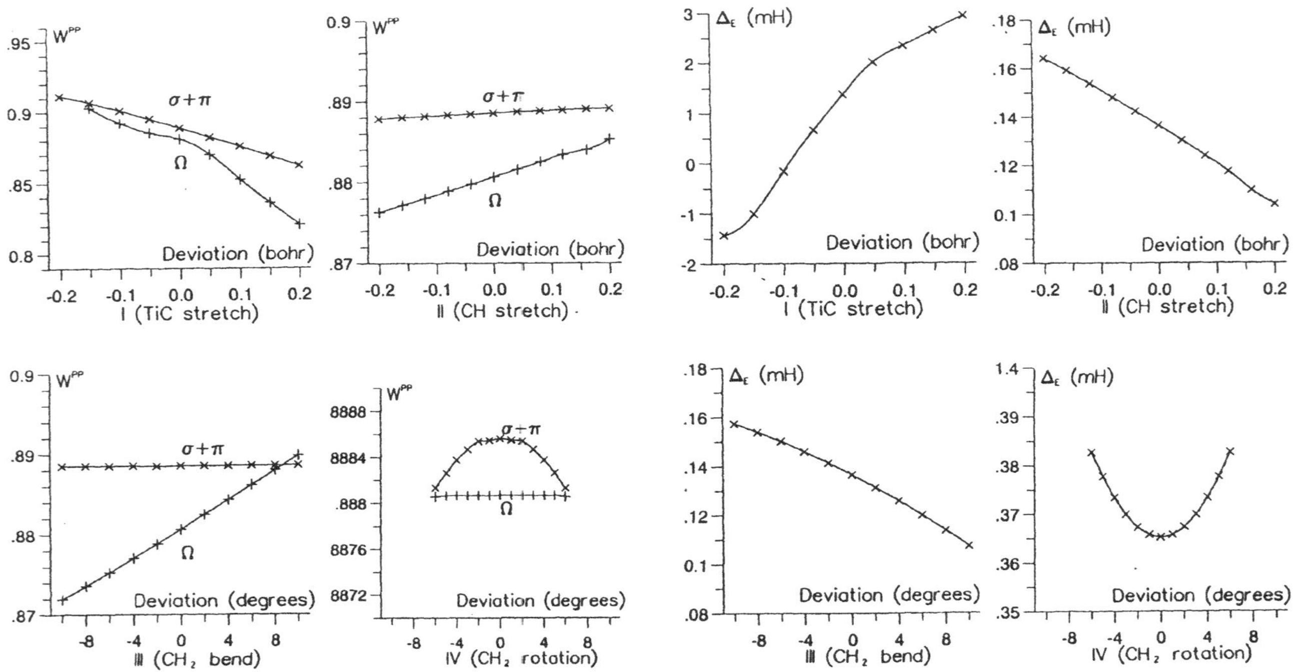

Figure 2. Evolution of the energy separation Δ_E for $TiCH_2^+$, and of the corresponding PP weight W^{PP} (expressed in the Serber basis) along several distortion paths (the zero on the $x_i^{\rightarrow}$ axis of each diagram indicates the equilibrium position, see text for further details).

The rotation of the methylene group **IV** takes place in the molecular plane $\sigma,\hat{}_{yz}$, about an axis perpendicular to this plane and passing through the methylene carbon. Since $\sigma,\hat{}_{yz}$ is preserved, this rotation does not change the difference between the orbital degrees of freedom in the two bonding representations observed at the C_{2v} geometry and, once again, $^{R}\Theta^{5}{}_{,\frac{1}{2};4}$ cannot participate in the active space spin function for the Ω bond solution. Thus, distortion **IV** preserves the differences between the orbital and spin flexibilities for the two models found at the C_{2v} symmetry. As a consequence, the impact of this distortion on the energy separation is very small (see Fig. 2-IV). A clockwise or anticlockwise rotation of the methylene group by 7° away from its equilibrium position changes Δ_E by only ~0.04 mH and brings about minor variations of weights of the PP terms.

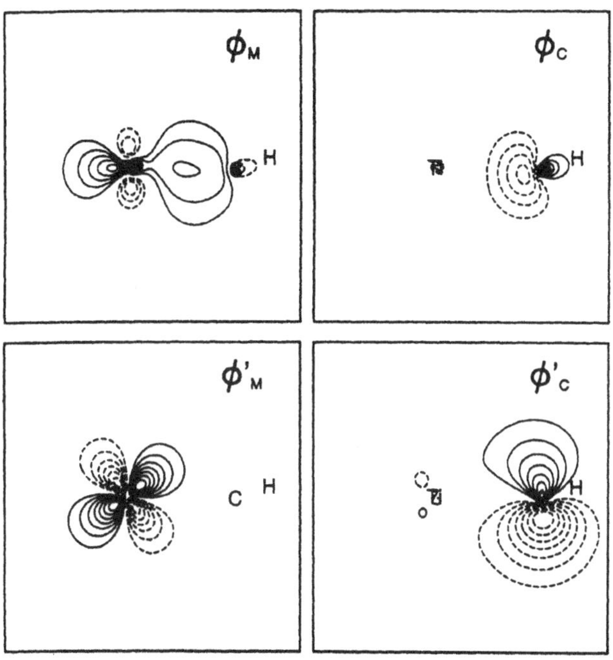

Figure 3. The four fully-variational SC orbitals from the metal-methylene bond for the distorted form **V** of $TiCH_2^{+}$ calculated within a *minimal* active space, without any symmetry constraints, and for a distortion angle of 30°.

In summary, our study of distortions **I** to **IV** shows that there is a direct correlation between the weights of the PP spin eigenfunctions (which demonstrate the extent to which the $\sigma+\pi$ and bent bond wavefunctions are capable of making use of spin flexibility) and the calculated energy separations between two models. In fact, a correlation of this type can be expected to apply to the whole metal-methylene series. Indeed, on moving from left to right in Table 1, the improved spin flexibility of the separated bond wavefunction, which leads to a uniform decrease of W^{PP}, is accompanied by a uniform increase of the energy separation between the two models. The only exception occurs for the chromium complex, for which Δ_E is slightly inferior to the value for $VCH^+{}_{,2}$. This small deviation is not very significant: For example, an increase of the CrC internuclear distance by just 0.1 bohr, which is not too much, bearing in mind that we are using theoretically optimized geometries of possibly limited accuracy, restores the uniformity of the Δ_E sequence.

Distortion **V** destroys the symmetry plane $\sigma,\hat{}_{yz}$, but retains $\sigma,\hat{}_{xz}$. With $\sigma,\hat{}_{yz}$ gone, it is not possible to impose a rigorous σ/π separation any longer, and all four orbitals describing the TiC interaction become different and can overlap freely between themselves. In these circumstances, it becomes meaningless to speak of separate $\sigma+\pi$ and bent bond wavefunctions. However, it is interesting to mention that the valence orbitals from the fully-optimized SC wavefunction, calculated at a relatively large value of the bending angle, 30°, retain a strong resemblance to those from the $\sigma+\pi$ solution at the C_{2v} equilibrium geometry (see Fig. 3).

4.2. TRIPLET CHARACTER OF THE π BOND

A. Significance of the non-PP terms. As mentioned in the introduction, the metal-carbon interactions are generally considered to involve a certain degree of triplet character, especially within the π bond. This is confirmed by the results of our calculations (see Table 1). It is obvious that the triplet characters of both the σ and π interactions increase significantly when moving towards the complexes in the right-hand side of Table 1. The values of the $\langle s,\hat{}_i \cdot s,\hat{}_j \rangle$ spin correlation matrix elements associated with σ_C and σ_M exhibit a moderate growth from −0.732 to −0.618, but their counterparts for π_C and π_M increase sharply from −0.732 to −0.422. There is no triplet character within the PP approximation since, by definition, the electrons participating in the bonds are entirely singlet-coupled. The partial triplet bond character depends entirely on the inclusion of non-PP spin eigenfunctions in the active space spin-coupling pattern. One particular non-PP spin coupling mode turns out to be specially important throughout the $MCH_2,^+$ series. It appears for the first time in the active space spin function for $TiCH_2,^+$ and corresponds to the third spin eigenfunction, with the second largest weight, within the Kotani and Serber spin bases (see Table 2). In the remaining part of this section we use the Serber spin basis, which shows explicitly the singlet and triplet pairs making up the individual spin eigenfunctions.

As we mentioned when discussing the structure of its symmetry-forbidden components, the spin coupling pattern in $VCH_2,^+$ is very similar to that for the titanium complex. There are two non-PP terms, ${}^S\Theta^6{}_{,1;4} \equiv ((01)1;0)$ and ${}^S\Theta^6{}_{,1;7} \equiv ((01)1;1)$ (see Table 3), which are direct descendants of the non-PP spin eigenfunction ${}^S\Theta^5{}_{,½;3} \equiv ((01)1;(½))$ for $TiCH_2,^+$. The weight of ${}^S\Theta^6{}_{,1;7}$ (0.160) makes it the second most important component in the active space spin-coupling pattern. The negligibly small contribution from ${}^S\Theta^6{}_{,1;4}$ follows from the fact that it couples the spins of the two nonbonding electrons to a sing-let, while $VCH_2,^+$ very much prefers to have them coupled to a triplet. Similar situations are also observed in the chromium and manganese complexes as a result of inheritance of the spin coupling pattern within the metal-methylene bond.

The most important non-PP spin eigenfunction can be shown to be responsible for nearly all of the energy gap separating solutions restricted to the PP term and solutions utilizing the full spin space. One way of doing this is to perform calculations involving only a few selected spin eigenfunctions with fixed active and core orbitals taken directly from the corresponding optimized full spin space wavefunction. Such calculations indicate, in the case of $TiCH_2,^+$, that including ${}^{S}\Theta^{5}{}_{,\frac{1}{2};3}$ with the PP term can account for approximately ¾ of the energy improvement. For $VCH_2,^+$, use of the PP term and the predominant non-PP term, ${}^{S}\Theta^{6}{}_{,1;7}$, captures approximately 83% of the correlation energy included in full spin space SC wavefunction. These results show that much of the stabilization energy of the metal-methylene complexes is due to an off-diagonal hamiltonian matrix element which corresponds to the interaction between two VB-style structures, one of which involves the PP spin eigenfunction, while the π bond orbitals within the spin-coupling pattern for the other one are coupled to a triplet.

B. Overall σ / π decoupling. As a next step, we need to assess the extent of coupling between the group of electrons involved in the σ metal-methylene bond, G_σ, and the group of electrons defined by the π bond and the metal unpaired electrons, $G_{\pi+\ell}$. One way to achieve this is to perform calculations in which the active space is deprived of the σ electrons, *i.e.* it consists only of the π and ℓ_i orbitals. In the case of $TiCH_2,^+$, for example, we find that the spin coupling situation depicted by a 3-electron calculation is in close agreement with that resulting from the corresponding 5-electron calculation. The $\langle \hat{s}_i \cdot \hat{s}_j \rangle$ spin correlation matrix elements for π_C and π_M, π_C and ℓ_1, and π_M and ℓ_1 are found to be equal to −0.668, −0.278 and 0.196, respectively; these values are very close to those observed for a *minimal* active space calculation (see Table 5). Alternatively, the weak coupling can be quantified with the help of the last term on the right-hand side of Eq. (5) (see Section 0). In the case of the titanium species, the $G_\sigma/G_{\pi+\ell}$ spin coupling is given by the sum of the six elements of the 3×2 block in the left lower corner of Table 5: the value is very close to zero, indicating that there is almost no interaction between the two groups. In order to compare the whole $MCH_2,^+$ (M=Ti–Mn) series, it is more convenient to report the $G_\sigma/G_{\pi+\ell}$ couplings as average spin couplings defined for one electron from each group. The $\langle \hat{s}(G_\sigma)\cdot\hat{s}(G_{\pi+\ell})\rangle$ values obtained in this way are −0.009, −0.008, −0.011, −0.014, −0.015, respectively. These values are all very small and indi-cate negligible coupling between the electrons from the G_σ and $G_{\pi+\ell}$ groups.

C. Overall triplet character of the π bonds. As a consequence of the fact that the two groups of electrons G_σ and $G_{\pi+\ell}$ are almost completely uncoupled, the triplet character of the π interaction can be rationalized using simple arguments involving only the electrons of the $G_{\pi+\ell}$ group. In the case of 3-electron treatment of $TiCH_2,^+$ ($f^3{}_{,\frac{1}{2}}=2$), analyzed in the Serber spin basis, it is possible to draw the following scheme (the numbers in parentheses are the corresponding $\langle \hat{s}_i \cdot \hat{s}_j \rangle$ spin correlation matrix elements):

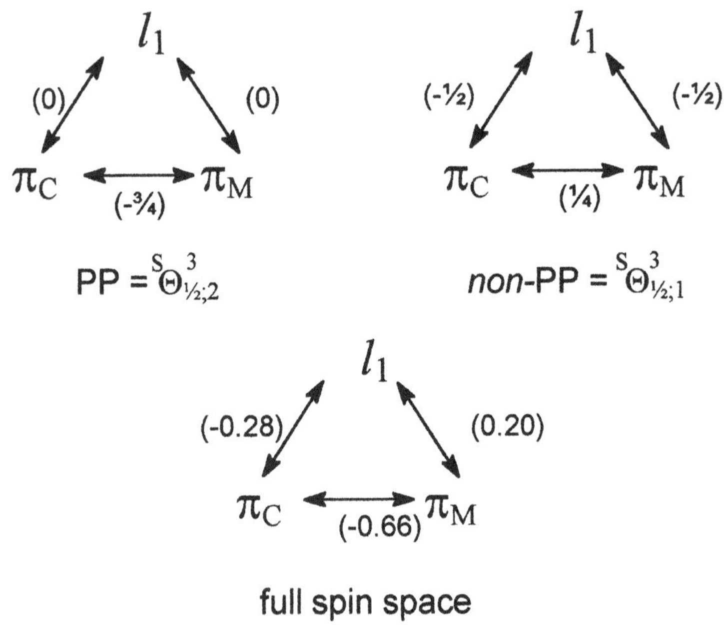

In the perfect-pairing case, electron spins associated with the three orbitals ℓ_1, π_M, and π_C participate in pure singlet and uncoupled interactions, so as to produce the required total net spin of 0.5. Within the full spin space the 3-electron system is allowed to relax through the mixing of ${}^{S}\Theta_{½,}{}^{3}{}_{;1}$ and ${}^{S}\Theta_{½,}{}^{3}{}_{;2}$. It is surprising, at first sight, to observe that the mixing of two spin patterns which are symmetric with respect to the π_C–ℓ_1 and π_M–ℓ_1 interactions does not result in a symmetric spin coupling mode. In fact, by writing correlation matrix elements as functions of the spin-coupling coefficients C_{sk} (Eq. (3)), it can be shown that this asymmetry arises from off-diagonal terms [46]. The full spin space calculation exhibits the following three interesting features: (*i*) a triplet coupling mode between π_M and ℓ_1; (*ii*) a unexpected singlet character in the π_C–ℓ_1 interaction; (*iii*) a certain triplet character in the metal-carbon π bond.

TABLE 5. Spin correlation matrix elements $\langle s,\hat{}_i \cdot s,\hat{}_j \rangle$ for the wavefunction of $TiCH^+$,2 ($f_{½,}{}^{5}$=5)[a]

	σ_C	σ_M	π_C	π_M	ℓ_1
σ_C	−0.750				
σ_M	−0.722	−0.750			
π_C	0.122	−0.146	−0.750		
π_M	−0.134	0.115	−0.645	−0.750	
ℓ_1	−0.099	0.093	−0.285	0.201	−0.750

[a] *Minimal* active space calculations

This picture has a clear physical interpretation. As hinted by Carter and Goddard [3], it seems plausible that the metal-methylene interaction in the $MCH_2{}^+$ (M=Ti–Mn) series is the result of a balance between two competitive effects, namely the preservation of the high-spin d–d exchange energy present in the isolated metal atom, and the formation of a singlet-coupled covalent bonding pair. Our simple scheme shows that, when comparing to the PP case, the only way in which the system could retain some triplet character between the bonding and the nonbonding metal electrons is to make use of the non-PP term, even if this weakens the π interaction by introducing some triplet character. The appearance of an unexpected singlet character in the interaction π_C–ℓ_1 is a side effect following from the requirement to the preserve the overall spin of the system.

Several observations provide additional confirmation of the validity of our descriptive model. In the first place, the triplet character of the π bond increases on moving towards the right-hand side of the periodic table, in parallel with an increase in the number of unpaired electrons. Indeed, as we have already demonstrated, the mainly triplet π_M–ℓ_i interactions are very similar irrespective of the choice of M and the nature of the unpaired electron. The related values of $\langle \hat{s}_i \cdot \hat{s}_j \rangle$ vary only between 0.189 and 0.201 throughout the whole $MCH_2{}^+$ series studied in this paper. On the basis of the arguments presented in the preceding paragraph we can expect that, with the increase of the number of unpaired electrons for which the exchange energy with the metal π_M bonding electron has to be preserved, the π bond will be forced to lose more of its singlet character. Indeed, the triplet character of the π bond decreases again beyond $MnCH_2{}^+$ [11], which is fully consistent with our model. For systems beyond $MnCH_2{}^+$, certain nonbonding orbitals on the metal centre become doubly occupied. The electrons in these orbitals are singlet coupled and, as a consequence, they do not contribute to the metal d–d exchange energy involving high-spin pairs. Further evidence comes from calculations carried out on the $^1A_1$ excited state of $VCH_2{}^+$. The triplet character of the π bond is found to be very small, $\langle \hat{s}_i \cdot \hat{s}_j \rangle = -0.708$, and much closer to the value for a species deprived of odd electrons, $ScCH_2{}^+$ (−0.732), than to those for the ground states of $TiCH_2{}^+$ (−0.644) and $VCH_2{}^+$ (−0.565) that incorporate one and two unpaired electrons, respectively. We will show in the second paper [11] that the differences between the triplet character present in the σ and in the π interactions can also be rationalized using this model.

5. Concluding remarks

The *ab initio* modern valence bond investigations of the nature of the metal-ligand double bond in the $MCH_2{}^+$ (M = Sc–Mn) series, performed in this paper using the formalism of spin-coupled theory, provide detailed information on several important issues:

i) The wavefunctions based on the two rival bond models using either σ+π or bent bond orbitals are found be essentially the same. They are also very close to the corresponding CASSCF wavefunctions, which is a general feature of the SC approach. This allows the use of frozen CASSCF core orbitals inside the SC wavefunction as an alternative to the simultaneous optimization of the core and valence subspaces without significant impact on the quantitative nature of the results.

ii) Our calculations indicate that, at the SC level of theory, the metal-methylene double bond is described by two predominantly singlet-coupled pairs of bonding orbitals. When the molecule contains nonbonding electrons on the metal centre [examples are provided by the $MCH_2,^+$ (M=Ti–Mn) series and by the $^1A_1$ singlet excited state of $VCH_2,^+$], the better description of the metal-methylene interaction is achieved through the $\sigma+\pi$ construction. On the other hand, complexes deprived of nonbonding electrons, such as $ScCH_2,^+$, $TiCH_2,^{2+}$ and $VCH_2,^{3+}$, opt for the bent bond scheme. These two trends can be rationalized on the basis of the differences between the numbers of free spin variational parameters allowed by the two bond models. The presence of nonbonding electrons on the metal centre reduces the spin flexibility of the bent bond solutions in comparison to their separated bond alternatives. Consequently, the latter become the lower in energy. This effect is independent of the coupling of the spins of the orbitals occupied by the nonbonding electrons, which can be singlet or triplet. From the viewpoint of the total energy, orbital flexibility proves to be a factor of very minor importance at the equilibrium geometries. The indepth analysis of the relationship between spin space flexibility and the preference for a bonding model also has to account for the fact that some spin-degrees of freedom, although allowed by symmetry, are not effectively used by the wavefunction.

iii) The study of small molecular distortions reveals the existence of a very good correlation between the magnitude of the energy gap separating the two bond models and the weights of the corresponding PP terms. Higher-weight PP terms favour the bent bond wavefunctions while, on the contrary, the presence of significant non-PP terms, which are associated with extra spin flexibility, is of greater benefit for the $\sigma+\pi$ construction. The bent bond description is found to be much more sensitive than is its $\sigma+\pi$ counterpart to small variations of the distance between the metal atom and the methylene carbon.

iv) Extension of the SC active space increases the difference between the spin flexibilities of the two bond models and, as a consequence, favours the $\sigma+\pi$ construction. As has been shown on the example of $ScCH_2,^+$, for systems in which the bent bond model is lower in energy, the use of a larger number of active orbitals can lead to a sharp decrease in the energy separation between the two models, which suggests that this difference could be completely eliminated, or even reversed upon further extension of the SC active space.

Apart from these findings, the $\sigma+\pi$ bond model offers one very significant conceptual and computational advantage over its bent bond counterpart: For symmetry reasons an MCSC or a non-orthogonal CI construction on top of the $\sigma+\pi$ wavefunction would involve a much smaller number of ionic configurations than an equivalent-quality construction based on the bent bond wavefunction. Of course, from a purely philosophical point of view, the question of which model is 'better' ultimately makes no sense at a full-CI limit, simply because such a wavefunction is invariant to nonsingular transformations of the active orbitals.

The present work provides the link between the weight of the PP term and the importance of spin flexibility for the stabilization of the wavefunction, which we could not establish on the basis of the results reported in our previous paper [13]. In fact, the previous study [13] involved a set of very different molecules, such as C_2H_2, C_2H_4, N_2, CN^-, $ScCH_2$,$^+$ and $TiCH_2$,$^+$. Within a set of unrelated molecules of this type, a subtle correlation can easily be obscured by larger differences between the wavefunctions for the individual molecules. For example, as the spin coupling pattern in each of these molecules was found to have little in common with those of the other molecules, the energy improvements related to non-PP terms could also be very different. The situation observed in the results from the present paper is entirely different: As we have shown, there is a well-defined inheritance of the spin-coupling patterns throughout the whole $ScCH_2$,$^+$–$MnCH_2$,$^+$ series. In particular, the symmetry constraint that prevents certain spin functions from contributing to the bent bond solution, which appears for the first time in $TiCH_2$,$^+$, is inherited by all remaining systems. The same is true of the most important non-PP term in $TiCH_2$,$^+$, which is responsible for the emergence of a VB-style resonance and thus has a strong stabilizing effect on the wavefunction.

Finally, we have also established the existence of a direct link between the triplet character of the metal-carbon π bond and the radical nature of complexes. Only unpaired electrons can instill some triplet character into the π interaction, and our results indicate that this triplet character increases in parallel with the number of unpaired electrons. The exact amount of triplet character depends on the interplay between two competing effects: the preservation of the high-spin d–d exchange energy present in the isolated metal atom, and the formation of a singlet-coupled covalent bonding pair between the metal- and carbon-centred π orbitals.

Based on our various findings in the present work, the second paper in this series [11] will concentrate on calculations for $ScCH_2$,$^+$–$CoCH_2$,$^+$ using only the symmetry-separated (σ+π) model. The particular order of sequentially filling the nonbonding orbitals ℓ_i, as utilized here, will be rationalized in terms of a compromise between minimizing repulsive electrostatic interactions and maximizing the exchange energy. The relative importance of metal→ligand and ligand→metal charge transfer in the σ and π interactions will be quantified, and trends in the dipole moments rationalized. We will also examine trends in the metal-ligand dissociation energy, and link the intrinsic bond strengths to the influence of unpaired electrons.

References

1. (*a*) Trost, B.M. *Chem. A Euro. J.* **1998**, *12*, 2405. (*b*) Chen, Y.M.; Armentrout, P.B. *J. Phys. Chem.* **1995**, *99*, 10775. (*c*) Haynes, C.L.; Chen, Y.M.; Armentrout, P.B. *J. Phys. Chem.* **1995**, *99*, 9110. (*d*) Haynes, C.L.; Armentrout, P.B.; Perry, J.K.; Goddard, W.A. *J. Phys. Chem.* **1995**, *99*, 6340. (*e*) Kemper, P.R.; Bushnell, J.; van Koppen, P.A.M.; Bowers, M.T. *J. Phys. Chem.* **1993**, *97*, 1810. (*f*) van Koppen, P.A.M.; Kemper, P.R.; Bowers, M.T. *J. Am. Chem. Soc.* **1992**, *114*, 1083.
2. (*a*) Li, J.H.; Feng, D.C.; Feng, S.Y. *Chem. J. of Chin. Universities* **1998**, *19*, 1495. (*b*) Abashkin, Y.G.; Burt, S.K.; Kusso, N. *J. Phys. Chem. A* **1997**, *101*, 8085. (*c*) Musaev, D.G.; Morokuma, K. *J. Phys. Chem.* **1996**, *100*, 11600. (*d*) Musaev, D.G.; Morokuma, K.; Koga, N.; Nguyen, K.A.; Gordon, M.S.; Cundari, T.R. *J. Phys. Chem.* **1995**, *97*, 11435. (*e*) Ricca, A.; Bauschlicher, C.W. *Chem. Phys. Lett.* **1995**, *245*, 150. (*f*) Blomberg, M.R.A.; Siegbahn, P.E.M.; Svenson, M. *J. Phys. Chem.* **1994**, *98*, 2062. (*g*) Musaev, D.G.; Koga, N.;

Morokuma, K. *J. Phys. Chem.* **1993**, *97*, 4964. (*h*) Bauschlicher, C.W.; Partridge, H.; Scuseria, G.E. *J. Chem. Phys.* **1992**, *97*, 7471. (*i*) Carter, E.A.; Goddard, W.A. *J. Am. Chem. Soc.* **1986**, *108*, 2180.
3. Carter, E.A.; Goddard, W.A. *J. Phys. Chem.* **1984**, *88*, 1485.
4. Ricca, A.; Bauschlicher, C.W. *Chem. Phys. Lett.* **1995**, *245*, 150.
5. Bauschlicher, C.W.; Partridge, H.; Sheehy, J.A.; Langhoff, S.R.; Rosi, M. *J. Phys. Chem.* **1992**, *96*, 6969.
6. Alvarado-Swaisgood, A.E.; Harrison, J.F. *J. Phys. Chem.* **1988**, *92*, 2757.
7. (*a*) Loades, S.D.; Cooper, D.L.; Gerratt, J.; Raimondi, M. *J. Chem. Soc., Chem. Comm.* **1989**, 1604. (*b*) Ohanessian, G.; Goddard, W.A. *Acc. Chem. Res.* **1990**, *23*, 386.
8. (*a)* Cooper, D.L.; Gerratt, J; Raimondi, M. *Chem. Rev.* **1991**, *91*, 929. (*b*) Gerratt, J; Cooper, D.L.; Karadakov, P.B.; Raimondi, M. *Chem. Soc. Rev.* **1997**, *26*, 87.
9. Cooper, D.L.; Thorsteinsson, T.; Gerratt, J. *Adv. in Quant. Chem.* **1999**, *32*, 51.
10. Loades, S.D. *Bonding to Transition Metal Atoms in Low Oxidation States* **1992**, Ph.D. Thesis, University of Liverpool: Liverpool, UK.
11. Ogliaro, F; Loades, S.D.; Cooper, D.L.; Karadakov, P.B. *in preparation.*
12. (*a*) Slater, J.C. *Phys. Rev.* **1931**, *37.* 481. (*b*) Pauling, L. *The Nature of the Chemical Bond*, **1960**, 3rd ed.; Cornell University Press: Ithaca, New-York.
13. Ogliaro, F.; Cooper, D.L.; Karadakov, P.B. *Int. J. Quant. Chem.* **1999**, *74*, 123, and references therein.
14. The CASSCF methodology is detailed in: (*a*) Werner, H.-J.; Knowles, P.J. *J. Chem. Phys.* **1985**, *82*, 5053. (*b*) Knowles, P.J.; Werner, H.-J. *J. Chem. Phys. Lett.* **1985**, *115*, 255.
15. See, for example: Cooper, D.L.; Thorsteinsson, T. ; Gerratt, J. *Int. J. Quant. Chem.* **1997**, 65, 439, and references therein.
16. Gerratt, J. *Adv. Atom. Mol. Phys.* **1971**, *7*, 141.
17. Goddard, W.A. *Acc. Chem. Res.* **1973**, *6*, 368.
18. Kotani, M.; Amemiya, A.; Ishiguro, E.; Kimura, T. *Tables of Molecular Integrals*, **1963**, 2nd ed., Maruzen: Tokyo, Japan.
19. Rumer, G. *Gottingen. Nachr.* **1932**, 337.
20. Serber, R. *Phys. Rev.* **1934**, *45*, 461.
21. Pauncz, R. *Spin Eigenfunctions* **1979**, Plenum Press: New York, and references therein.
22. Chirgwin, B.H.; Coulson, C.A. *Proc. Roy. Soc. Lond. A* **1950**, *201*, 196.
23. Gallup, G.A.; Norbeck, J.M. *Chem. Phys. Lett.* **1970**, *21*, 495.
24. Löwdin, P.-O. *Ark. Mat. Astr. Fysik* **1947**, *35A*, 9.
25. Thorsteinsson, T.; Cooper D.L. *J. Math. Chem.* **1998**, *23*, 105.
26. The methodology and a few applications are detailed in: (*a*) Raos, G.; Gerratt, J.; Cooper, D.L.; Raimondi, M. *Chem. Phys.* **1994**, *186*, 233, 251. (*b*) Cooper, D.L.; Ponec, R., Thorsteinsson, T.; Raos, G. *Int. J. Quant. Chem.* **1996**, *57*, 501.
27. (*a*) Raos, G.; McNicholas, S.J.; Gerratt, J.; Cooper, D.L.; Karadakov, P.B. *J. Phys. Chem. B* **1997**, *101*, 6688. (*b*) Friis-Jensen, B.; Cooper, D.L.; Rettrup, S.; Karadakov, P.B. *J. Chem. Soc. Faraday Trans.* **1998**, *94*, 3301.
28. Penotti, F. *Int. J. Quant. Chem* **1996**, *59*, 349.
29. Clarke, N.J.; Raimondi, M.; Sironi, M.; Gerratt, J.; Cooper, D.L. *Theor. Chem. Acc.* **1998**, *99*, 8.
30. Thorsteinsson, T.; Cooper, D.L. *Prog. Theor. Chem. and Phys.* (in press).
31. Gerratt, J.; Raimondi, M. *Proc. Roy. Soc. Lond. A* **1980**, *371*, 525.
32. Watchers, A.J.H. *J. Chem. Phys.* **1970**, *52*, 1033.
33. Bauschlicher, C.W.; Langhoff, S.R.; Barnes, L.A. *J. Chem. Phys.* **1989**, *91*, 2399.
34. Dunning, T.H. *J. Chem. Phys.* **1989**, *90*, 1007.

35. MOLPRO is a package of *ab initio* programs written by Werner, H.–J.; Knowles, P.J. with contributions from Amos, R.D.; Berning, A.; Cooper, D.L.; Deegan, M.J.O.; Dobbyn, A.J.; Eckert, F.; Hampel, C.; Leininger, T.; Lindh, R.; Lloyd, A.W.; Meyer, W.; Mura, M.E.; Nicklaß, A.; Palmieri, P.; Peterson, K.; Pitzer, R.; Pulay, P.; Rauhut, G.; Schütz, M.; Stoll, H.; Stone, A.J.; Thorsteinsson, T..
36. Karadakov, P.B.; Gerratt, J.; Cooper, D.L.; Raimondi, M. *Theor. Chim. Acta.* **1995**, *90*, 51.
37. Thorsteinsson, T.; Cooper, D.L.; Gerratt, J.; Karadakov, P.B.; Raimondi, M. *Theor. Chim. Acta* **1996**, *93*, 343.
38. Karadakov, P.B.; Gerratt, J.; Cooper, D.L.; Raimondi; M. *J. Am. Chem. Soc.* **1993**, *115*, 6863.
39. Palke, W.E. *J. Am. Chem. Soc.* **1994**, *108*, 6543.
40. Messmer, R.P.; Schultz, P.A.; Tatar, R.C.; Freund, H.-J. *Chem. Phys. Lett.* **1986** *126*, 176.
41. Messmer, R.P.; Schultz, P.A. *Phys. Rev. Lett.* **1986**, *57*, 2653.
42. Schultz, P.A.; Messmer, R.P. *Phys. Rev. Lett.* **1987**, *58*, 2416.
43. Schultz, P.A.; Messmer, R.P. *J. Am. Chem. Soc.* **1993**, *115*, 10925, 10938, 10943.
44. Karadakov, P.B.; Gerratt J.; Cooper, D.L.; Raimondi, M. *J. Am. Chem. Soc.* **1993**, *115*, 6863.
45. Cunningham, T.P.; Cooper, D.L.; Gerratt, J.; Karadakov, P.B.; Raimondi, M. *J. Chem. Soc. Faraday Trans.* **1997**, *93*, 2247.
46. Cooper, D.L. *unpublished results*.

HARTREE-FOCK STUDY OF HYDROGEN-BONDED SYSTEMS IN THE ABSENCE OF BASIS-SET SUPERPOSITION ERROR: THE NUCLEIC-ACID BASE PAIRS

A. FAMULARI, M. SIRONI, E. GIANINETTI AND M. RAIMONDI

Dipartimento di Chimica Fisica ed Elettrochimica and Centro CNR-CSRSRC, Università degli Studi di Milano via Golgi 19, 20133 Milano, Italy
E-mail: antonino.famulari@unimi.it

Abstract. The Roothaan equations have been recently modified for computing molecular interactions between weakly bonded systems at the SCF level in order to avoid the introduction of the basis-set superposition error (BSSE). Due to the importance that nucleic-acid base interactions play in DNA and RNA, which are 3-D structures, we present applications of this approach of Self Consistent Field for Molecular Interactions (SCF-MI) to the study of several hydrogen-bonded DNA bases. Nucleic-acid pairs are extensively investigated: structures and energies for both isolated and paired molecules are thoroughly studied. Cs-symmetry equilibrium geometries and binding energies are calculated in the framework of the SCF-MI formalism by using standard basis sets. SCF-MI/3-21G stabilisation energies for the studied base pairs lie in the range - 22.5 / -8.0 kcal/mol, in good agreement with previous, high-level theoretical values. The hydrogen bonding potential energy surface (PES) and propeller twist potentials are also calculated for some of the molecular complexes. The SCF-MI interaction density is used to interpret the nature of the interactions involved in the hydrogen bond formation: structure and stabilisation of the base pairs turn out to be determined mostly by electrostatic interactions. Preliminary calculations on stacked cytosine dimer are also reported. The SCF-MI/3-21G results show agreement with the counterpoise corrected SCF/6-31G** calculations.

1. Introduction

Accurate *ab initio* calculations on large van der Waals molecular clusters are one of the most challenging tasks of theoretical chemistry. In spite of a number of theoretical and experimental studies [1] employing more and more sophisticated

J. Maruani et al. (eds.), New Trends in Quantum Systems in Chemistry and Physics, Volume 1, 313–333.

techniques, the unambiguous characterisation of these systems is still not practical. The significance of the theoretical contribution has been widely recognised, emphasising that only a close cooperation between theory and experiment can provide the full description of van der Waals molecule properties [2]. The study of these systems can be considered as one of the fields of chemistry where quantum chemical methods have increased our understanding both in a quantitative and qualitative way. In fact, while experimental techniques can provide information on average equilibrium geometries and corresponding binding energies of the molecular complexes and clusters, quantum chemical ab initio methods can cover the entire potential surface, providing valuable information for a deeper understanding of the nature of these interactions.

Hydrogen-bonded van der Waals molecular clusters represent undoubtedly one of the most important class of van der Waals systems. The great interest to the subject is emphasised by the impressive number of reviews, monographs and books that have recently appeared [3]. Hydrogen bond interactions play a fundamental role in the life sciences. These interactions are in fact responsible of biomolecular structures and related chemical processes: it would be practically impossible to find important biological phenomena in which these interactions do not play an important role [2].

The structure of one of the most important biomolecules, the DNA molecule, is determined by nucleic acid base interactions. In fact, although the unique 3-D double helix DNA structure is influenced by various contributions, the hydrogen bonding and stacking interactions of DNA bases are prominent. A complete elucidation of these interactions as well as the comprehension of the mechanistic aspects of genetic information encoding and trasduction is undoubtedly one of the most fascinating and ambitious goals of scientific research. Due to the difficulties connected to obtaining gas phase experimental data for isolated bases and base pairs characterisation (only a limited number of reliable experimental studies are available [4]) quantum chemical calculations can represent a useful tool to obtain reference data on the structure, properties and interactions of nucleic acid pairs. It is necessary to mention that there are not experimental results on the structure of hydrogen bonded or stacked DNA base pairs in the gas phase. Theoretical studies can provide an help to properly understand the functions and properties of nucleic acids and are fundamental for verification (validation) and parameterisation of empirical potential for molecular modelling of larger biomacromulecules and their interactions.

Although a considerable computational effort is required to describe the fundamental components of nucleic acids by *ab initio* methods [5], the application of quantum chemistry to biodisciplines is nowadays feasible and reliable [2]. A number of systematic and accurate studies have been accomplished on hydrogen bonding of DNA bases [6-8]. Among these, the paper by Sponer, Leszczynski and Hobza [7] represents the most extended study. In the present work we wish to compare these studies with the SCF-MI (Self Consistent Field for Molecular

Interactions) [9-13] results when a small basis set, namely 3-21G, is employed to realise a computational saving for such extended molecular systems.

Differences with standard SCF/3-21G calculations are also reported. Special attention is paid to the basis-set superposition error (BSSE) [14-15], which can be of the same order of magnitude of the interaction itself and represents a well known problem in the study of the potential of weakly interacting fragments. Although several approaches for avoiding this error have been discussed in the literature [15-16], the Boys and Bernardi [17] formulation of the function counterpoise principle (CP) is the most prominent mean for correcting BSSE. The function counterpoise has been a subject for debate and a considerable amount of literature concerning the accuracy of CP correction is available [2,15,18-26]. Different results and conclusions have been reported which will not be discussed here in detail. It is out of questions that the CP correction requires extra computational costs because n+1 calculations in the composite basis (or full basis set), rather than a single one, have to be performed for a cluster consisting of n molecules. Moreover, since the composite basis set does not possess the symmetry of the individual monomers, the partner functions not only lower the energy, in accordance with the variational principle, but also introduce spurious multipole components [18]. The contamination of the resulting energy by the so-called "secondary superposition error" can be particularly significant for charged systems [19]. It should also be pointed out that geometry relaxation effects require a more complex definition of the interaction energy [27-29]. Very recently, a general straightforward method for computing geometries on the corrected potential energy surface (PES) has been developed [28-29]. However, normally, CP correction is added as a single point calculation on geometry optimised in the presence of the BSSE [30]. Recent results show that BSSE removal is essential to correctly predict the "quasi linear" structure of the $(HF)_2$ system [29] and the experimentally observed planar dimer of pyrimidine and *p*-benzoquinone [31].

The SCF-MI method used here is a modification of the Roothaan equations for computing molecular interactions between weakly-bonded systems at the SCF level. The BSSE is avoided in an *a priori* fashion and the resulting scheme is compatible with the usual formulation of the analytic derivatives of the SCF energy; this allows an efficient implementation of BSSE free gradient optimisation algorithms in both the direct and conventional SCF approaches into several quantum chemistry packages (GAMESS-US, PC-GAMESS and MOLPRO). In the next section we report an introduction to the most relevant elements of the SCF-MI approach; a more detailed account can be found elsewhere [9-13]. The validity of the method extends from the long range to the region of the minimum and of short distances.

2. Theoretical background

According to the SCF-MI strategy, the supersystem AB formed by two interacting monomers A and B, consisting of $N=2N_A+2N_B$ electrons, is described by the one-determinant wavefunction:

$$\Psi^0 = (N!)^{-\frac{1}{2}} \boldsymbol{A} \left[\Phi_1^A(1)\overline{\Phi}_1^A(2)\ldots\overline{\Phi}_{N_A}^A(2N_A)\Phi_1^B(2N_A+1)\overline{\Phi}_1^B(2N_A+2)\ldots \right.$$
$$\left. \ldots\overline{\Phi}_{N_B}^B(2N_A+2N_B)\right], \tag{1}$$

where $\boldsymbol{A}$ is the usual total antisymmetrizer operator. The key of the method is the partitioning of the total basis set $\chi = \{\chi_k\}_{k=1}^{M}$, so that the MOs of fragments A ($\Phi_1^A \ldots \Phi_{N_A}^A$) and B ($\Phi_1^B \ldots \Phi_{N_B}^B$) are expanded in two different subsets $\chi^A = \{\chi_p^A\}_{p=1}^{M_A}$ and $\chi^B = \{\chi_q^B\}_{q=1}^{M_B}$ centred respectively on fragments A and B:

$$\Phi_a^A = \sum_{p=1}^{M_A} \chi_p^A T_{pa}^A, \quad a = 1\ldots N_A \tag{2a}$$

$$\Phi_b^B = \sum_{q=1}^{M_B} \chi_q^B T_{qb}^B, \quad b = 1\ldots N_B \tag{2b}$$

where $M=M_A+M_B$ is the basis-set size. The orbitals of different fragments are free to overlap with each other, although this non orthogonality does not involve any particularly severe computational problem. With this partitioning the (MxN) matrix of molecular orbital coefficients assumes a block diagonal form:

$$\mathbf{T} = \begin{bmatrix} \mathbf{T}_A & \mathbf{0} \\ \mathbf{0} & \mathbf{T}_B \end{bmatrix} \tag{3}$$

The energy corresponding to the SCF-MI wavefunction is

$$E = \mathrm{Tr}[\mathbf{D}\cdot\mathbf{h}] + \mathrm{Tr}[\mathbf{D}\cdot\mathbf{F}(\mathbf{D})] \tag{4}$$

while its variation becomes

$$\delta E = 2\mathrm{Tr}[\mathbf{F}(\mathbf{D})\delta\mathbf{D}] \tag{5}$$

where **F** and **h** are the usual Fock and one-electron integral matrices expressed in the atomic orbitals basis set. The density matrix defined as

$$\mathbf{D} = \mathbf{T}\left(\mathbf{T}^{\dagger}\mathbf{S}\mathbf{T}\right)^{-1}\mathbf{T}^{\dagger} \tag{6}$$

satisfies the general idempotency condition

$$\mathbf{DSD} = \mathbf{D} \tag{7}$$

The occurrence of BSSE is avoided by assuming and maintaining the orbital coefficient variation matrix in block diagonal form:

$$\delta\mathbf{T} = \begin{bmatrix} \delta\mathbf{T}_{\mathrm{A}} & \mathbf{0} \\ \mathbf{0} & \delta\mathbf{T}_{\mathrm{B}} \end{bmatrix} \tag{8}$$

As a result, the general stationary condition $\delta E = 0$ yields the following coupled secular problems

$$\begin{cases} \mathbf{F}_{\mathrm{A}}'\mathbf{T}_{\mathrm{A}} = \mathbf{S}_{\mathrm{A}}'\mathbf{T}_{\mathrm{A}}\mathbf{L}_{\mathrm{A}} \\ \mathbf{T}_{\mathrm{A}}^{\dagger}\mathbf{S}_{\mathrm{A}}'\mathbf{T}_{\mathrm{A}} = \mathbf{I}_{\mathrm{A}} \end{cases} \qquad \begin{cases} \mathbf{F}_{\mathrm{B}}'\mathbf{T}_{\mathrm{B}} = \mathbf{S}_{\mathrm{B}}'\mathbf{T}_{\mathrm{B}}\mathbf{L}_{\mathrm{B}} \\ \mathbf{T}_{\mathrm{B}}^{\dagger}\mathbf{S}_{\mathrm{B}}'\mathbf{T}_{\mathrm{B}} = \mathbf{I}_{\mathrm{B}} \end{cases} \tag{9}$$

in terms of "effective" Fock and overlap matrices $\mathbf{F}_{\mathrm{A}}'$, $\mathbf{F}_{\mathrm{B}}'$ and $\mathbf{S}_{\mathrm{A}}'$, $\mathbf{S}_{\mathrm{B}}'$. It is substantial to recognise that matrices $\mathbf{F}_{\mathrm{A}}'$, $\mathbf{F}_{\mathrm{B}}'$ and $\mathbf{S}_{\mathrm{A}}'$, $\mathbf{S}_{\mathrm{B}}'$ are Hermitian and have the correct asymptotic behaviour: that is, in the limit of infinite separation of the fragments A and B, $\mathbf{F}_{\mathrm{A}}'$, $\mathbf{F}_{\mathrm{B}}'$ and $\mathbf{S}_{\mathrm{A}}'$, $\mathbf{S}_{\mathrm{B}}'$ become exactly the Fock and overlap matrices of the individual systems A and B.

The dimerisation energy is

$$\Delta E_{\mathrm{SCF-MI}} = E_{\mathrm{SCF-MI}}^{\mathrm{AB}} - E_{\mathrm{SCF}}^{\mathrm{A}} - E_{\mathrm{SCF}}^{\mathrm{B}} \tag{10}$$

and takes properly account of the geometry relaxation effects. Following the scheme proposed by Gerratt and Mills [32] (see also Pulay [33,34] and Yamaguchi et al. [35]) the calculation of first and second derivatives was easily implemented [10-11,36].

The version of the SCF-MI code implemented into GAMESS-US package [11,36] can perform single point conventional and direct SCF-MI energy calculation, analytic gradient, numerical Hessian evaluation and geometry optimisation; vibrational analysis is also available. Increase in complication and computation time with respect to standard SCF algorithms is minimal. The SCF-MI option is also incorporated in the particularly efficient PC GAMESS version

[37] of the GAMESS-US [36] quantum chemistry package. Another version [38] adopting the CASVB algorithm [39] is being incorporated into MOLPRO code [40].

3. Computational details

All calculations were performed *ab initio* with the SCF and SCF-MI procedures implemented in the GAMESS-US package [11,36]. Several standard split valence Cartesian basis sets (namely 3-21G, 4-31G, 6-31G and 6-31G(p,s) [36]) were employed, with all the electrons considered explicitly. The structures were fully gradient-optimised by imposing severe convergence criteria (norm of gradient less than $3*10^{-6}$ a.u.). The single bases turn out planar in all optimisations; calculations performed including polarisation functions (6-31G**) revealed a slight piramidalisation of the exocyclic nitrogen of isolated guanine confirming previous literature results [7]. It is to be noted, however, that the energy difference between this distorted minimum structure and the planar conformation is very small, around 0.35 kcal/mol. We conclude that the use of planar structures does not introduce significant errors from a numerical point of view; in addition, the question on the planarity of isolated bases remains undecided, even experimentally [7,41].

4. Calculations and results

4.1. SCF-MI/3-21G EQUILIBRIUM GEOMETRIES AND BINDING ENERGIES

To cover a sufficiently large range of DNA base pairs, the calculations were carried out for several possible combinations of the neutral major tautomers of adenine (A), cytosine (C), guanine (G), thymine (T) and uracil (U) bases. Figure 1 reports the atomic structure of DNA base pairs optimised at the SCF-MI/3-21G level. The abbreviations WC and H are used for Watson-Crick and Hoogsteen respectively. The structures of the isolated systems were found to possess Cs symmetry; the planarity of the aromatic rings of the bases was maintained upon hydrogen bond formation. Table 1 summarises the SCF-MI/3-21G interaction energies of the pairs. SCF-MI results are compared with those obtained by standard SCF/3-21G calculations. The values reported demonstrate that standard SCF approach predict too high binding energies if BSSE is not properly taken into account. The stabilisation energies are in the range -22.5/-8.0 kcal/mol. The GCWC pair with three hydrogen bonds is the most stable pair while the AA pair with two hydrogen bonds is the weakest one. SCF-MI interaction energies are consistent with the MP2 results reported by Sponer, Leszczynski and Hobza [7], confirming the electrostatic nature of the base pairs.

TABLE 1. Interaction energies (in kcal/mol) of hydrogen-bonded DNA base pairs obtained for the SCF-MI/3-21G optimised geometries (see figure 1)

Base pairs	ΔE SCF	ΔE SCF-MI	Ref. [7]
GCWC	-39.8	-22.5	-23.4
ATWC	-22.0	-9.5	-9.6
CC	-25.5	-15.6	-15.9
GCH	-35.1	-19.1	-20.6
AA	-16.9	-8.0	-8.1
GG	-37.4	-20.7	-22.5
UA	-22.0	-9.5	-
TT	-17.6	-8.9	-8.5
AG	-22.1	-10.7	-11.5
TAH	-20.2	-9.5	-10.2

Selected SCF-MI geometrical parameters for the investigated base pairs are reported in Table 2

TABLE 2. Selected optimised SCF-MI/3-21G intermolecular geometry parameters - distances in Å / angles in degrees - for the investigated base pairs (see figure 1)

Base pairs	parameters	SCF-MI	Literature (*)
GCWC	$N_2(H)...O_{2'}$	2.97/171.9	3.02/178.1
	$N_1(H)...N_{3'}$	3.04/173.8	3.04/176.1
	$O_6...(H)N_{4'}$	2.94/172.4	2.92/177.0
ATWC	$N_6(H)...O_{4'}$	3.09/170.2	3.09/172.0
	$N_{3'}(H)...N_{1'}$	3.01/176.7	2.99/178.8
CC	$N_3...(H)N_{4'}$	3.05/174.2	3.05/173.2
	$N_4(H)...N_{3'}$	3.05/174.2	3.05/173.2
GCH	$N_{1'}(H)...O_2$	2.83/177.2	2.82/175.0
	$O_{6'}...(H)N_1$	2.95/170.1	2.92/179.0
AA	$N_6(H)...N_{1'}$	3.14/179.1	3.16/179.4
	$N_1...(H)N_{6'}$	3.14/179.2	3.16/179.4
GG	$N_1(H)...O_{6'}$	2.87/177.7	2.87/178.1
	$O_6...(H)N_{1'}$	2.90/177.7	2.87/178.1
UA	$N_6(H)...O_{4'}$	3.09/170.2	-
	$N_{3'}(H)...N_{1'}$	3.01/176.7	-
TT	$O_{2'}...(H)N_3$	2.95/159.1	2.98/166.2
	$O_2...(H)N_{3'}$	2.96/158.8	2.98/166.6
AG	$O_{6'}...(H)N_6$	2.95/175.2	2.95/179.9
	$N_1...(H)N_{1'}$	3.19/176.4	3.19/179.3
TAH	$O_{4'}...(H)N_6$	3.11/171.8	3.14/170.1
	$N_{3'}(H)...N_7$	2.96/176.5	2.95/175.6

(*) The data reported here are obtained at the SCF/6-31G** level by adding the counterpoise correction and the fragment relaxation terms [7]

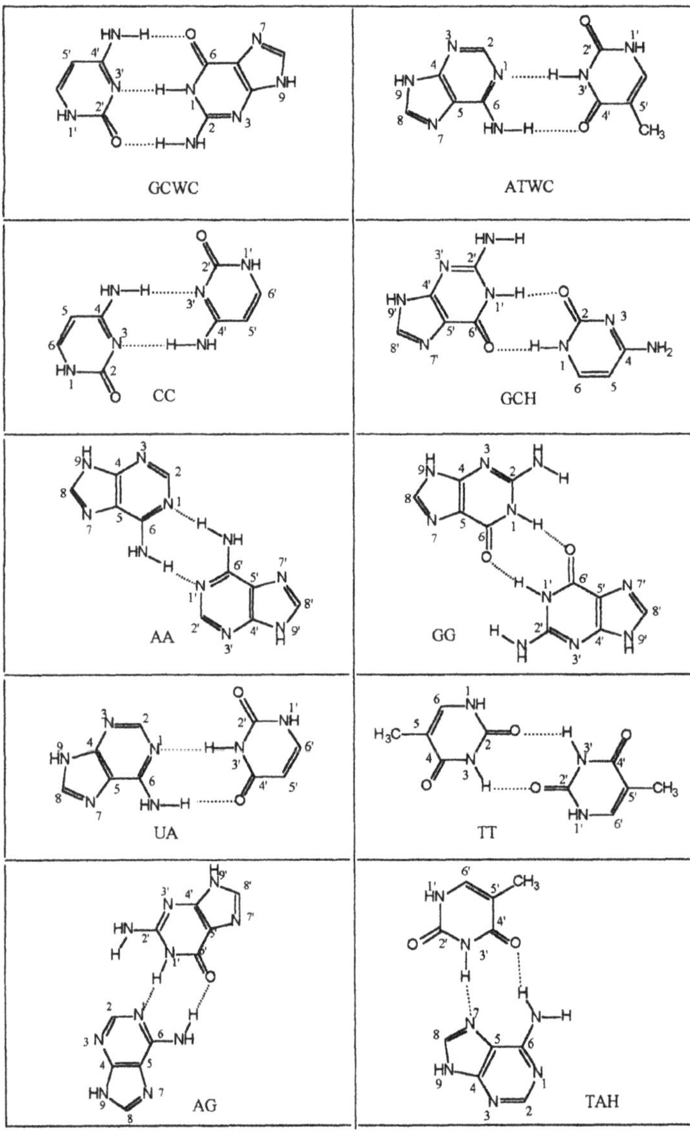

Figure 1. SCF-MI/3-21G optimised geometries of hydrogen bonded DNA base pairs. For GCWC and ATWC pairs, broken lines represent hydrogen-bond bridges.

In Tables 3 and 4, additional, extensive SCF-MI calculations on ATWC and GCWC, performed with 6-31G and 6-31G(p,s) basis sets, confirm the accuracy and reliability of this approach, both for interaction energies and geometric parameters.

TABLE 3. Selected SCF and SCF-MI geometrical parameters and binding energies of ATWC base pair for several standard basis sets

Basis set	$N_{3'}$(H)...N_1 Å	$N_{3'}$(H)...N_1 deg.	ΔE SCF-MI kcal/mol	ΔE SCF kcal/mol
3-21G	3.01	176.7	-9.50	-22.01
6-31G	3.04	176.9	-10.32	-15.19
6-31G(ps)	3.08	176.9	-9.78	-

TABLE 4. Selected SCF and SCF-MI geometrical parameters and binding energies of GCWC base pair for several standard basis sets

Basis set	N_1(H)...$N_{3'}$ Å	N_1(H)...$N_{3'}$ deg.	ΔE SCF-MI kcal/mol	ΔE SCF kcal/mol
3-21G	3.04	173.8	-22.51	-39.81
6-31G	3.09	178.4	-24.80	-31.61
6-31G(ps)	3.10	172.9	-24.56	-

4.2. POTENTIAL ENERGY CURVES

The SCF and SCF-MI potential energy curves for thymine approaching adenine was computed as a function of the distance R between the H($N_{3'}$) of thymine and N_1 of adenine. The same kind of calculations have been performed for uracil approaching adenine (with R equal to the distance between the H($N_{3'}$) of uracil and N_1 of adenine) and for guanine approaching cytosine (with R equal to the distance between $N_{3'}$ of cytosine and H(N_1) of guanine). Figures 2-4 show the effect of the basis-set quality on the potential energy surfaces.

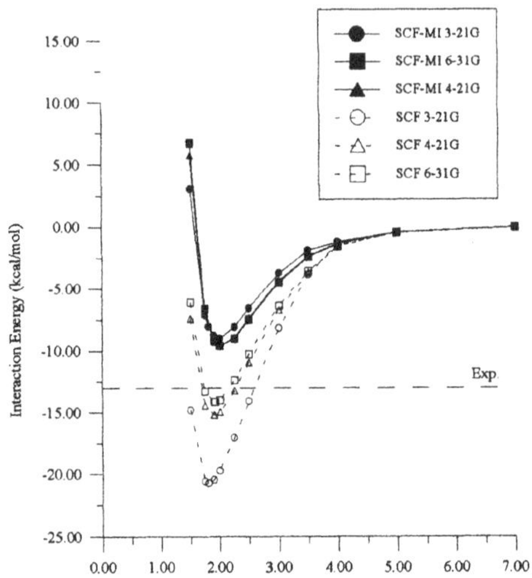

Figure 2. SCF and SCF-MI potential interaction energy surfaces for ATWC system. The standard 3-21G, 4-31G and 6-31G basis sets are used. Distances are in Å and energies in kcal/mol. The experimental interaction energy is also reported.

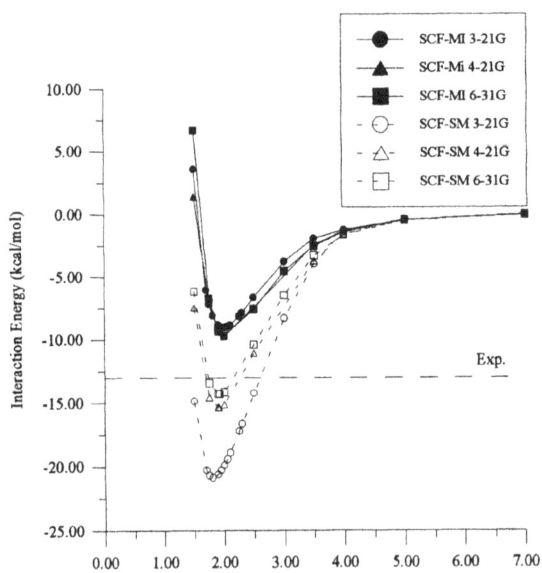

Figure 3. SCF and SCF-MI potential interaction energy surfaces for UA the system. The standard 3-21G, 4-31G and 6-31G basis sets are used. Distances are in Å and energies in kcal/mol. The experimental interaction energy is also reported.

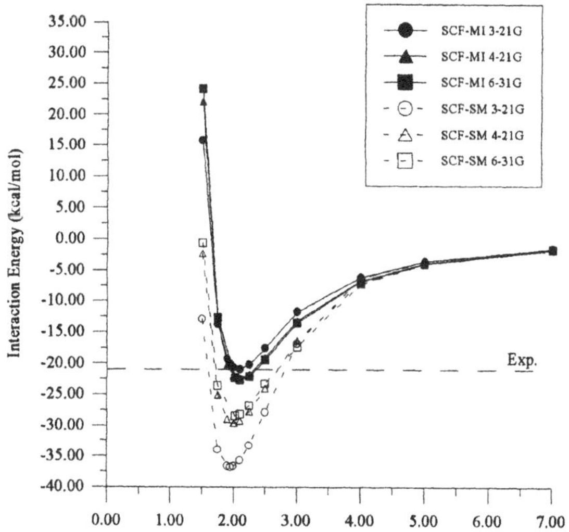

Figure 4. SCF and SCF-MI potential interaction energy surfaces for GCWC system. The standard 3-21G, 4-31G and 6-31G basis sets are used. Distances are in Å and energies in kcal/mol. The experimental interaction energy is also reported.

4.3. SCF-MI INTERACTION DENSITY

The SCF-MI orbitals and interaction density maps for ATWC and GCWC base pairs have been used to interpret the nature of the interactions involved in the hydrogen bond formation. The SCF-MI orbitals describing the electrons directly involved in the interaction are shown in figure 5 for the ATWC and in figure 6 for the GCWC base pairs and can be clearly identified. In figures 7 and 8 we report the plots of the SCF-MI interaction density. The density difference $\Delta\rho = \rho_{dimer} - \rho_{monomers}$, plotted in the plane of the molecules, shows that the effect of the hydrogen bond is not restricted to the bonding region alone. The hydrogen bond increases the polarity of the monomers involving intramolecular as well as intermolecular rearrangements: the hydrogen atom becomes more positive, while the electron density on the heteroatom increases. See [43] for a more detailed analysis of this point.

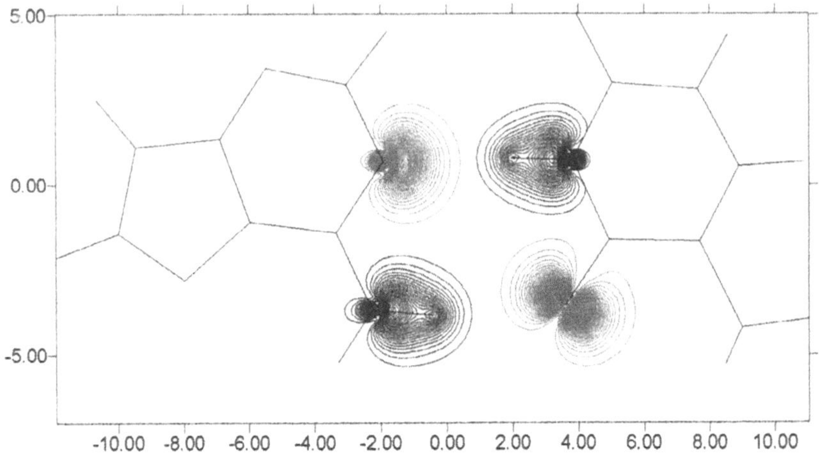

Figure 5. SCF-MI/6-31G orbitals for ATWC base pair shown as contours of $\left|\varphi_{\mu}(\mathbf{r})\right|^2$ in the plane of the interacting molecules.

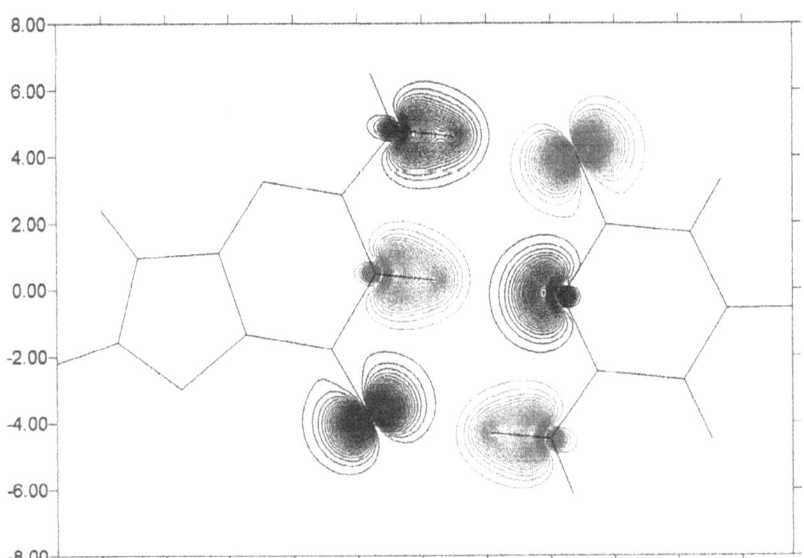

Figure 6. SCF-MI/6-31G orbitals for GCWC base pair shown as contours of $\left|\varphi_{\mu}(\mathbf{r})\right|^2$ in the plane of the interacting molecules.

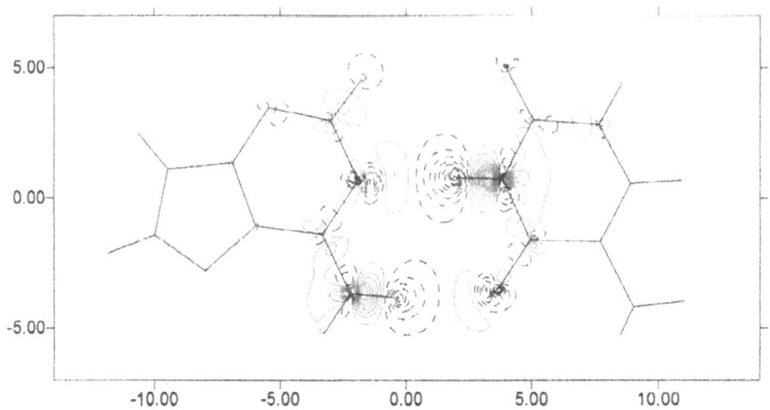

Figure 7. SCF-MI/6-31G interaction densities ($\Delta\rho=\rho_{dimer}-\rho_{monomers}$) in ATWC base pair. Dashed contours denote negative and solid contours positive values of $\Delta\rho$.

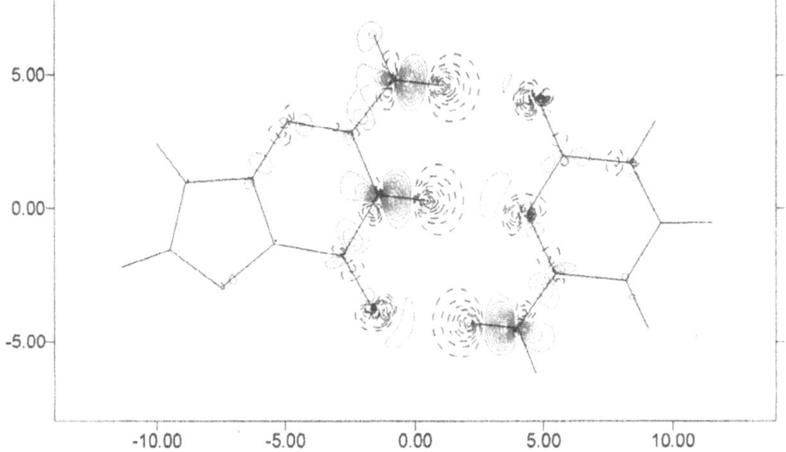

Figure 8. SCF-MI/6-31G interaction densities ($\Delta\rho=\rho_{dimer}-\rho_{monomers}$) in GCWC base pair. Dashed contours denote negative, solid contours positive values of $\Delta\rho$.

4.4. THE PROPELLER TWIST ANGLE

Single crystal X-ray analysis of nucleic acids shows that base pairs in DNA are not planar. Arnott, Dover and Wanacott [44] found that the angle of rotation of one molecular plane with respect to the other around the roll axis of the reference system - *propeller twist angle* - is 15 degrees. As hydrogen bond interaction favours planarity, packing effects probably accounts for the deviation. We have evaluated such destabilisation by the SCF-MI method and compared it with the corresponding value obtained with the standard SCF procedure.

Starting from the most stable planar dimer structure SCF-MI and SCF destabilisation energies were calculated for different values of the Arnott, Dover and Wanacott's propeller twist angle. Figures 9-11 report the results obtained with the 3-21G, 4-31G and 6-31G basis sets for ATWC, UA, and GCWC base pairs. Due to the relative low rigidity of the potentials around the minimum it is confirmed that small deviations from planarity [45] are allowed.

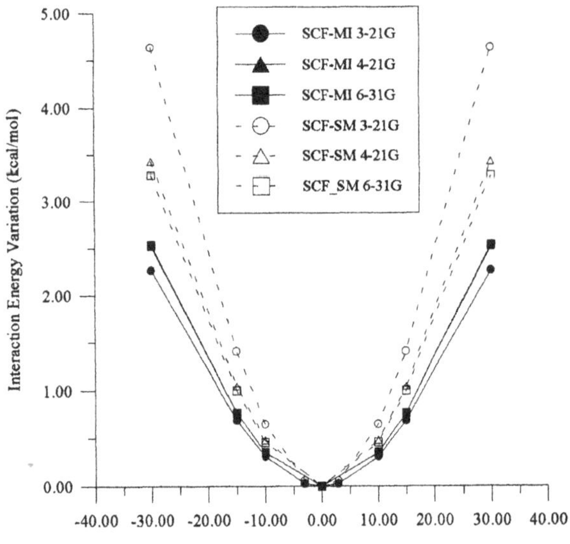

Figure 9. SCF and SCF-MI interaction energy variations of the ATWC hydrogen base pair as a function of propeller twist angle α. The standard 3-21G, 4-31G and 6-31G basis sets are used. Angles are in degrees and energies in kcal/mol.

4.5. STACKING EFFECTS: PARALLEL INTERACTION BETWEEN NUCLEIC-ACID BASES

In the double helix structure of real nucleic acids, bases are parallelly stacked, at a distance of about 3.4 Å; as a complete turn of the double helix is 34 Å long and contains 10 base pairs, each one is found to be rotated by 36 degrees around the double helix axis. The stabilisation of stacked complexes of DNA bases originates from both dispersion effects and electrostatic dipole-dipole interactions. The importance of the use of a proper geometry optimisation procedure is demonstrated by the PES reported in figure 12 where a fixed ring approach of two cytosine bases perfectly superimposed turn out fully repulsive. A proper stationary point was found only after a full optimisation of the CC stacked base pair without any constraint. The SCF-MI/3-21G equilibrium structure (see figure 13) was found at an average interring distance of 3.9 Å with an interaction energy of -3.0 kcal/mol, in accordance with literature results [7] obtained at the MP2 correlated level. It is to be noted that stacking interactions induce pyramidalisation of the ammonia groups of the bases so that the planar geometry of cytosine is deformed.

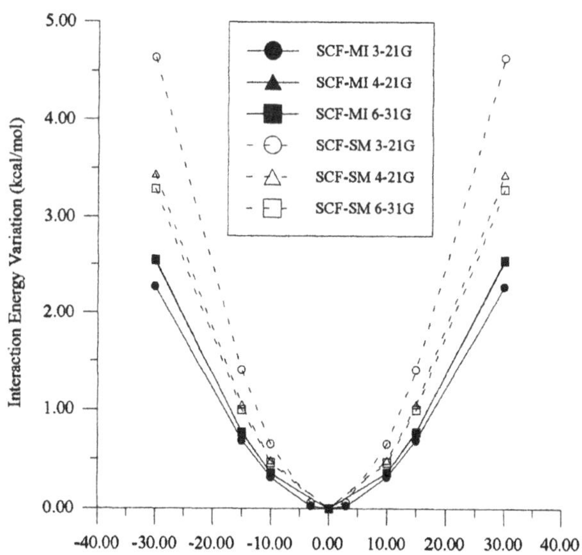

Figure 10. SCF and SCF-MI interaction energy variations of the UA hydrogen base pair as a function of propeller twist angle α. The standard 3-21G, 4-31G and 6-31G basis sets are used. Angles are in degrees and energies in kcal/mol.

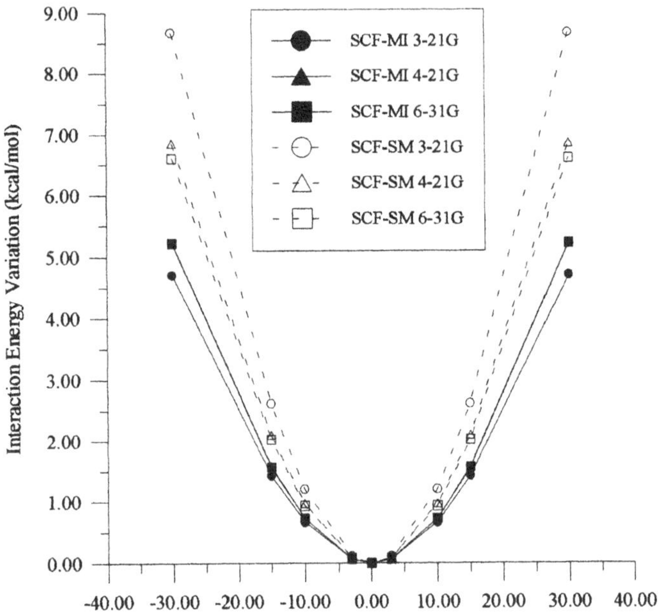

Figure 11. SCF and SCF-MI interaction energy variations of the GCWC hydrogen base pair as a function of propeller twist angle α. The standard 3-21G, 4-31G and 6-31G basis sets are used. Angles are in degrees and energies in kcal/mol.

4.6. BADER ANALYSIS AND CHARGE TRANSFER EFFECTS

The ATWC and GCWC systems in their minimum-energy planar geometries obtained with the 6-31G basis set at the SCF and SCF-MI levels (Tables 3 and 4) were analysed using Bader's procedure. The analysis of the charge density confirms that two H-bonds are formed in the ATWC and three in the GCWC system. The data regarding net atomic charges reported in Table 5 provide evidence that the SCF-MI method, while eliminating the BSSE contributions to the interaction potential, does not prevent a proper description of all main physical effects even when employing modest 6-31G basis sets.

TABLE 5. Bader's net charges on each fragment for nucleic-acid base pairs ATWC and GCWC at the equilibrium geometries. The 6-31G basis set is used

Base pairs	Bases	SCF	SCF-MI
ATWC	A	+0.021	+0.016
	T	-0.021	-0.016
GCWC	G	-0.030	-0.018
	C	+0.030	+0.018

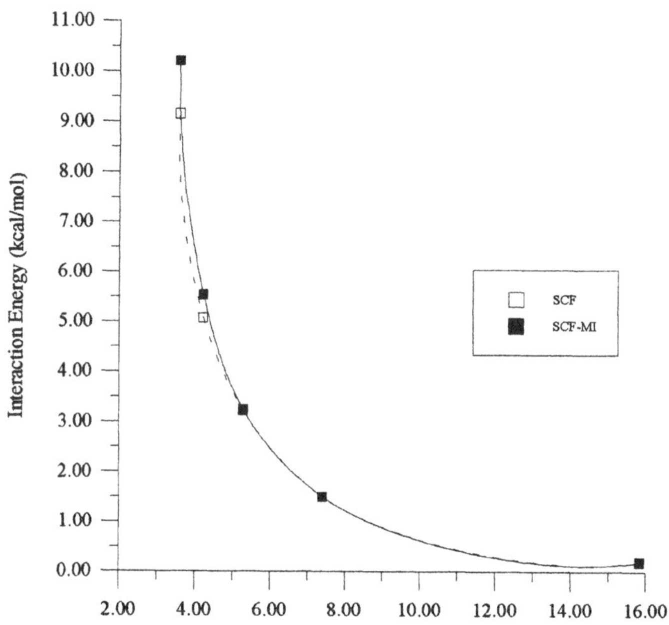

Figure 12. SCF and SCF-MI parallel interaction energies between nucleic-acid bases: Cytosine-Cytosine case. The standard 6-31G basis set is used. Interring distances (d) are in Å and energies in kcal/mol.

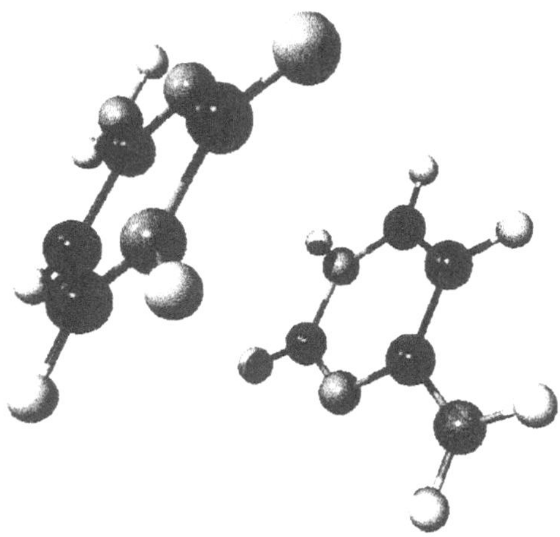

Figure 13. Optimised SCF-MI/3-21G structure of stacked CC pair.

5. Conclusions

The present paper provides the first geometry determination for several hydrogen-bonded DNA base pairs performed using an *a priori* BSSE-free gradient optimisation SCF algorithm. The effect of the overlap between the orbitals of the fragments is naturally taken into account. SCF-MI/3-21G stabilisation energies range from -22.5 to -8.0 kcal/mol, in good agreement with reliable literature theoretical values. It was gratifying to find that, for the systems studied, the *a priori* exclusion of BSSE reproduces accurate results by employing a small basis set. The hydrogen bonding PES and the propeller twist potentials are also calculated for some of the molecular complexes. The SCF-MI interaction density is used to interpret the nature of the interactions involved in hydrogen-bond formation in adenine...thymine and guanine...cytosine pairs in the Watson-Crick configuration, confirming that the structure and stabilisation of the base pairs are determined mostly by electrostatic interactions. Preliminary calculations on the stacked cytosine dimer are also reported.

It may be questioned whether the SCF-MI procedure is able to take into consideration the possibility of a charge transfer. As the basis functions of the two fragments A and B are kept strictly partitioned, there is the justifiable doubt that the electrons of one fragment cannot delocalise over the other. However, as the functions centred on fragment A have tails extending to the space of fragment B and vice versa, it is to be expected that electronic transfer will not be strictly forbidden. Bader analysis has confirmed this, demonstrating that, even in the unfavourable case of employing the 6-31G basis set, the computed total charge located on A and B is of the same order of magnitude for both the SCF and SCF-MI wave functions.

The natural development of the present work, which is in progress in our laboratory, is the investigation of the interaction of the DNA base pairs with ionic molecules or fragments that show citotoxic effects on biological systems. In particular, we have studied the coordination of metal cations to the N7 and O6 sites GCWC pair that can generate a non-Watson-Crick hydrogen-bonding pattern [46]. Inclusion of dispersion terms will be attempted along the BSSE-free procedure already tested in the case of water properties [47], with the aim of verifying the present results when correlation is included.

References

1. a) *Advances in Chemical Physics*, Ed. J.O. Hirschfelder, Interscience Publisher, New York (1967), vol. 12; b) *Discuss. Faraday Soc.*, 40 (1965), 62 (1977), 73 (1982); c) G.C. Maitland, M. Rigby, E.B. Smith and W.A. Wakeham, 'Intermolecular Forces', Claredon Press, Oxford (1997); e) Chem. Rev. 88 (1988), 94 (1994); f) S. Scheiner Ed: 'Molecular Interactions, from van der Waals to strongly Bound Complexes', Wiley, Chichester (1997); g) A.J. Stone, "The theory of Intermolecular Forces', Claredon Press, Oxford 1996).
2. P. Hobza, Prog. Phys. Chem. 93, 257 (1997).
3. a) P. Hobza and R. Zahradnik, 'Intermolecular Complexes, Elsevier, Amsterdam (1988); b) P. Hobza and R. Zahradnik, Chem. Rev. 88, 871 (1988); c) S. Scheiner: 'Calculating the Properties of Hydrogen Bonds by *ab initio* Methods', in K.B. Lipkowitz and D.B. Boyd (Eds.), Reviews in Computational Chemistry II, VCH Publisher, New York, 165 (1991); b) S. Scheiner: Annu. Rev. Phys. Chem. 45, 23 (1994); c) A.W. Castleman Jr. and P. Hobza, Guest Editors, "van der Waals Molecules II", Chem. Rev. 94 (1994); d) S. Scheiner: 'Molecular Orbital Theory of Hydrogen Bonded Systems and Proton Transfer Reactions', Oxford University Press, New York (1997); e) D. Hadzi, Theoretical Treatments of Hydrogen Bonding, Wiley Research Series in Theoretical Chemistry, John Wiley & Sons Ltd Publishers, Chichester-England (1997).
4. a) I. K. Yanson, A. B. Teplitsky, and L. F. Sukhodub, Biopolymers 18, 1149 (1970); b) M. Dey, F. Moritz, J. Grotemeyer and E. W. Schloag, J. Am. Chem. Soc. 116, 9211 (1994); c) W. Nerdal, D. Hare, and B. Reid, Nuc. Acid Res. 28, 10008 (1989).
5. R. Rein in *Perspectives in Quantum Chemistry and Biochemistry Vol. II.* "Intermolecular Interactions: from Diatomics to Biopolymers", Ed. B. Pullman, Wiley (1978).
6. a) P. Hobza and C. Sandorfy, J. Am. Chem. Soc. 109, 1302 (1987); b) P. Hobza, J. Sponer, and M. Polasek, J. Am. Chem. Soc. 1117, 792 (1995).
7. a) J. Sponer, J. Leszczynski, and P. Hobza, J. Phys. Chem. 100, 1965 (1996); b) J. Sponer, J. Leszczynski, and P. Hobza, J. Phys. Chem. 100, 5590 (1996).
8. a) J. Sponer, J. Leszczynski, and P. Hobza, J. Biomol. Struct. Dyn. 14, 117 (1996); b) P. Hobza, J. Sponer, Chem Phys. Lett. 288, 7 (1997); c) P. Hobza, Prog. Phys. Chem. 93, 257

(1997); d) J. Sponer and P. Hobza, Encyclopedia of Computational Chemistry, P. v R. Schleyer, Editor in Chief. Wiley, Chirchester (1998) pag. 777.
9. E. Gianinetti, M. Raimondi and E. Tornaghi, Int. J. Quantum Chem. 60, 157 (1996).
10. Famulari PhD Thesis, University of Milan (1997).
11. Famulari, E. Gianinetti, M. Raimondi and M. Sironi, Int. J. Quantum Chem. 69, 151 (1998).
12. Famulari, E. Gianinetti, M. Raimondi, M. Sironi and I. Vandoni, Theor. Chim. Acc. 99, 358-365 (1998). *Electronic version*: DOI 10.1007/s002149800m20.
13. E. Gianinetti, I. Vandoni, A. Famulari and M. Raimondi, Adv. Quantum Chem. 31, 251 (1998).
14. E. Clementi, J. Chem. Phys. 46, 3851 (1967); b) N. R. Kestner, J. Chem. Phys. 48, 252 (1968).
15. a) J. H. van Lenthe, J. C. G. M. van Duijneveldt-van de Rijdt, and D. B. van Duijneveldt, Adv. Chem. Phys. 69, 521 (1987); b) F. B. van Duijneveldt, J. G. C. M. van Duijneveldt-van de Rijdt, and J. H. van Lenthe, Chem. Rev. 94, 1873 (1994).
16. a) I. Mayer, P. R. Surjan, In. J. Quantum Chem. 36, 225 (1989); b) J. Almlöf, P. R. Taylor, J. Chem. Phys. 86, 553 (1987); c) W. Saebø, P. Pulay, J. Chem. Phys. 105, 1884 (1988).
17. a) S.F. Boys and F. Bernardi, Mol. Phys. 19, 553 (1970); b) A. Meunier, B. Levy, G. Berthier, Theor. Chim. Acta 29, 49 (1973); c) H. Janses, P. Ross, Chem. Phys. Lett. 3, 40 (1969).
18. G. Karlström and A.J. Sadlej, Theor. Chim. Acta 61, 1 (1982).
19. Z. Latajka, S. Scheiner, Chem. Phys. Lett. 140, 338 (1987).
20. D.W. Schwenke and D.G. Truhlar, J. Chem. Phys. 82, 2418 (1985).
21. S. K. Louishin, S. Liu, C. E. Dykstra, J. Chem. Phys. 84, 2720 (1986).
22. M. Gutowski, J. H. van Lenthe, J. Verbeek, F. B. van Duijneveldt, Chem. Phys. Lett. 124, 370 (1986).
23. M.M. Szczesniak, S. Scheiner, J. Chem. Phys. 84, 6328 (1986).
24. Liou, A. D. McLean, J. Chem. Phys. 91, 2348 (1989).
25. F. Tao, Y. Pan, J. Phys. Chem. 95, 3582 (1991).
26. G. Chalasinski and M.M. Szczesniak, Chem. Rev. 94 (1994) 1723.
27. S.S. Xantheas, J. Chem. Phys. 104, 8821 (1996).
28. a) S. Simon, M. Duran, J.J. Dannenberg, J. Chem. Phys. 105, 11024 (1996); b) S. Simon, M. Duran, J.J. Dannenberg, J. Phys. Chem. A 103, 1640 (1999).
29. P. Hobza, and Z. Havles, Theor. Chim. Acc. 99, 372 (1998). *Electronic version*: DOI 10.1007 /s002149800m21.
30. a) J. E. Del Bene, and H. D. Mettee, J. Phys. Chem. 95, 5387 (1991); b) Y. Bouteiller, H. Behrouz, J. Chem. Phys.96, 6033 (1992); c) J. M. Leclercq, M. Allavena, Y. Bouteiller, J. Chem. Phys. 78, 4606 (1983).
31. W. McCarthy, A. M. Plokhotnichenko, E. D. Radchenko, J. Smets, D. M. A. Smith, S. G. Stepanian, L. Adamowicz, J. Phys. Chem. A 101, 7208 (1997).
32. J. Gerratt and I. M. Mills, J. Chem. Phys. 49, 1719 (1968).
33. P. Pulay, Mol. Phys. 17, 197 (1969).
34. P. Pulay, Adv. Chem. Phys. 69, 241 (1987).
35. Y. Yamaguchi, Y. Osamura, J. D. Goddard and H. F. Schaefer, in A New Dimension to Quantum Chemistry: Analytic Derivative Methods in Ab Initio Molecular Electronic Structure Theory, Oxford University Press, Oxford, UK (1994).
36. a) M. W. Schmidt, K. K. Baldridge, J. A. Boatz, S. T. Elbert, M. S. Gordon, J. Jensen, S. Koseki, N. Matsunaga, K. A. Nguyen, S. J. Su, T. L. Windus, M. Dupuis, J. A. Montgomery, J. Comput. Chem. 14, 1347 (1993); b) GAMESS User's Manual (1997).
37. A. Granowsky, www http://classic.chem.msu.su/gran/gamess/index.html
38. T. Thorsteinsonn, A. Famulari and M. Raimondi (1999). Manuscript in preparation.
39. a) T. Thorsteinsonn, D.L. Cooper, J. Gerratt, P. B. Karadakov and M. Raimondi, Theor. Chim. Acta 93, 343 (1996); b) T. Thorsteinsonn and D.L. Cooper, Theor. Chim. Acta 94, 233 (1996); c) T. Thorsteinsonn, D.L. Cooper, J. Gerratt and M. Raimondi, Theor. Chim. Acta 95,

131 (1997); d) D.L. Cooper, T. Thorsteinsonn and J. Gerratt, Adv. Quantum Chem. 32, 51 (1998).

40. MOLPRO is a package of programs written by H.-J.Werner and P.J Knowles, with contributions from R. D. Amos, A. Berning, D. L. Cooper, M. J. O. Deegan, A. J. Dobbyn, F. Eckert, C. Hampel, T. Leininger, R. Lindh, A. W. Lloyd, W. Meyer, M. E. Mura, A. Nicklass, P. Palmieri, K. Peterson, R. Pitzer, P. Pulay, G. Rauhut, M. Schütz, H. Stoll, A. J. Stone and T. Thorsteinsson, University of Birmingham (1998).
41. J. Šponer, J. Leszczynski, V. Vetterl, P. Hobza, J. Biomol. Struct. Dyn. 13, 695 (1996).
42. a) J. D. Watson and F. H. Crick, Nature 171, 757 (1953); b) J. D. Watson and F. H. Crick, Nature 171, 964 (1953).
43. a) C. Gatti and A. Famulari, «Interaction Energies and Densities. A Quantum Theory of Atom in Molecules insight on the Effect of Basis-Set Superposition Error Removal». Kluwer book series, Understanding Chemical Reactivity: Electron, Spin and Momentum Densities and Chemical Reactivity Vol. 2 (1998). P.G. Mezey and B. Robertson Editors. In press; b) M. Raimondi, A. Famulari and E. Gianinetti, Int. J. Quantum. Chem. 74, 259 (1999).
44. S. Arnott, S. D. Dover, and A. J. Wanacott, Acta Cristallogr. B25, 2192 (1969).
45. M. Aida, J. Comput. Chem. 9, 362 (1988).
46. a) M. Raimondi, A. Famulari, E. Gianinetti, M. Sironi, and F. Moroni, "Modification of Roothaan equations for the ab-initio calculation of Interactions in Large Molecular Systems in the absence of Basis-Set Superposition Error", Large-Scale Scientific Computations" (WLSSC'-99). Ed. J. Leszczynski, 1999. In press; b) A. Pelmenshchikov, I.L. Zilberberg, J. Leszczynski, A. Famulari, M. Sironi, and M. Raimondi, Chem. Phys. Lett. (1999). In press; c) A. Famulari, M. Raimondi, M. Sironi, Computer & Chemistry (1999). In press.
47. a) A. Famulari, R. Specchio, M. Sironi and M. Raimondi, J. Chem. Phys. 108, 3296 (1998); b) A. Famulari, M. Raimondi, M. Sironi and E. Gianinetti, Chem. Phys. 232, 289 (1998); c) R. Specchio, A. Famulari, M. Sironi, M. Raimondi, J. Chem. Phys. 111, 6204 (1999).

PROTON TRANSFER AND NON-DYNAMICAL CORRELATION ENERGY IN MODEL MOLECULAR SYSTEMS

HENRYK CHOJNACKI
Institute of Physical and Theoretical Chemistry,
Wroclaw University of Technology,
Wyb. Wyspianskiego 27, PL 50-370 Wroclaw, Poland

Abstract. Hypersurfaces for proton transfer processes have been studied at the non-empirical level for model molecules, including hydrogen-bonded systems. Non-dynamical correlation energies were evaluated and analyzed with different basis sets. It has been shown that the barrier heights calculated with CASPT2 are not very different from those of single-determinantal methods, e.g. MP2. Therefore we conclude that hydrogen-bonded systems may be treated at a good accuracy with single-determinantal wavefunctions. The inclusion of zero-point energy corrections and crystal-field effects enables for a reasonable interpretation of the low potential barriers observed in NMR experimental studies on some molecular crystals involving carboxylic dimers.

1. Introduction

Proton transfer reactions are the simplest but very important in many chemical problems as well as in some biological processes. It appears that the low-barrier hydrogen bond (LBHB) for this displacement may play a fundamental role in stabilizing intermediates in enzymatic reactions and in energy lowering of transition states [1]. There is considerable evidence that a LBHB may be important in the reaction catalyzed by Δ^5-3-ketosteroid isomerase [2]. Recent computational and gas phase experimental studies [3] have also shown that LBHB can exist in gas phase systems. On the other hand, the multiple proton transfer seems to play an important role in quantum chemical interpretation at molecular level of some biological processes like mutations, aging and cancerogenic action [4]. As the evaluation of the potential energy surface with good accuracy is of great importance for dynamics and interpretation of the proton transfer mechanism, these processes have recently been intensively studied in many groups, by using both semi-empirical and non-empirical quantum chemical methods [5-9]. Unfortunately, in spite of recent progress in computational chemistry and more rigorous

J. Maruani et al. (eds.), New Trends in Quantum Systems in Chemistry and Physics, Volume 1, 335–344.

treatment of correlation energy corrections, reliable results are still very difficult to obtain by quantum mechanical calculations on hydrogen-bonded systems. The aim of our calculations is to improve the accuracy in the evaluation of potential energy surfaces and to estimate the role of basis set, geometry optimization and correlation energy. It should be emphasized that an appropriate accuracy in the evaluation of potential hypersurfaces is of crucial importance in quantum chemical studies of the mechanism and dynamics of proton transfer.

2. Basis-set problem

It was found that the lowering of the potential barrier for the proton transfer is dependent on the correlation energy taken into account in the calculation and its some limiting value is reached when more than 60 % of this energy is taken into account [10]. On the other hand, experimental geometry of the formic acid [11] and other carboxylic acids may be well reproduced [12] within the 6-31G** basis at the MP2 correlation level (Table 1) where theoretical results concerning hydrogen bond length very well resemble those of experimental ones.

TABLE 1. Hydrogen bond length [Å] for the formic acid dimer within the different basis sets evaluated at the MP2 level with full geometry optimization

Basis set	Hydrogen bond length
6-31G*	2.704
6-311G**	2.704
CBS	2.670

It should be noted that not so good results were obtained even with the complete basis set (CBS) [13]. In this situation we decided to perform extensive studies on the proton transfer for a number of carboxylic acid dimers within the 6-31G** basis set at the second-order Moeller-Plesset level.

The effect of basis set superposition error (BSSE) in the proton displacement potential has been studied by Latajka et al. [14]. It was concluded that within the counterpoise scheme, the BSSE is comparable for the endpoint and midpoint of the proton position and therefore has only a negligible effect upon the barrier to proton movement.

3. Non-dynamical correlation energy

According to the generally accepted Löwdin's definition, the electron correlation energy E_{CORR} is the difference between the exact non-relativistic energy eigenvalue of the electronic Schrödinger equation E_{EXACT} and the basis limit energy of the single configuration state function approximation, i.e. the Hartree-Fock energy E_{HF}:

$$E_{CORR} = E_{EXACT} - E_{HF}$$

While this definition is satisfactory near molecular equilibrium, it becomes less adequate as molecular bonds are stretched when a single configuration is not sufficient.

TABLE 2. Non-dynamical correlation energy (a.u.) for the water monomer in different basis sets

Basis set	No. of functions	Non-dynamical correlation energy
DZV	13	-0.05440
TZV	20	-0.05509
TZV+P	60	-0.05362
CBS	80	-0.05420
CBS+F	100	-0.05355

Non-dynamical correlation is associated with the lowering of the molecular energy as a result of interaction of the Hartree-Fock configuration with low-lying excited states. It may be evaluated from the relationship

$$E_{ND} = E_{CASSCF} - E_{HF}$$

where E_{CASSCF} is the total energy with the relevant multireference wavefunction. An unambiguous definition is to include in the secular matrix CFS's which arise from all possible occupancies of the valence orbitals. The number of such orbitals is the same as that of basis functions within a minimal basis set. However, in order to obtain a unique definition of the non-dynamical correlation energy, the orbitals should be optimized to self-consistency [15].

Non-dynamical correlation energies for the water monomer, calculated according to the above-mentioned definition, are given in Table 2 for different basis sets. All atomic orbitals are taken into account without freezing. It appears that, in general, the respective values are not strongly dependent on the basis choice and, therefore, it is difficult to estimate the basis set quality in that way.

TABLE 3. Non-dynamical correlation energy (a.u.) for the two extremal protons positions within the hydrogen bond

System	Basis set	Equilibrium	Middle
HF_2^-	(TZV)	-0.0221147	-0.0275612
$(H_2O)_2$	(Duijneveldt)	-0.0528120	-0.0836097
$H_5O_2^+$	(TZV)	-0.1068028	-0.1016403
$(HF)_3$	(TZV)	-0.0335594	-0.0355322
$(HCOOH)_2$	(DZV)	-0.0955092	-0.0850460
NH_4^+	(CBS)	-0.3777727	-0.1306017

4. Model molecular systems with possible proton transfer

In *Figure 1*, there are depicted the model hydrogen-bonded systems for which the calculations of potential hypersurfaces for proton transfer have been carried out. In the case of the formic acid dimer, the reactant C_{2h} is transformed to its symmetry related product via a transition state D_{2h} (*Figure 2*). In this case, it is assumed that in the gaseous state the hydrogen-bond protons displace between the two symmetrical potential minima of the equivalent tautomers. In condensed phases, however, because of interactions with the environment, the initial and final states may be trapped in one configuration [16]. Thus, the role of crystal field in proton transfer reactions may be important, but it is still not clearly understood. In this situation, a comparison of the potential energy hypersurfaces for isolated gaseous dimers with those for the crystal lattice may enable deeper understanding of the proton transfer mechanism.

In all cases full geometry optimization has been carried out, at the MP2 level, for both the equilibrium and the transition states. It is worth noticing that the optimization essentially influences the final results of the calculations.

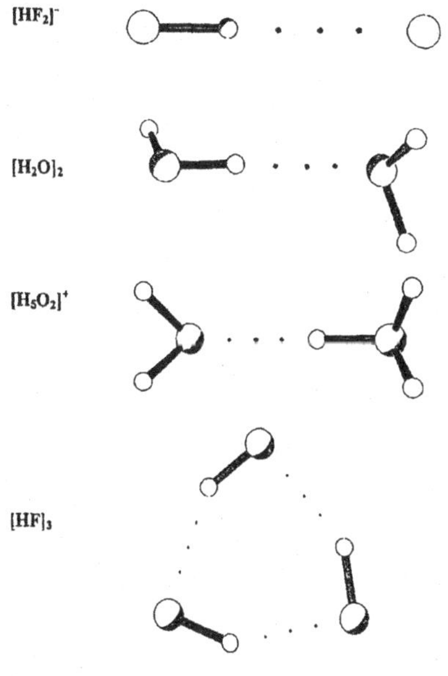

Figure 1. Geometry of model hydrogen-bonded systems: the dotted lines denote the respective hydrogen bonds.

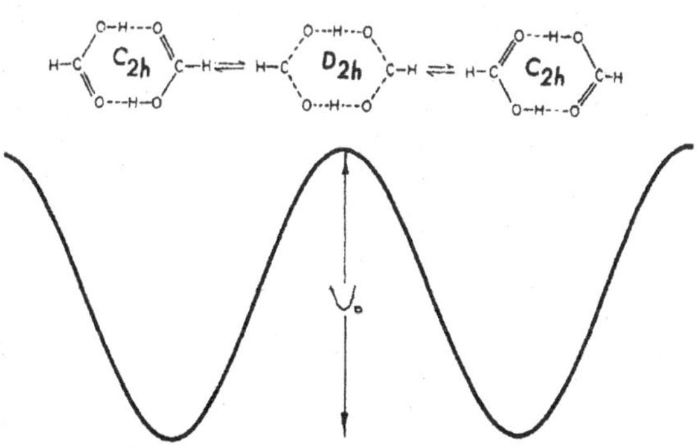

Figure 2. Stationary points: C_{2V} (equilibrium geometry) and D_{2h} (transition state) on the potential energy hypersurface for synchronous double proton transfer in the formic acid dimer. The potential barrier height is denoted by V_o.

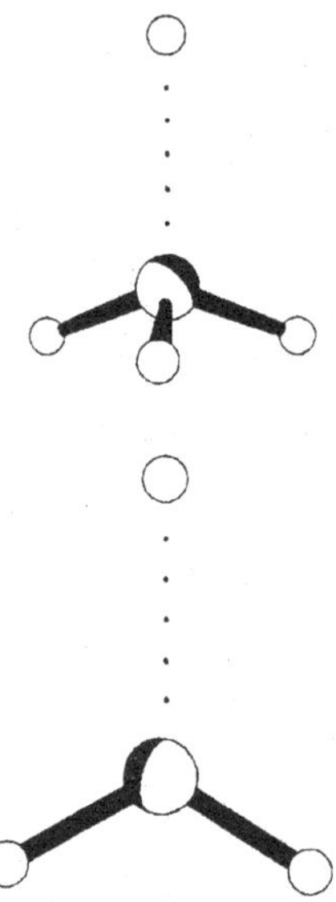

Figure 3. Geometry of dissociating H_3O^+ and NH_4^+ systems. The dotted lines denote the interatomic O ... O and N ... N distances, respectively.

5. Results and Conclusions

For the above-described, model hydrogen-bonded systems (*Figure 1*), all CAS-SCF and CASPT2 calculations were performed with frozen 1s and 2s orbitals. The non-dynamical correlation energies were evaluated for the two extremal, i.e. equilibrium and middle, proton positions of the model systems within the hydrogen bond (Table 3).

It is only in the NH_4^+ case that the non-dynamical correlation energy value is much greater than those for hydrogen-bonded systems. This fact is in agreement with the Hartree-Fock potential curve behaviour when stretching the N...H bond (*Figure 4*). The dependence of the dynamical and non-dynamical correlation energies for the H_3O^+ cation is shown in *Figure 5*, where the latter is rather small for the equilibrium proton position and much larger for shorter and longer O...H distances.

As non-dynamical correlation energies are rather small for the equilibrium geometry of the systems (*Figure 5*), the difference in these correlation energies for the two extremal proton positions is not very large, except for NH_4^+ where this quantity assumes its largest value. This result, given in Table 4, is in agreement with our previous calculations, which showed that this system is a typical multiconfigurational one [17].

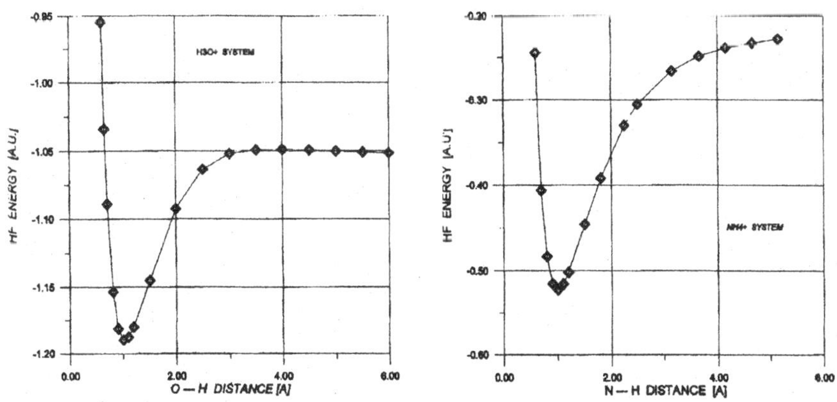

Figure 4. Dependence of the potential energy of the H_3O^+ system at the Hartree-Fock level (E_{HF}+75.0) [a.u.] on the O ... H distance [A] (to the left) and for the NH_4^+ cation (E_{HF}+56.0) [a.u] on the N ... H distance [A] (to the right) within the aug-cc-pVTZ basis set.

TABLE 4. Hydrogen bond lengths [A] and potential barriers [kcal/mole] for the double proton transfer in carboxylic acid dimers within the 6-31G** basis set without (BH) and with (BH-ZPE) zero-point energy corrections (full geometry optimization at the MP2 level)

Dimer	HB length	BH	BH-ZPE
Formic acid	2.704	8.08	4.62
Acetic acid	2.694	9.43	2.76
Oxalic acid	2.674	8.98	2.51
Malonic acid	2.700	9.85	1.41
Benzoic acid	2.675	7.07	1.10
o-Cl-Benzoic acid	2.682	14.98	8.56

On the other hand, this means that for hydrogen-bonded systems calculations with one-determinantal functions may be reliable. In fact, there is reasonable agreement of potential barriers for proton transfer between MP2 and CASPT2 calculations (Table 5). Thus, the ground state of hydrogen-bonded systems may be, with good accuracy, considered as single-determinantal. These results are consistent with Roszak et al. calculations [18] for the $[H_3N...H...OH_2]^+$ cation. However, this is not the case when the improper dissociation limit takes place, e.g. for NH_4^+, characteristic for computational schemes based on the restricted Hartree-Fock wavefunction.

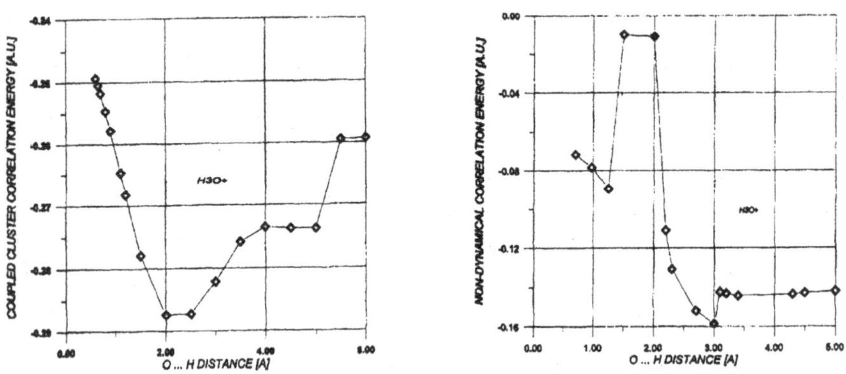

Figure 5. Dependence of the CCSD(T) correlation energy (left side) and non-dynamical correlation energy (right side) on the O ... H distance of the H_3O^+ system within the aug-cc-pVTZ basis set.

With the above encouraging results we extended our calculations to more complicated systems. In the case of the simultaneous double proton transfer the potential energy barrier for the gaseous dimeric system of the malonic acid calculated within the 6-31G** basis set at the MP2 correlation level amounts to 9.85 kcal/mole [19].

The difference in the zero-point energies resulting from our calculations is of the order of 8.44 kcal/mole, lowering the potential barriers by this value. With this correction the potential barrier for the synchronous double proton transfer in the malonic acid crystal was found to be 2.21 kcal/mole.

The role of crystal lattice effects in molecular lattices is still a moot point. In the case of the malonic acid dimer the simulation of the surrounding effects have been taken into account by using monopole atomic charges taken from ab initio calculations. It should be noted that the estimated field effect on the potential barrier in case of the simultaneous double proton transfer is of the order of 0.80 kcal/mole (lowering the barrier). Thus, the final barrier for the above-mentioned mechanism is 1.41 kcal/mole, respectively. Therefore, taking into account the zero-point energy, it shows that the barrier height of the malonic acid dimer is quite close to the experimental result of 1.33 kcal/mole [20].

TABLE 5. Potential barriers [kcal/mole] for the proton transfer in model systems calculated at the MP2 level and with the CASPT2 method

System	MP2	CASPT2
HF_2^-	21.41	22.19
$(H_2O)_2$	46.12	51.27
$(HF)_3$	29.19	31.11
$H_5O_2^+$	11.11	8.87
Formic acid	8.08	8.02

The following final conclusions can be drawn from our calculations:

- An important influence of correlation effects is documented and the most accurate results are obtained by using multiconfigurational wavefunctions (coupled-cluster methods lead to correct results as well).
- An estimation of the non-dynamical correlation energy may be a test for the quality of proton-transfer quantum-chemical calculations.
- The ground state of simple hydrogen-bonded systems is approximately single-determinantal.
- Results of DFT calculations depend essentially on the functional used. In general, they give a qualitatively correct potential energy hypersurface. However, they underestimate the proton transfer barrier height [21].
- Except for basis set and correlation energy, full geometry optimization is essential in potential energy hypersurface calculations.

The numerical calculations were performed using the GAMESS package.

Acknowledgments

This work has been sponsored by the Research State Committee (KBN) under Contracts SPUB390017 and 3TO9A06416 and in part by Wroclaw University of Technology as well as Wroclaw Networking and Supercomputing Center.

References

1. Frey, P.A., Whitt, S.A. and Tobin, J.B. (1994), A low-barrier hydrogen bond in the catalytic triad of serine proteases, *Science*, **264**, 1927-1930; Warshel, A. and Papazyan, A. (1995) On low-barrier hydrogen bonds and enzyme catalysis, **269**, 102-103.

2. Zhao, Q. Abeygunawardana, C., Talalay, P. and Mildvan, A. (1996) Low barrier in hydrogen-bonded systems, *Proc. Natl. Acad. Sci.*, **93**, 8220-8225.
3. Garcia-Viloca, M., Gonzales-Lafont, A. and Lluch, J. M. (1997) Theoretical study of the low-barrier hydrogen bond in the maleate anion in the gas phase. Comparison with normal hydrogen bonds, *J. Am. Chem. Soc.*, **119**, 1081-1086.
4. Löwdin, P. O. (1965) Quantum genetics and the aperiodic solid. Some aspects on the biological problems of heredity, mutations, aging, and tumors in view of the quantum theory of the DNA molecule, Adv. Quantum Chem., Academic Press, New York, Vol. 2, pp. 215-360.
5. Hayashi, S. Umemura, J. Kato, S. and Morokuma, K. (1980) Ab inito molecular orbital study on the formic acid dimer, *J. Phys. Chem.*, **88**, 1330-1334.
6. Shida, N. Barbara, P.F. and Almloef, J. (1991) A reactive surface Hamiltonian treatment of the double proton transfer of formic acid dimer, *J. Chem. Phys.*, **94**, 3633-3643.
7. Chojnacki, H., Lipi□ski, J. and Sokalski. W. A. (1981) Potential energy curves in complementary bases and model hydrogen-bonded systems, *Int. J. Quantum Chem.*, **19**, 339-346.
8. Kim, Y. (1996) Non-empirical quantum chemical studies on the proton transfer processes, *J. Am. Chem. Soc.*: **118**, 1522-1529.
9. Chojnacki, H. and Pyka, M.J. (1990*) Modelling of Molecular Structures and Properties*, Ed. J.-L. Rivail, Elsevier, Amsterdam 1990, pp. 351-358.
10. Chojnacki, H. (1997) Correlation effects in the proton transfer of the $[FHF]^-$ system, *J. Mol. Struct.*, **404**, 83-85.
11. Almenningen, A., Bastiansen, O. and Motzfeldt, T. (1970) A study of the influence of deuterium substitution on the hydrogen bond of dimeric formic acid, *Acta Chem. Scan.*, **24**, 747-748.
12. Chojnacki, H. (1999) Quantum chemical studies of the double proton transfer in carboxylic acid dimers, in preparation.
13. Petersson, G.A., Tenfeldt, T.G., and Ochterski, J.W. (1994) A complete basis set model chemistry. IV. An improved atomic pair natural orbital method, *J. Chem. Phys.*, **101**, 5900-5909.
14. Latajka, Z., Scheiner, S. and Cha□asi□ski, G. (1992) Basis set superposition error in proton transfer potentials, *Chem. Phys. Lett.*, **196**, 384-389.
15. Mok, D.K.W., Neumann, R. and Handy, N.C. (1996) Dynamical and nondynamical correlation, *J. Phys. Chem.*, **100**, 6225-6230.
16. Roszak, S. and Chojnacki, H. (1998) The performance of the density functional theory on reaction pathways requiring the multideterminantal description, *Comput. Chem.*, **22**, 3-6.
17. Chojnacki, H. (1998) Quantum chemical studies of the double proton transfer in oxalic acid dimer, *Polish J. Chem.*, **72**, 421-425.
18. Roszak, S., Kaldor, U., Chapman, D.A. and Kaufman, J.J. (1992) Ab initio multireference configuration interaction and coupled cluster studies of potential surfaces for proton transfer in (H_3N--H--OH_2), *J. Phys. Chem.*, 96, 2123-2129.
19. Chojnacki, H. (1999) Quantum chemical studies of the double proton transfer in malonic acid, unpublished.
20. Idziak, S. and Pislewski, N. (1987) An NMR relaxation study on the hydrogen dynamics in malonic acid, *Chem. Phys.*, **111**, 439-443.
21. Sadhukhan, S., Munoz, D., Adamo, C. and Scuseria, G.E. (1999) Predicting proton transfer barriers with density functional methods, *Chem. Phys. Lett.*, **306**, 83-87.

Part V

Nuclear Motion

LARGE AMPLITUDE MOTIONS IN ELECTRONICALLY EXCITED STATES: A STUDY OF THE S_1 EXCITED STATE OF FORMIC ACID

LEANNE M. BEATY-TRAVIS[1], DAVID C. MOULE[1]
CAMELIA MUÑOZ-CARO[2] AND ALFONSO NIÑO[2]

*1) Department of Chemistry, Brock University,
St. Catharines, Ontario, L2S3A1 Canada
2) E.U. Informática de Ciudad Real,
Universidad de Castilla la Mancha,
Ronda de Calatrava s/n 13071, Ciudad Real, Spain*

Abstract. The $S_1 \leftarrow S_0$ electronic band system in the formic acid was simulated from RHF / UHF *ab initio* calculations for the two electronic states. The torsion-wagging energy levels were evaluated by the variational method using free-rotor basis functions for torsional coordinates and harmonic-oscillator basis functions for wagging coordinates. A comparison of the calculated band spectrum to the jet-cooled excitation spectrum allowed for the assignments of a number of clearly defined bands. Along with these simulations of the overall spectrum there is also the prediction that the individual bands contain a complex mixture of rotational hybrid bands. For the 0^0 origin band the calculations predict that there should be three components within the band cluster. The allowed component of the electronic transition attaches to the 0^+, v=0 zero-point level of the S_1 state as a *c*-type band, and is referred to as the Franck-Condon component, whereas the electronically forbidden but vibronically allowed *a* / *b*-type Herzberg-Teller bands terminate on the 0^- first excited level of the torsion-wagging manifold. The calculated separation between these bands is very small, 0.00 cm^{-1}. Relative intensities of the *a* / *b* / *c* components within the 0^0 origin band are predicted to appear with ratios of about 0.02 / 1.24 / 0.67.

1. Introduction

In general, higher electronic states of molecules exhibit greater structural flexibility than do the structures of the corresponding ground electronic states. This nonrigidity is a direct consequence of the excitation process, whereby an electron is lifted from a bonding or nonbonding orbital to a molecular orbital that is ostensibly antibonding. Thus, the $n \rightarrow \pi^*$ excitation process in the carbonyl chromophore that places an electron in the π^* orbital of the C=O group has the effect

J. Maruani et al. (eds.), New Trends in Quantum Systems in Chemistry and Physics, Volume 1, 347–358.

of reducing the C=O bond order from 2 to 1.5 while at the same time increasing the length of the bond by 0.08-0.11 Å. Even more dramatic changes occur in the bond angle relationships. In the case of the molecular prototype [1] constituted by formaldehyde, CH_2O, the rigid planar conformation of the S_0 ground electronic state converts into a pyramidal structure on excitation to the S_1 excited state. The out-of-plane motion inverting the pyramidal S_1 structure is described by a double minimum potential function that contains a central barrier of 350 cm^{-1}. The barriers to molecular inversion are found to be sensitive to the nature of the attached group. For example, while the first excited singlet and triplet states of the sulphur analogue, CH_2S, are found to be pseudo-planar, the fully fluorinated species, CF_2O is observed to have a barrier of 3100 cm^{-1}.

Additional large-amplitude information is introduced into these systems when more complex groups are attached to the carbonyl center. In the case of formic acid, the HCOO frame is rigidly planar in the S_0 state, and the internal rotation of the hydroxy group is the sole large-amplitude mode [2-7]. As would be expected, the molecular frame in the singlet S_1 excited state is pyramidal and, as a result, the molecular structure is highly flexible in both aldehyde wagging and hydroxy torsion coordinates [8]. The low frequency vibrational dynamics of the excited state are thus governed by two large amplitude modes, a torsion of the hydroxy group and a wagging-inversion of the aldehyde hydrogen.

It is the Franck-Condon principle that makes electronic spectroscopy an ideal tool for investigating large amplitude motions in polyatomic molecules. This principle relates the activity of the observed band progressions to the normal coordinates that most closely correspond to those changes in molecular conformations that occur on electronic excitation. It would be expected that the CH wagging and the OH torsion would be active in forming band progressions in the spectrum. As each band in a given progression is a suborigin for every other progression, the band spectrum very quickly becomes complex at ever increasing energies from the electronic origin. As a result, the UV spectra of these simple systems are highly congested and satisfactory vibrational assignments are often difficult to achieve.

The present studies were undertaken to provide information about the structure and dynamics of the S_1 first excited state of formic acid. In particular, a knowledge of the inversion-torsion hypersurface for the two combining states, along with maps of the electronic transition moments should be of value in the interpretation assignments of the 260-200 nm spectrum that is attributed to the $n \rightarrow \pi^*$ excitation process within the monomer molecule.

2. The S_0 ground electronic state

To describe the torsion-wagging motion, it is necessary to express the potential energy as a function of the two internal coordinates θ (torsion) and α (wagging) that are defined in Figure 1. The two-dimensional potential may be described as a V(θ, α) surface. The form of this plot, Fig. 2, follows without difficulty. The S_0 ground state contains two minima, corresponding to the wells created by the planar *syn* and *anti* conformers. In the α direction the hydrogen wagging potential has a single minimum and is harmonic (quadratic in α). The very slight tilt to the elliptical contours at the bottom of the wells is the result of a small coupling between the θ and α internal coordinates.

Fully optimized MP2 / 6-31G (d,p) calculations with the GAUSSIAN program were used to establish the potential surfaces [9]. In this method, the total energy was calculated for a set of molecular conformations defined by selected values of θ and α, forming grid points on the V(θ, α) surface. These *ab initio* data points were reduced to analytical form by fitting the energy calculated at the grid points to a series expression containing polynomic, exponential and Fourier terms:

$$V = \sum_{k}^{Nv} V_k^0 \prod_{j}^{2} f_{kj} \tag{1}$$

where the V's are the coefficients in the expansion for the potential energy obtained from the fitting procedure and the f_{kj} represent trigonometric (torsion) / polynomic (wagging) terms. N_v is the number of terms in the series expansion.

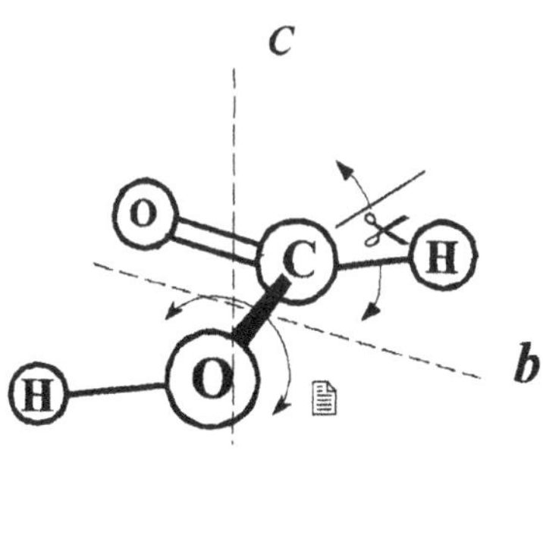

Figure 1. The formic acid structure, the b and c principal axes and definitions of the θ and α internal coordinates.

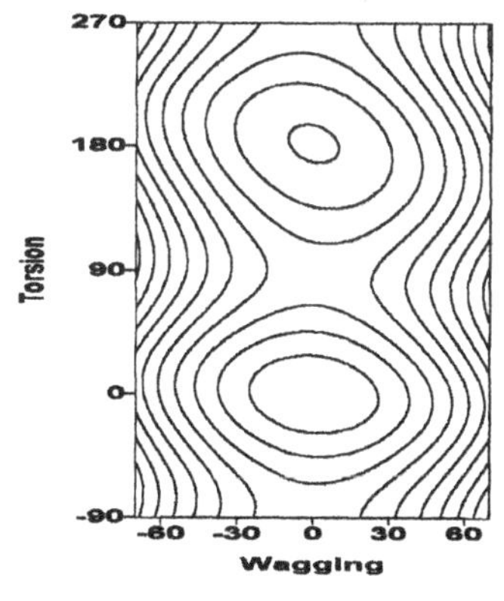

Figure 2. The morphed potential energy surface V(θ, α) for the S_0 ground state. Contour intervals at 2000 cm^{-1}.

Expansion coefficients f_{kj} for the S_0 electronic state were obtained from fitting the energy calculated at 70 different conformations to the analytical potential given by Eq. (1). A Fourier series of 3 terms was used for the sinusoidal potential generated by the OH internal rotation, and powers α^2 and α^4 for the anharmonic wagging potential. The coupling between the two internal coordinates was accounted for by 4 cross terms for $N_v = 10$.

The kinetic terms, B_{ij}, were treated in a somewhat similar way. Kinetic energies were evaluated from fully optimized molecular geometries calculated at grid points in θ and α as elements of the rovibrational G matrix, using the KICO program [10]. In this method, the kinetic energy parameters were obtained by inversion of the inertial matrix

$$\begin{bmatrix} I & X \\ X^t & Y \end{bmatrix} \tag{2}$$

where I is the inertial tensor corresponding to the overall rotation, Y is the vibrational submatrix, X represents the interaction between the external and internal motions, and

$$X_i = \sum_a m_a \left(\vec{r}_a \times \frac{\partial \vec{r}_a}{\partial \alpha_i} \right)_i \tag{3}$$

$$Y_{ij} = \sum_a m_a \left(\frac{\partial \vec{r}_a}{\partial \alpha_i} \right) \left(\frac{\partial \vec{r}_a}{\partial \alpha_j} \right) \tag{4}$$

Here the mass of atom *a* is given by m_a and its displacement vector by r_a. The B_{ij} are the coefficients obtained from fitting Eq. (2) to points on the grid. N_B is the number of terms in the expansion for the kinetic energy, while

$$B_{ij} = \sum_k^{N_B} B^0_{ijk} \prod_j^2 f_{ijkl} \tag{5}$$

A general Hamiltonian [11] was used for the treatment of the multidimensional vibrational problem:

$$\hat{H} = \sum_i^n \sum_j^n \left(-B_{ij} \frac{\partial^2}{\partial q_i \partial q_j} - \frac{\partial B_{ij}}{\partial q_i} \frac{\partial}{\partial q_j} \right) + \hat{V} \tag{6}$$

where the number of vibrations, n, is two.

The two-dimensional Hamiltonian is solved variationally for the eigenvalues and eigenvectors, using a hybrid free-rotor + harmonic-oscillator basis set for the hydroxy-torsion and hydrogen-wagging coordinates, respectively. Nonrigid group theory was used for the labeling of the energy levels, factorization of the Hamiltonian matrix and generation of the selection rules. The existence of a single plane of symmetry allows the S_0 and S_1 structures to be classified by the S switch operation:

$$\hat{S} f(z,\theta) = f(-z,-\theta) \tag{7}$$

creating a nonrigid group G_2 that is isomorphous to the point group C_S [12]. Thus, the torsion-wagging functions classify into the symmetry species a' (in-plane) and a" (out-of-plane). The calculated energy levels are adjusted to fit the observed levels by morphing the potential energy surface that was initially obtained from the *ab initio* procedure. The refinement was carried out by minimizing the differences between the calculated and observed levels with a quasi-Newton method [13]. Throughout the fitting procedure, the kinetic energy coefficients were fixed at their optimized values. The S_0 expansion coefficients for the morphed potential and kinetic energy functions are collected together in Table 1.

TABLE 1. Potential[a] and kinetic[b] energy expansion coefficients for the S_0 ground electronic state (in cm^{-1})

coeff.	V	$B_{\theta,\theta}$	$B_{\theta,\alpha}$	$B_{\alpha,\alpha}$
constant	0.36549+04	0.24447+02	0.65430+01	0.24585+02
α^2	0.25424+01	-0.10620-02	-0.40400-07	-0.81244-04
α^4	0.19043-05	0.66000-07	0.22500-07	0.81173-08
cos(θ)	-0.40268+03	-0.19010-02	0.17939+02	0.23183+01
cos(2θ)	-0.29300+04	0.26649+00	0.19074+00	0.42686+00
cos(3θ)	-0.32224+03			
αsin(θ)	-0.13143+02		-0.46790-02	
αsin(2θ)	0.22303+02			
α^2cos(θ)	0.57425-01		-0.63085-04	
α^2cos(2θ)	-0.44351-02			

a) from the morphed potential surfaces.
b) from the MP2 / 6-31G (d,p) optimized structures.

3. The S_1 excited electronic state

On an n→π* electron excitation, antibonding density is introduced into the C=O bond, with the result that the HCOO group distorts into a pyramidal conformation with the aldehyde hydrogen projecting out of the plane. Thus, the T_1 and S_1 states are nonplanar in the frame, and the potential function describing the wagging motion contains two minima separated by a central barrier. To further complicate the picture the OH group undergoes a conformational change on excitation, and rotates from its planar *syn* and *anti* forms to a staggered conformation. Thus, for the excited state, the positions of the minima on the S_0 surface now become maxima for the S_1 potential that is illustrated in Fig. 3. For one full revolution of the hydroxy group, the $V(\theta, \alpha)$ surface contains four separate wells. The double minimum nature of the S_1 wagging potential for the upper state required the use of the Coon function.[14] This is constructed from a Gaussian term that describes the central barrier while the outer edges of the two wells are formed from the addition of quadratic-quartic terms.

To obtain a starting point for the analysis of the torsion-wagging level structure of the S_1 state, MP2 / 6-31G (d,p) calculations were carried out on the companion nπ* triplet state, T_1. As only a few thousand wavenumbers separate the singlet and triplet states, it would be expected that the $V(\theta, \alpha)$ potential surface of the T_1 state would resemble that of the S_1 state. Thus, the T_1 surface was used to generate a manifold of levels that became a starting point for the fitting procedure. In all, 15 expansion coefficients yield an adequate fit to the *ab initio* data points. The excited-state expansion coefficients for the morphed potential and kinetic energy terms are given in Table 2.

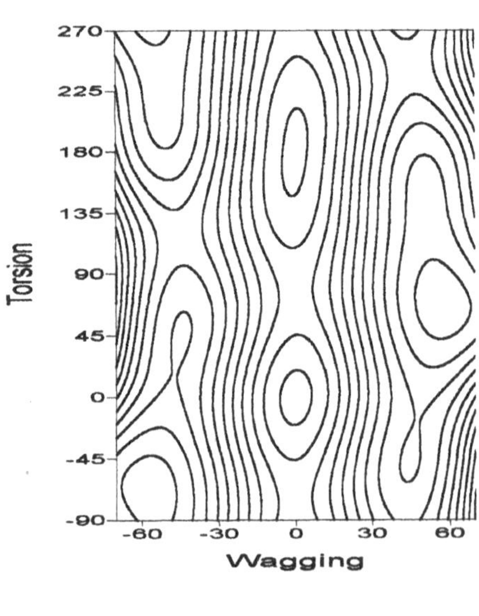

Figure 3. The morphed potential energy surface $V(\theta, \alpha)$ for the S_1 first electronic state. Contour intervals: 500 cm^{-1}.

TABLE 2. Potential[a] and kinetic[b] energy expansion coefficients for the S_1 excited electronic state (in cm^{-1})

coeff.[c]	V	$B_{\theta,\theta}$	$B_{\theta,\alpha}$	$B_{\alpha,\alpha}$
constant	-0.29182+04	0.21900+02	0.41527+01	0.23799+02
α^2	0.69532+00	-0.89100-04	-0.28600-03	-0.43900-04
α^4	0.67900-04	0.56300-07	0.15100-07	
$\cos(\theta)$	-0.12436+02	-0.17450-02	0.16290+01	0.15609+01
$\cos(2\theta)$	-0.13731+03	0.20287+00	0.12251+00	0.41449+00
$\cos(3\theta)$	-0.22865+03			
$\alpha\sin(\theta)$	-0.26475+01		-0.52862-02	
$\alpha\sin(2\theta)$	0.94358+01			
$\exp(-cx^2)$	0.76105+04			
$\exp(-cx^2)\cos(\theta)$	-0.97771+02			
$\exp(-cx^2)\cos(2\theta)$	0.58683+03			
$\exp(-cx^2)\cos(3\theta)$	-0.85855+02			
$\alpha^3\sin(\theta)$	-0.54576-02			
$\alpha^3\sin(2\theta)$	0.31249-03			
$\alpha^2\cos(\theta)$			-0.50800-04	

(a) from the morphed potential surfaces.
(b) from the MP2 / 6-31G (d,p) optimized structures.
(c) c was fixed at 6.5×10^{-4}.

4. Vibronic transitions

The intensities of the various transitions were determined from the transition dipole moments between an *n* and *m* pair of torsion-wagging vibronic states belonging to electronic states *e'* and *e"*:

$$\mu_{nm} = \langle \psi_n \psi_{e'} \mid \hat{\mu} \mid \psi_{e''} \psi_m \rangle, \tag{8}$$

where μ represents the dipole moment operator. The electronic transition moments are obtained directly from the GAMESS package with the CI / 6-31G (d,p) basis set and the length approximation, using double excitations for the lower state and triple excitations for the upper state [15]:

$$\mu_{e'e''} = \langle \psi_{e'} \mid \hat{\mu} \mid \psi_{e''} \rangle. \tag{9}$$

The electronic transition moments were then expanded as a series expression derived from the same θ and α grid points as those selected for the potential surface:

$$\mu_{e'e''} = \sum_{j}^{N_\mu} \mu_j^0 \prod_{k}^{n} f_{kj}. \tag{10}$$

Transition dipole moments were calculated with the above procedure from the geometry of the ground state, using the assumption that the absorption of the photon and excitation to the upper state is an adiabatic process. The results are shown, in Fig. 4, as contour maps that depend on the torsional and wagging angles for each projection on the *a, b* and *c* principal axes. The μ_c component is perpendicular to the O-C=O frame and belongs to the a' representation, while μ_a and μ_b lie in the molecular plane and are of a" species. The selection rules were derived from the symmetry of the torsion-wagging wave functions and the components of the transition dipole moment. The transition moment expansion coefficients along the *a, b* and *c* directions are displayed in Table 3. The relative intensities of the transitions within the manifold of S_0 and S_1 levels were obtained from

$$I = (g_n - g_m)\mu_{nm}^2 \tag{11}$$

where the g_n represent the populations of the levels.

The allowed a' ↔ a' component of the transition between the S_0 (v = 0) and S_1 (0^+, v = 0) levels of formic acid is directed out of the O-C=O molecular plane and entails *c*-type polarization. Transitions to the higher inversion level, 0^-, are electronically forbidden, but indeed allowed through a Herzberg-Teller vibronic coupling generating *a* and *b* in-plane polarized components.

TABLE 3. Transition moment expansion coefficients resolved along the principal a, b and c-type axes (in Debye)

coeff.	μ_a	μ_b	μ_c
constant			0.29058+00
$\sin(\theta)$	-0.76768-01	0.90750-02	
$\sin(2\theta)$	0.53163-01	-0.54520-02	
$\cos(\theta)$			-0.18930-02
$\cos(2\theta)$			-0.54658-01
α	-0.48400-02	0.20216-01	
α^2			-0.91580-04
α^3	0.34334-06	-0.11034-04	
α^4			0.56157-07
α^5		0.22940-08	
$\alpha\cos(\theta)$	0.56579-03	0.70768-03	
$\alpha^2\cos(\theta)$			0.13263-04
$\alpha^3\cos(\theta)$	-0.15935-07	-0.18944-06	
$\alpha^4\cos(\theta)$			-0.92085-08
$\alpha\cos(2\theta)$	0.19490-02	0.32450-02	
$\alpha^2\cos(2\theta)$			0.11028-03
$\alpha^3\cos(2\theta)$	0.94377-06	-0.10842-05	
$\alpha^4\cos(2\theta)$			-0.25997-07

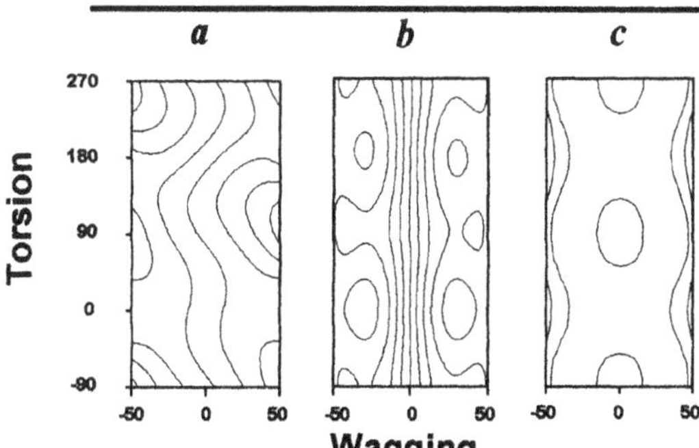

Figure 4. Transition moment maps resolved along the a, b and c principal axes. Contour lines are spaced by 0.1 Debye Transition moments are calculated for S_0-state geometry.

TABLE 4. Relative band positions and band intensities for the a, b and c-type transitions from the morphed S_0 and S_1 potential energy surfaces

		a-type		*b*-type		*c*-type	
v'S_1	v"S_0	displ.[a]	int.[b]	displ.[a]	int.[b]	displ.[a]	int.[b]
0	0	0.0	0.02	0.00	1.24	0.00	0.67
1	0	250.49	0.07	250.49	5.50	250.49	2.91
2	0	376.56	0.07	376.56	5.95	376.41	3.10
3	0	512.94	0.22	512.94	22.05	512.71	11.40
4	0	670.10	0.51	670.10	53.13	669.85	27.35
5	0	808.70	1.04	808.70	100.00	808.54	51.66
6	0	865.19	0.00	865.19	1.38	865.33	0.63
7	0	983.79	0.70	983.79	76.72	983.73	39.26
8	0	1037.0	0.22	1037.0	14.45	1037.1	9.27
9	0	1094.9	0.20	1094.9	11.71	1095.0	6.24

a) in cm^{-1}.
b) scaled from 100.

Figure 5 shows the simulated and observed UV spectra of HCOOH. The upper panel shows the computed spectra where the calculated positions and intensities of the vibronic bands are illustrated as histograms. The intensities of the *c*-type bands are given by the heights of the open rectangles and the *a*/*b*-hybrid bands as hatched rectangles. While the location of the bands came from the energy data of Table 4, the rectangles were offset from each other for clarity. The lower panel is a low resolution excitation spectrum of HCOOH that we have recently recorded under jet-cooled conditions [16]. The notation given to the vibronic bands labels the active mode Q_9 (OH torsion) as upper case 9, with the subscripts and superscripts designating the vibrational excitation. It is clear that the gross vibrational features in the spectrum are accounted for. The interval between the *c* and *a*/*b*-type bands measures the inversion-doubling splitting, 0^--0^+, imposed by the conditions of the *S* switch operation. This interval attaches to the quanta of the torsional levels to form sets of doublets in the spectrum. From the calculations of Table 5 these splittings are vanishingly small for the 0_0^0, 9_0^1 and 9_0^2 torsional bands. Thus, under this resolution, each band is predicted to have a mixed *a*/*b*/*c* rotational band character. The contributions of each band type to the overall band profile can be obtained from Table 4. For the 0_0^0 origin band, the *a/b/c* contributions to the intensity based on the foregoing calculations are predicted to appear in the ratio 0.02 / 1.24 / 0.67. Similar ratios are predicted for the 9_0^1 and 9_0^2 torsional bands. Thus, the rotational structure within the origin band should be complicated by the appearance of three separate bands. Very high rotational resolution will be required to disentangle the different contributions. The torsion-wagging bands within the envelopes of the Franck-Condon and Herzberg-Teller progressions have very similar intensity profiles. This is perhaps not too surprising since the separation between the (-) and (+) levels is very small and the associated vibrational wavefunctions are similar, but for their symmetry characteristics.

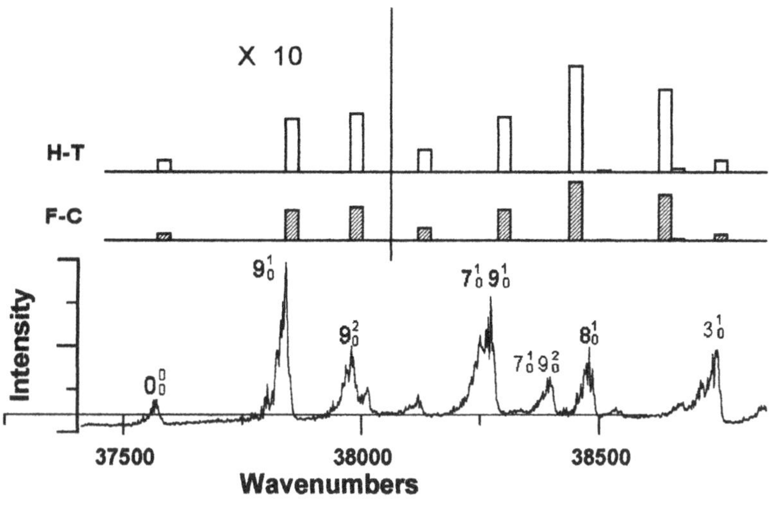

Figure 5. The observed and calculated band progressions in the $S_1 \leftarrow S_0$ electronic transition in formic acid. The upper panel shows histograms of the FC (Franck-Condon) and HT (Herzberg-Teller) induced transitions. The lower panel is the LIF (laser-induced excitation) spectrum of formic acid under jet-cooled conditions. The calculated intensities of the first three bands are magnified 10x for the sake of clarity.

The *c*-type component of the vibronic transition moment results from the out-of-plane electronic transition that connects the combining $S_0(a')$ and $S_1(a'')$ electronic states. These transitions are allowed by electronic selection rules and form the Franck-Condon components to the overall electronic transition. The *a/b*-band types come from the in-plane moments and they are forbidden as electric dipole transitions but are allowed as vibronic transitions. The sum of the two components *a/b* is referred to as the Herzberg-Teller transition. For formic acid, the allowed and forbidden components of the $S_1 \leftarrow S_0$ transition roughly have the same strength. This result can be traced back to the excitation process. An examination of the molecular orbitals of formic acid shows that the n and π^* orbitals project at right angles from the sides of the oxygen atom. As a result the promotion of the n-orbital electron involves more of a rotation than a translation of the electronic charge. The consequence of a lack of charge translation is that the system has low electric dipole strength. Formic acid is thus a textbook example of an electronic transition that is allowed by the overall selection rules but forbidden by local symmetry of the C=O chromophore. The relative intensity of the vibronically induced Herzberg-Teller bands in the spectrum is not so much a consequence of their absolute strength, but rather is related to the weakness of the allowed Franck-Condon bands.

Acknowledgments

LMBT and DCM would like to thank the National Sciences and Engineering Council of Canada for financial support. CMC and AN acknowledge financial support from the Universidad de Castilla la Mancha.

References

1. Clouthier, D. J., and Moule, D. C. (1989) *Topics in Current Chemistry*, edited by K. Niedenzu (Springer, Berlin), vol. 150, pp 167-247.
2. Bellet, J., Deldalle, A., Samson, C., Steenbeckeliers, G., and Wertheimer, R. (1971) *J. Mol. Struct.* **9**, 65.
3. Bellet, J., Samson, C., Steenbeckeliers, G., and Werthheimer, R. (1971) *J. Mol. Struct.* **9**, 49.
4. Bjarnov, E., and Hocking, W.H. (1978) *Z. Naturforsch.* **33a**, 610.
5. Hocking, W.H. (1976) *Z. Naturforsch.* **31a**, 1113.
6. Bertie, J. E., and Michaelian, K. H. (1982) *J. Chem. Phys.* **76**, 886.
7. Hisatsune, I. C., and Heicklen, K. (1982) *Can. J. Spectrosc.* **8**, 135.
8. Ioannoni, F., Moule, D. C., and Clouthier, D. J. (1990) *J. Phys. Chem.* **94**, 2290.
9. Frisch, M. J., Trucks, G. E., Schegel, H. B., Gill, P. M. W., Johnson, B. G., Robb, M. A., Cheeseman, J. R., Keith, T. A., Petersson, G. A., Montgomery, J. A., Raghavachari, K., Al-Laham, M. A., Zakzewski, V. G., Ortiz, J. V., Foresman, J. B., Ciolowski, J., Stefanov, B. B., Nanayakkara, A., Challacombe, M., Peng, C. Y., Ayala, P. Y., Chen, W., Wong, M. W., Andres, J. L., Replogle, E. S., Gomperts, R., Martin, R. L., Fox, D. J., Brinkly, J. S., Defrees, D. J., Baker, J., Steward, J. P., Head-Gordon, M., Gonzalez, C., and Pople, J. A. (1995) *Gaussian, Inc.*, Pittsburgh, PA, USA.
10. Niño, A., and Muñoz-Caro, C. (1994) *Comput. Chem.* **56**, 27.
11. Pickett, H. M. (1972) *J. Chem. Phys.* **56**, 1715.
12. Smeyers, Y. G. (1992) *Advances in Quantum Chemistry* **24**, 1.
13. Fletcher, R. (1987) *Practical Methods of Optimization* (Wiley), ch 3.
14. Coon, J. B., Naugle, N. W., and McKenzie, R. D. (1966) *J. Mol. Spectrosc.* **20**, 107.
15. Schmidt, M. W., Balderidge, K. K., Boatz, J. A., Elbert, S. T., Gordon, M. S., Jensen, J. H., Koseki, S., Matsunaga, N., Nguyen, K. A., Su, S., Theresa, L. W., Dupuis, M., and Montgomery, J. A. (1993) *Comput. J. Chem.* **14**, 1347.
16. Beaty-Travis, L. M., Moule, D. C., Judge, R. H., Liu, H., and Lim, E. C. (to be published).

AB-INITIO HARMONIC ANALYSIS OF LARGE-AMPLITUDE MOTIONS IN ETHANOL DIMERS

M. LUISA SENENT[1], YVES G. SMEYERS[1] AND ROSA DOMÍNGUEZ-GÓMEZ[2]

[1]*Departamento de Química y Física Teóricas, Instituto de Estructura de la Materia, C.S.I.C., c/ Serrano 113 bis, 28006 Madrid, Spain*
[2]*Departamento de Ingeniería Civil, Cátedra de Química, E.U.I.T. Obras Públicas, Universidad Politécnica de Madrid, c/ Alfonso XIII 3-5, 28014 Madrid, Spain*

Abstract. At relatively low pressures, dimerization of ethanol yields three different structures trans-gauche, trans-trans and gauche-gauche, each of them with different minimum energy conformations. The energy differences among the stable structures are relatively low. All of them may be present in the same sample.Dimerization shifts the whole spectrum to higher frequencies. The six new intermolecular modes push up the remaining vibrational modes which become constrained by the presence of the second molecule. The normal modes involving the hydrogen bonded atoms show the largest vibrational shifts. Five of the additional modes lies below 100 cm^{-1} and confer some non-rigidity to the dimer. The harmonic fundamental of the hydrogen bond stretching is located at 173.6, 155.3 and 189.4 cm^{-1} for the different conformers of the trans-gauche structure, i.e., quite below the OH torsion that located at 305.7 cm^{-1} in the molecule and moves up to 700 cm^{-1} in the dimers. For the trans-trans and gauche-gauche, this transition lies at 164.8 and 188.7 cm^{-1}, respectively. The pattern observed between 150 and 190 cm^{-1} may be assigned to this stretching.

1. Introduction

Ethanol is a well known molecule of chemical and astrophysical interest. Relatively high concentrations of the isolated species have been detected in interstellar clouds. At high pressures, it forms dimers and even polymers bonded by hydrogen bonds. The structure of the dimer was recently studied experimentally [1-2] and theoretically [3,4].

The full vibrational spectrum of ethanol has been recorded in gas phase at low pressures or trapped in argon or nitrogen matrices [5-9]. In the electronic ground

J. Maruani et al. (eds.), New Trends in Quantum Systems in Chemistry and Physics, Volume 1, 359–374.

state, the molecule shows two stable conformers, trans and gauche, which are almost isoenergetic and interconvert to each other by the internal rotation of the hydroxyl group. In addition, the molecule presents a second large amplitude mode that also confers non-rigidity, the torsion of the methyl group. In gas phase, the two fundamental torsion frequencies of the OH and CH_3 groups for the non-deuterated isotopic variety have been observed at 305.7 cm^{-1} and 244.4 cm^{-1}, respectively [5].

In argon matrices, some bands corresponding to the dimers and polymers can be observed near to the torsional bands of the isolated molecule [7]. Dimerization provides six additional transitions involving hydrogen bonds, that produce a displacement of the whole spectrum to higher frequencies. The six additional transitions lie in the far infrared zone close to the molecular torsion bands give rise to patterns that are not easy to assign. For example, Barnes and Hallam [7] have assigned the low resolution branch observed at 213 cm^{-1} to a superposition of hydrogen-bond stretching and OH torsion. Both modes may lie approximately at the same frequencies making the spectrum analysis arduous. For this reason, in this paper we compare the structures and the vibrations of the two conformers trans and gauche as well as those of the dimer by using ab initio calculations. We hope that this analysis may be of assistance in the assignment of ethanol spectra.

The analysis of the Fourier infrared (FIR) vibrational spectra of the dimer is also of interest for establishing the properties of ethanol in the condensed phase [10-13]. The crystal packing shows hydrogen bonds in which the molecules are linked forming infinite chains. The unit cell of the monoclinic crystal phase shows four molecules per unit cell. There are two independent molecular configurations in the unit cell which are not related by symmetry. Half of the molecules show a gauche structure and half a trans structure. The large isotopic effects in the excess contributions to the specific heat of the disordered phases of ethanol may be correlated with the low-frequency spectra [13]. As it appears on the solid phase, we shall put special emphasis on the trans-gauche dimer.

2. Computational details

The isolated ethanol molecule exhibits two minimal energy conformations, trans and gauche, of C_s and C_1 symmetries. From the spectroscopic experimental data [14,15], it may be inferred that the most stable geometry is the planar trans-conformer where the two dihedral angles HCCO and CCOH are equal to 180°.

The energies and geometries of the two conformers have been determined with the MP2 approach (Möller-Plesset perturbation theory up to second order) with the program Gaussian 94 [16]. All the core and valence electrons have been taken into account. The calculations have been performed with Dunning's double-zeta correlation-consistent basis set, cc-pVDZ [17], adding some diffuse functions for describing the hydrogen bond of the dimers (*s*, *p* and *d* diffuse functions on the oxygen atom and *s* and *p* on the hydroxyl group hydrogen, AUG on OH). They appear to be indispensable for reproducing the experimental relative energies. The structural parameters of both conformations are shown in Table 1. Table 2 shows the total energies and several spectroscopic and electrostatic properties. With MP2(full) / cc-pVDZ (AUG on OH), the energy difference between conformers has been determined to be ΔH=73.1 cm^{-1}, which is in the range of experimental data (45.0 cm^{-1} in Ref [5]; 39.2 cm^{-1} in Ref. [15]). The total energies of the two conformers are -154.600852 a.u. (trans-ethanol) and -154.600519 a.u. (gauche-ethanol). Without diffuse functions the order of stability is reversed [18].

TABLE 1. Structural parameters for the two conformers of ethanol[a]

	trans-ethanol	*gauche-ethanol*
C1C2	1.516815	1.522131
O3C2	1.441018	1.439301
H4C1	1.102177	1.103343
H5C1	1.101425	1.104057
H6C1	1.101420	1.101582
H7C2	1.106010	1.106392
H8C2	1.106010	1.100944
H9O3	0.966233	0.967708
O3C2C1	106.981	112.346
H4C1C2	114.413	110.550
H5C1C2	110.171	110.764
H6C1C2	110.219	110.451
H7C2C1	110.514	110.730
H8C2C1	110.487	110.703
H9O3C2	108.020	107.406
H4C1C2O3	180.0	178.0
H5C1C2H4	120.2	119.5
H6C1C2H4	-120.2	-120.4
H7C2C1O3	120.1	123.8
H8C2C1O3	-120.1	-116.7
H9O3C2C1	180.0	60.4

a) distances in Å; angles in degrees

The two minima (Figs. 1 and 2) correspond to <H9O3C2C1= 180° and 60.4°. O3C2C1 is the internal coordinate that shows the largest OH internal rotation dependence. It shifts approximately 5° during the torsion (O3C2C1 is 106.981° in the trans and 112.346° in the gauche geometries), in good agreement with microwave measurements [14]. With torsion the methyl group loses the C_{3v} symmetry and the central bond C1C2 lengthens from 1.51682 Å (trans) to 1.52213 Å (gauche), minimizing the steric interactions. The rotational constants have been calculated to be

34.3110568, 9.3587363 and 8.1158306 MHz in trans-ethanol, and 33.8185519, 9.1259165 and 8.0515401 MHz in gauche-ethanol (Table 2). The dipole moment displays a significant fluctuation from the gauche form (1.9292 Debyes) to the trans form (1.7861 Debyes), with possible consequences on band intensities. Hyperconjugation induces a negative charge that increases on the carbon of the methyl group during the OH-torsion.

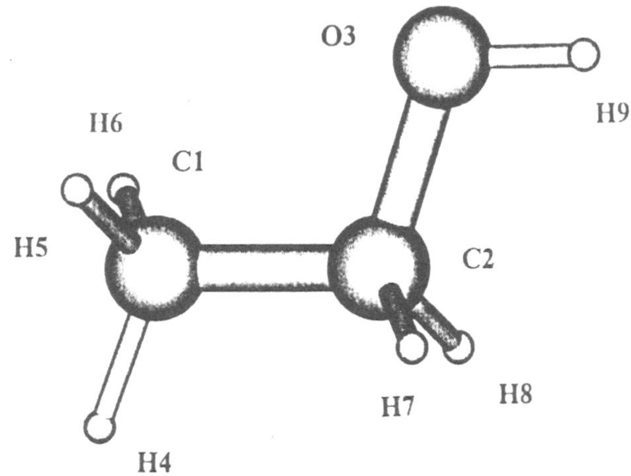

Figure 1. The trans-conformer of ethanol.

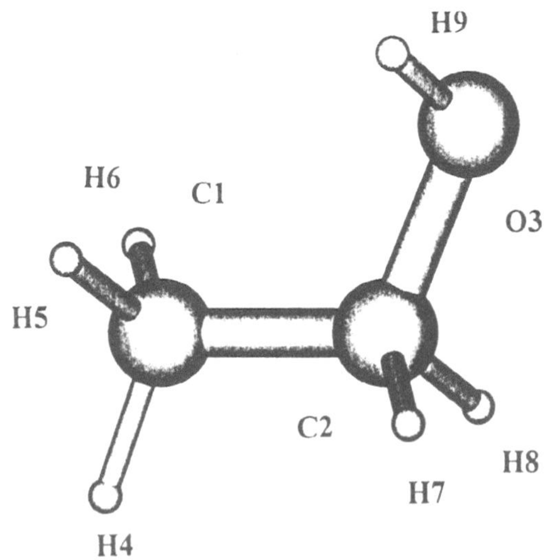

Figure 2. The gauche-conformer of ethanol.

TABLE 2. Total energies, O-H distances, atomic charges and dipole moment of trans and gauche-ethanol, and the dimers

			Trans-Gauche Dimers				
	trans	*gauche*	*A-dimer*	*B-dimer*	*C-dimer*	*trans-trans*	*gauche-gauche*
Energy	-154.600852	-154.600519	-309.229335	-309.228216	-309.229240	-309.228289	-309.229469
Structural parameters (Å / degrees)[a]							
O^g-H^t			1.8670	1.8814	1.8664	1.8951	1.8698
O^g-H^t-O^t			167.1	165.4	167.9	169.6	168.8
H^t-O^t	0.9662		0.9750	0.9743	0.9753	0.9740	0.9751
Rotational Constants (MHz)							
A	34.3110568	33.8185519	6.9841463	12.7934180	6.1258724	5.8766496	5.1464216
B	9.3587363	9.1259165	1.0888570	0.8485763	1.2306146	1.2237822	1.3760997
C	8.1158306	8.0515401	0.9940642	0.8204907	1.1158718	1.1482287	1.2785301
Atomic charges (a.u.)[a]							
H^t	0.1649		0.2662	0.2634	0.2753	0.2573	0.2666
O^t	-0.5226		-0.5453	-0.5544	-0.5704	-0.5540	-0.5731
C_2^t	0.2546		0.2269	0.2328	0.2416	0.2192	0.2453
C_1^t	-0.0626		-0.0583	-0.0558	-0.0888	-0.0351	-0.0894
H^g		0.1658	0.1979	0.1904	0.2019	0.1932	0.2072
O^g		-0.5288	-0.6204	-0.5973	-0.6313	-0.5997	-0.6373
C_2^g		0.2566	0.2187	0.1968	0.2163	0.2312	0.2169
C_1^g		-0.0829	-0.0783	-0.0511	-0.0765	-0.0790	-0.0751
μ *(Debyes)*	1.7861	1.9292	2.1594	2.1894	2.2122	3.8328	2.3758

a) X^t are the atoms of the molecule that confers the H for the hydrogen bond; X^g are the second molecule atoms

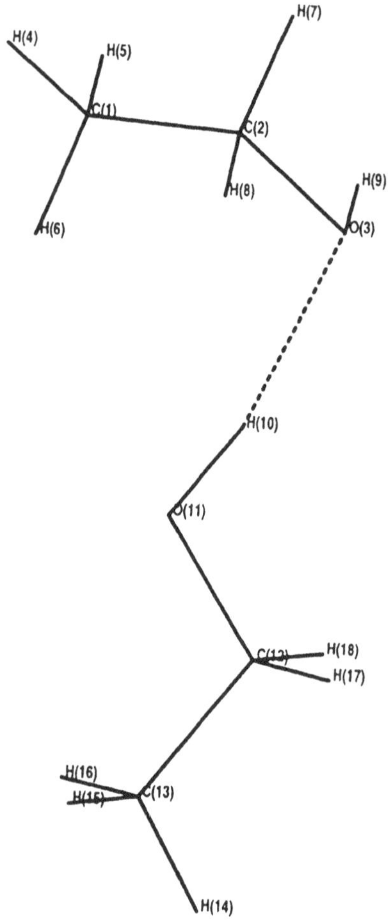

Figure 3. The trans-gauche-dimer of ethanol.

The MP2 total energies of the three possible dimers, trans-gauche (E = -309.229335 a.u.), trans-trans (E = -309.228289 a.u.) and gauche-gauche (E = -309.229469 a.u.), are shown on Table 2. The relative energies are 0.08, 0.74 and 0.00 kcal/mol. Surprisingly, the gauche-gauche dimer appears as the most stable conformation, although in the solid phase the trans-gauche conformer is the most stable form [12]. In the case of the trans-gauche form the hydrogen bond connects the OH hydrogen of the trans form through the oxygen atom of the gauche form (Figure 3). The trans-trans conformer appears to be less stable than it was reported by Ehbrecht and Huisken [4]. In that paper, the gauche-gauche conformer appears to be less stable.

All the forms show a rather similar stability. For this reason it is difficult to assert which is the absolute minimal structure on the dimer potential energy surface. Taking into account that the energy difference between the trans and gauche monomers is very small ($\Delta G(\textit{trans} \rightarrow \textit{gauche})$=0.17 kcal/mol), the possible existence of gauche-gauche and trans-trans forms of lower energy was carefully investigated.

On the basis of the relevance of the trans-gauche conformer, a careful search for all the possible minimal energy conformation was performed with the MP2(full) / cc-pVDZ (AUG on OH) procedure. As a result, three stable geometries were localized that we call

A-dimer, B-dimer and C-dimer (Figure 4). They have been determined by optimizing the geometry from different starting points, in which the relative positions of the two components, the trans and gauche molecules, were different. The initial positions were defined using different distances between the two centers of mass and different relative orientations of the components. The orientations were defined by the three Euler angles connecting the principal axis of the two molecules.

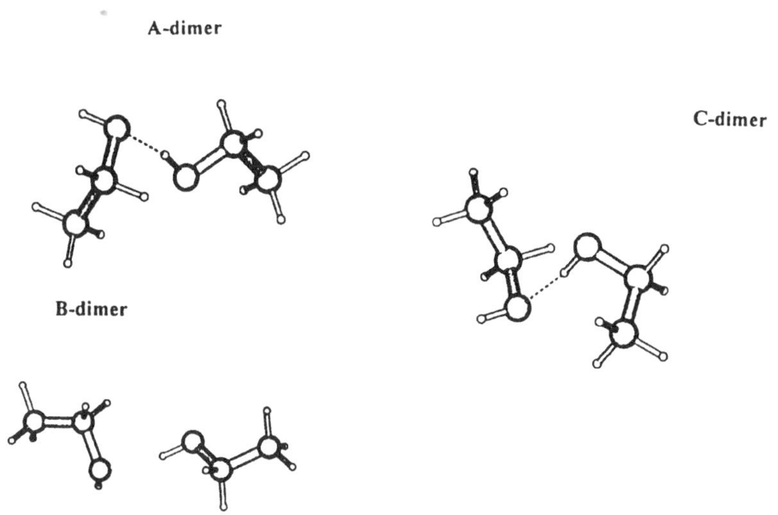

Figure 4. The three stable trans-gauche dimers of ethanol.

The three total electronic energies: E (A-dimer) = -309.229335 a.u.; E (B-dimer) = -309.228216 a.u., E (C-dimer) = -309.229240 a.u., are shown in Table 2. Table 2 also shows the rotational constants, dipole moments and Mulliken atomic charges and the structural parameters related to the hydrogen bond.

The A-dimer represents the most stable conformer we have found. The corresponding rotational constants are 6.1258724, 1.2306146 and 1.1158718 MHz and the dipole moment, 2.1594 Debyes, is the lowest of the three trans-gauche dimers (Table 2). The hydrogen-bond distance is 1.8670 Å. The O^g-H-O^t angle and the O^t-H^t distance are 167.1° and 0.9753 Å. The O^t-H^t bond is slightly larger than in the isolated molecule (0.9662 Å).

The relative energies of the B and C forms with respect to the A form were calculated to be about 0.70 and 0.06 kcal/mol. The B-dimer is the least stable and shows the largest hydrogen bond, 1.8814 Å. The structural parameters <O^g-H-O^t and O^t-H^t of the C-dimer approximately coincide with those of the A-form, although the orientation of the two molecules is different in the two conformers, which have different rotational constants (Table 2).

The molecular properties of the trans-gauche, trans-trans and gauche-gauche dimers are also given in Table 2. X^t (O^t or H^t) represents the atom of the molecule that supplies the H atom to the hydrogen bond. O^g is the atom of the second molecule involved in the hydrogen bond. The trans-trans dimer displays the largest dipole moment of 3.8328 Debyes, whereas the gauche-gauche dimer exhibits 2.3758 Debyes. The largest separation between the two molecules is found in the trans-gauche dimer (O^g-H^t).

The formation probability of the dimers was evaluated from the Gibbs energies. The thermodynamic potentials for the dimer formation process

$$x\text{-}ethanol + y\text{-}ethanol \rightarrow dimer \quad (x=t,g;\ y=t,g)$$

are reported in Table 3.

The enthalpies of the trans-gauche dimers have been calculated to be -19.5988 kcal / mol (A-dimer), -18.7871 kcal/mol (B-dimer) and -19.5247 kcal/mol (C-dimer), which is in good agreement with previous quantum cluster equilibrium (QCE) results of Ludwig et al. [1,2]. In the case of the trans-trans and gauche-gauche dimers, they are -18.600 kcal/mol and -19.967 kcal/mol respectively. The free energies of the trans-gauche, trans-trans and gauche-gauche dimers were determined to be -11.252, -10.4063 and -11.3025 kcal/mol, respectively, using the ideal-gas model to determine the entropies and volume increments at P=1 atm and T=298.15°K. The formation probabilities of the A and C dimers appear to be similar (Table 3) while those of the trans-trans and gauche-gauche dimers are slightly lower.

TABLE 3. Enthalpy, entropy and free-energy variations of the dimer formation processes (in kcal/mol; T=298°K)

	Trans-Gauche			**Trans-Trans**	**Gauche-Gauche**
	A-dimer	*B-dimer*	*C-dimer*		
ΔH	-19.5988	-18.7871	-19.5247	-18.6000	-19.9670
TΔS	-8.3366	-7.4678	-8.3032	-8.1937	-8.6645
ΔG	-11.2622	-11.3193	-11.2215	-10.4063	-11.3025

It may be concluded that dimerization of ethanol yields three different structures: trans-gauche, trans-trans and gauche-gauche, each of them showing several conformers of minimum energy. Although the gauche-gauche dimer appears to be the most stable, the energy differences between the conformers are relatively small. Thus, the formation of all of them is plausible. They can all coexist in the same sample, and have to be considered simultaneously for an understanding of the spectra. In solution, the probabilities may change as a result of the different electric dipole moments.

3. Assignments

Tables 4, 5, 6 and 7 show the harmonic frequencies corresponding to the 21 vibrational modes of the monomers and to the 48 modes of the dimers. They have been calculated with MP2(full) / cc-pVDZ (AUG on OH).

The harmonic frequencies of the trans- and gauche-ethanol are shown in Table 3. In the case of the trans-conformer, the frequencies are assigned to the symmetry vibrations and are classified in the C_s point group representation. Five vibrational modes lie below 1000 cm^{-1}, the two torsions, the CCO bending, the CCO stretching and that of the asymmetric rocking mode. The calculated frequencies of the skeletal modes have been calculated to be 413.3, 822.4 and 907.6 cm^{-1}. The experimental bands are observed nearby at 419, 801 and 885 cm^{-1}, which attest the harmonic character of these vibrations. However there is a significant disagreement between the experimental and calculated values for the torsions. The torsional normal modes appear as combinations of the two torsional internal coordinates. The weight of each internal coordinate in the normal mode fluctuates between 40% to 60%. The contribution of the methyl group to the mode at 250.4 cm^{-1} is slightly large than that of the OH torsion, whereas it is smaller for the mode at 305.5 cm^{-1}. In the case of the gauche conformer, the relative order of these modes is reversed.

The band calculated at 250.4 cm^{-1} can be assigned to the band observed at 243 cm^{-1} but the difference between the calculated position for the OH-torsion (305.5 cm^{-1}) and the observed frequency (202.6 cm^{-1}) is unacceptable. The rigorous study of the torsional spectra of ethanol requires a more sophisticated two dimensional model for non-rigid molecules that is able to describe the interactions between the torsional coordinates and torsional motions hindered by low barriers [19-20]. The harmonic normal coordinates model represents only a first approach and is more suitable for the methyl torsion than for the OH torsion in taking into account the barrier heights (402.8 and 399.1 cm^{-1} for the OH group and 1185 and 1251 cm^{-1} for the CH_3 group [5]).

Dimerization generates six additional modes of low frequency from the external modes. They represent relative motions of the two molecules, the torsion around the hydrogen bond (υ_{48}), the hydrogen-bond stretching (υ_{43}), and four skeletal modes. The calculated harmonic frequencies, υ_{48}, υ_{47}, υ_{46}, υ_{45}, υ_{44} and υ_{43} of the A-dimer are 30.2, 39.9, 51.4, 80.9, 97.9 and 173.6 cm^{-1} (Table 5). Most of the remaining frequencies can be assigned to internal motions of each one of the molecules trans (t) and gauche (g) of the dimer. With dimerization, the whole ethanol spectrum shifts up to higher frequencies.

In the dimer the two torsion modes of trans-ethanol have been calculated at 269 cm^{-1} (methyl group) and 725 cm^{-1} (OH group). The methyl torsion shifts by 16.8 cm^{-1} with dimerization. The OH torsion appears at 725.0 cm^{-1}, above the COC bending modes. The formation of the hydrogen bond hinders the internal rotation, which becomes an oscillation. It may be drawn that the OH trans-torsion of the dimer lies around 700 cm^{-1} (it was found at 721.9 and 720.2 cm^{-1} for the B and C-dimers) since the harmonic model appears to be more adequate. In the trans-trans and gauche-gauche dimers, the OH torsion corresponding to the H atom involved in the hydrogen bond appears at 709.6 and 716.5 cm^{-1}, respectively.

The stretching of the hydrogen bond was calculated at 173.6, 155.3 and 189.4 cm^{-1} for the trans-gauche dimers A, B and C. It lies at 164.8 and 188.7 cm^{-1} for the trans-trans and gauche-gauche dimers, respectively. The correct assignment of this band is essential for a proper assignment of the ethanol spectrum, because it is close to the OH torsion. Barnes et al. [7] have assigned the low resolution band observed at 213 cm^{-1} in an argon matrix to a superposition of the OH trans-torsion and stretching. It should be noted that the argon matrix spectrum is displaced to higher frequencies when compared with the pure-gas spectrum. The OH-torsion shifts to 211 cm^{-1}.

However, the 213 cm^{-1} value is too large to be assigned to the stretching hydrogen bond. The stretching can be associated with the pattern observed between 160 and 180 cm^{-1} [7]. In addition, the harmonic analysis used in these calculations overestimates the frequency values.

The theoretical calculations for the low frequencies of several isotopic species may help in the assignment. Table 7 shows the low frequencies of four isotopic species: tEtOH-gEtOH, tEtOD-gEtOH, tEtOH-gEtOD and tEtOD-gEtOD. The isotopic effect allows the torsional frequencies in trans and gauche modes to be classified. The stretching of the hydrogen bond appears at 173.6, 172.3, 167.3 and 166.0 cm^{-1}, respectively. Although Barnes et al. [7] place the stretching of Et-OH and Et-OD at 215 and 214 cm^{-1}, we found them at 173.6 and 166.0 cm^{-1}. The isotopic effect predicts a shift of only 7.6 cm^{-1}.

TABLE 4. Harmonic analysis of trans- and gauche-ethanol (units are cm^{-1})

	trans-ethanol			*gauche-ethanol*
	assignments	calc.	exp.	calc.
υ_{21}	*OH tor A''*	305.5	201	273.5
υ_{20}	*CH3 tor A''*	250.4	243	287.6
υ_{19}	*CCO bend A'*	413.3	419	416.6
υ_{18}	*CH2 roc A''*	822.4	801	805.9
υ_{17}	*CCO stretch A'*	907.6	884.6	894.8
υ_{16}	*CH3 roc A'*	1061.2	1032.6	1065.6
υ_{15}	*CCO stretch A'*	1099.5	1089.2	1096.0
υ_{14}	*CH3 roc A''*	1179.2	1062.1	1133.4
υ_{13}	*OH bend A'*	1270.0	1241.3	1280.0
υ_{12}	*CH2 tor A''*	1298.1		1364.8
υ_{11}	*CH3 def A'*	1395.5	1393.7	1395.7
υ_{10}	*CH2 roc A'*	1454.8		1425.0
υ_{9}	*CH3 def A''*	1480.6	1451.6	1484.5
υ_{8}	*CH3 def A'*	1500.3	1451.6	1492.6
υ_{7}	*CH2 def A'*	1524.1	1490	1511.7
υ_{6}	*CH2 stretch A'*	3061.4	2900.5	3074.6
υ_{5}	*CH3 stretch A'*	3092.6	2943.4	3079.2
υ_{4}	*CH2 stretch A''*	3116.3	2948.5	3160.6
υ_{3}	*CH3 stretch A'*	3194.2	2989.4	3178.1
υ_{2}	*CH3 stretch A'*	3202.4	3676.1	3195.5
υ_{1}	*OH stretch A''*	3836.3	2989.4	3822.7

TABLE 5. Harmonic analysis. Low frequency modes (units are cm^{-1})

	A-tg	B-tg	C-tg	tg	gg	trans	gauche	assignment
υ_{48}	30.2	19.4	20.7	28.0	25.3			
υ_{47}	39.4	29.9	43.7	41.0	47.3			
υ_{46}	51.4	34.5	58.5	54.6	63.7			
υ_{45}	80.9	62.3	77.8	75.3	88.5			
υ_{44}	97.9	92.1	94.1	86.7	100.8			
υ_{43}	**173.6**	**155.3**	**189.4**	**164.8**	**188.7**			**H-bond streching**
υ_{42}	269.0	268.2	283.3	264.5	287.0	250.4		*CH_3 torsion*
υ_{41}	285.8	277.2	286.4	268.9	288.1		287.6	*CH_3 torsion*
υ_{40}	329.7	337.4	333.8	346.0	333.6		273.5	*OH torsion*
υ_{39}	423.5	419.9	422.9	415.1	420.4	413.3		*COC bending*
υ_{38}	438.9	441.9	431.1	427.5	434.6		416.6	*COC bending*
υ_{37}	**725.0**	**721.9**	**720.2**	**709.6**	**716.5**	305.7		*OH torsion*

TABLE 6. Harmonic analysis of the dimers (units are cm^{-1})

	A-dimer		B-dimer		C-dimer		t-t	g-g
υ_{36}	810.9	g-CH_2 roc	807.0	g	808.1	g	820.9	809.5
υ_{35}	820.0	t-CH_2 roc	820.3	t	811.0	t	824.3	810.3
υ_{34}	891.9	g-CCO stretch	892.6	g	891.8	g	903.3	891.9
υ_{33}	913.7	t-CCO stretch	914.1	t	900.8	t	915.3	900.6
υ_{32}	1066.6	g-CH_3 roc	1065.5	g	1065.9	g	1067.0	1065.4
υ_{31}	1087.3	g-CCO stretch	1086.4	tg	1087.0	g	1086.9	1087.0
υ_{30}	1090.4	t-CH_3 roc	1094.6	tg	1102.9	t	1088.8	1101.6
υ_{29}	1116.7	t-CCO stretch	1117.2	t	1104.7	t	1117.9	1104.1
υ_{28}	1142.4	g-CH_3 roc	1137.1	g	1140.8	g	1179.3	1140.6
υ_{27}	1182.3	t-CH_3 roc	1182.2	t	1148.8	t	1182.0	1147.8
υ_{26}	1278.4	g-OH bend	1277.1	g	1278.1	g	1269.5	1277.9
υ_{25}	1298.0	t-OH bend	1299.5	t	1299.9	t	1297.7	1299.1
υ_{24}	1337.7	t-CH_2 tor	1338.0	t	1360.8	g	1299.9	1360.5
υ_{23}	1361.0	g-CH_2 tor	1365.2	g	1381.6	t	1333.6	1380.7
υ_{22}	1398.6	tg-CH_3 def	1399.0	t	1399.8	g	1398.9	1399.5
υ_{21}	1401.2	tg-CH_3 def	1399.7	g	1412.5	t	1399.5	1411.3
υ_{20}	1430.5	g-CH_2 roc	1430.8	g	1429.3	g	1453.4	1428.9
υ_{19}	1480.4	t-CH_2 roc	1481.7	t	1454.5	t	1478.0	1454.9
υ_{18}	1482.1	tg-CH_3 def	1482.2	t	1487.5	tg	1482.0	1487.5
υ_{17}	1491.8	tg-CH_3 def	1488.6	g	1493.0	tg	1484.5	1492.5
υ_{16}	1498.2	tg-CH_3 def	1495.2	g	1494.9	tg	1500.7	1496.8
υ_{15}	1501.2	tg-CH_3 def	1501.3	t	1498.8	tg	1504.0	1499.2
υ_{14}	1513.5	g-CH_2 def	1514.3	g	1512.6	tg	1523.0	1512.7
υ_{13}	1525.1	t-CH_2 def	1525.5	t	1513.3	tg	1525.6	1514.4
υ_{12}	3052.6	t-CH_2 stretch	3051.4	t	3063.8	t	3036.1	3065.3
υ_{11}	3082.4	g-CH_2 stretch	3085.8	g	3081.4	t	3076.5	3080.3
υ_{10}	3093.6	t-CH_3 stretch	3093.7	t	3082.4	g	3086.5	3082.6
υ_{9}	3095.9	g-CH_3 stretch	3099.7	g	3096.0	g	3094.4	3096.0
υ_{8}	3104.0	t-CH_2 stretch	3102.7	t	3153.0	t	3097.4	3153.7
υ_{7}	3171.6	g-CH_2 stretch	3174.4	g	3171.7	g	3135.2	3171.7
υ_{6}	3188.1	g-CH_3 stretch	3190.7	g	3180.1	t	3195.9	3179.9
υ_{5}	3194.9	t-CH_3 stretch	3195.1	t	3188.0	g	3200.8	3187.8
υ_{4}	3203.4	t-CH_3 stretch	3203.7	t	3195.8	t	3205.1	3192.9
υ_{3}	3207.4	g-CH_3 stretch	3204.9	g	3207.1	g	3209.9	3206.6
υ_{2}	3651.3	t-OH stretch	3661.0	t	3642.2	t	3672.7	3647.1
υ_{1}	3812.6	g-OH stretch	3808.5	g	3811.4	g	3828.2	3812.1

TABLE 7. Large amplitude frequencies of the *A-dimer* isotopic varieties (units are cm^{-1})

	tEtOH-gEtOH	tEtOD-gEtOH	tEtOH-gEtOD	tEtOD-gEtOD
υ_{48}	30.2	30.0	30.2	30.0
υ_{47}	39.4	39.1	38.9	38.7
υ_{46}	51.4	50.9	51.2	50.8
υ_{45}	80.9	80.9	80.3	80.2
υ_{44}	97.9	96.8	97.6	96.6
υ_{43}	**173.6**	**172.3**	**167.3**	**166.0**
υ_{42}	269.0	268.9	248.8	246.3
υ_{41}	285.8	285.0	269.9	269.9
υ_{40}	329.7	322.0	284.8	284.8
υ_{39}	423.5	417.3	417.4	413.6
υ_{38}	438.9	432.7	435.9	429.3
υ_{38}	**725.0**	**542.6**	**715.6**	**525.3**

The remaining modes of the trans-gauche, trans-trans and gauche-gauche dimers are shown in Table 6. In the case of the trans-gauche dimer they may be assigned to the gauche and trans components. With the formation of the dimer, the shifts of the frequencies in the gauche component are below 10 cm^{-1}. In the case of the trans-component the most significant changes correspond to the OH bending (υ_{25}) and OH stretching (υ_1) modes where the contribution of the trans-OH hydrogen atom is quite important. Both modes lie at 1270.0 and 3826.3 cm^{-1} in the isolated molecule and at 1298.0 and 3651.3 cm^{-1} in the dimer. Likewise, when the trans and gauche frequencies are close together, a large shift is observed in the trans-modes. In this case, both molecules show similar contributions. In the trans-trans and gauche-gauche dimers the largest shifts affect the component containing the H atom of the hydrogen bond.

It may be concluded that dimerization produces a displacement of the spectrum to high frequencies. The additional 6 modes push up the remaining vibrations which are constrained by the presence of the second molecule. The normal modes associated with the hydrogen bond show the largest shifts. Five of the additional modes lie below 100 cm^{-1} and confer non-rigidity to the dimer. The harmonic fundamental of the hydrogen bond stretching lies at 173.6, 155.3 and 189.4 cm^{-1} in the different conformers of the trans-gauche structure, below the OH torsion which lies at 202.6 cm^{-1} in the molecule and moves up to 700 cm^{-1} in the dimers. For the trans-trans and gauche-gauche dimers, this transition lies at 164.8 and 188.7 cm^{-1}. These results have to be considered simultaneously as all conformers coexist in the same sample.

Acknowledgments

This work has been supported by the "Comision Interministerial de Ciencias y Tecnologia" of Spain through Grant PB 96-0882. M.L.S. wishes to acknowledge Prof. Bermejo for his collaboration and help.

References

[1] R. Ludwog, F. Weinhold and T. C. Farrar, *Mol. Phys.*, **97**, 465 (1999).

[2] R. Ludwog, F. Weinhold and T. C. Farrar, *Mol. Phys.*, **97**, 479 (1999).

[3] W. O. George, T. Has, M. F. Hossain, B. F. Jones and R. Lewis, *J. Chem. Soc., Faraday Trans.*, **94**, 2701 (1998).

[4] M. Ehbrecht and F. Huisken, *J. Chem. Phys. A*, **101**, 7768 (1997).

[5] J. R. Durig and R. A. Larsen, *J. Mol. Struct.*, **238**, 195 (1989).

[6] J. P. Perchard and M. L. Josien, *J. Chim. Phys.*, **65**, 1834 (1968).

[7] A. J. Barnes and H. E. Hallam, *Trans. Faraday Soc.*, **66**, 1932 (1970).

[8] J. R. Durig, W. E. Bucy, C. J. Wurrey and L. A. Carreira, *J. Phys. Chem*, **79**, 988 (1975).

[9] S. Coussan, Y. Bouteiller, J. P. Perchard and W. Q. Zheng, *J. Phys. Chem.*, **102**, 5789 (1998).

[10] J. P. Perchard and M. L. Josien, *J. Chim. Phys.*, **65**, 1856 (1968).

[11] J. R. Durig and C. W. Hawley, *J. Phys. Chem.*, **75**, 3993 (1971).

[12] P. G. Jönsson, *Acta Crystallogr., Sect. B: Struct. Crystallogr. Cryst. Chem.*, **32**, 232 (1976).

[10] A. Anderson and W. Smith, *Chem. Phys. Lett.*, **257**, 143 (1996).

[13] C. Talón, M. A. Ramos, S. Vieria, G. J. Cuello, F. J. Bermejo, A. Criado, M. L. Senent, S. M. Bennington, H. E. Fischer and H. Schober, *Phys. Rev. B*, **58**, 745 (1998).

[14] Y. Sasada, M. Takano and T. Satoh, *J. Mol. Struct.*, **38**, 33 (1971).

[15] J. C. Pearson, K. V. L. N. Sastry, E. Herbst and F. C. de Lucia, *J. Mol. Spectrosc.*, **175**, 246 (1996).

[16] M. J. Frisch, G. W. Trucks, H. B. Schlegel, P. M. W. Gill, B. G. Johnson, M. A. Robb, J. R. Cheeseman, T. Keith, G. A. Petersson, J. A. Montgomery, K. Raghavachari, M. A. Al-Laham, V. G. Zakrzewski, J. V. Ortiz, J. B. Foresman, J. Cioslowski, B. B. Stefanov, A. Nanayakkara, M. Challacombe, C. Y. Peng, P. Y. Ayala, W. Chen, M. W. Wong, J. L. Andres, E. S. Replogle, R. Gomperts, R. L. Martin, D. J. Fox, J. S. Binkley, D. J. Defrees, J. Baker, J. P. Stewart, M. Head-Gordon, C. Gonzalez, and J. A. Pople, *Gaussian 94, Revision E.2,* Gaussian, Inc., Pittsburgh PA, 1995.

[17] D. E. Woon and T. H. Dunning Jr., *J. Chem. Phys.*, **98** (1993) 98, 1358; R. A. Kendal, T. H. Dunning, Jr and R. J. Harrison, *J. Chem. Phys.*, **96** (1992) 6796; T. H. Dunning, Jr, *J. Chem. Phys.*, **90** (1989) 1007.

[18] C. H. Görbitz, *J. Mol. Struct. (TEOCHEM)*, **262**, 209 (1992).

[19] A. Vivier-Bunge, V. H. Uc and Y. G. Smeyers, *J. Chem. Phys.*,**109**, 2279 (1998); Y. G. Smeyers, M. L. Senent, V. Botella and D. C. Moule, *J. Chem. Phys.*, **98**, 2754 (1993); M. L. Senent, D. C. Moule and Y. G. Smeyers, *J. Chem. Phys.*, **102**, 5952 (1995); M. L. Senent and Y. G. Smeyers, *J. Chem. Phys.*, **105**, 2789 (1996).

[20] M. L. Senent, *J. Mol. Spectrosc.*,**191**, 265 (1998); M. L. Senent, *Int. J. Quant. Chem.*, **399**, 58 (1995).

VIBRATIONAL FIRST HYPERPOLARIZABILITY OF METHANE AND ITS FLUORINATED ANALOGS

O. QUINET AND B. CHAMPAGNE[1]
Laboratoire de Chimie Théorique Appliquée
Facultés Universitaires Notre-Dame de la Paix
rue de Bruxelles, 61, B-5000 Namur, Belgium

Abstract. The vibrational first hyperpolarizability of methane and its fluorinated analogs has been computed *ab initio* at the Hartree-Fock and Møller-Plesset second-order levels of approximation by adopting the perturbation approach due to Bishop and Kirtman. Both the pure vibrational and the zero-point vibrational averaging contributions have been determined. In the static limit, it turns out that the pure vibrational term is at least of the same order of magnitude as its electronic counterpart and the ratio $|\beta^{\mathrm{v}}/\beta^{\mathrm{e}}|$ increases with the fluorine content. The first-order anharmonicity correction to this pure vibrational term increases also with the fluorine content whereas the zero-point vibrational averaging is one/two order of magnitude smaller and decreases with the fluorine content. The suitability of the bond additivity scheme and of the infinite optical frequency approximation is assessed as well as the consistency between experimentally derived dc-Pockels and second harmonic generation first hyperpolarizability values.

1. Introduction

Methane and its fluorinated derivatives have been the subject of several experimental [1-10] and theoretical [11-25] determinations of their nonlinear optical (NLO) properties. On the one hand, their simple chemical structure is well-suited for comparing these complementary approaches. On the other hand, as early as thirty years ago, these five molecules have been used to assess the validity of the bond additivity model [1], the simplest model to establish structure-property relationships.

[1]Research Associate of the National Fund for Scientific Research (Belgium)

J. Maruani et al. (eds.), New Trends in Quantum Systems in Chemistry and Physics, Volume 1, 375–392.

The linear polarizability (α), first (β) and second (γ) hyperpolarizabilities describe the linear and nonlinear dependences of the dipole moment with respect to static and dynamic external electric fields. The amplitudes of α, β, and γ are related to the importance of the field-induced rearrangements of both the electronic and nuclear charges. Whereas rotations, phase transitions and temperature effects are slow and of little interest for our purpose, the electronic and vibrational effects can be substantial and of considerable interest when optimizing the NLO responses [26]. The clamped-nucleus (CN) approximation [27] is generally adopted to calculate the (hyper)polarizabilities. In this approximation, one assumes that the external electric fields act sequentially rather than simultaneously upon the electronic and nuclear motions : the field-dependence of the electronic distribution - which is associated to the electronic contribution to the (hyper)polarizabilities, (P^{e}, with $P = \alpha$, β or γ) - modifies the potential energy surface (curvature contribution) and provokes nuclear displacements (nuclear relaxation (NR) contribution). In the time-dependent perturbative approach due to Bishop and Kirtman [28, 29] the separation into electronic and vibrational terms leads to expressions given in terms of electrical and mechanical anharmonicities. The electronic (P^{e}) and zero-point vibrational average (ΔP^{ZPVA}) contributions are treated together whereas the remaining terms form the so-called pure vibrational contribution (P^{v}). Such separation into electronic ($P^{e} + \Delta P^{ZPVA}$) and vibrational contributions is appealing when analyzing the frequency dispersion. On the one hand, the dispersion of the $P^{e} + \Delta P^{ZPVA}$ contribution follows a power series expansion in the square of the optical frequencies :

$$\beta\left(-\omega_\sigma;\omega_1,\omega_2\right) = \beta\left(0;0,0\right)\left[1 + A\omega_L^2 + B\omega_L^4 + \ldots\right] \qquad (1)$$

$$\gamma\left(-\omega_\sigma;\omega_1,\omega_2,\omega_3\right) = \gamma\left(0;0,0,0\right)\left[1 + A'\omega_L^2 + B'\omega_L^4 + \ldots\right] \qquad (2)$$

where ω_i ($i = 1,2,3$) are the circular frequencies of the ingoing waves, $\omega_\sigma = \sum_i \omega_i$ the frequency of the outgoing wave and $\omega_L^2 = \omega_\sigma^2 + \sum_i \omega_i^2$. The expansion coefficients (A, B, ..., A', B', ...) are molecule-dependent but, for all-diagonal or orientationally-averaged components, A and A' do not depend upon the NLO process. For other A and A' tensor components as well as for the B and B' coefficients, additional relations have been highlighted for some NLO processes (see for instance Ref. [30]). On the other hand, since their poles are linear combinations of normal mode vibrational frequencies, the vibrational contribution presents large variations in the infrared region. As a result, when the frequency increases it becomes less and less frequency-dependent. This fact justifies the interest for the simple infinite optical frequency approximation ($\omega \to \infty$) [31].

The separation between $P^{\rm v}$ and $P^{\rm e} + \Delta P^{\rm ZPVA}$ combined with their different dispersion relations enable therefore to deduce the response associated with a given NLO process from another process or to check the consistency between different NLO measurements [10]. For instance, in order to deduce either mean electric field-induced second harmonic generation [ESHG ; $\gamma_{//}(-2\omega;\omega,\omega,0)$] or anisotropic dc-Kerr [$\gamma^{\rm K}(-\omega;\omega,0,0)$] total second hyperpolarizability values of CH_4 from the dispersion curve of the other one, both $\gamma^{\rm v,K}(-\omega;\omega,0,0)$ and $\gamma^{\rm v}_{//}(-2\omega;\omega,\omega,0)$ are necessary. Since the A' coefficient for $\gamma_{//}$ and $\gamma_{\perp}$ are not exactly identical and $\gamma^{\rm K} = \frac{3}{2}\left(\gamma_{//} - \gamma_{\perp}\right)$, other informations could also be needed to reach sufficient accuracy. If, like in the infinite optical frequency approximation, one can assume negligible dispersion for $\gamma^{\rm v,K}(-\omega;\omega,0,0)$ and $\gamma^{\rm v}_{//}(-2\omega;\omega,\omega,0)$ only their difference $\Delta_{\omega\to\infty} = \gamma^{\rm v,K}(-\omega;\omega,0,0)_{\omega\to\infty} - \gamma^{\rm v}_{//}(-2\omega;\omega,\omega,0)_{\omega\to\infty}$ is needed. Similarly, Δ can be determined from the knowledge of both ESHG and dc-Kerr dispersion curves while taking care, if necessary, of its dispersion. In the same way, the mean second harmonic generation (SHG ; $\beta_{//}(-2\omega;\omega,\omega)$) and the anisotropic dc-Pockels ($\beta^{\rm K}(-\omega;\omega,0)$) total first hyperpolarizabilities can be related to each other through frequency dispersion of their electronic + ZPVA component (Eq. (1)) and the knowledge of their vibrational counterpart. In particular, $\beta^{\rm v}_{//}(-2\omega;\omega,\omega)$ tends towards zero in the infinite optical frequency limit. Consequently, $\beta^{\rm v,K}(-\omega;\omega,0)_{\omega\to\infty}$ is expected to explain most of the difference between $\beta^{\rm K}(-\omega;\omega,0)$ and $\beta_{//}(-2\omega;\omega,\omega)$ after removing the frequency dispersion of the electronic part.

In a recent study [25] we have determined the static and dynamic vibrational second hyperpolarizabilities for methane and its fluorinated analogs by including electron correlation effects and taking into account the first-order electrical and mechanical anharmonicity corrections. Systematic variations of the different γ contributions have been obtained w.r.t. the number of hydrogen (fluorine) atoms and analysis of the potential energy surface of the C-H and C-F bonds has enabled us to understand some of these γ variations. In particular, it permits us to improve the agreement with the experimental data for $\Delta = \gamma^{\rm v,K}(-\omega;\omega,0,0) - \gamma^{\rm v}_{//}(-2\omega;\omega,\omega,0)$ of CF_4 [10]. On the other hand, in the case of methane, the inclusion of electron correlation and first-order anharmonicity corrections did not help to close the gap between theory and experiment.

In this paper we determine the vibrational (pure vibrational + ZPVA) first hyperpolarizabilities of these five compounds and compare them to their static electronic counterpart as a function of the NLO process. Frequency - dispersion is only considered for the pure vibrational term. As in Ref. [25], electron correlation effects are considered within the Møller-

Plesset perturbation theory limited to second order (MP2) and first-order electrical and mechanical anharmonicity contributions are evaluated. By comparing these results to available [1, 2] experimental data, we discuss the consistency between these values obtained for different NLO processes. For the pure vibrational component, we also explore the replacement of the optical frequency (ω) by $\omega \to \infty$, *i.e.* within the so-called enhanced [5] or infinite optical frequency [31] approximation. Then, we address the validity of the bond additivity scheme for the different β contributions. Section 2 describes the methodological and computational approaches we have adopted. The results and subsequent discussions are given in Sections 3 and 4 while our conclusions are drawn in Section 5.

2. Method

Static/dynamic electric fields applied on a molecule force the system to rearrange its charges in order to minimize its energy. The field-induced dipole moment is described by a Taylor series expansion in the static and/or dynamic (of circular frequency ω_i) electric fields (E),

$$\begin{aligned} \mu_\zeta(\omega_\sigma) &= \mu_\zeta^0 + \sum_\eta \alpha_{\zeta\eta}(-\omega_\sigma;\omega_1) E_\eta(\omega_1) \\ &+ \frac{1}{2}K^{(2)} \sum_{\eta\xi} \beta_{\zeta\eta\xi}(-\omega_\sigma;\omega_1,\omega_2) E_\eta(\omega_1) E_\xi(\omega_2) \\ &+ \ldots \end{aligned} \quad (3)$$

where the $K^{(2)}$ factor is such that the first-order nonlinear responses converge towards the same static limit, and the summations run over the field indices η and ξ associated with the Cartesian coordinates. The total CN first hyperpolarizability is given by the sum of the electronic contribution, the ZPVA correction and the pure vibrational contribution,

$$\beta_{\zeta\eta\xi} = \beta^{\mathrm{e}}_{\zeta\eta\xi} + \Delta\beta^{\mathrm{ZPVA}}_{\zeta\eta\xi} + \beta^{\mathrm{v}}_{\zeta\eta\xi} \quad (4)$$

From the exact sommation-over-states (SOS) expressions, Bishop and Kirtman [28, 29, 32] have derived the CN expressions of β^{v} and $\Delta\beta^{\mathrm{ZPVA}}$ as summations of harmonic and anharmonic terms,

$$\begin{aligned} \beta^{\mathrm{v}}_{\zeta\eta\xi} &= [\mu\alpha]^0 + [\mu\alpha]^{\mathrm{II}} + \ldots \\ &\quad [\mu^3]^{\mathrm{I}} + [\mu^3]^{\mathrm{III}} + \ldots \end{aligned} \quad (5)$$

$$\Delta\beta^{\mathrm{ZPVA}}_{\zeta\eta\xi} = [\beta]^{\mathrm{I}} + [\beta]^{\mathrm{III}} + \ldots \quad (6)$$

where, for instance, $[\mu\alpha]^{\mathrm{II}} = [\mu\alpha]^{2,0} + [\mu\alpha]^{1,1} + [\mu\alpha]^{0,2}$. The notation $[X]^{n,m}$ means n^{th} order electrical anharmonicity [$(n-j)$ is the number of times a $(2+j)^{th}$ derivatives of an electric property appears, with j ranging between 0 and $n-1$] and m^{th} order mechanical anharmonicity [$(m-j)$ is the number of times a cubic ($j=0$) (F_{abc}), a quartic ($j=1$) (F_{abcd}), ... force constant appears, with j ranging between 0 and $m-1$]. In this study, we have only considered the lowest-order non-vanishing terms of each type, *i.e.* $[\mu\alpha]^0 = [\mu\alpha]^{0,0}$, $[\mu^3]^{\mathrm{I}} = [\mu^3]^{1,0} + [\mu^3]^{0,1}$ and $[\beta]^{\mathrm{I}} = [\beta]^{1,0} + [\beta]^{0,1}$. When neglecting the electrical and mechanical anharmonicity corrections, the only $[\mu\alpha]^{0,0}$ term remains : it defines the so-called double harmonic oscillator approximation. In addition to $[\mu\alpha]^{0,0}$, the static nuclear relaxation contribution to $\beta^{\mathrm{v}} = (\beta^{\mathrm{NR}})$ includes also the $[\mu^3]^{\mathrm{I}}$ term [23, 31]. Expressions for the square bracketed contributions are given in Refs. [28, 29] under the form of summations over modes (SOM),

$$[\mu\alpha]^{0,0} = \frac{1}{2}\sum P_{-\sigma,1,2}\sum_{a}\frac{\left(\frac{\partial\mu^{\mathrm{e}}_{\zeta}}{\partial Q_a}\right)_0\left(\frac{\partial\alpha^{\mathrm{e}}_{\eta\xi}}{\partial Q_a}\right)_0}{(\omega_a^2-\omega_\sigma^2)} \tag{7}$$

$$[\mu^3]^{1,0} = \frac{1}{2}\sum P_{-\sigma,1,2}\sum_{a,b}\frac{\left(\frac{\partial\mu^{\mathrm{e}}_{\zeta}}{\partial Q_a}\right)_0\left(\frac{\partial^2\mu^{\mathrm{e}}_{\eta}}{\partial Q_a\partial Q_b}\right)_0\left(\frac{\partial\mu^{\mathrm{e}}_{\xi}}{\partial Q_b}\right)_0}{(\omega_a^2-\omega_\sigma^2)(\omega_b^2-\omega_2^2)} \tag{8}$$

$$[\mu^3]^{0,1} = -\frac{1}{6}\sum P_{-\sigma,1,2}\sum_{a,b,c}F_{abc}\frac{\left(\frac{\partial\mu^{\mathrm{e}}_{\zeta}}{\partial Q_a}\right)_0\left(\frac{\partial\mu^{\mathrm{e}}_{\eta}}{\partial Q_b}\right)_0\left(\frac{\partial\mu^{\mathrm{e}}_{\xi}}{\partial Q_c}\right)_0}{(\omega_a^2-\omega_\sigma^2)(\omega_b^2-\omega_1^2)(\omega_c^2-\omega_2^2)} \tag{9}$$

where, the sums run over the 3N-6 (3N-5 for linear molecules) vibrational normal modes, $\sum P_{-\sigma,1,2}$ is the summation over the 6 permutations of the pairs $(-\omega_\sigma,\zeta)$, $(-\omega_1,\eta)$ and $(-\omega_2,\xi)$ and Q_a is the normal mode coordinate having the frequency $\omega_a = 2\pi\nu_a$. Higher-order terms involve higher-order derivatives of the electronic properties and higher-order anharmonic force constants. Like the pure vibrational terms, the ZPVA terms is also described by SOM expressions,

$$[\beta]^{0,1} = -\frac{1}{4}\sum_a\frac{1}{\omega_a^2}\left(\sum_b\frac{F_{abb}}{\omega_b}\right)\left(\frac{\partial\beta^{\mathrm{e}}}{\partial Q_a}\right)_0 \tag{10}$$

$$[\beta]^{1,0} = \frac{1}{4}\sum_a\frac{1}{\omega_a}\left(\frac{\partial^2\beta^{\mathrm{e}}}{\partial Q_a^2}\right)_0 \tag{11}$$

The evaluation of Eqs. (7)-(11) requires the determination of several partial derivatives of the energy with respect to the electric field and/or the normal coordinates. These are performed either analytically by adopting coupled-perturbed procedures [33] or numerically by using finite difference methods. Table 1 summarizes how the different intermediate properties have been computed.

TABLE 1. Methods used to compute the different quantities involved in the evaluation of the first hyperpolarizability contributions

Properties	RHF	MP2
F_a	A	A
ω_a	A	A
F_{abc}	N	N
μ	A	A
α^e	A	A
β^e	A	N
$\partial\mu/\partial Q$	A	A
$\partial\alpha^e/\partial Q$	A	N
$\partial\beta^e/\partial Q$	N	N
$\partial^2\mu/\partial Q^2$	N	N
$\partial^2\alpha^e/\partial Q^2$	N	N
$\partial^2\beta^e/\partial Q^2$	N	N

A = analytic determination
N = numerical determination

For instance, at the RHF levels the $(\partial\beta^e/\partial Q_a)_0$ quantities were evaluated by calculating β^e for different structures obtained by the addition of different fractions of the normal coordinate Q_a to the equilibrium geometry. In order to reach a sufficient accuracy (of the order of 0.01 - 0.1 a.u.) it was sufficient to use distortions such that the amplitude of the Cartesian displacements is 0.050 Å and 0.025 Å [34] together with one Romberg iteration [35]. This Romberg procedure is used to remove higher order contamination. At the MP2 level, in addition to the finite distortion procedure used to perform the derivative w.r.t Q, the β^e values are obtained numerically from the field-dependent α^e. All the calculations have been done by using the GAUSSIAN94 program [36]. Since the various dipole and (hyper)polarizability derivatives are very sensitive to the geometry, a very tight convergence threshold was chosen in the geometry optimization: limiting the residual forces on the atoms at 1.5×10^{-6} hartree/bohr or hartree/rad. The RHF and MP2 calculations were performed by using the

Sadlej atomic basis set[37]. Its use provides results that are in good agreement with other theoretical results obtained with more extended basis sets. Indeed, by using the series of the aug-cc-pvdz, aug-cc-pvtz and aug-cc-pvqz basis sets, one can see, for instance, that RHF/Sadlej values for CH_4 are always within 10% of the largest basis set values; the variation on the total $\beta^e + \beta^v$ being less than 2% [38].

3. Results and discussions

3.1. PRELIMINARY REMARKS

The ground state optimized geometrical parameters and the vibrational normal mode frequencies have been taken from Ref. [25]. Because CH_4 and CF_4 are of T_d symmetry, β_{xyz} (x, y and z are the coordinates of the cube which circumscribes the tetrahedron) is the unique non-zero first hyperpolarizability tensor component, and moreover only two vibrational modes of F_2 representation (degenerate three times) contribute to the pure vibrational terms. For the sake of compactness, we have only listed experiment-related quantities. In SHG experiment, all applied fields have parallel polarization and the measured quantity is $\beta_{//}$. It corresponds to the vector component of the tensor β in the direction of the permanent dipole moment μ which defines the molecular z axis. It is given by,

$$\beta_{//}(-\omega_\sigma;\omega_1,\omega_2) = \frac{1}{5}\left[\beta_{\xi\xi z}(-\omega_\sigma;\omega_1,\omega_2) + \beta_{\xi z\xi}(-\omega_\sigma;\omega_1,\omega_2) + \beta_{z\xi\xi}(-\omega_\sigma;\omega_1,\omega_2)\right] \quad (12)$$

which for SHG reduces to

$$\beta_{//}(-2\omega;\omega,\omega) = \frac{1}{5}\left[2\beta_{\xi z\xi}(-2\omega;\omega,\omega) + \beta_{z\xi\xi}(-2\omega;\omega,\omega)\right] \quad (13)$$

where we have assumed Einstein summation. In the static limit, this reduces further to,

$$\beta_{//} = \frac{3}{5}\beta_{\xi\xi z} \quad (14)$$

For the dc-Pockels effect — dc-Kerr experiment — the measured quantity is the anisotropy of the refractive index and the related NLO parameter is $\beta^K = \frac{3}{2}(\beta^K_{//} - \beta^K_\perp)$. Since for dc-Pockels effect $\beta_{\zeta\eta\xi}(-\omega;\omega,0) = \beta_{\eta\zeta\xi}(-\omega;\omega,0)$,

$$\beta^K_{//}(-\omega;\omega,0) = \frac{1}{5}\left(\beta_{\xi\xi z}(-\omega;\omega,0) + 2\beta_{z\xi\xi}(-\omega;\omega,0)\right) \quad (15)$$

$$\beta^K_\perp(-\omega;\omega,0) = \frac{1}{5}\left(2\beta_{\xi\xi z}(-\omega;\omega,0) - \beta_{\xi z\xi}(-\omega;\omega,0)\right) \quad (16)$$

and therefore,

$$\beta^{K}(-\omega;\omega,0)=\frac{3}{10}\left(3\beta_{\xi\xi z}(-\omega;\omega,0)-\beta_{\xi z\xi}(-\omega;\omega,0)\right) \tag{17}$$

Again, in the static limit,

$$\beta^{K}=\frac{3}{5}\beta_{\xi\xi z}=\beta_{//} \tag{18}$$

Tables 2-4 list the RHF and MP2 the electronic and vibrational static first hyperpolarizability contributions for the CH_3F, CH_2F_2 and CHF_3 molecules whereas the RHF and MP2 xyz tensor component of the electronic and vibrational static first hyperpolarizability contributions for methane and tetrafluoromethane are given in Tables 5 and 6, respectively. Since $\beta_{//}$ and $\beta_{\perp}$ are zero, only hyper-Rayleigh scattering measurements can probe β_{xyz} [39]. Figs. 1(2) shows at the RHF(MP2) level of approximation the evolution with the number of fluorine atoms of the electronic and vibrational $\beta_{//}$ contributions as well as of the dipole moment.

3.2. CH_3F, CH_2F_2 AND CHF_3

In the static limit, the electronic and the pure vibrational contributions ($[\mu\alpha]^0_{//}+[\mu^3]^I_{//}$) are of similar amplitude but of opposite sign. The RHF(MP2) $\left|\beta^{v}_{//}/\beta^{e}_{//}\right|$ ratio is 0.96(0.81), 1.47(1.37) and 2.12(2.07) for CH_3F, CH_2F_2 and CHF_3, respectively. In fact, the $\left|\beta^{v}_{//}/\beta^{e}_{//}\right|$ ratio increases with the number of fluorine atoms. In any case the double harmonic term dominates the vibrational reponse but the first-order anharmonic contributions is not negligible. $[\mu^3]^{1,0}_{//}$ is smaller than $[\mu^3]^{0,1}_{//}$ and of opposite sign to $[\mu\alpha]^0_{//}$. The importance of $[\mu^3]^{0,1}_{//}$ ranges between 49% and 72% of the double harmonic contribution. As observed in Ref. [25] for the second hyperpolarizability, the higher the fluorine content, the higher the anharmonicity contributions. The ZPVA correction is small compared to the pure vibrational term but still represents 10-14 percents of the electronic term.

We observe similar trends among the studied quantities at both levels of approximation (Figs. 1-2). μ and $[\mu\alpha]^0_{//;\omega=0}$ show a maximum amplitude for the CH_2F_2 molecule. In fact, there is a good linear relationship between μ and $[\mu\alpha]^0_{//;\omega=0}$ (the square of the correlation coefficient (R^2) = 0.997 at the RHF level and 0.996 at the MP2 level). On the other hand, the electronic and the ZPVA contributions present a minimum amplitude for the CH_3F molecule whereas for the anharmonic contributions the maximum is obtained for the CH_2F_2 (CHF_3) molecule at the RHF(MP2) level of approximation.

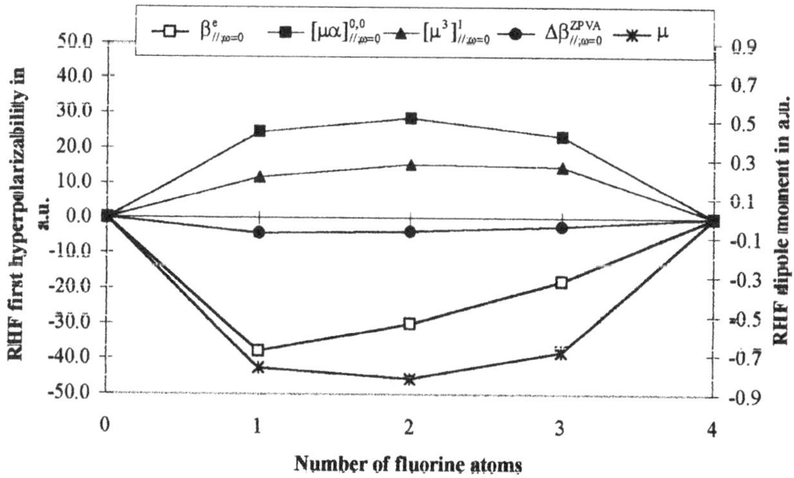

Figure 1. Evolution with the number of fluorine atoms of the electronic and vibrational contributions to $\beta_{//}$ as well as of the dipole moment computed at the RHF level of approximation.

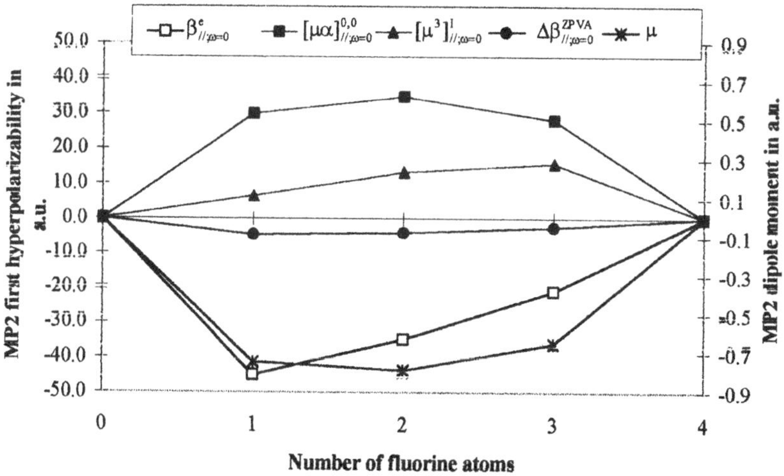

Figure 2. Evolution with the number of fluorine atoms of the electronic and vibrational contributions to $\beta_{//}$ as well as of the dipole moment computed at the MP2 level of approximation.

TABLE 2. Dipole moment, electronic and vibrational static first hyperpolarizability contributions of CH_3F calculated at the RHF and MP2 levels of approximation. All the values are given in a.u. (1.0 a.u. of electric dipole moment = 8.4784358 × 10^{-30} Cm = 2.5418 × 10^{-18} esu ; 1.0 a.u. of first hyperpolarizability = 3.206361 × 10^{-53} $C^3m^3J^{-2}$ = 8.6392 × 10^{-33} esu)

CH_3F	RHF	MP2	Other theoretical work
μ	-0.77	-0.74	
$\beta^{e}_{//;\omega=0}$	-37.9	-45.0	-25.6[a] -40.3[b]
$[\mu\alpha]^{0,0}_{//;\omega=0}$	24.6	29.9	
$[\mu^3]^{1,0}_{//;\omega=0}$	-4.1	-8.3	
$[\mu^3]^{0,1}_{//;\omega=0}$	15.9	14.8	
$\beta^{v}_{//;\omega=0}$	36.4	36.4	
$[\beta]^{0,1}_{//;\omega=0}$	-1.1	-1.0	
$[\beta]^{1,0}_{//;\omega=0}$	-3.3	-3.6	
$\Delta\beta^{ZPVA}_{//;\omega=0}$	-4.4	-4.6	

[a] RHF results due to Sekino and Bartlett [12].
[b] MP2 results due to Rice *et al* [13].

TABLE 3. Dipole moment, electronic and vibrational static first hyperpolarizability contributions of CH_2F_2 calculated at the RHF and MP2 levels of approximation. All the values are given in a.u.

CH_2F_2	RHF	MP2	Other theoretical work
μ	-0.82	-0.79	
$\beta^{e}_{//;\omega=0}$	-30.0	-35.0	-25.1[a]
$[\mu\alpha]^{0,0}_{//;\omega=0}$	28.6	34.7	
$[\mu^3]^{1,0}_{//;\omega=0}$	-3.7	-9.4	
$[\mu^3]^{0,1}_{//;\omega=0}$	19.1	22.6	
$\beta^{v}_{//;\omega=0}$	44.0	48.0	
$[\beta]^{0,1}_{//;\omega=0}$	-0.9	-0.8	
$[\beta]^{1,0}_{//;\omega=0}$	-3.0	-3.3	
$\Delta\beta^{ZPVA}_{//;\omega=0}$	-3.9	-4.2	

[a] RHF results due to Sekino and Bartlett [12].

TABLE 4. Dipole moment, electronic and vibrational static first hyperpolarizability contributions of CHF_3 calculated at the RHF and MP2 levels of approximation. All the values are given in a.u.

CHF_3	RHF	MP2	Other theoretical work
μ	-0.69	-0.65	
$\beta^{e}_{//;\omega=0}$	-18.0	-21.1	-20.0[a]
			-17.2[b]
$[\mu\alpha]^{0,0}_{//;\omega=0}$	23.4	28.0	
$[\mu^3]^{1,0}_{//;\omega=0}$	-0.8	-4.8	
$[\mu^3]^{0,1}_{//;\omega=0}$	15.5	20.4	
$\beta^{v}_{//;\omega=0}$	38.2	43.7	
$[\beta]^{0,1}_{//;\omega=0}$	-0.6	-0.6	
$[\beta]^{1,0}_{//;\omega=0}$	-1.8	-1.9	
$\Delta\beta^{ZPVA}_{//;\omega=0}$	-2.4	-2.5	

[a] RHF results due to Sekino and Bartlett [12].
[b] RHF results due to Karna and Dupuis [14].

3.3. CH_4 AND CF_4

When including the first-order anharmonicity correction, the pure vibrational contribution to the static β_{xyz} of CH_4 amounts to 53.3% and 39.2% of the static electronic part at the RHF and MP2 levels of approximation, respectively. The double harmonic term dominates the pure vibrational response. On the other hand, due to the substantial $[\mu^3]^{0,1}_{xyz}$ term, β^{v}_{xyz} of CF_4 is 7.9 times larger than β^{e}_{xyz} in the static limit. However, in the dynamic limit the β^{v}_{xyz} of both CH_4 and CF_4 will become more similar (see next paragraph). Indeed, in the infinite optical frequency limit, the β^{v}_{xyz} associated to the dc-Pockels effect tends towards $[\mu\alpha]^{0,0}_{xyz;\omega=0}/3$ and amounts to about -2 au. (-3 au.) for CH_4 (CF_4). The ZPVA term is small for CH_4 and negligible for CF_4. It is interesting to point out that the main effect of including electron correlation corrections is a general decrease (increase) of all the contributions for CH_4 (CF_4) (except $[\mu^3]^{1,0}_{xyz}$ of CH_4). Taking into account differences in basis set and the inclusion of $[\mu\alpha]^{II}_{xyz}$ at the RHF level, a general good agreement between our results and those of Bishop *et al.*[24] is obtained both at RHF and MP2 levels.

TABLE 5. xyz tensor component of the electronic and vibrational static first hyperpolarizability contributions of methane calculated at the RHF and MP2 levels of approximation. All the values are given in a.u.

CH_4	RHF	MP2	Other theoretical work
$\beta^{e}_{xyz;\omega=0}$	-11.8	-9.7	-5.9[a] -10.9[b] -8.1[c]
$[\mu\alpha]^{0,0}_{xyz;\omega=0}$	-7.0	-4.7	-7.7[b] -7.3[c]
$[\mu^3]^{1,0}_{xyz;\omega=0}$	0.8	1.0	0.8[b]
$[\mu^3]^{0,1}_{xyz;\omega=0}$	-0.1	-3.8×10^{-2}	-0.1[b]
$\beta^{v}_{xyz;\omega=0}$	-6.3	-3.7	-9.1[b,d] -7.3[c,e]
$[\beta]^{0,1}_{xyz;\omega=0}$	-0.6	-0.5	
$[\beta]^{1,0}_{xyz;\omega=0}$	-0.9	-0.7	
$\Delta\beta^{ZPVA}_{xyz;\omega=0}$	-1.5	-1.2	-1.5[b] -1.1[c]

[a] RHF results due to Sekino and Bartlett [12]. [b] RHF results due to Bishop *et al.* [24]. [c] MP2 results due to Bishop *et al.* [24]. [d] $\beta^{v} = [\mu\alpha]^{0} + [\mu\alpha]^{II} + [\mu^3]^{I}$. [e] $\beta^{v} = [\mu\alpha]^{0}$.

TABLE 6. xyz tensor component of the electronic and vibrational static first hyperpolarizability contributions of tetafluoromethane calculated at the RHF and MP2 levels of approximation. All the values are given in a.u.

CF_4	RHF	MP2	Other theoretical work
$\beta^{e}_{xyz;\omega=0}$	-4.8	-6.1	-3.1[a] -3.7[b] -4.2[c]
$[\mu\alpha]^{0,0}_{xyz;\omega=0}$	-9.8	-11.4	-10.7[b] -10.7[c]
$[\mu^3]^{1,0}_{xyz;\omega=0}$	3.5	8.0	3.2[b]
$[\mu^3]^{0,1}_{xyz;\omega=0}$	-31.3	-44.4	-25.9[b]
$\beta^{v}_{xyz;\omega=0}$	-37.7	-47.9	-34.4[b,d] -10.7[c,e]
$[\beta]^{0,1}_{xyz;\omega=0}$	2×10^{-2}	-1×10^{-2}	
$[\beta]^{1,0}_{xyz;\omega=0}$	9×10^{-2}	3×10^{-2}	
$\Delta\beta^{ZPVA}_{xyz;\omega=0}$	1.1×10^{-1}	2×10^{-2}	0.1[b] -0.1[c]

[a] RHF results due to Sekino and Bartlett [12]. [b] RHF results due to Bishop *et al.* [24]. [c] MP2 results due to Bishop *et al.* [24]. [d] $\beta^{v} = [\mu\alpha]^{0} + [\mu\alpha]^{II} + [\mu^3]^{I}$. [e] $\beta^{v} = [\mu\alpha]^{0}$.

3.4. FREQUENCY-DEPENDENT VIBRATIONAL FIRST HYPERPOLARIZABILITY

dc Pockels and SHG vibrational first hyperpolarizabilities are given in Tables 7-11 for four common laser wavelengths and for the infinite optical frequency approximation ($\omega \to \infty$). As expected, when ω increases, β^{v} tends toward $[\mu\alpha]^{0,0}_{\omega=0}/3$ and zero for the dc-Pockels and SHG phenomena, respectively. For the all five molecules the difference between the true frequency-dependent and the $\omega \to \infty$ values is of the order or less than 1 au. The maximum relative differences appear for the CH_4 molecule due to its larger normal mode vibrational frequencies whereas they get much smaller for the CF_4 of which all the frequencies are smaller than 1500 cm^{-1}. Consequently, it turns out that the infinite optical frequency approximation is also a suitable approximation scheme for evaluating the β of $CH_{4-n}F_n$ ($n = 0 - 4$) in the UV-visible frequency domain.

Our best estimates for the $\beta^{\mathrm{v,K}}(-\omega;\omega,0)$ of CH_2F_2 (17.3 au.) and CHF_3 (13.7 au.) can be compared with the values determined by Bishop[11] (CHF_3 : ± 31 au.) and by Elliot and Ward [5] (CH_2F_2 : 28.1 au. ; CHF_3 : 50.2 au.) using spectroscopic data. Since the sign of the individual normal mode contributions to β^{v} is undetermined in these experiment-based calculations, the agreement is rather good.

3.5. BOND ADDITIVITY SCHEME

Together with the atom additivity scheme, the bond additivity scheme aims at providing transferable parameters for describing any new set of compounds[1, 2, 40, 41]. Their success has rather been limited to the polarizability whereas for the hyperpolarizabilities, it failed in several instances [2, 25]. Within the bond additivity scheme, the $\beta_{//}$ of CHF_3, CH_2F_2 and CHF_3 satisfies the relation,

$$\beta_{//}(CH_3F) = (\sqrt{3}/2)\beta_{//}(CH_2F_2) = \beta_{//}(CHF_3) = \beta_{//}(\text{C-H}) - \beta_{//}(\text{C-F}) \quad (19)$$

which accounts for the vector nature of $\beta_{//}$ and assumes exact tetrahedral angles. Analysis of the results in Tables 2-6 shows that at the exception of μ and $[\mu\alpha]^{0,0}_{\omega=0}$, this scheme fails: the contributions from the bond interactions can not be neglected. Similar conclusion was also drawn for the corresponding $[\mu\beta]^{0}_{//;\omega=0}$ term whereas for $[\alpha^2]^{0}_{//;\omega=0}$ and $[\mu^2\alpha]^{\mathrm{I}}_{//;\omega=0}$ (the three are $\gamma^{\mathrm{v}}_{//}$ contributions), it performs well [25].

TABLE 7. Vibrational contribution to the dynamic first hyperpolarizability (dc-Pockels and SHG) of CH_4 calculated at the RHF and MP2 levels of approximation for common laser wavelengths and for the infinite optical frequency approximation ($\omega \to \infty$). All the values are given in a.u.

CH_4	$\beta^{v}_{xyz}(-\omega;\omega,0)$		$\beta^{v}_{xyz}(-2\omega;\omega,\omega)$	
	RHF	MP2	RHF	MP2
$\lambda = 1064$ nm, $\hbar\omega = 1.165$ eV	-1.7	-1.1	0.8	0.5
$\lambda = 694.3$ nm, $\hbar\omega = 1.786$ eV	-2.1	-1.4	0.3	0.2
$\lambda = 632.8$ nm, $\hbar\omega = 1.959$ eV	-2.1	-1.4	0.3	0.2
$\lambda = 514.5$ nm, $\hbar\omega = 2.410$ eV	-2.2	-1.5	0.2	0.1
$\omega \to \infty$	-2.4	-1.6	0.0	0.0

TABLE 8. Vibrational contribution to the dynamic first hyperpolarizability (dc-Pockels and SHG) of CH_3F calculated at the RHF and MP2 levels of approximation for common laser wavelengths and for the infinite optical frequency approximation ($\omega \to \infty$). All the values are given in a.u.

CH_3F	$\beta^{v,K}(-\omega;\omega,0)$		$\beta^{v}_{//}(-2\omega;\omega,\omega)$	
	RHF	MP2	RHF	MP2
$\lambda = 1064$ nm, $\hbar\omega = 1.165$ eV	15.5	15.7	0.7	0.7
$\lambda = 694.3$ nm, $\hbar\omega = 1.786$ eV	14.8	15.1	0.3	0.2
$\lambda = 632.8$ nm, $\hbar\omega = 1.959$ eV	14.7	15.1	0.2	0.2
$\lambda = 514.5$ nm, $\hbar\omega = 2.410$ eV	14.6	15.0	0.1	0.1
$\omega \to \infty$	14.4	14.8	0.0	0.0

TABLE 9. Vibrational contribution to the dynamic first hyperpolarizability (dc-Pockels and SHG) of CH_2F_2 calculated at the RHF and MP2 levels of approximation for common laser wavelengths and for the infinite optical frequency approximation ($\omega \to \infty$). All the values are given in a.u.

CH_2F_2	$\beta^{v,K}(-\omega;\omega,0)$		$\beta^{v}_{//}(-2\omega;\omega,\omega)$	
	RHF	MP2	RHF	MP2
$\lambda = 1064$ nm, $\hbar\omega = 1.165$ eV	16.0	17.8	0.7	0.5
$\lambda = 694.3$ nm, $\hbar\omega = 1.786$ eV	15.5	17.4	0.2	0.2
$\lambda = 632.8$ nm, $\hbar\omega = 1.959$ eV	15.4	17.3	0.2	0.2
$\lambda = 514.5$ nm, $\hbar\omega = 2.410$ eV	15.3	17.3	0.1	0.1
$\omega \to \infty$	15.2	17.3	0.0	0.0

TABLE 10. Vibrational contribution to the dynamic first hyperpolarizability (dc-Pockels and SHG) of CHF_3 calculated at the RHF and MP2 levels of approximation for common laser wavelengths and for the infinite optical frequency approximation ($\omega \rightarrow \infty$). All the values are given in a.u.

CHF_3	$\beta^{v,K}(-\omega;\omega,0)$		$\beta^{v}_{//}(-2\omega;\omega,\omega)$	
	RHF	MP2	RHF	MP2
$\lambda = 1064$ nm, $\hbar\omega = 1.165$ eV	12.2	13.8	0.3	0.2
$\lambda = 694.3$ nm, $\hbar\omega = 1.786$ eV	12.0	13.7	0.1	0.1
$\lambda = 632.8$ nm, $\hbar\omega = 1.959$ eV	11.9	13.7	0.1	0.1
$\lambda = 514.5$ nm, $\hbar\omega = 2.410$ eV	11.9	13.7	0.1	0.1
$\omega \rightarrow \infty$	11.8	13.6	0.0	0.0

TABLE 11. Vibrational contribution to the dynamic first hyperpolarizability (dc-Pockels and SHG) of CF_4 calculated at the RHF and MP2 levels of approximation for common laser wavelengths and for the infinite optical frequency approximation ($\omega \rightarrow \infty$). All the values are given in a.u.

CF_4	$\beta^{v}_{xyz}(-\omega;\omega,0)$		$\beta^{v}_{xyz}(-2\omega;\omega,\omega)$	
	RHF	MP2	RHF	MP2
$\lambda = 1064$ nm, $\hbar\omega = 1.165$ eV	-3.1	-3.6	0.3	0.3
$\lambda = 694.3$ nm, $\hbar\omega = 1.786$ eV	-3.2	-3.7	0.1	0.1
$\lambda = 632.8$ nm, $\hbar\omega = 1.959$ eV	-3.2	-3.7	0.1	0.1
$\lambda = 514.5$ nm, $\hbar\omega = 2.410$ eV	-3.3	-3.7	0.1	0.1
$\omega \rightarrow \infty$	-3.3	-3.8	0.0	0.0

4. Comparison with experiment

Table 12 lists β values deduced from Kerr ($\beta^{K}(-\omega;\omega,0)$, $\lambda = 632.8$ nm) and SHG ($\beta_{//}(-2\omega;\omega,\omega)$, $\lambda = 694.3$ nm) gas phase experiments on CH_3F, CH_2F_2 and CHF_3 together with the theoretical estimates for the static electronic + ZPVA and dynamic vibrational responses. Provided one approximately accounts for the frequency-dispersion of $\beta^{e}_{//} + \Delta\beta^{ZPVA}_{//}$ (larger for $\beta(-2\omega;\omega,\omega)$ than $\beta(-\omega;\omega,0)$), the negligible $\beta^{v}_{//}(-2\omega;\omega,\omega)$ value, and one considers the error bars on the experimental values as well as the limit of our theoretical estimates, one can check the consistency between the different NLO measurements. In the case of the CH_3F, the values obtained for the two NLO processes are consistent. For CH_2F_2 the much smaller

$\beta^{K}(-\omega;\omega,0)$ value w.r.t. $\beta_{//}(-2\omega;\omega,\omega)$ can partly be explained by the β^{K} vibrational contribution which is of opposite sign to $\beta^{e}_{//}$. However, it is not sufficient whereas $\beta^{e}_{//}(CH_2F_2)$ is in close agreement with the experimental $\beta_{//}(-2\omega;\omega,\omega)$ values. In the case of CHF_3, our theoretical values cannot explain the large (and of different sign) experimentally-derived $\beta^{K}(-\omega;\omega,0)$ value whereas the SHG result matches again our theoretical estimates. Refinement of the Kerr experimental data, which has to be isolated from several other electrical properties (μ, α and γ) defining the measured Kerr constant, is probably needed for a better agreement between theory and experiment.

TABLE 12. Comparison between theory and experiment. Experimental results are provided for the dc-Pockels effect (λ = 632.8 nm) and the second harmonic generation (λ = 694.3 nm). The theoretical results are given for the electronic + ZPVA static first hyperpolarizability and the vibrational dynamic first hyperpolarizability for the same processes and the same wavelengths. All the values are given in a.u. [Note that in the static limit, $\beta^{e,K} + \Delta\beta^{ZPVA,K} = \beta^{e}_{//} + \Delta\beta^{ZPVA}_{//}$]

	$\beta^{K}(-\omega;\omega,0)$ (632.8 nm)	$\beta_{//}(-2\omega;\omega,\omega)$ (694.3 nm)	$\beta^{e}_{//}(0;0,0)+$ $\Delta\beta^{ZPVA}_{//}(0;0,0)$	$\beta^{v,K}(-\omega;\omega,0)$ (632.8 nm)	$\beta^{v}_{//}(-2\omega;\omega,\omega)$ (694.3 nm)
CH_3F	-59 ± 31^{a}	-57.0 ± 4.2^{b}	-49.6	15.1	0.2
		-58.2 ± 1.2^{c}			
CH_2F_2	-12.8 ± 3.1^{a}	-42.1 ± 1.9^{b}	-38.9	17.3	0.2
CHF_3	84 ± 31^{a}	-25.2 ± 0.9^{b}	-23.6	13.7	0.1
		-27.8 ± 0.6^{c}			

[a] experimental results due to Buckingham and Orr [1].
[b] experimental results due to Ward and Bigio [2].
[c] experimental results due to Shelton and Buckingham [4].

5. Conclusions

The vibrational first hyperpolarizability of methane and its fluorinated analogs has been computed *ab initio* at the Hartree-Fock and MP2 levels of approximation by adopting the perturbation approach due to Bishop and Kirtman. By including the lowest-order non vanishing terms of each type, both the pure vibrational and the zero-point vibrational averaging contributions have been determined. In the static limit, it turns out that the pure vibrational term is at least of the same order of magnitude as its electronic counterpart and the ratio $|\beta^{v}/\beta^{e}|$ increases with the fluorine content. The first-order anharmonicity correction to this pure vibrational term increases also with the fluorine content whereas the zero-point vibrational averaging is one/two order of magnitude smaller and decreases with the fluorine content. The study of the frequency dispersion for the five molecules reveals that the infinite optical frequency approximation is a satisfactory approximation for evaluating the vibrational dc-Pockels responses.

Since the bond additivity scheme is only valid for the vibrational part within the double harmonic oscillator approximation, it turns out that bond interactions are important for the anharmonicity corrections and the electronic counterpart.

These theoretical determinations of $\beta^{\mathrm{v,K}}(-\omega;\omega,0)$ and $\beta^{\mathrm{v}}_{//}(-2\omega;\omega,\omega)$ on one side and of $\beta^{\mathrm{e}}_{//} + \Delta\beta^{\mathrm{ZPVA}}_{//}(0;0,0)$ on the other side have enabled to address the consistency between dc-Pockels- and SHG-derived experimental β values for CH_3F, CH_2F_2 and CHF_3. It turns out that for CH_2F_2 and, more obviously, for CHF_3, refined experimental data are needed.

Acknowledgements

The authors are pleased to acknowledge Prof. J.M. André, Prof. D.M. Bishop, Dr. E.A. Perpète, and Dr. D. Jacquemin for stimulating discussions. O.Q. thanks the 'Fonds pour la Formation et la Recherche dans l'Industrie et dans l'Agriculture' (FRIA) for financial support. B.C. thanks the Belgian National Fund for Scientific Research for his Research Associate position. The calculations have been performed on the IBM SP2 of the Namur Scientific Computing Facility (Namur-SCF). The authors gratefully acknowledge the financial support of the FNRS-FRFC, the 'Loterie Nationale' for the convention N° 2.4519.97 and the Belgian National Interuniversity Research Program on 'Sciences of Interfacial and Mesoscopic Structures' (PAI/IUAP N^o P4/10).

References

1. A.D. Buckingham and B.J. Orr, *Trans. Faraday Soc.* **65**, 673 (1969).
2. J.F. Ward and I.J. Bigio, *Phys. Rev. A* **11**, 60 (1975).
3. C.K. Miller and J.F. Ward, *Phys. Rev. A* **16**, 1179 (1977).
4. D.P. Shelton and A.D. Buckingham, *Phys. Rev. A* **26**, 2787 (1982).
5. D.S. Elliott and J.F. Ward, *Mol. Phys.* **51**, 45 (1984).
6. J.F. Ward and D.S. Elliott, *J. Chem. Phys.* **80**, 1003 (1984).
7. D.P. Shelton, *Phys. Rev. A* **34**, 304 (1986).
8. Z. Lu and D.P. Shelton, *J. Chem. Phys.* **87**, 1969 (1987).
9. D.P. Shelton, *Phys. Rev. A* **42**, 2578 (1990).
10. D.P. Shelton and J.J. Palubinskas, *J. Chem. Phys.* **104**, 2482 (1996).
11. D.M. Bishop, *Mol. Phys.* **42**, 1219 (1981).
12. H. Sekino and R.J. Bartlett, *J. Chem. Phys.* **85**, 976 (1986).
13. J.E. Rice, R.D. Amos, S.M. Colwell, N.C. Handy, and J. Sanz, *J. Chem. Phys.* **171**, 201 (1990).
14. S.P. Karna and M. Dupuis, *J. Chem. Phys* **92**, 7418 (1990).

15. J. Martí, J.L. Andrés, J. Bertrán, and M. Duran, *Mol. Phys.* **80**, 625 (1993).
16. G. Maroulis, *Chem. Phys. Lett.* **226**, 420 (1994).
17. D.M. Bishop and J. Pipin, *J. Chem. Phys.* **103**, 4980 (1995).
18. G. Maroulis, *Chem. Phys. Lett.* **259**, 654 (1996).
19. D.M. Bishop and E.K. Dalskov, *J. Chem. Phys.* **104**, 1004 (1996).
20. D.M. Bishop and S.P.A. Sauer, *J. Chem. Phys.* **107**, 8502 (1997).
21. J.M. Luis, J. Martí, M. Duran, and J.L. Andrés, *Chem. Phys.* **217**, 29 (1997).
22. P. Norman, *PhD Thesis Nonlinear Optical Properties of Fullerenes, Oligomers, and Solutions*, chapter 19, Linköping, 1998.
23. J.M. Luis, J. Martí, M. Duran, J.L. Andrés, and B. Kirtman, *J. Chem. Phys.* **108**, 4123 (1998).
24. D.M. Bishop, F.L. Gu, and S.M. Cybulski, *J. Chem. Phys.* **109**, 8407 (1998).
25. O. Quinet and B. Champagne, *J. Chem. Phys.* **109**, 10594 (1998).
26. B. Kirtman and B. Champagne, *Int. Rev. Phys. Chem.* **16**, 389 (1997).
27. D.M. Bishop, B. Kirtman, and B. Champagne, *J. Chem. Phys.* **107**, 5780 (1997).
28. D.M. Bishop and B. Kirtman, *J. Chem. Phys.* **95**, 2646 (1991), *ibidem*, **97**, 5255 (1992).
29. D.M. Bishop, J.M. Luis, and B. Kirtman, *J. Chem. Phys.* **108**, 10013 (1998).
30. C. Hättig, *Mol. Phys.* **94**, 455 (1998).
31. D.M. Bishop, M. Hasan, and B. Kirtman, *J. Chem. Phys.* **103**, 4157 (1995).
32. D.M. Bishop, *Adv. Chem. Phys.* **104**, 1 (1998).
33. for example, see Y. Yamaguchi, Y. Osamura, J.D. Goddard, and H.F. Schaefer III, *A New Dimension to Quantum Chemistry : Analytic Derivative Methods in Ab Initio Molecular Electronic Structure Theory*, Oxford University Press, Oxford, 1994.
34. B. Champagne, *Chem. Phys. Lett.* **261**, 57 (1996).
35. P.J. Davis and P. Rabinowitz, *Numerical Integration*, p. 166, Blaisdell Publishing Company, London, 1967.
36. GAUSSIAN 94, Revision B.1, M.J. Frisch, G.W. Trucks, H.B. Schlegel, P.M.W. Gill, B.G. Johnson, M.A. Robb, J.R. Cheeseman, T. Keith, G.A. Petersson, J.A. Montgomery, K. Raghavachari, M.A. Al-Laham, V.G. Zakrezewski, J.V. Ortiz, J.B. Foresman, J. Cioslowski, B.B. Stefanov, A. Nanayakkara, M. Challacombe, C.Y. Peng, P.Y. Ayala, W. Chen, M.W. Wong, J.L. Andres, E.S. Repolgle, R. Gomperts, R.L. Martin, D.J. Fox, J.S. Binkley, D.J. Defrees, J. Baker, J.P. Stewart, M. Head-Gordon, C. Gonzalez, and J.A. Pople, Gaussian Inc., Carnegie-Mellon University, Pittsburg, PA, 1995.
37. A.J. Sadlej, *Coll. Czech. Chem. Commun.* **53**, 1995 (1992). K. Andersson and A.J. Sadlej, *Phys. Rev. A* **46**, 2356 (1992).
38. Basis set effect upon the different contributions to the static first hyperpolarizability of methane.

Basis sets	β^{e}_{xyz}	$[\mu\alpha]^{0,0}_{xyz}$	$[\mu^3]^{1,0}_{xyz}$	$[\mu^3]^{0,1}_{xyz}$	Total static
aug-cc-pvdz	-12.2	-8.1	0.9	-0.1	-19.5
aug-cc-pvtz	-11.4	-7.8	0.8	-0.1	-18.5
aug-cc-pvqz	-11.0	-7.8			(-18.8)
Sadlej	-11.8	-7.0	0.8	-0.1	-18.1

39. E. Hendrickx, K. Clays, and A. Persoons, *Acc. Chem. Res.* **31**, 675 (1998).
40. J.M. Stout and C.E. Dykstra, *J. Phys. Chem. A* **102**, 1576 (1998).
41. K.J. Miller, *J. Am. Chem. Soc.* **112**, 8533 (1990).

STAGGERING EFFECTS IN NUCLEAR AND MOLECULAR SPECTRA

DENNIS BONATSOS, N. KAROUSSOS
Institute of Nuclear Physics, N.C.S.R. "Demokritos"
GR-15310 Aghia Paraskevi, Attiki, Greece
C. DASKALOYANNIS
Department of Physics, Aristotle University of Thessaloniki
GR-54006 Thessaloniki, Greece
S. B. DRENSKA, N. MINKOV, P. P. RAYCHEV, R. P. ROUSSEV
Institute for Nuclear Research and Nuclear Energy, Bulgarian Academy of Sciences
72 Tzarigrad Road, BG-1784 Sofia, Bulgaria
AND
J. MARUANI
Laboratoire de Chimie Physique, CNRS and UPMC
11, rue Pierre et Marie Curie, F-75005 Paris, France

Abstract. It is shown that the recently observed $\Delta J = 2$ staggering effect (i.e. the relative displacement of the levels with angular momenta J, $J+4$, $J+8, \ldots$, relatively to the levels with angular momenta $J+2$, $J+6$, $J+10$, $\ldots$) seen in superdeformed nuclear bands is also occurring in certain electronically excited rotational bands of diatomic molecules (YD, CrD, CrH, CoH), in which it is attributed to interband interactions (bandcrossings). In addition, the $\Delta J = 1$ staggering effect (i.e. the relative displacement of the levels with even angular momentum J with respect to the levels of the same band with odd J) is studied in molecular bands free from $\Delta J = 2$ staggering (i.e. free from interband interactions/bandcrossings). Bands of YD offer evidence for the absence of any $\Delta J = 1$ staggering effect due to the disparity of nuclear masses, while bands of sextet electronic states of CrD demonstrate that $\Delta J = 1$ staggering is a sensitive probe of deviations from rotational behaviour, due in this particular case to the spin–rotation and spin–spin interactions.

J. Maruani et al. (eds.), New Trends in Quantum Systems in Chemistry and Physics, Volume 1, 393–416.

1. Introduction

Several *staggering* effects are known in nuclear spectroscopy [1]:

1) In rotational γ bands of even nuclei the energy levels with odd angular momentum I (I=3, 5, 7, 9, ...) are slightly displaced relatively to the levels with even I (I=2, 4, 6, 8, ...), i.e. the odd levels do not lie at the energies predicted by an $E(I) = AI(I+1)$ fit to the even levels, but all of them lie systematically above or all of them lie systematically below the predicted energies [2].

2) In octupole bands of even nuclei the levels with odd I and negative parity (I^π=1^-, 3^-, 5^-, 7^-, ...) are displaced relatively to the levels with even I and positive parity (I^π=0^+, 2^+, 4^+, 6^+, ...) [3, 4, 5, 6].

3) In odd nuclei, rotational bands (with $K = 1/2$) separate into signature partners, i.e. the levels with I=3/2, 7/2, 11/2, 15/2, ... are displaced relatively to the levels with I=1/2, 5/2, 9/2, 13/2, ... [7].

In all of the above mentioned cases each level with angular momentum I is displaced relatively to its neighbours with angular momentum $I \pm 1$. The effect is then called $\Delta I = 1$ *staggering*. In all cases the effect has been seen in several nuclei and its magnitude is clearly larger than the experimental errors. In cases 1) and 3) the relative displacement of the neighbours increases in general as a function of the angular momentum I [2, 7], while in case 2) (octupole bands), the relevant models [8, 9, 10, 11, 12] predict constant displacement of the odd levels with respect to the even levels as a function of I, i.e. all the odd levels are raised (or lowered) by the same amount of energy.

A new kind of staggering ($\Delta I = 2$ *staggering*) has been recently observed [13, 14] in superdeformed nuclear bands [15, 16, 17]. In the case in which $\Delta I = 2$ staggering is present, the levels with I=2, 6, 10, 14, ..., for example, are displaced relatively to the levels with I=0, 4, 8, 12, ..., i.e. the level with angular momentum I is displaced relatively to its neighbours with angular momentum $I \pm 2$.

Although $\Delta I = 1$ staggering of the types mentioned above has been observed in several nuclei and certainly is an effect larger than the relevant experimental uncertainties, $\Delta I = 2$ staggering has been seen in only a few cases [13, 14, 18, 19] and, in addition, the effect is not clearly larger than the relevant experimental errors.

There have been by now several theoretical works related to the possible physical origin of the $\Delta I = 2$ staggering effect [20, 21, 22, 23, 24, 25, 26], some of them [27, 28, 29, 30, 31, 32] using symmetry arguments which could be of applicability to other physical systems as well.

On the other hand, rotational spectra of diatomic molecules [33] are known to show great similarities to nuclear rotational spectra, having in addition the advantage that observed rotational bands in several diatomic molecules [34, 35, 36, 37] are much longer than the usual rotational nuclear bands. We have been therefore motivated to make a search for $\Delta J = 1$ and $\Delta J = 2$ staggering in rotational bands of diatomic molecules, where by J we denote the total angular momentum of the molecule, while I has been used above for denoting the angular momentum of the nucleus. The questions to which we have hoped to provide answers are:

1) Is there $\Delta J = 1$ and/or $\Delta J = 2$ staggering in rotational bands of diatomic molecules?

2) If there are staggering effects, what are their possible physical origins?

In Sections 2 and 3 the $\Delta J = 2$ staggering and $\Delta J = 1$ staggering will be considered respectively, while in Section 4 the final conclusions and plans for further work will be presented.

2. $\Delta J = 2$ staggering

In this section the $\Delta J = 2$ staggering will be considered. In subsection 2.1 the $\Delta I = 2$ staggering in superdeformed nuclear bands will be briefly reviewed. Evidence from existing experimental data for $\Delta J = 2$ staggering in rotational bands of diatomic molecules will be presented in subsection 2.2 and discussed in subsection 2.3, while subsection 2.4 will contain the relevant conclusions. We mention once more that by J we denote the total angular momentum of the molecule, while by I the angular momentum of the nucleus is denoted.

2.1. $\Delta I = 2$ STAGGERING IN SUPERDEFORMED NUCLEAR BANDS

In nuclear physics the experimentally determined quantities are the γ-ray transition energies between levels differing by two units of angular momentum ($\Delta I = 2$). For these the symbol

$$E_{2,\gamma}(I) = E(I+2) - E(I) \tag{1}$$

is used, where $E(I)$ denotes the energy of the level with angular momentum I. The deviation of the γ-ray transition energies from the rigid rotator behavior can be measured by the quantity [14]

$$\Delta E_{2,\gamma}(I) = \frac{1}{16}(6E_{2,\gamma}(I) - 4E_{2,\gamma}(I-2) - 4E_{2,\gamma}(I+2) + E_{2,\gamma}(I-4) + E_{2,\gamma}(I+4)). \tag{2}$$

Using the rigid rotator expression

$$E(I) = AI(I+1), \tag{3}$$

one can easily see that in this case $\Delta E_{2,\gamma}(I)$ vanishes. In addition the perturbed rigid rotator expression

$$E(I) = AI(I+1) + B(I(I+1))^2, \tag{4}$$

gives vanishing $\Delta E_{2,\gamma}(I)$. These properties are due to the fact that Eq. (2) is a (normalized) discrete approximation of the fourth derivative of the function $E_{2,\gamma}(I)$, i.e. essentially the fifth derivative of the function $E(I)$.

In superdeformed nuclear bands the angular momentum of the observed states is in most cases unknown. To avoid this difficulty, the quantity $\Delta E_{2,\gamma}$ is usually plotted not versus the angular momentum I, but versus the angular frequency

$$\omega = \frac{dE(I)}{dI}, \tag{5}$$

which for discrete states takes the approximate form

$$\omega = \frac{E(I+2) - E(I)}{\sqrt{(I+2)(I+3)} - \sqrt{I(I+1)}}. \tag{6}$$

For large I one can take the Taylor expansions of the square roots in the denominator, thus obtaining

$$\omega = \frac{E(I+2) - E(I)}{2} \doteq \frac{E_{2,\gamma}(I)}{2}. \tag{7}$$

Examples of superdeformed nuclear bands exhibiting staggering are shown in Figs 1–2 [13, 14]. We say that $\Delta I = 2$ staggering is observed if the quantity $\Delta E_2(I)$ exhibits alternating signs with increasing ω (i.e. with increasing I, according to Eq. (7)). The following observations can be made:

1) The magnitude of $\Delta E_2(I)$ is of the order of 10^{-4}–10^{-5} times the size of the gamma transition energies.

2) The best example of $\Delta I = 2$ staggering is given by the first superdeformed band of ^{149}Gd, shown in Fig. 1a. In this case the effect is almost larger than the experimental error.

3) In most cases the $\Delta I = 2$ staggering is smaller than the experimental error (see Figs 1b, 2a, 2b), with the exception of a few points in Fig. 1b.

2.2. $\Delta J = 2$ STAGGERING IN ROTATIONAL BANDS OF DIATOMIC MOLECULES

In the case of molecules [38] the experimentally determined quantities regard the R branch ($(v_{lower}, J) \rightarrow (v_{upper}, J+1)$) and the P branch ($(v_{lower}, J) \rightarrow (v_{upper}, J-1)$), where v_{lower} is the vibrational quantum

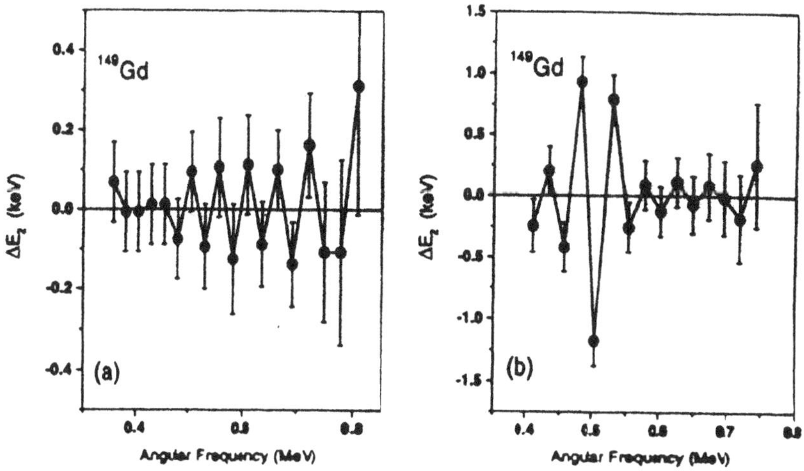

Figure 1. $\Delta E_2(I)$ (in keV), calculated from Eq. (2), versus the angular frequency ω (in MeV), calculated from Eq. (7), for various superdeformed bands in the nucleus ^{149}Gd [13]. a) Band (a) of Ref. [13]. b) Band (d) of Ref. [13].

number of the initial state, while v_{upper} is the vibrational quantum number of the final state. They are related to transition energies through the equations [38]

$$E^R(J) - E^P(J) = E_{v_{upper}}(J+1) - E_{v_{upper}}(J-1) = DE_{2,v_{upper}}(J), \quad (8)$$

$$E^R(J-1) - E^P(J+1) = E_{v_{lower}}(J+1) - E_{v_{lower}}(J-1) = DE_{2,v_{lower}}(J), \quad (9)$$

where in general

$$DE_{2,v}(J) = E_v(J+1) - E_v(J-1). \quad (10)$$

$\Delta J = 2$ staggering can then be estimated by using Eq. (2), with $E_{2,\gamma}(I)$ replaced by $DE_{2,v}(J)$:

$$\Delta E_{2,v}(J) = \frac{1}{16}(6DE_{2,v}(J) - 4DE_{2,v}(J-2) - 4DE_{2,v}(J+2) + DE_{2,v}(J-4) + DE_{2,v}(J+4)). \quad (11)$$

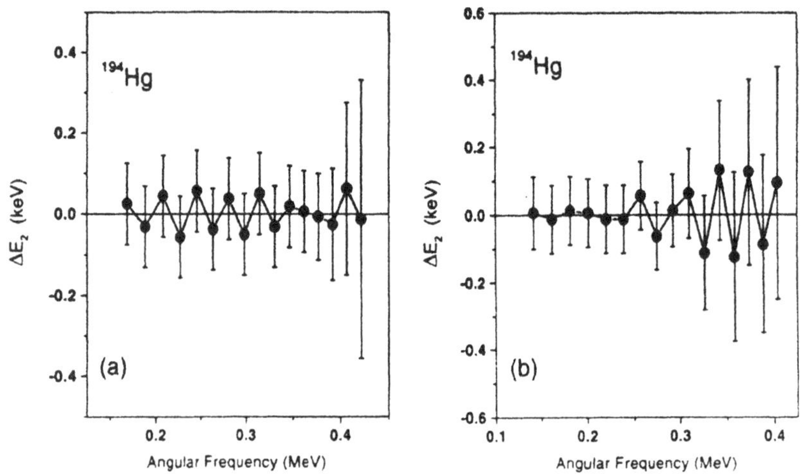

Figure 2. $\Delta E_2(I)$ (in keV), calculated from Eq. (2), versus the angular frequency ω (in MeV), calculated from Eq. (7), for various superdeformed bands in the nucleus ^{194}Hg [14]. a) Band 1 of Ref. [14]. b) Band 2 of Ref. [14].

Results for several rotational bands in different electronic and vibrational states of various diatomic molecules are shown in Figs 3–9. We say that $\Delta J = 2$ staggering is observed if the quantity $\Delta E_2(J)$ exhibits alternating signs with increasing J (J is increased by 2 units each time). The magnitude of $\Delta E_2(J)$ is usually of the order of 10^{-3}–10^{-5} times the size of the interlevel separation energy. In Figs 7 and 8, which correspond to sextet electronic states, the rotational angular momentum N is used instead of the total angular momentum J, the two quantities been connected by the relation $\mathbf{J} = \mathbf{N} + \mathbf{S}$, where S is the spin. Several observations can be made:

1) In all cases shown, the "upper" bands (which happen to be electronically excited) exhibit (Figs 3, 4, 7-9) $\Delta J = 2$ staggering (or $\Delta N = 2$ staggering) which is 2 to 3 orders of magnitude larger than the experimental error, while the corresponding "lower" bands (which, in the cases studied, correspond to the electronic ground state of each molecule), show (Figs 5, 6) some effect smaller than the experimental error.

2) There is no uniform dependence of the $\Delta J = 2$ staggering on the angular momentum J. In some cases of long bands, though, it appears that

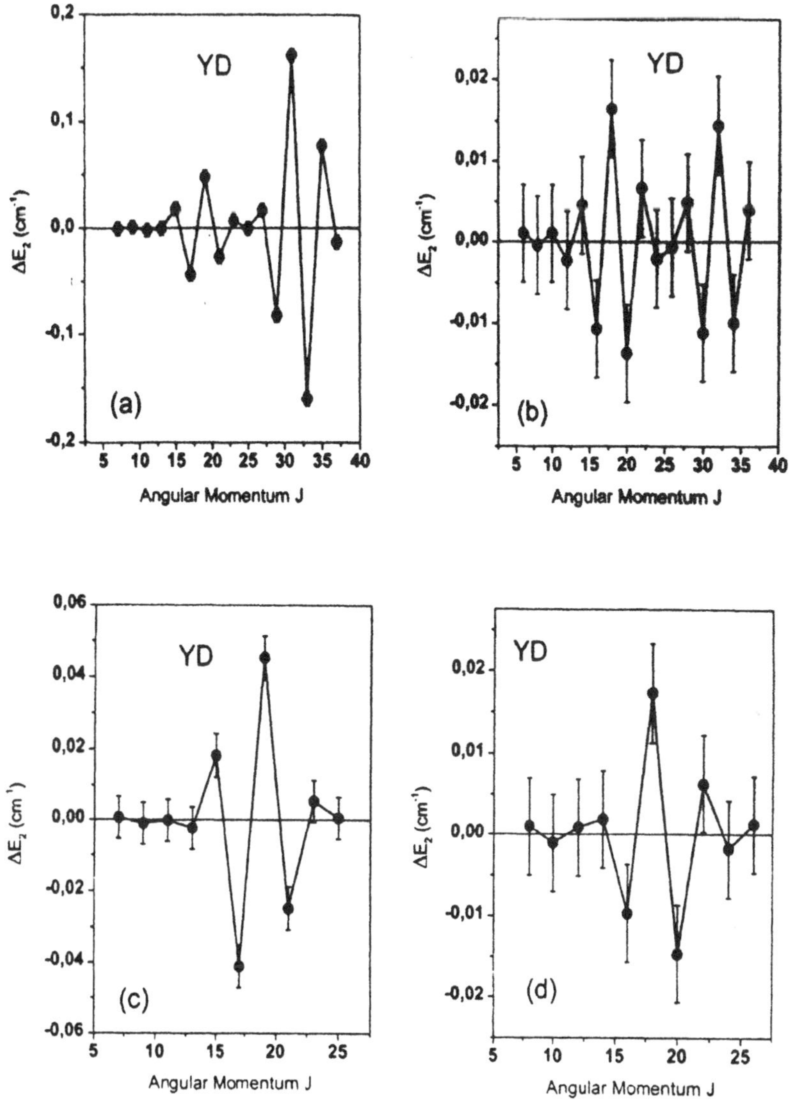

Figure 3. $\Delta E_2(J)$ (in cm^{-1}), calculated from Eq. (11), for various bands of the YD molecule [34]. a) Odd levels of the $v = 1$ $C^1\Sigma^+$ band calculated from the data of the 1–1 $C^1\Sigma^+$–$X^1\Sigma^+$ transitions. b) Even levels of the previous band. c) Odd levels of the $v = 1$ $C^1\Sigma^+$ band calculated from the 1–2 $C^1\Sigma^+$–$X^1\Sigma^+$ transitions. d) Even levels of the previous band.

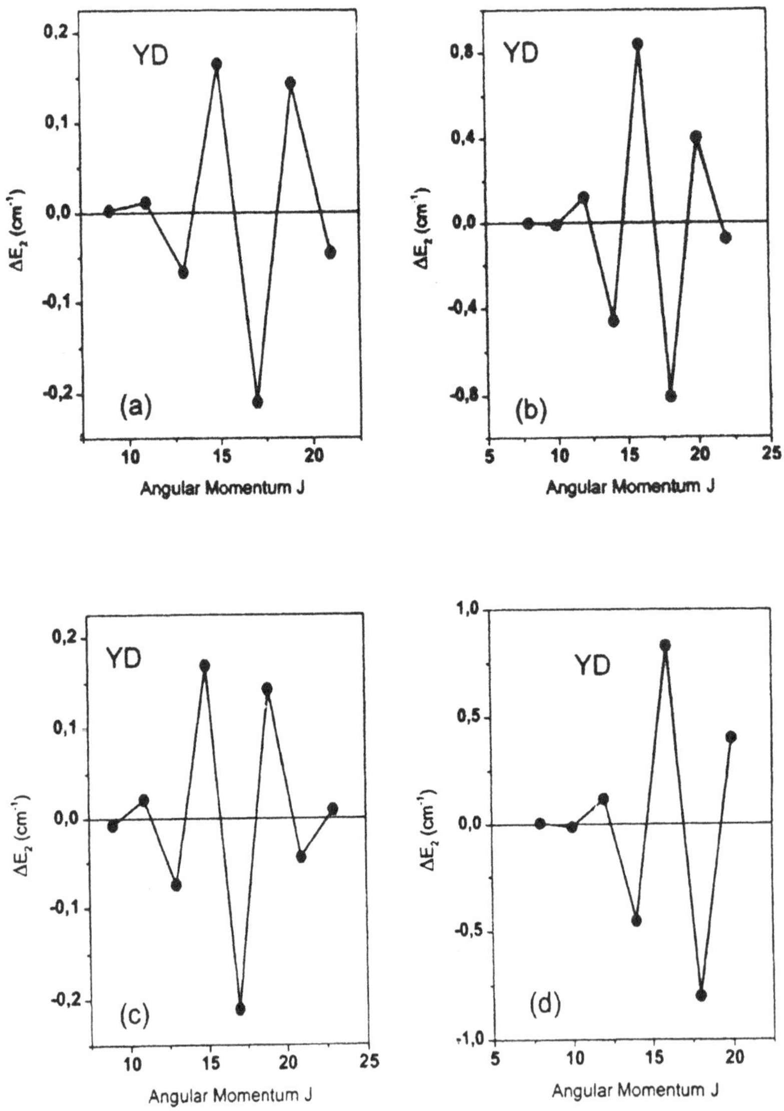

Figure 4. $\Delta E_2(J)$ (in cm^{-1}), calculated from Eq. (11), for various bands of the YD molecule [34]. a) Odd levels of the $v = 2$ $C^1\Sigma^+$ band calculated from the data of the 2–2 $C^1\Sigma^+$–$X^1\Sigma^+$ transitions. b) Even levels of the previous band. c) Odd levels of the $v = 2$ $C^1\Sigma^+$ band calculated from the 2–3 $C^1\Sigma^+$–$X^1\Sigma^+$ transitions. d) Even levels of the previous band. The experimental error in all cases is ± 0.006 cm^{-1} and therefore is hardly or not seen.

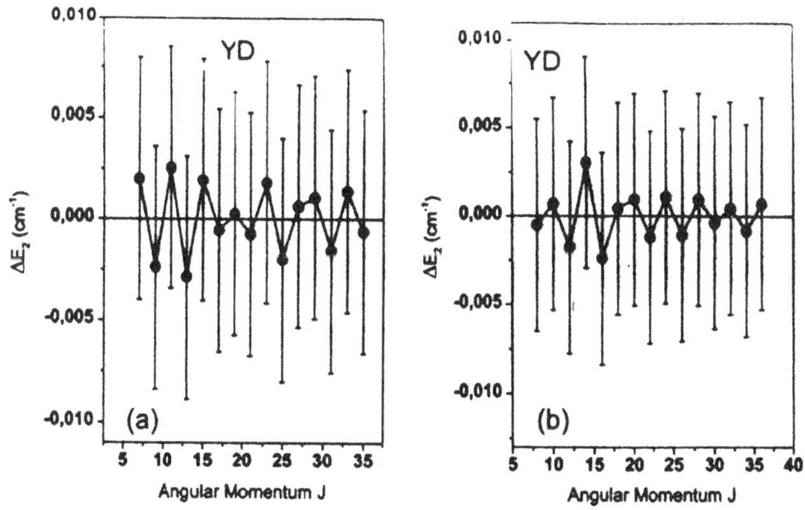

Figure 5. $\Delta E_2(J)$ (in cm^{-1}), calculated from Eq. (11), for various bands of the YD molecule [34]. a) Odd levels of the $v = 1$ $X^1\Sigma^+$ band calculated from the data of the 1–1 $C^1\Sigma^+$–$X^1\Sigma^+$ transitions. b) Even levels of the previous band.

the pattern is a sequence of points exhibiting small staggering, interrupted by groups of 6 points each time showing large staggering. The best examples can be seen in Figs 3a, 3b, 7a, 7b. In Fig. 3a (odd levels of the $v = 1$ $C^1\Sigma^+$ band of YD)) the first group of points showing appreciable $\Delta J = 2$ staggering appears at $J = 13$–23, while the second group appears at $J = 27$–37. In Fig. 3b (even levels of the $v = 1$ $C^1\Sigma^+$ band of YD) the first group appears at $J = 12$–22, while the second group at $J = 26$–36. In Fig. 7a (odd levels of the $v = 0$ $A^6\Sigma^+$ band of CrD) the first group appears at $N = 15$–25, while the second at $N = 27$–37. Similarly in Fig. 7b (even levels of the $v = 0$ $A^6\Sigma^+$ band of CrD) the first group appears at $N = 14$–24, while the second group at $N = 26$–36.

3) In all cases shown, the results obtained for the odd levels of a band are in good agreement with the results obtained for the even levels of the same band. For example, the regions showing appreciable staggering are approximately the same in both cases (compare Fig. 3a with Fig. 3b and Fig. 7a with Fig. 7b, already discussed in 2)). In addition, the positions of the local staggering maxima in each pair of figures are closely related. In Fig. 3a, for example, maximum staggering appears at $J = 19$ and $J = 31$,

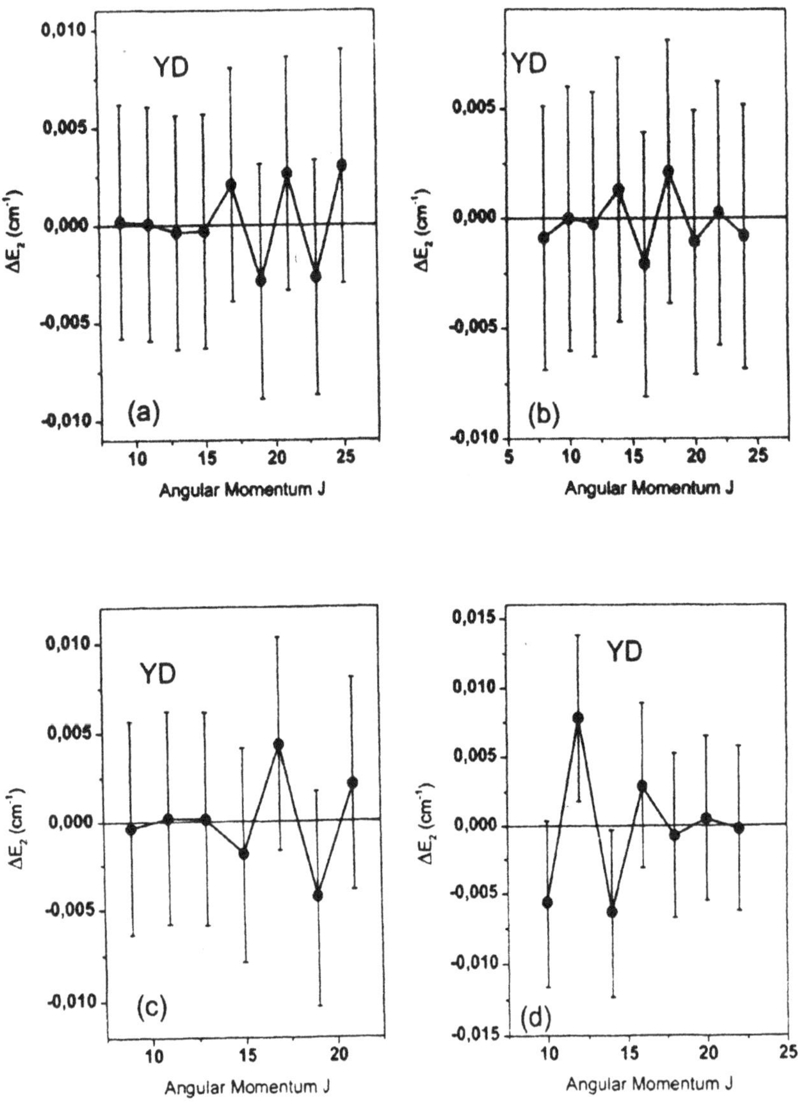

Figure 6. $\Delta E_2(J)$ (in cm^{-1}), calculated from Eq. (11), for various bands of the YD molecule [34]. a) Odd levels of the $v = 2$ $X^1\Sigma^+$ band calculated from the data of the 1–2 $C^1\Sigma^+$–$X^1\Sigma^+$ transitions. b) Even levels of the previous band. c) Odd levels of the $v = 2$ $X^1\Sigma^+$ band calculated from the 2–2 $C^1\Sigma^+$–$X^1\Sigma^+$ transitions. d) Even levels of the previous band.

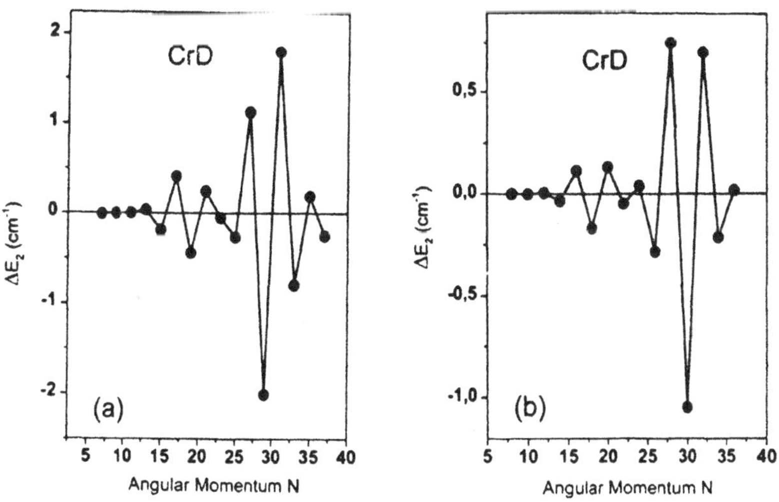

Figure 7. $\Delta E_2(N)$ (in cm^{-1}), calculated from Eq. (11), for various bands of the CrD molecule [35]. a) Odd levels of the $v = 0$ $A^6\Sigma^+$ band calculated from the data (R2, P2 branches) of the 0–0 $A^6\Sigma^+$–$X^6\Sigma^+$ transitions. b) Even levels of the previous band. The experimental error in all cases is ± 0.006 cm^{-1} and therefore is not seen.

while in Fig. 3b the maxima appear at $J = 18$ and $J = 32$.

4) In several cases the $\Delta J = 2$ staggering of a band can be calculated from two different sets of data. For example, Figs 3a, 3b show the $\Delta J = 2$ staggering of the $v = 1$ $C^1\Sigma^+$ band of YD calculated from the data on the 1–1 $C^1\Sigma^+$–$X^1\Sigma^+$ transitions, while Figs 3c, 3d show the staggering of the same band calculated from the data on the 1–2 $C^1\Sigma^+$–$X^1\Sigma^+$ transition. We remark that the results concerning points showing staggering larger than the experimental error come out completely consistently from the two calculations (region with $J = 13$–23 in Figs 3a, 3c; region with $J = 12$–22 in Figs 3b, 3d), while the results concerning points exhibiting staggering of the order of the experimental error come out randomly (in Fig. 3a, for example, $J = 11$ corresponds to a local minimum, while in Fig. 3c it corresponds to a local maximum). Similar results are seen in the pairs of figures (3b, 3d), (4a, 4c), (4b, 4d), (6a, 6c), (6b, 6d), (9a, 9c), (9b, 9d). The best example of disagreement between staggering pictures of the same band calculated from two different sets of data is offered by Figs 6b, 6d, which concern the $v = 2$ $X^1\Sigma^+$ band of YD, which shows staggering of the order of the experimental

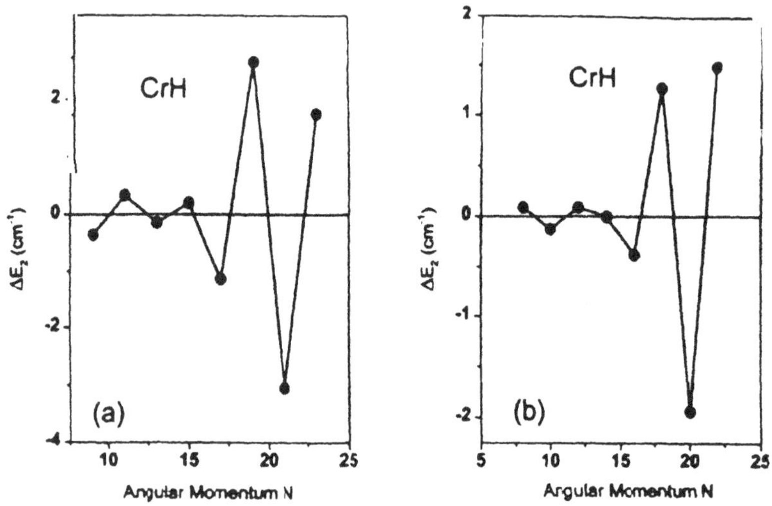

Figure 8. $\Delta E_2(N)$ (in cm^{-1}), calculated from Eq. (11), for various bands of the CrH molecule [36]. a) Odd levels of the $v = 0$ $A^6\Sigma^+$ band calculated from the data (R2, P2 branches) of the 0–0 $A^6\Sigma^+$–$X^6\Sigma^+$ transitions. b) Even levels of the previous band. The experimental error in all cases is ± 0.004 cm^{-1} and therefore is not seen.

error.

5) When considering levels of the same band, in some cases the odd levels exhibit larger staggering than the even levels, while in other cases the opposite is true. In the $v = 1$ $C^1\Sigma^+$ band of YD, for example, the odd levels (shown in Fig. 3a, corroborated by Fig. 3c) show staggering larger than that of the even levels (shown in Fig. 3b, corroborated by Fig. 3d), while in the $v = 2$ $C^1\Sigma^+$ band of YD the odd levels (shown in Fig. 4a, corroborated by Fig. 4c) exhibit staggering smaller than that of the even levels (shown in Fig. 4b, corroborated by Fig. 4d).

2.3. DISCUSSION

The observations made above can be explained by the assumption that the staggering observed is due to the presence of one or more bandcrossings [39, 40]. The following points support this assumption:

1) It is known [41] that bandcrossing occurs in cases in which the interband interaction is weak. In such cases only the one or two levels closest to the crossing point are affected [42]. However, if one level is influenced by the

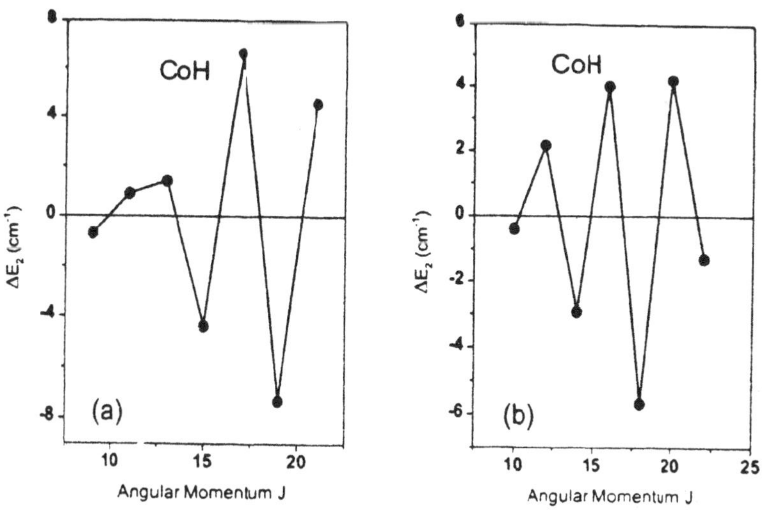

Figure 9. $\Delta E_2(J)$ (in cm^{-1}), calculated from Eq. (11), for various bands of the CoH molecule [37]. a) Odd levels of the $v = 0$ $A'^3\Phi_4$ band calculated from the data (Ree, Pee branches) of the 0–1 $A'^3\Phi_4$–$X^3\Phi_4$ transitions. b) Even levels of the previous band. The experimental error in all cases is ± 0.01 cm^{-1} and therefore is not seen.

crossing, in the corresponding staggering figure six points get influenced. For example, if E(16) is influenced by the crossing, the quantities $DE_2(15)$ and $DE_2(17)$ are influenced (see Eq. (10)), so that in the corresponding figure the points $\Delta E_2(J)$ with J = 11, 13, 15, 17, 19, 21 are influenced, as seen from Eq. (11). This fact explains why points showing appreciable staggering appear in groups of 6 at a time.

2) It is clear that if bandcrossing occurs, large staggering should appear in approximately the same angular momentum regions of both even levels and odd levels. As we have already seen, this is indeed the case.

3) It is clear that when two bands cross each other, maximum staggering will appear at the angular momentum for which the energies of the relevant levels of each band are approximately equal [42]. If this angular momentum value happens to be odd, then $\Delta E_2(J)$ for even values of J in this region (the group of 6 points centered at this J) will show larger staggering than the $\Delta E_2(J)$ for odd values of J in the corresponding region, and vice versa. For example, if the closest approach of two bands occurs for $J = 31$, then $\Delta E_2(J)$ for even values of J in the J = 26–36 region will show larger

staggering than $\Delta E_2(J)$ for odd values of J in the same region. This is in agreement with the empirical observation that in some cases the odd levels show larger staggering than the even levels, while in other cases the opposite holds.

4) The presence of staggering in the "upper" (electronically excited) bands and the lack of staggering in the "lower" (electronic ground state) bands can be attributed to the fact that the electronically excited bands have several neighbours with which they can interact, while the bands built on the electronic ground state are relatively isolated, and therefore no bandcrossings occur in this case. In the case of the CrD molecule, in particular, it is known [35] that there are many strong Cr atomic lines present, which frequently overlap the relatively weaker (electronically excited) molecular lines. In addition, Ne atomic lines are present [35]. Similarly, in the case of the YD molecule the observed spectra are influenced by Y and Ne atomic lines [34], while in the case of the CrH molecule there are Ne and Cr atomic lines influencing the molecular spectra [36].

5) The fact that consistency between results for the same band calculated from two different sets of data is observed only in the cases in which the staggering is much larger than the experimental error, corroborates the bandcrossing explanation. The fact that the results obtained in areas in which the staggering is of the order of the experimantal error, or even smaller, appear to be random, points towards the absence of any real effect in these regions.

It should be noticed that bandcrossing has been proposed [43, 44, 45] as a possible explanation for the appearance of $\Delta I = 2$ staggering effects in normally deformed nuclear bands [26, 43, 45] and superdeformed nuclear bands [44].

The presence of two subsequent bandcrossings can also provide an explanation for the effect of mid-band disappearance of $\Delta I = 2$ staggering observed in superdeformed bands of some Ce isotopes [18]. The effect seen in the Ce isotopes is very similar to the mid-band disappearance of staggering seen, for example, in Fig. 3a.

2.4. CONCLUSION

In conclusion, we have found several examples of $\Delta J = 2$ staggering in electronically excited bands of diatomic molecules. The details of the observed effect are in agreement with the assumption that it is due to one or more bandcrossings. In these cases the magnitude of the effect is clearly larger than the experimental error. In cases in which an effect of the order of the experimental error appears, we have shown that this is an artifact of the method used, since different sets of data from the same experiment and for the same molecule lead to different staggering results for the same rotational band. The present work emphasizes the need to ensure in all cases (including staggering candidates in nuclear physics) that the effect is larger than the experimental error and, in order to make assumptions about any new symmetry, that it is not due to a series of bandcrossings.

3. $\Delta J = 1$ staggering

In this section the $\Delta J = 1$ staggering effect (i.e. the relative displacement of the levels with even angular momentum J with respect to the levels of the same band with odd J) will be considered in molecular bands free from $\Delta J = 2$ staggering (i.e. free from interband interactions/bandcrossings), in order to make sure that $\Delta J = 1$ staggering is not an effect due to the same cause as $\Delta J = 2$ staggering.

The formalism of the $\Delta J = 1$ staggering will be described in subsection 3.1 and applied to experimental molecular spectra in subsection 3.2. Finally, subsection 3.3 will contain a discussion of the present results and plans for further work.

3.1. FORMALISM

By analogy to Eq. (2), $\Delta I = 1$ staggering in nuclei can be measured by the quantity

$$\Delta E_{1,\gamma}(I) = \frac{1}{16}(6E_{1,\gamma}(I) - 4E_{1,\gamma}(I-1) - 4E_{1,\gamma}(I+1) + E_{1,\gamma}(I-2) + E_{1,\gamma}(I+2)), \tag{12}$$

where

$$E_{1,\gamma}(I) = E(I+1) - E(I). \tag{13}$$

The transition energies $E_{1,\gamma}(I)$ are determined directly from experiment.

In order to be able to use an expression similar to that of Eq. (12) for the study of $\Delta J = 1$ staggering in molecular bands we need transition energies similar to those of Eq. (13), i.e. transition energies between levels differing by one unit of angular momentum. However, Eqs (8) and (9) can provide us only with transition energies between levels differing by two units of angular momentum. In order to be able to determine the levels with even J from Eqs (8) or (9), one needs the bandhead energy $E(0)$. Then one has

$$E_{v_{upper}}(2) = E_{v_{upper}}(0) + E^R(1) - E^P(1), \tag{14}$$

$$E_{v_{upper}}(4) = E_{v_{upper}}(2) + E^R(3) - E^P(3), \ldots \tag{15}$$

$$E_{v_{lower}}(2) = E_{v_{lower}}(0) + E^R(0) - E^P(2), \tag{16}$$

$$E_{v_{lower}}(4) = E_{v_{lower}}(2) + E^R(2) - E^P(4), \dots \tag{17}$$

In order to be able to determine the levels with odd J from Eqs (8) and (9) in an analogous way, one needs $E(1)$. Then

$$E_{v_{upper}}(3) = E_{v_{upper}}(1) + E^R(2) - E^P(2), \tag{18}$$

$$E_{v_{upper}}(5) = E_{v_{upper}}(3) + E^R(4) - E^P(4), \dots \tag{19}$$

$$E_{v_{lower}}(3) = E_{v_{lower}}(1) + E^R(1) - E^P(3), \tag{20}$$

$$E_{v_{lower}}(5) = E_{v_{lower}}(3) + E^R(3) - E^P(5), \dots \tag{21}$$

For the determination of $E(0)$ and $E(1)$ one can use the overall fit of the experimental data (for the R and P branches) by a Dunham expansion [46]

$$E(J) = T_v + B_v J(J+1) - D_v[J(J+1)]^2 + H_v[J(J+1)]^3 + L_v[J(J+1)]^4, \tag{22}$$

which is usually given in the experimental papers.

After determining the energy levels by the procedure described above, we estimate $\Delta J = 1$ staggering by using the following analogue of Eq. (12),

$$\Delta E_{1,v}(J) = \frac{1}{16}(6DE_{1,v}(J) - 4DE_{1,v}(J-1) - 4DE_{1,v}(J+1) + DE_{1,v}(J-2) + DE_{1,v}(J+2)), \tag{23}$$

where

$$DE_{1,v}(J) = E_v(J) - E_v(J-1). \tag{24}$$

Using Eq. (24) one can put Eq. (23) in the sometimes more convenient form

$$\Delta E_{1,v}(J) = \frac{1}{16}(10E_v(J) - 10E_v(J-1) + 5E_v(J-2) - 5E_v(J+1) + E_v(J+2) - E_v(J-3)). \tag{25}$$

In realistic cases the first few values of $E^R(J)$ and $E^P(J)$ might be experimentally unknown. In this case one is forced to determine the first few values of $E(J)$ using the Dunham expansion of Eq. (22) and then continue by using the Eqs (14)–(21) from the appropriate point on. Denoting by J_{io} the "initial" value of odd J, on which we are building through the series of equations starting with Eqs (18)–(21) the energy levels of odd J, and by J_{ie} the "initial" value of even J, on which we are building through the series of equations starting with Eqs (14)–(17) the energy levels of even J, we find that the error for the levels with odd J is

$$Err(E(J)) = D(J_{io}) + (J - J_{io})\epsilon, \tag{26}$$

while the error for the levels with even J is

$$Err(E(J)) = D(J_{ie}) + (J - J_{ie})\epsilon, \tag{27}$$

where $D(J_{io})$ and $D(J_{ie})$ are the uncertainties of the levels $E(J_{io})$ and $E(J_{ie})$ respectively, which are determined through the Dunham expansion of Eq. (22), while ϵ is the error accompanying each $E^R(J)$ or $E^P(J)$ level, which in most experimental works has a constant value for all levels.

Using Eqs (26) and (27) in Eq. (25) one easily sees that the uncertainty of the $\Delta J = 1$ staggering measure $\Delta E_{1,v}(J)$ is

$$Err(\Delta E_{1,v}(J)) = D(J_{io}) + D(J_{ie}) + (2J - J_{io} - J_{ie} - 1)\epsilon. \tag{28}$$

This equation is valid for $J \geq \max\{J_{io}, J_{ie}\} + 3$. For smaller values of J one has to calculate the uncertainty directly from Eq. (25).

3.2. ANALYSIS OF EXPERIMENTAL DATA

3.2.1. *YD*

We have applied the formalism described above to the 0–1, 1–1, 1–2, 2–2 transitions of the $C^1\Sigma^+$–$X^1\Sigma^+$ system of YD [34]. We have focused attention on the ground state $X^1\Sigma^+$, which is known to be free from $\Delta J = 2$ staggering effects (see subsection 2.2), while the $C^1\Sigma^+$ state is known to exhibit $\Delta J = 2$ staggering effects, which are fingerprints of interband interactions (bandcrossings), as we have seen in subsection 2.2. Using the formalism of subsection 3.1, we calculated the $\Delta J = 1$ staggering measure $\Delta E_1(J)$ of Eq. (23) for the $v = 1$ band of the $X^1\Sigma^+$ state (Fig. 10a, 10b) and for the $v = 2$ band of the $X^1\Sigma^+$ state (Fig. 10c, 10d). At this point the following comments are in place:

1) In all cases the levels $E(0)$, $E(1)$, $E(2)$, $E(3)$ have been determined using the Dunham expansion of Eq. (22) and the Dunham coefficients given in Table II of Ref. [34]. This has been done because $E^R(1)$ is missing in the tables of the 1–1 and 2–2 transitions [34], so that Eq. (20) cannot be used for the determination of $E(3)$. In the cases of the 0–1 and 1–2 transitions, $E^R(1)$ is known, but we prefered to calculate $E(3)$ from the Dunham expansion in these cases as well, in order to treat the pairs of cases 0–1, 1–1 and 1–2, 2–2 on equal footing, since we intend to make comparisons between them.

2) For the calculation of errors we have taken into account the errors of the Dunham coefficients given in Table II of Ref. [34], as well as the fact that the accuracy of the members of the R- and P-branches is $\epsilon = \pm 0.002$ cm^{-1} [34]. It is clear that the large size of the error bars is due to the accumulation of errors caused by Eqs (14)–(21), as seen in Eqs (25)–(28).

3) In Figs 10a and 10b the $\Delta J = 1$ staggering measure $\Delta E_1(J)$ for the $v = 1$ band of the $X^1\Sigma^+$ state of YD is shown, calculated from two different

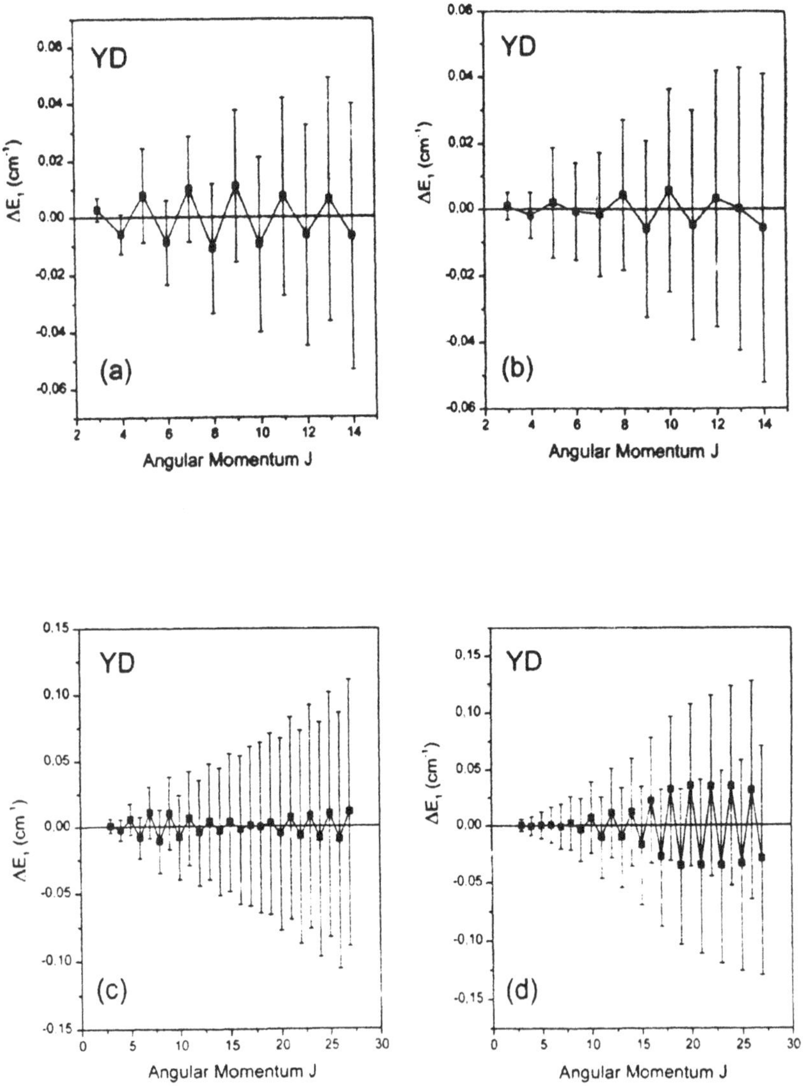

Figure 10. $\Delta E_1(J)$ (in cm^{-1}), calculated from Eq. (23), for various bands of the YD molecule [34]. a) Levels of the $v = 1$ $X^1\Sigma^+$ band calculated from the data of the 0–1 $C^1\Sigma^+$–$X^1\Sigma^+$ transitions. b) Levels of the $v = 1$ $X^1\Sigma^+$ band calculated from the data of the 1–1 $C^1\Sigma^+$–$X^1\Sigma^+$ transitions. c) Levels of the $v = 2$ $X^1\Sigma^+$ band calculated from the data of the 1–2 $C^1\Sigma^+$–$X^1\Sigma^+$ transitions. d) Levels of the $v = 2$ $X^1\Sigma^+$ band calculated from the data of the 2–2 $C^1\Sigma^+$–$X^1\Sigma^+$ transitions.

sources, the 0–1 and 1–1 transitions. If a real $\Delta J = 1$ staggering effect were present, the two figures should have been identical, or at least consistent with each other. However, they are completely different (even the maxima and the minima appear at different values of J in each figure), indicating that what is seen is not a real physical effect, but random experimental errors (buried in the large error bars, anyway).

4) Exactly the same comments as in 3) apply to Figs 10c and 10d, where the $\Delta J = 1$ staggering measure for the $v = 2$ band of the $X^1\Sigma^+$ state of YD is shown, calculated from two different sources, the 1–2 and 2–2 transitions.

We conclude therefore that no $\Delta J = 1$ staggering effect appears in the $v = 1$ and $v = 2$ bands of the $X^1\Sigma^+$ state of YD, which are free from $\Delta J = 2$ staggering, as proved in subsection 2.2.

This negative result has the following physical implications. It is known in nuclear spectroscopy that reflection asymmetric (pear-like) shapes give rise to octupole bands, in which the positive parity states ($I^\pi = 0^+$, 2^+, 4^+, ...) are displaced reletively to the negative parity states ($I_\pi = 1^-$, 3^-, 5^-, ...) [3, 4, 5, 6, 47, 48, 49]. Since a diatomic molecule consisting of two different atoms possesses the same reflection asymmetry, one might think that $\Delta J = 1$ staggering might be present in the rotational bands of such molecules. Then YD, because of its large mass asymmetry, is a good testing ground for this effect. The negative result obtained above can, however, be readily explained. Nuclei with octupole deformation are supposed to be described by double well potentials, the relative displacement of the negative parity levels and the positive parity levels being attributed to the tunneling through the barrier separating the wells [47, 48, 49]. (The relative displacement vanishes in the limit in which the barrier separating the two wells becomes infinitely high.) In the case of diatomic molecules the relevant potential is well known [33] to consist of a single well. Therefore no tunneling effect is possible and, as a result, no relative displacement of the positive parity levels and the negative parity levels is seen.

3.2.2. *CrD*

The formalism of subsection 3.1 has in addition been applied to a more complicated case, the one of the 0–0 and 1–0 transitions of the $A^6\Sigma^+$–$X^6\Sigma^+$ system of CrD [35]. We have focused our attention on the ground state $X^6\Sigma^+$, which is known to be free from $\Delta N = 2$ staggering effects (see subsection 2.2), while the $A^6\Sigma^+$ state is known from subsection 2.2 to exhibit $\Delta N = 2$ staggering effects, which are fingerprints of interband interactions (bandcrossings). The CrD system considered here has several differences from the YD system considered in the previous subsection, which are briefly listed here:

1) The present system of CrD involves sextet electronic states. As a

result, each band of the $A^6\Sigma^+$-$X^6\Sigma^+$ transition consists of six R- and six P-branches, labelled as R1, R2, ..., R6 and P1, P2, ..., P6 respectively [35]. In the present study we use the R3 and P3 branches, but similar results are obtained for the other branches as well.

2) Because of the presence of spin–rotation interactions and spin–spin interactions, the energy levels cannot be fitted by a Dunham expansion in terms of the total angular momentum J, but by a more complicated Hamiltonian, the N^2 Hamiltonian for a $^6\Sigma$ state [50, 51]. This Hamiltonian, in addition to a Dunham expansion in terms of N (the rotational angular momentum, which in this case is different from the total angular momentum $\mathbf{J} = \mathbf{N} + \mathbf{S}$, where S the spin), contains terms describing the spin–rotation interactions (preceded by three γ coefficients), as well as terms describing the spin–spin interactions (preceded by two λ coefficients [35, 50]).

In the present study we have calculated the staggering measure of Eq. (23) for the $v = 0$ band of the $X^6\Sigma^+$ state of CrD, using the R3 and P3 branches of the 0–0 (Fig. 11a) and 1–0 (Fig. 11b) transitions of the $A^6\Sigma^+$-$X^6\Sigma^+$ system. Since in this case the Dunham expansion involves the rotational angular momentum N, and not the total angular momentum J, the formalism of subsection 3.1 has been used with J replaced by N everywhere. This is why the calculated staggering measure of Eq. (23) is in this case denoted by $\Delta E_1(N)$ and not by $\Delta E_1(J)$, the relevant effect being called $\Delta N = 1$ staggering instead of $\Delta J = 1$ staggering. At this point the following comments are in place:

1) In both cases the levels $E(0)$, $E(1)$, $E(2)$, $E(3)$, $E(4)$ have been determined using the Dunham expansion of Eq. (22) (with J replaced by N) and the Dunham coefficients given in Table V of Ref. [35]. This has been done because $E^R(2)$ is missing in the tables of the 0–0 transitions [35], so that Eq. (17) cannot be used for the determination of $E(4)$. In the case of the 1–0 transitions, $E^R(2)$ is known, but we prefered to calculate $E(4)$ from the Dunham expansion in this case as well, in order to treat the cases 0–0 and 1–0 on equal footing, since we intend to make comparisons between them.

2) For the calculation of errors we have taken into account the errors of the Dunham coefficients given in Table V of Ref. [35], as well as the fact that the accuracy of the members of the R- and P- branches is $\epsilon = \pm 0.001$ cm^{-1} for the 0–0 transitions and $\epsilon = \pm 0.003$ cm^{-1} for the 1–0 transitions [35]. In this case it is clear, as in the previous one, that the large size of the error bars is due to the accumulation of errors caused by Eqs (14)–(21), as seen in Eqs (25)–(28).

3) In Figs 11a and 11b the $\Delta N = 1$ staggering measure $\Delta E_1(N)$ for the $v = 0$ band of the $X^6\Sigma^+$ state of CrD is shown, calculated from two different sources, the 0–0 and 1–0 transitions. The two figures are nearly

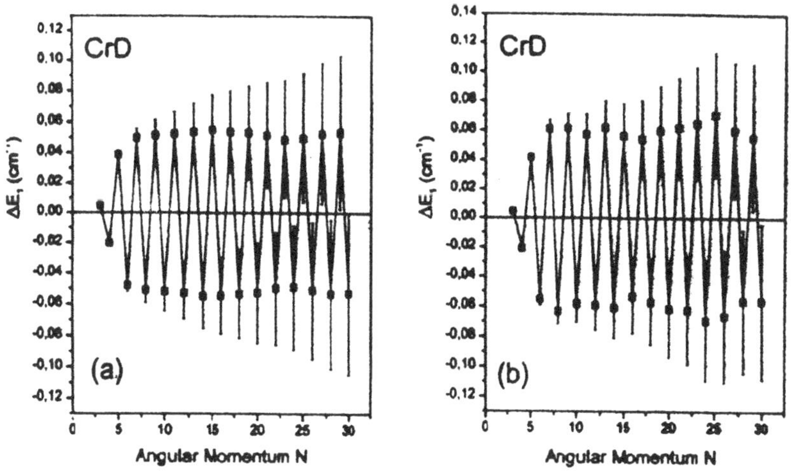

Figure 11. $\Delta E_1(N)$ (in cm^{-1}), calculated from Eq. (23), for various bands of the CrD molecule [35]. a) Levels of the $v = 0$ $X^6\Sigma^+$ band calculated from the data of the 0–0 $A^6\Sigma^+$–$X^6\Sigma^+$ transitions (R3, P3 branches). b) Levels of the $v = 0$ $X^6\Sigma^+$ band calculated from the data of the 1–0 $A^6\Sigma^+$–$X^6\Sigma^+$ transitions (R3, P3 branches). The error bars in case (b) have been divided by a factor of 3, in order to be accommodated within the figure.

identical. The maxima and the minima appear at the same values of N in both figures, while even the amplitude of the effect is almost the same in both figures. It should be pointed out, however, that the error bars in Fig. 11b have been made smaller by a factor of three, in order to the accommodated in the figure.

We conclude therefore that in the $v = 0$ band of the $X^6\Sigma^+$ state of CrD the two different calculations give consistent results, despite the error accumulation mentioned above. The result looks like $\Delta N = 1$ staggering of almost constant amplitude. The reason behind the appearance of this staggering is, however clear: It is due to the omission of the spin-rotation and spin-spin terms of the N^2 Hamiltonian mentioned above [35, 50, 51]. As a result, we have not discovered any new physical effect. What we have demonstated, is that Eq. (23) is a very sensitive probe, which can uncover small deviations from the pure rotational behaviour. However, special care should be taken when using it, because of the accumulation of errors, which is inherent in this method. This problem is avoided by producing results for the same band from two different sets of data, as done above. If both sets lead to consistent results, some effect is present. If the two sets give randomly different results, it is clear that no effect is present.

It should be pointed out at this point that the appearance of $\Delta J = 1$ staggering (or $\Delta N = 1$ staggering) does *not* mean that an effect with oscillatory behaviour is present. Indeed, suppose that the energy levels of a band follow the $E(J) = AJ(J+1)$ rule, but to the odd levels a constant term c is added. It is then clear from Eq. (25) that we are going to obtain $\Delta E_1(J) = +c$ for odd values of J, and $\Delta E_1(J) = -c$ for even values of J, obtaining in this way perfect $\Delta J = 1$ staggering of constant amplitude c, without the presence of any oscillatory effect. This comment directly applies to the results presented in Fig. 11. The presence of $\Delta N = 1$ staggering of almost constant amplitude is essentially due to the omission of the rotation–spin and spin–spin interactions in the calculation of the $E(3)$ and $E(4)$ levels. The difference of the omitted terms in the $N = 3$ and $N = 4$ cases plays the role of c in Fig. 11.

3.3. DISCUSSION

In this section we have addressed the question of the possible existence of $\Delta J = 1$ staggering (i.e. of a relative displacement of the odd levels with respect to the even levels) in rotational bands of diatomic molecules, which are free from $\Delta J = 2$ staggering (i.e. free from interband interactions/bandcrossings). The main conclusions drawn are:

1) The YD bands studied indicate that there is no $\Delta J = 1$ staggering, which could be due to the mass asymmetry of this molecule.

2) The CrD bands studied indicate that there is $\Delta N = 1$ staggering, which is, however, due to the spin–rotation and spin–spin interactions present in the relevant states.

3) Based on the above results, we see that $\Delta J = 1$ staggering is a sensitive probe of deviations from the pure rotational behaviour. Since the method of its calculation from the experimental data leads, however, to error accumulation, one should always calculate the $\Delta J = 1$ staggering measure for the same band from two different sets of data and check the consistency of the results, absence of consistency meaning absence of any real effect.

It is desirable to corroborate the above conclusions by studying rotational bands of several additional molecules.

4. Conclusions

In this work we have examined if the effects of $\Delta J = 2$ staggering and $\Delta J = 1$ staggering, which appear in nuclear spectroscopy, appear also in rotational bands of diatomic molecules. For the $\Delta J = 2$ staggering it has been found that it appears in certain electronically excited rotational bands of diatomic molecules (YD, CrD, CrH, CoH), in which it is attributed to interband interactions (bandcrossings). The $\Delta J = 1$ staggering has been examined in rotational bands free from $\Delta J = 2$ staggering, i.e. free from interband interactions (bandcrossings). Bands of YD offer evidence for the absence of any $\Delta J = 1$ staggering effect due to the disparity of nuclear masses, while bands of sextet electronic states of CrD demonstrate that $\Delta J = 1$ staggering is a sensitive probe of deviations from rotational behaviour, due in this particular case to the spin-rotation and spin-spin interactions. We conclude therefore that both $\Delta J = 2$ staggering and $\Delta J = 1$ staggering are sensitive probes of perturbations in rotational bands of diatomic molecules and do not constitute any new physical effect.

The number of rotational bands of diatomic molecules examined in the case of the $\Delta J = 2$ staggering is satisfactory. For the case of the $\Delta J = 1$ staggering it is desirable to corroborate the findings of the present work through the examination of rotational bands of more diatomic molecules.

Acknowledgements

One of the authors (PPR) acknowledges support from the Bulgarian Ministry of Science and Education under contract Φ-547. Another author (NM) has been supported by the Bulgarian National Fund for Scientific Research under contract no MU-F-02/98. Three authors (DB,CD,NK) have been supported by the Greek Secretariat of Research and Technology under contract PENED 95/1981.

References

1. A. Bohr and B. R. Mottelson, *Nuclear Structure Vol. II: Nuclear Deformations* (World Scientific, Singapore, 1998).
2. D. Bonatsos, *Phys. Lett. B* **200** (1988) 1.
3. W. R. Phillips, I. Ahmad, H. Emling, R. Holzmann, R. V. F. Janssens, T. L. Khoo and M. W. Drigert, *Phys. Rev. Lett.* **57** (1986) 3257.
4. P. Schüler *et al.*, *Phys. Lett. B* **174** (1986) 241.
5. I. Ahmad and P. A. Butler, *Annu. Rev. Nucl. Part. Sci.* **43** (1993) 71.
6. P. A. Butler and W. Nazarewicz, *Rev. Mod. Phys.* **68** (1996) 349.
7. C. S. Wu and Z. N. Zhou, *Phys. Rev. C* **56** (1997) 1814.
8. J. Engel and F. Iachello, *Phys. Rev. Lett.* **54** (1985) 1126.
9. J. Engel and F. Iachello, *Nucl. Phys. A* **472** (1987) 61.
10. A. Georgieva, P. Raychev and R. Roussev, *J. Phys. G* **8** (1982) 1377.
11. A. Georgieva, P. Raychev and R. Roussev, *J. Phys. G* **9** (1983) 521.
12. A. Georgieva, P. Raychev and R. Roussev, *Bulg. J. Phys.* **12** (1985) 147.
13. S. Flibotte *et al.*, *Phys. Rev. Lett.* **71** (1993) 4299; *Nucl. Phys. A* **584** (1995) 373.
14. B. Cederwall *et al.*, *Phys. Rev. Lett.* **72** (1994) 3150.
15. P. J. Twin *et al.*, *Phys. Rev. Lett.* **57** (1986) 811.
16. P. J. Nolan and P. J. Twin, *Ann. Rev. Nucl. Part. Sci.* **38** (1988) 533.
17. R. V. F. Janssens and T. L. Khoo, *Ann. Rev. Nucl. Part. Sci.* **41** (1991) 321.

18. A. T. Semple *et al.*, *Phys. Rev. Lett.* **76** (1996) 3671.
19. R. Krücken *et al.*, *Phys. Rev. C* **54** (1996) R2109.
20. Y. Sun, J.-Y. Zhang and M. Guidry, *Phys. Rev. Lett.* **75** (1995) 3398; *Phys. Rev. C* **54** (1996) 2967.
21. I. N. Mikhailov and P. Quentin, *Phys. Rev. Lett.* **74** (1995) 3336.
22. P. Magierski, K. Burzyński, E. Perlińska, J. Dobaczewski and W. Nazarewicz, *Phys. Rev. C* **55** (1997) 1236.
23. V. K. B. Kota, *Phys. Rev. C* **53** (1996) 2550.
24. Y.-X. Liu, J.-G. Song, H.-Z. Sun and E.-G. Zhao, *Phys. Rev. C* **56** (1997) 1370.
25. I. M. Pavlichenkov, *Phys. Rev. C* **55** (1997) 1275.
26. H. Toki and L.-A. Wu, *Phys. Rev. Lett.* **79** (1997) 2006; L.-A. Wu and H. Toki, *Phys. Rev. C* **56** (1997) 1821.
27. I. Hamamoto and B. Mottelson, *Phys. Lett. B* **333** (1994) 294.
28. A. O. Macchiavelli, B. Cederwall, R. M. Clark, M. A. Deleplanque, R. M. Diamond, P. Fallon, I. Y. Lee, F. S. Stephens and S. Asztalos, *Phys. Rev. C* **51** (1995) R1.
29. I. M. Pavlichenkov and S. Flibotte, *Phys. Rev. C* **51** (1995) R460.
30. F. Dönau, S. Frauendorf and J. Meng, *Phys. Lett. B* **387** (1996) 667.
31. W. D. Luo, A. Bouguettoucha, J. Dobaczewski, J. Dudek and X. Li, *Phys. Rev. C* **52** (1995) 2989.
32. P. Magierski, Warsaw University of Technology preprint nucl-th/9512004, to appear in *Acta Physica Polonica B*.
33. G. Herzberg, *Molecular Spectra and Molecular Structure*, Vol. I: *Spectra of Diatomic Molecules* (Van Nostrand, Toronto, 1950).
34. R. S. Ram and P. F. Bernath, *J. Molec. Spectr.* **171** (1995) 169.
35. R. S. Ram and P. F. Bernath, *J. Molec. Spectr.* **172** (1995) 91.
36. R. S. Ram, C. N. Jarman and P. F. Bernath, *J. Molec. Spectr.* **161** (1993) 445.
37. R. S. Ram, P. F. Bernath and S. P. Davis, *J. Molec. Spectr.* **175** (1996) 1.
38. G. M. Barrow, *Introduction to Molecular Spectroscopy* (McGraw-Hill, London, 1962).
39. I. M. Pavlichenkov, *Phys. Lett. B* **53** (1974) 35.
40. L. P. Marinova, P. P. Raychev and J. Maruani, *Molec. Phys.* **82** (1994) 1115.
41. M. J. A. de Voigt, J. Dudek and Z. Szymanski, *Rev. Mod. Phys.* **55** (1983) 949.
42. D. Bonatsos, *Phys. Rev. C* **31** (1985) 2256.
43. W. Reviol, H.-Q. Jin and L. L. Riedinger, *Phys. Lett. B* **371** (1996) 19.
44. K. Hara and Y. Sun, *Int. J. Mod. Phys. E* **4** (1995) 637.
45. K. Hara and G. A. Lalazissis, *Phys. Rev. C* **55** (1997) 1789.
46. J. L. Dunham, *Phys. Rev.* **41** (1932) 721.
47. G. A. Leander, R. K. Sheline, P. Möller, P. Olanders, I. Ragnarsson and A. J. Sierk, *Nucl. Phys. A* **388** (1982) 452.
48. G. A. Leander and R. K. Sheline, *Nucl. Phys. A* **413** (1984) 375.
49. H. J. Krappe and U. Wille, *Nucl. Phys. A* **124** (1969) 641.
50. J. M. Brown and D. J. Milton, *Molec. Phys.* **31** (1976) 409.
51. R. M. Gordon and A. J. Merer, *Can. J. Phys.* **58** (1980) 642.

Contents of Volume 2

Part VIII. Condensed Matter

Combined Index to Volumes 1 and 2

(Entries are in the form [volume number]:[page number])

Progress in Theoretical Chemistry and Physics

1. S. Durand-Vidal, J.-P. Simonin and P. Turq: *Electrolytes at Interfaces.* 2000
ISBN 0-7923-5922-4
2. A. Hernandez-Laguna, J. Maruani, R. McWeeny and S. Wilson (eds.): *Quantum Systems in Chemistry and Physics.* Volume 1: Basic Problems and Model Systems, Granada, Spain, 1997. 2000 ISBN 0-7923-5969-0; Set 0-7923-5971-2
3. A. Hernandez-Laguna, J. Maruani, R. McWeeny and S. Wilson (eds.): *Quantum Systems in Chemistry and Physics.* Volume 2: Advanced Problems and Complex Systems, Granada, Spain, 1998. 2000 ISBN 0-7923-5970-4; Set 0-7923-5971-2
4. J.S. Avery: *Hyperspherical Harmonics and Generalized Sturmians.* 1999
ISBN 0-7923-6087-7
5. S.D. Schwartz (ed.): *Theoretical Methods in Condensed Phase Chemistry.* 2000
ISBN 0-7923-6687-5
6. J. Maruani, C. Minot, R. McWeeny, Y.G. Smeyers and S. Wilson (eds.): *New Trends in Quantum Systems in Chemistry and Physics.* Volume 1: Basic Problems and Model Systems. 2001 ISBN 0-7923-6708-1; Set: 0-7923-6710-3
7. J. Maruani, C. Minot, R. McWeeny, Y.G. Smeyers and S. Wilson (eds.): *New Trends in Quantum Systems in Chemistry and Physics.* Volume 2: Advanced Problems and Complex Systems. 2001 ISBN 0-7923-6709-X; Set: 0-7923-6710-3

KLUWER ACADEMIC PUBLISHERS – DORDRECHT / LONDON / BOSTON

FSC
www.fsc.org
MIX
Papier | Fördert gute Waldnutzung
FSC® C083411

Kolibri 360 GmbH
Ferdinand-Jühlke-Straße 7
99095 Erfurt, Deutschland
produktsicherheit@kolibri360.de